International Review of

Cytology

A Survey of

Cell Biology

MITOCHONDRIAL
GENOMES

VOLUME 141

International Review of Cytology

A Survey of Cell Biology

Guest Edited by

David R. Wolstenholme
Department of Biology
University of Utah
Salt Lake City, Utah

Kwang W. Jeon
Department of Zoology
The University of Tennessee
Knoxville, Tennessee

MITOCHONDRIAL GENOMES

VOLUME 141

Academic Press, Inc.
Harcourt Brace Jovanovich, Publishers
San Diego   New York   Boston   London   Sydney   Tokyo   Toronto

Dustjacket micrograph: This figure was first published in an article by C.R. Wolstenholme, J.M. Goddard, and C.M.-R. Fauron *in* "The ICN–UCLA Symposium on Molecular and Cellular Biology," Volume XV, "Extrachromosomal DNA" (D.J. Cummings, I.B. Dawid, P. Borst, S.M. Weissman, and C.F. Fox, eds.), pp. 409–425, Academic Press, San Diego (1979).

This book is printed on acid-free paper. ∞

Academic Press, Inc.
1250 Sixth Avenue, San Diego, California 92101-4311

United Kingdom Edition published by
Academic Press Limited
24–28 Oval Road, London NW1 7DX

Library of Congress Catalog Number: 52-5203

International Standard Book Number: 0-12-364544-1

PRINTED IN THE UNITED STATES OF AMERICA
92 93 94 95 96 97   EB   9 8 7 6 5 4 3 2 1

CONTENTS

Mitochondrial Genomes of the Ciliate

Donald J. Cummings

Mitochondrial DNA of Kinetoplastids

Kenneth Stuart and Jean E. Feagin

Evolution of Mitochondrial Genomes in Fungi

G. D. Clark-Walker

Structure and Function of the Higher Plant Mitochondrial Genome

Maureen R. Hanson and Otto Folkerts

Animal Mitochondrial DNA: Structure and Evolution

David R. Wolstenholme

Transcription and Replication of Animal Mitochondrial DNAs

David A. Clayton

The Endosymbiont Hypothesis Revisited

Michael W. Gray

CONTRIBUTORS

Numbers in parentheses indicate the pages on which the authors' contributions begin.

G. D. Clark-Walker (89), *Molecular and Population Genetics Group, Research School of Biological Sciences, Australian National University, Canberra City, 2601 Australia*

David A. Clayton (217), *Department of Developmental Biology, Stanford University School of Medicine, Stanford, California 94305*

Donald J. Cummings (1), *Department of Microbiology and Immunology, University of Colorado School of Medicine, Denver, Colorado 80262*

Jean E. Feagin (65), *Seattle Biomedical Research Institute, Seattle, Washington 98109*

Otto Folkerts (129), *Agricultural Biotechnology Laboratory, DowElanco, Midland, Missouri 48674*

Michael W. Gray (233), *Program in Evolutionary Biology, Canadian Institute for Advanced Research, Department of Biochemistry, Dalhousie University, Halifax, Nova Scotia B3H 4H7, Canada*

Maureen R. Hanson (129), *Section of Genetics and Development, Cornell University, Ithaca, New York 14853*

Kenneth Stuart (65), *Seattle Biomedical Research Institute, Seattle, Washington 98109*

David R. Wolstenholme (173), *Department of Biology, University of Utah, Salt Lake City, Utah 84112*

PREFACE

In 1924, E. Bresslau and L. Scremin, using the then-new DNA-specific cytochemical technique of R. Feulgen and H. Rosenbach, demonstrated that the basophilic body situated in the cytoplasm of trypanosome flagellates contained DNA. However, it was not until 1958 that H. Meyer and co-workers, using the electron microscope, showed that this DNA (kinetoplast DNA) is located in a modified region of a mitochondrion, and thus provided the first unequivocal evidence for DNA in mitochondria. That mitochondria of many other eukaryotes contain DNA was established over the next several years, mainly from studies carried out by P. Borst, I.B. Dawid, A.W. Linnane, D.J.L. Luck, M.M.K. Nass, S. Nass, M. Rabinowitz, P.P. Slonimski, H.H. Swift, and D. Wilkie, and their associates. The occurrence of DNA genomes in mitochondria became generally accepted only in the late 1960s.

The application in many laboratories of the powerful molecular techniques developed over the last two decades has resulted in a detailed understanding of the coding capacity, expression, and replication of mitochondrial genetic systems in a wide variety of organisms. These genetic systems have been shown to exhibit many unusual features that collectively include modified genetic codes, structurally different transfer RNAs and ribosomal RNAs, post-transcriptional modification of messenger RNAs (RNA editing), a uniquely asymmetrical mode of replication, chimeric genes, and complex protein gene–intron arrangements.

Since the mitochondrial genomes of yeast and other fungi are particularly amenable to genetic analysis, they are the best understood of all such genomes at this time. Animal mitochondrial genomes are being used with an ever-increasing frequency in studies of populations, systematics, and evolution, principally because entire sequences of many of these mitochondrial genomes are known, they rearrange at very low frequency, and they are maternally inherited.

This volume contains seven chapters that are designed to bring readers up to date on the present state of knowledge of some of the major structural, functional, and evolutionary aspects of mitochondrial genetic systems of various eukaryotes. The first two chapters by D.J. Cummings, and K. Stuart and J.E. Feagin deal with the mitochondrial genomes of ciliates and kinetoplast-containing protozoa. These are followed by a chapter by G.D. Clark-Walker on the structure and evolution of fungal mitochondrial genomes. The next chapter by M.R. Hanson and O. Folkerts is concerned with the structure and function of plant mitochondrial genomes, and chapters by D.R. Wolstenholme and D.A. Clayton cover the structure, evolution, replication, and transcription of animal mitochondrial genomes. The final chapter by M.W. Gray is a comprehensive treatise on the intriguing question of whether or not mitochondria and other organelles are the products of ancient endosymbiotic relationships.

Many groups of eukaryotes remain whose mitochondrial genomes have yet to be examined. There is reasonable expectation that studies of these genomes will uncover further genetic novelties, while at the same time increasing the pool of molecular information that has great promise for significantly advancing our understanding of the evolution of eukaryotes. It is hoped that the present volume will act as a stimulus for further studies on mitochondrial genomes.

We thank the authors for their contributions and patience. We also thank Dr. Nicholas W. Gillham for his suggestions and input during the initial planning stage of this volume. The excellent cooperation of the editorial staff at Academic Press is gratefully acknowledged.

David R. Wolstenholme
Kwang W. Jeon

Mitochondrial Genomes of the Ciliates

Donald J. Cummings

Department of Microbiology and Immunology, University of Colorado School of
Medicine, Denver, Colorado, 80262

I. Introduction

According to Corliss (1979), the Ciliophora are a group of protozoans that
possess cilia at some point in their life cycle and contain macro (somatic)-
and micro (germ line)-nuclei. There are three major ciliate classes: the
Kinetofragminophora, the Polyhymenophora, and the Oligohymenophora.
Hypotrichous ciliates, such as *Euplotes* and *Stylonichia*, are members of
the Polyhymenophora, whereas *Paramecium aurelia* and *Tetrahymena
pyriformis* or *thermophila* are members of the Oligohymenophora, order
Hymenostomatida. On the basis of phylogenetic tree construction utilizing
small subunit cytoplasmic ribosomal RNAs, Sogin and Elwood (1986)
have proposed that all these organisms constitute only a loose phylogenetic
grouping that had an ancient common ancestor. In fact, the depth of
branching in this tree between *Euplotes* and *Tetrahymena* was greater than
the phylogenetic distance between *Saccharomyces cerevisiae* and any of
the ciliates analyzed. Although in Corliss' groupings, *Tetrahymena* and
Paramecium are in the same order, the relatedness in their RNA sequence
homologies is not much greater than that between different classes of
ciliates. These phylogenetic groupings are emphasized here because in
describing ciliate mitochondrial genomes I will focus solely on *Paramecium* and *Tetrahymena*, the only two ciliates whose mitochondria have
been studied in any detail. Because of their relative phylogenetic distance,
no general conclusions about other ciliate mitochondrial genomes should
be drawn. We will see, however, that remarkable similarities (and differences) between these mitochondrial genomes do exist.

   The study of ciliate mitochondria began in earnest with the discovery
of cytoplasmically inherited drug-resistant mutants in *P. aurelia* (Beale,
1969; Adoutte and Beisson, 1970). Elegant microinjection experiments
with *Paramecium* were possible because unlike *S. cerevisiae* mitochondria, which can, under certain conditions, fuse into a single convoluted

1

structure (Hoffman and Avers, 1973), *Paramecium* mitochondria appear to be unique entities at all times. They are about 2–5 μm in size and thus are amenable to direct microinjection. Moreover, like other ciliates, paramecia are quite large (100–200 μm), and able to withstand the insult of a 5-μm needle (Fig. 1). Utilizing several antibiotic-resistant cells, Beale's group (Beale *et al.*, 1972; Knowles, 1974; Beale and Knowles, 1976) demonstrated that "sensitive" cells could be transformed into resistant cells by microinjection of mitochondria isolated from resistant cells. Under selective pressure, the recipient paramecia undergo a gradual transformation through intermediate stages characterized by increasing proportions of "resistant" mitochondria. Differences in the restriction enzyme patterns of species 1, 4, 5, and 7 (Sonneborn, 1975, cataloged the *P. aurelia* complex into a collection of species, all with characteristic genetic and morphological properties; Nanney and McCoy, 1976, have done the same for *T. pyriformis*) allowed Maki and Cummings (1977) to demonstrate directly that it is the DNA within mitochondria that determines antibiotic

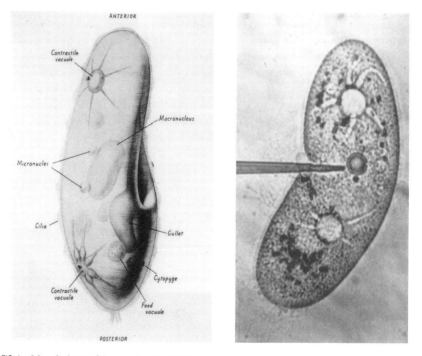

FIG. 1 Morphology of *P. aurelia*. (Left) The most important structures; (right) a 5-μm needle penetrating a recipient cell (kindly provided by J. K. C. Knowles).

resistance when they found that in transformed recipient paramecia the mitochondrial DNA was indistinguishable from the donor mitochondrial DNA, rather than that contained in the original recipient cells. Adoutte *et al.* (1979), using a variety of genetic and microinjection protocols, showed quite convincingly that *P. aurelia* mitochondria do not undergo recombination, as is so prominent in fungal mitochondria. This failure to undergo mitochondrial recombination may be related to the fact that *Paramecium* mitochondria are discrete entities and do not fuse.

The results of Maki and Cummings (1977) are important in two other respects. First, they confirm clearly that the recipient and donor mitochondria DNAs do not recombine, at least not at a level necessary to illustrate restriction enzyme pattern differences. Second, they also may explain the so-called "incompatibility" phenomenon discovered by Beale and co-workers in that in order for antibiotic resistant cells to multiply, the injected mitochondria themselves must multiply. Too much nuclear–mitochondrial incompatibility may arise if the species are too distantly related to allow mitochondrial multiplication. Incompatibility was defined by Beale and Knowles (1976) as an inability to support antibiotic transformation between distant species. Successful transfer occurs at essentially 100%, even when only one mitochondrion is injected (Knowles, 1974), when the same species is used as donor and recipient or between species 1, 5, or 7. Transformation between species 4 and 1, 5, or 7 is not possible. Interspecies transfer provided a nice counterbalance to the demonstration by Maki and Cummings (1977) that the mitochondrial DNA of the donor persisted, in that species 1 and 7 show different electrophoretic forms of the fumarase enzyme and it is the recipient species fumarase that prevails in the transformation. This shows that variation of this enzyme is under nuclear control (Knowles and Tait, 1972). Evolutionary divergence measurements, based on 6- and 4-bp endonuclease recognition sites, also demonstrate the relatedness of species 1, 5, and 7 and the distant divergence of species 4 mitochondrial DNA (Cummings, 1980). Further details of the genetics and microinjection experiments as well as some information on the unusual respiratory properties of ciliate mitochondria can be obtained from Beale and Tait (1981) and Sainsard-Chanet and Cummings (1988).

Although the study of the cytoplasmic genetics of ciliates has revealed some remarkable characteristics, the purpose of this chapter is to discuss the properties of their mitochondrial genomes. In this respect as well, the ciliates harbor distinctive properties. Except for the fungus *Hansenula mrakii* (Wesolowski and Fukuhara, 1979), all mitochondrial genomes are circular, ranging in size from about 12 to over 600 kbp in some plants (see Wallace, 1982, for a review). Both *Paramecium* and *Tetrahymena* have linear genomes with unique replicative pathways (Goddard and Cummings, 1975; Suyama and Miura, 1968; Goldbach *et al.*, 1979). Their genomes are about the same size, 41 kbp for *Paramecium* and about 50 kbp for *Tetrahymena*, but they differ substantially in G-C content with estimates of 40% for *Paramecium* and only 25% for *Tetrahymena* (Suyama and Preer, 1965; Flavell and Jones, 1971). The G-C content (41%) of *Paramecium* mitochondrial DNA has been confirmed by direct sequence analysis of the entire genome (Pritchard *et al.*, 1990b).

Computer analyses of the known sequences for each genome show that the ciliate mitochondrial structural genes are among the most divergent known (Pritchard *et al.*, 1990a). Remarkably, each also contains a surprising number of ribosomal protein genes, and *Paramecium* shows genes with good sequence similarity to chloroplast genes. These and other properties will be dealt with in detail.

II. Mitochondrial Genome Structure and Replication

Our knowledge of the structure and replication pathways for the linear genomes of both *P. aurelia* and *T. pyriformis* is incomplete. But, as we will see, in important aspects what we know of each complements the other.

A. Paramecium aurelia

The mitochondrial genome of *P. aurelia* was shown to be a linear duplex by Goddard and Cummings (1975). Examination in the electron microscope demonstrated that 65% of the isolated molecules were 13–15 μm in length, 10% were linear dimers (25–28 μm), and 3–6% were lariats, i.e., a circle plus a tail. For the lariat molecules, the length of the tail plus half the contour length of the circle corresponded to the length of the linear monomeric molecules. These molecules are depicted as A through C in the replication scheme presented in Fig. 2. Considerable evidence has been accumulated to justify all the steps in this pathway. Much of this evidence was made possible by the fact that lariat molecules were enriched to about 30% of the population by growth of the paramecia in the presence of ethidium bromide and that when the protozoa were grown in the presence of chloramphenicol (or erythromycin), more than half of the molecules had the dimer length. Partial denaturation mapping of the linear monomers indicated that one-half of the molecule was more sensitive to denaturation than the other and that about 30% of the molecules contained a small bubble at the end of the more resistant half. In lariat molecules this small bubble was exactly opposite the replication fork. For dimer molecules,

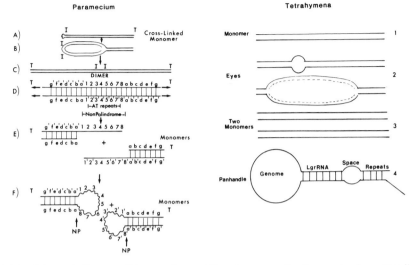

FIG. 2 Schematic diagram of *Paramecium* and *Tetrahymena* replication mechanisms. Details are discussed in the text. I, initiation; T, termination; NP, nonpalindromic.

the denaturation pattern showed two mirror images with the small bubble in the center. This denaturation study strongly suggested that replication was initiated at a unique end (labeled I) and proceeded undirectionally, resulting in a head-to-head dimer intermediate (Goddard and Cummings, 1977). The most unusual feature of this mechanistic scheme is the apparent linkage of the complementary strands at the initiation end of the molecule. This was studied further by determining whether monomer and dimer molecules could "snap-back" after denaturation. In a preparation in which 59% of the molecules were originally of dimer length, after allowing rapid renaturation, 39% of the molecules were completely double-stranded monomers and 19% contained both double-stranded and single-stranded regions with a total length of 15 μm or less. There was not a single molecule containing a double-stranded molecule of greater than monomer length. DNA molecules isolated from cells in the stationary growth state have a much lower proportion of dimer and lariat molecules. From a population where 92% of the molecules were of monomer length (or less), denaturation–renaturation yielded a population containing 32% completely or partially double-stranded molecules of monomer length. It thus appeared that a terminal linkage was present in a significant fraction of the monomer population. To determine the nature of this linkage, both monomers and dimers were treated with S1, a single-strand-specific endonuclease, and the snap-back experiment was repeated. The proportion of monomer length molecules resulting from denaturation–renaturation of the dimer population was not affected but few if any completely double-stranded monomer molecules were present in the monomer renatured population. Pretreatment of the DNA population with either proteolytic enzymes or RNase had no effect on these results. The terminal linkage of monomer length molecules consisted of a single-stranded length of DNA. Because of this property of snap-back renaturation, the question arose whether in the replicative pathway dimers were processed to monomers by enzymatic breaking and rejoining steps or whether the entire dimer molecule unraveled and then renatured, yielding two monomer length cross-linked molecules. This question was resolved by examining 5-Br Ura-labeled mitochondrial DNA populations and showing that replication was semi-conservative, as predicted by the direct breaking–rejoining pathway (Cummings, 1977).

Having established this general replication scheme, we then proceeded to study this cross-linked initiation region. By taking advantage of the availability of dimer molecules, we were able to clone the double-stranded initiation region (Pritchard et al., 1980) and sequence the entire region from five different species (Pritchard and Cummings, 1981; Pritchard et al., 1983). For all species, this dimer initiation region contains a series of A T-rich repeats, specific for each species. These repeats will be presented in Section II,C. Now, we wish only to use them for the replication scheme shown in Figs. 2D–2F, where the monomer length molecule is shown with the AT-rich repeats (1, 2, 3, 4, etc.) in the cross-link connecting the predominant double-stranded linear monomer. After the unidirectional replication of the entire molecule is completed, the dimer length molecule must be processed to two cross-linked monomers. The model shown in steps E and F suggests that staggered nicks of the dimer somewhere in the spacer region separating the repeats from the palindromic mitochondrial DNA followed by ligation of the single strands would yield the desired monomer molecules. This model predicts that each monomer would be a "flip-flop" isomer of the other. Using P. primaurelia, direct evidence for these monomer isomers was obtained by taking advantage of the anomalous electrophoretic behavior of such monomers and isolating them for DNA sequence analysis. The results showed quite clearly that both isomers contained sequences spanning the entire nonpalindromic region and were inverted complements of each other, as predicted (Pritchard and Cummings, 1984). The presence of sequence isomers in P. tetraurelia was suggested earlier by the finding that dimer molecules contain different numbers of the repeating element in the same preparation of DNA (Cummings and Laping, 1981).

The presence of nonpalindromic repeats is not unique to Paramecium. In T. thermophila (Kiss and Pearlman, 1981) as well as T. pigmentosa (Kan and Gall, 1981), the otherwise palindromic extrachromosomal ribosomal RNA dimer also has such a region. Vaccinia virus DNA possesses a terminal sequence linking both strands of the genome that consists of an AT-rich series of repeats (Baroudy *et al.*, 1982) which may be a replication origin. Similarly, parvovirus H1 has a 55-bp AT-rich terminal repeat that may be required for replication (Rhode and Klaasen, 1982; Rhode, 1982). Despite these many different approaches in characterizing the role of AT-rich repeats in the initiation end of the *Paramecium* mitochondrial DNA, as yet we do not have any information on how the termination of replication is accomplished. Understanding completion of replication of linear DNAs has been a nagging problem for some time (Watson, 1972; Cavalier-Smith, 1974). In the next section, we will describe the termination of the linear molecule of *T. pyriformis* mitochondrial DNA, but we have no information on how replication is initiated.

B. Tetrahymena pyriformis

The mitochondrial genome of *T. pyriformis* (or *thermophila*) provides an interesting complement to that from *P. aurelia*. It also is a linear duplex molecule of about 50 kbp (slightly larger than *P. aurelia*), but the replication mechanism is quite different. The DNAs of all strains differ in size and the size variation is entirely dependent on the size of a duplication–inversion present at both ends (Arnberg *et al.*, 1975; Goldbach *et al.*, 1977). As we will discuss more completely in later sections, the duplication–inversion has two components, a duplicate large subunit ribosomal RNA gene and a set of repeated sequences (Arnberg *et al.*, 1977; Goldbach *et al.*, 1976). Both of these properties have been utilized in the replication scheme proposed (Goldbach *et al.*, 1979), as depicted in Fig. 2, panels 1–4. As indicated, the mitochondrial genome of *T. pyriformis* is a linear duplex. Replication is initiated near the left center of the molecule and proceeds bidirectionally toward both termini (Arnberg *et al.*, 1974). Molecules showing the replication bubble, termed Eyes, are enriched by growth of the protozoan in the presence of ethidium bromide (Upholt and Borst, 1974). Recall that for *P. aurelia* lariat replicative intermediates accumulated in the presence of ethidium bromide; unlike *P. aurelia*, dimer length molecules are not observed in *T. pyriformis* mitochondrial DNA molecules and growth in the presence of chloramphenicol has no effect on the proportion of replicative intermediates.

The essence of the model for termination of replication is offered in Fig. 2, panel 4. As indicated earlier, termination of linear duplexes presents a special problem in that all known DNA polymerases synthesize DNA in a 5' to 3' direction. Since initiation of DNA synthesis requires an RNA primer (see Gefter, 1975), a newly replicated linear DNA will have difficulty completing replication following the removal of the RNA primer. Electron microscope analysis of partially denatured mitochondrial DNA from *T. pyriformis* showed "panhandle" structures; i.e., molecules with a single-stranded loop and a double-stranded handle. Further analysis indicated that the single-stranded region represented the bulk of the molecule and that the entire double-stranded region contained the variable duplicate inversion located at each end. The first 2400 bp of the handle consisted of the inverted duplicate large subunit ribosomal RNA followed by a nonhomologous bubble and then a region at the end of the entire molecule representing a set of direct repeats. Temperature-controlled experiments indicated that these direct repeats are GC-rich. Taking advantage of a tandem repeat model put forth by Heumann (1976), Goldbach *et al.* (1979) proposed that replication of the linear *T. pyriformis* mitochondrial DNA is completed by the introduction of a specific endonuclease single-stranded nick between two repeating units at a variable distance from the terminus. Displacement synthesis would then lead to a molecule with a single-stranded tail; DNA ligase, possibly in combination with DNA polymerase, would then complete the replication of the 3' tail. This model also provides an explanation for the variable length of the duplicated inversion since pairing of the single-stranded terminus with the displaced strand would lead to an increase in the number of repeats, and internal recombination or occasional loss of single-stranded ends could lead to a decrease. Remarkably, as we will see in the next section, Morin and Cech (1986) for *T. thermophila* and Middleton and Jones (1987) for *T. pyriformis* showed that there was a variable length GC-rich repeat at both termini. In *T. thermophila* and *T. malaccensis*, Morin and Cech (1986, 1988) found that the number of repeats could vary greatly. They speculated that unequal homologous recombination within the repeats could maintain the telometric structure. Although no other details of the replication model proposed by Goldbach *et al.* (1979) have been forthcoming, the finding of GC-rich repeats at both termini lends great support to the model.

What we know about the replication mechanisms of these two linear mitochondrial genomes is in some ways complementary. In *Paramecium* much is known about the initiation of replication, whereas for *Tetrahymena* more is currently understood about termination. Because chloramphenicol had no effect on the proportion of replication intermediates in *Tetrahymena*, Goldbach *et al.* (1979) speculated that the mechanism for *Paramecium* mitochondrial DNA must be fundamentally different from that for *Tetrahymena*. But this need not be the case. In *Paramecium*, inhibition of mitochondrial protein synthesis with chloramphenicol prevented the processing of dimers to monomers, a step affecting the initiation region. The effect on termination may well be the same as in *Tetrahymena*.

Nevertheless, no information is available as yet on termination of replication in *Paramecium* or on initiation in *Tetrahymena*.

C. Initiation and Termination Repeats

The initiation and termination regions of *Paramecium* and *Tetrahymena*, respectively, were determined by two fundamentally different methods. In *Paramecium*, restriction enzyme digestion of dimer length molecules yielded a new fragment not present in monomer molecules. This head-to-head initiation region was then cloned into a plasmid and amplified. For *Tetrahymena*, advantage was taken of the fact that the termination restriction fragment was heterogeneous in size and gave rise to a "fuzzy" appearance on gels. This fragment was isolated and filled in with Klenow polymerase, blunt-ended, and cloned into a plasmid. In principle, these methods should be applicable for both linear molecules. For *Tetrahymena*, restriction enzyme digestion of the "Eyed" replicative intermediates should produce an enriched I–I fragment. We do not know whether this experiment has been attempted. In *Paramecium* we found that in four species the termination fragment was also fuzzy (Cummings and Laping, 1981). Despite repeated efforts in our own laboratory as well as Gregg Morin's, we were not able to clone the termination fragment. Thus, we do not know whether the initiation and termination regions of these two ciliate mitochondrial DNAs are similar.

The initiation and termination repeats for *Paramecium* and *Tetrahymena* are given in Fig. 3. For *Paramecium*, it can be noted that the

```
4,51     6 to 14 repeats of 34 bp   TTAATATATTTATA TTTTTTTATTTTAATAAATA
4,172    8 or 9 repeats of 35 bp    TTAGTATATTTATAATTTTTTTATTTTAATAAATA

8,299    ATAATAATAATTTATTATTTATATTTATAAT       2 repeats of 28 bp
                                               42 repeats of TAA

1,513    ATATAATTAATTATAAATTTAATTTTTTAATAAATTATATTATTTATATA
1,168    ACATAATTAATTATAAATTTAATTTTTTAATAAATTATATTATTTATATA
5,87                  ATATATTATAAATAATTTAATATTAAAATATA
5,311                ATATATTATAAATAATTTAATATTAAAATATA
7,227                      ATATTATAATTAAATATTTATAAT
7,325                      ATATTTTAATTAAATATTTATATT

1,513    CATATATTATAATTATTTTT      6 repeats of 72 bp
1,168    CATGTATTATAATTATTTTT
5,87     TTTATATAATTTAAATATTATTATTTTTAAATAC      8 repeats of 66 bp
5,311    TTTATATAATTTAAATATTATTATTTTTAAATAC
7,227    AATTTATTTTTACAATTAAATTATTTTTATTTTATTTATATAAAT      7 repeats of 69 bp
7,325    AAATTGTTTTTGTAATTGTATTATTTTTATCTTACTTATATAATC

T. ther.  TCTTAGAGGTATGTTAGCTATTAGTGTTGTTTAGGCTTGTTATGGTATGTGTA      15 repeats of 53 bp
T. pyri.  ACTACCCTCGTGTCTCTTTAGTTTCATATCT      8 repeats of 31 bp
```

FIG. 3   The DNA sequences for the initiation region repeat for five species of *P. aurelia* and two termination repeats for *Tetrahymena*. The paramecia sequences were aligned to show the TATA box (see text).

sequence of the AT-rich repeat is species specific. The only sequence common to each repeat is a so-called Goldberg–Hogness box, TA-TAAATA (Corden et al., 1980). Earlier we showed that the dimer initiation regions of species 1, 5, and 7 cross-hybridized but no hybridization to the species 4 dimer was noted (Cummings et al., 1979; Pritchard et al., 1980). This is borne out by the sequences of the repeats illustrated in Fig. 3. Species 1, 5, and 7, in addition to the Goldberg–Hogness box, show additional homologies, especially a T-rich element near the end of the repeat. It may also be significant that the general size and structure of each species initiation region are similar. The first repeat is bordered by a GC-rich unique sequence and the last repeat is followed by an AT-rich unique sequence, whereas the larger the repeat, the smaller the number (Pritchard et al., 1983). As alluded to earlier, species 4 with a repeat size of only 34 bases exhibits polymorphism in its number of repeats, ranging from 6 to 14 repeats depending on the mutant or species examined (Cummings and Laping, 1981; Pritchard et al., 1983). We proposed that this variation in the number of repeats is due to an unequal crossover recombinantional event. Although no intermitochondrial recombination has ever been detected in *Paramecium* (Adoutte et al., 1979), we have no evidence for or against intramitochondrial recombination. In fact, the variation in the number of repeats could be considered support for such an event.

For the *Tetrahymena* termination sequence, the situation is the same. The GC-rich repeat is species specific and the number is greatly variable ranging from 4 to 30 in *T. thermophila* (Morin and Cech, 1986). In terms of being GC-rich, these repeats are not at all as rich as the AT-rich repeat in *P. aurelia*, where the repeats were almost all AT. But relative to the average 25% GC content of the genome, the *Tetrahymena* repeat is about 36%. It can also be noted that this telomere repeat is quite different from the short tandem repeats found in nuclear telomeres, C_4A_2 (Blackburn and Gall, 1978). Morin and Cech (1988) also proposed a mechanism involving unequal crossover recombination for the maintenance of these mitochondrial telomeres. There is no evidence for or against inter- or intramitochondrial recombination in *Tetrahymena*.

D. Conservation of Sequences Adjacent to Initiation Region in *P. aurelia*

Because of the lack of sequence homology between the initiation region of species 1 and 4 and the general lack of cross-hybridization between these two *P. aurelia* species (Cummings et al., 1979), we were surprised to find that the sequence adjacent to the dimer initiation region for all the species examined was quite homologous (Pritchard *et al.*, 1983). This is shown in Fig. 4 for most of this sequence, which immediately abuts the end of the single-stranded cross-link. Like with the sequence of the repeat itself, the homology is best between species 1, 5, and 7 or between species 4 and 8, but the similarity of all the sequences is evident. The exact reason for this sequence homology is unknown but it is clear that this region is transcribed (Pritchard *et al.*, 1986). We considered the possibility that because the sequence is so similar in all the species, this might be the replication origin and the nonpalindromic repeat region provided only a cross-link. This possibility was excluded when we demonstrated that the dimer initiation regions from species 1 and 4 served as autonomously replicating sequences in a yeast transformation system, even for subfragments that retained only the repeated sequences themselves (Lazdins and Cummings, 1984). The adjacent sequence containing the homologous region did not function as an autonomously replicating sequence. This region could simply encode a common structural gene, perhaps a gene-encoding protein necessary for the initiation of replication.

```
4, 51     CGCTGCACGGGCTCGTTTTTTGTATGGGGGGGCATAAAATTAGGGGGTAC
4, 172    CGCTGCACGGGCCCGTTTTTTGTATGGGGGGGCATAAAATTAGGGGGTAC
8, 299    GGA GG CGGGCTCGCTTTTTGTATGGTGGGGCATAAAATTGGGGGGTAC
1, 513    GGG G  CGGGCTCGAAGTTTGTATGGCGGGGCATAAAATTAGCAAATAC
5, 87     GGG GG CGGGCTCGAAGTTTGTATGGTGGGGCATAAAATTAGCAAATAC
7, 227    GGG G  CGGGCTCGAAGTTTGTATGGCGGGGCATAAAATTAGCAAATAC

4, 51     CCAGGTGCTGTTTAATTCTCTAGCTTTAAGGATAAGCCTCAATCAAATCC
4, 172    CCAGGTGCTGTTTAATTCTCTAGCTTTAAGGATAAGCCTCAATCAAATCC
8, 299    CCACCTGCTGTTTAATTCTCTAGCTTGAAGGATAAGCCTCAATCAAATCC
1, 513    CCAGGTGTTGTTTAATTCTCTAGCTTTAAAAGTAAGCCCCCAACGGCCCC
5, 87     CCATGTGTTGTTTAATTCTCTAGCTTTAAAAGTAAGCCCCCAGCGGTCCC
7, 227    CCAGGTGTTGTTTAATTCTCTAGCTTTAAAAGTAAGCCCCCAACGGCCCC

4, 51     AACGCGCCTGCTTCTGCAGCGGGGGGGGTTAAAAAAAAAAAGAAGCCCAAG
4, 172    AACGCTCCTGCTCC
8, 299    AACGCTCCTGTTTTTGTAGCGACGAGGGCAAAAAAAAAAAAAAGCCCAAG
1, 513    AGCGCGCCTTTTTTTGTAGCGAGGGGGGCTAAGAAGAAGAAGAAGCCCAAG
5, 87     AGCGCGCCTTTTTTTGTAGCGAGGGGGGCTAAGAAGAAGAAGAAGCCCAAG
7, 227    AGCGCGCCTTTTTTTGTAGCGAGGGGGGCTAAGAAGAAGAAGAAGCCCAAG

4, 51     TACC GCTTTGGCTTCCCCATGGCCCTTTATTTTTTTTTTTGAGAAGCTAAA
4, 172
8, 299    TACCAGCTTTGGGTTCCCCATGGCCCTTTATTTCTTTTTCGAGAAGCTAAA
1, 513    TACCAGCTTCGGCTTCCCCATGGCCCTGTATTTTTTTTTTTGAGAAGCTAAA
5, 87     TATCGCTTCGGCTTTCCCCATGGCCCTGTATTTTTTTTTTTGAGAAGCTAAG
7, 227    TACCAGCTTCGGCTTCCCCATGGCCCTGTATTTTTTTTTTTGAGAAGCTAAA
```

FIG. 4   DNA sequence of the P1 region abutting the initiation repeat region in five different *P. aurelia* species. The sequence homology continues for a few hundred bases downstream.

III. DNA Sequence of *P. aurelia*

Although the complete DNA sequences of some plant chloroplast genomes (greater than 100 kbp) are known (Shinozaki *et al.*, 1986; Hiratsuka *et al.*, 1989), very few large mitochondrial genomes have been sequenced in their entirety. In general, complete sequences are known only for the highly compacted 16-kbp genomes (human, Anderson *et al.*, 1981; mouse, Bibb *et al.*, 1981; *Drosophila*, Clary and Wolstenholme, 1984; *Xenopus*, Roe *et al.*, 1985; and sea urchin, Jacobs *et al.*, 1988) from multicellular organisms. Until now, no protozoan mitochondrial DNA sequences were known and only one fungal genome (*Podospora anserina*, 94 and 100 kbp for two different races) was completed (Cummings *et al.*, 1990). For all of these mitochondrial genomes, despite the range in size, the same known genes have been identified: cytochrome *c* oxidase subunits I, II, and III; NADH (ndh) subunits 1, 2, 3, 4L, 5, and 6; Fo-ATPase subunits 6, 8, and 9 (with some exceptions where the gene is apparently encoded in the nucleus); apocytochrome *b* (cyt *b*); the large and small ribosomal RNAs; a full complement (22–25) of tRNA genes (after wobble has been taken into account, Crick, 1966); and several open reading frames (ORFs), which could encode important mitochondrial proteins of unknown function. With a few exceptions, mitochondrial genomes do not encode the protein components of the mitochondrial translation machinery. Ribosomal proteins Var 1 in *S. cerevisiae* (Hudspeth *et al.*, 1982), S-5 in *Neurospora crassa* (Burke and RajBhandary, 1982; LaPolla and Lambowitz, 1981) and *P. anserina* (Cummings *et al.*, 1989a), S-14 in broad bean (Wahleithner and Wolstenholme, 1988), and S-13 in plants (Bland *et al.*, 1986) are thought to be unusual in not being encoded in the nucleus. Similarly, most organisms contain a full set of mitochondrially encoded tRNA genes. But in the bean *P. vulgaris*, at least four mitochondrial tRNAs are imported from the cytoplasm (Marechal-Drouard *et al.*, 1988); in *Chlamydomonas reinhardtii*, Boer and Gray (1988) have sequenced 80% of the 15.8-kbp mitochondrial genome and have found only three tRNA genes; and for the 20-kbp trypanosoid protozoans, mitochondrially encoded tRNAs appear to be completely absent (Simpson, 1987). As we will see, *P. aurelia* and *T. pyriformis* have several unique genes in their mitochondrial genomes and are deficient in the number of tRNA genes.

A. Organization and Types of Mitochondrial Genes

We have completed the major structural sequence of the mitochondrial genome of *P. aurelia* (Pritchard *et al.*, 1990b). The only region still unsequenced is that containing the extreme termination sequence. If this is similar to the termination sequence of *Tetrahymena*, as was discussed in Section II,C, then it would amount to only a few hundred bases. The 40,469 bp completed from *P. tetraurelia* is presented in Fig. 5, and Fig. 6 contains the organizational features of both *P. aurelia* and *T. pyriformis*. We present the DNA sequence here in a format different from that published to emphasize certain characteristics. The sequence commences at the initiation single-stranded crosslink that contains 11 copies of the 34-bp AT-rich repeat. As discussed earlier, the next gene downstream of this loop is P1, which is present at the same location in at least five separate *P. aurelia* species. It is possible that, except for the position of the small and large subunit ribosomal RNA genes on lines 24500 and 37500, respectively, the organization of the genes in *P. tetraurelia* is not the same for some *P. aurelia* species. Earlier we found that whereas species 1, 5, and 7 cross-hybridized indistinguishably, species 4 demonstrated only about 50% cross-hybridization. In particular, the 12-kbp region downstream of the P1 gene region showed no cross-hybridization between species 1 and 4 (Cummings *et al.*, 1979). A similar situation prevails for *T. pyriformis*. Goldbach *et al.* (1977) analyzed six *Tetrahymena* strains and found they not only differed in the length of their duplicate inversions, but five of the strains demonstrated less than 20% cross-hybridization. The ribosomal RNA genes at the same location were the primary regions of identity. It is therefore curious that the organization of the identified genes in *Tetrahymena* (about half of the genome has been sequenced, Y. Suyama, personal communication) is almost identical to that of *P. tetraurelia*. The single large subunit RNA gene in *Paramecium* is also quite near the termination region and the location of the small subunit ribosomal RNA gene is indistinguishable in the two ciliates with the 5' flank containing COII and ndh5 in the same order as well as tRNA[Trp] and COI on the 3' flank. The only identified gene apparently misplaced is ndh3, which is positioned near the 5' end in *P. aurelia* just downstream of P1, and in *Tetrahymena* is located approximately 10–15 kbp from the 5' end. This difference in position may not be relevant, however. As discussed earlier, *Tetrahymena* has a large duplicate inversion, including a second large subunit ribosomal RNA gene at the 5' end. These additional sequences

---

FIG. 5   Nucleotide sequence of the noncoding strand of *P. tetraurelia*, beginning with the set of 34-bp repeats that constitute the initiation loop. All genes, tRNAs, and ORFs are marked by square brackets. Genes and ORFs, on the coding strand are delineated by parentheses (see Fig. 6). 5' and 3' ends of known mRNAs are indicated by arrows. *Eco*RI and *Pst*I clones used for sequencing are noted as are other significant restriction sites. The terminal 200 bp are an estimate. (Overleaf, pages 14–34)

Initiation loop start
[ATATATTAATAATATTATGCTACTTTACAATATCTTATTTGTTAAATATTGTGTACTTACATTTCTTGATATTAATAAAGATCATATTAATATTATTGTA   100
                                             first repeat of 11   200
TATACACAAATTTTACATCTTCTTTTTTTAATGAGTTTTATCATAATTATGGTTTATCTTATTAGTAAAATTAATATATTTTTTATTTTTAAT   300
AAATATTAATATATTATATTTTTTTTTTATTTTTTAATAAATATTAATAAAATATTTAATATATATTTTTTATTTTTATTTA   400
ATAAATATTAATAATATTTTTATATTTTTTAATTTTTAATAAATATTAATAAAATATTAATATATTTTTTATTTTTTATTTT   500
TAATAAATATTAATAATATTTTTTTTTTATTTTTTTATTTTTTATTTTTATTTTTTATTTTTATTTTTATTTTTATTTTTTAT   600
       last complete repeat
TTTAATAAATAATAATATTTTTTTATTTTTAATAATATTAGTATATATATACAAATTATATATTTTTTTATTTTTTATTAAAATATTAA   700
     Initiation loop end   KpnI   ↓ 5'mRNA 1
TATATAATTCTTTTTATTTCAATAAATTA]TTTTAATAAATGTATTATTACTATATATTATTTTTATGGGGCGGGGCCTCAAGTGCTGCTGAAACCGCTG   800
                                                    P1 start
CACGGGCTCGTTTTTGTATGGGGGGCATAAAATTAGGGGTACCCAGGTGCTGTTAATCTCTAGCTTTAAGG[ATAAGCCTCAATCAAATCCAACGC   900
                                         NcoI
GCCTGCTTCTGCAGCGGGGGGGGTAAAAAAAAAAGAGAGCCCAAGTACCGCTTGGCTTCCCATGGCCCTTTATTTTTTTTTTGAGAAGCTAAACTTCT   1000
                    ↓ 5'mRNA 2   SstI
CGTACTGGACCTCGACGACGGTGAACCCAACCACTACGTTTGCGTCCTCCCCGAGGCGGCCAAGCCCTGCCAAGCGTCTCGGGGGGGAGCTCTT   1100
                                    AflII
CTTAGCCAAGTCCCAGCTCGTGGAGGCCACCGCCTTGACCTGACGGGGCAGGGCGGCTGAGGCCGTGGAACCTGCGTCGTCTTCTTAAGGAACAACGGGGTC   1200
                                     XhoI
GTCCTTAGCTATAGCTTTTACTTCTTTCTCCTTAAAAAGCGCATAACTTTTTCTTGCACGGGGGCGACAAGGTCAGCTCGCTCGAGTCTTTCTACAGCA   1300
      XhoI   SstI
ATGCTAACTGGCTGAGCGCGAGATTCCGAGATGTTCGGGGCTCCAACCTCCTCAAGAAGAATCCCGCAACCTCCTCCGACTACGGAGCTCGTT   1400
                                                P1 end
CAACCCCTTCCTGAAGAAGTTCCCTTCCACGGGCCACGCCGAGTCGAGTCGTCTTCAACCTCCTTCTTAAAGACGACGCGCCGCGGGCGTCGAG   1500
   SstI   3'mRNA 1, 2↓   ↓ ORF1 (113) and 5' mRNA 3
CTC]AGCTTAAAGACGTTTTTTTTTTTT[ATGTTAATTGGCTCAACTTCTTCAACTTCTTCTTCTTCTTCTCTTCGTCTTTCTGACCACGTCGTT   1600
                                   SstI
TAATTTTTTTGTCTAGTTTGACCTGGACCTCATGTGGGCCTCAGCGGTCCATCCTCAGCGGTCCGGCTCCATCCTCGACGACTTCTTCCTGGCC   1700
                                   AflII
TCCTTCGCCTTCCTCCTCCGGAGCTTTGCTTCGGTAGAGCTCTCGATCGGCCTCACCTTAAGACCACAACCTCTCCCGAACCTCG   1800
      XhoI   ORF1 (113) end   5'mRNA 4 ↓   NDH3 (ORF2)
CGGCTCATAAGGGCTCGAGCTACGCAACGCTCTTCTTGCTCAAGAACGGGGCCTCTAAAAAAAAATC]TAGGCTTGGGATAC[ATGGGTAGTATGCACATTG   1900
                                                           SstI
CTTTTCTTCGTGTGGAGCACGTTTTATTTTTTGCATGATTTTTTGGCTCCTCACCTGAGTCGCCGAGTATTCTTTAAGTCTAAAAACAACAAGCAGAAGC   2000
ACCAGTTTACCAGTGCGGAATCAGGGCCCTCAACATCCAGATAAACTTGAACTTTCTATTGTTGCGTCTTCCTCATCCTCCTACGACGT

```
GGAGTTCATCTTTATGTACCCCTTCTTTTTTTAACTTCTTTTAGTCAATGCGGGTGCTTCCTGCGTCTTCTTTGTCTTCCTCTTCGTCTTCGTCTTCTACTCC    2100
CTCGTGTACGATAGCGTCCAGAACTCGCTGGCCCTCCAGCTC[AATCCTGATTTCTTTTTTTTTTTTTGAAAAAGTCATCGCAGAAGTACCCGCATAAGG    2200
                       NDH3(ORF2) end  3'mRNA 3, 4 ↓
GTCGGCACCTTCTTAAATTTTTTTTTATCGATCCCTAGGCTCTATCTCTCCACTGGGAGATCGCCAGCTTTTTTACGCCTTTAAGCACTTAAGT    2300
                            ClaI
ATTTTTTGAAAACATCTACCTTTCGACCCGCTTCATCTTCTTTACTAGCACAACTACCTGAGCTTAATCTTAAGAACGCCAGGTTTTTTTAACCT    2400
                                                                        AflII
ACTCGACCCCTTTAGCGTTCGAAAAGATTCTTCTCGAATGGGGGGATCCTCCGCTTCCTAAAAAAACGAGAAAACGAAGCAAGTACAAGCAAAGCAAC    2500
                                          BamHI
GGGGTGGCCTTCATCTTGTTCCTGCGGAGGTCTTCTTTAGGGTCTTCTTGTATAAAAACCTCGTCCTCATCTTTAAGTCTTTTAGGCTCCGCCACAGTACT    2600
TTTTGAAGCTCGCTATCGAGCTCTTGGCGACGAGTCGTCCGGCATGTACGTCGTGCCCGGGCCCGCTACTGCAGGGTACTTTTTAGGCGCGTTAAGGC    2700
          SstI                                  SmaI     XhoI
CGTTAAGAGGAAGATTAAGCGCAAGCTTAAGTAGCATAAAGTAAAAATAAAAAGAGTAGCTCTTGTTCCTTAGTCTCCTATATTTTGTGTTTTAGGGG    2800
                 HindIII  AflII
GGCGGCCCAGGCCATGAGCCTGGGGGCTTAACGCTCTACCTCTACTTCTTCTTTTACCTGAAGCGGAAGGCTCTCTGTATGCCCTGAATTATGAAAACTTC    2900
TTCTTCTTCCTGCGTTCGAACTTTTTGATGTGAAGCTAAGCGCCTTCCTCCTCGCCTTCTTTTAGGAACTTCCTCTGGCCTTTTACAACGTGAGGGTGG    3000
CCCTCCTCTTTTGTTACCCCAGCTTCCTACTTCTTCTTCTTCTTTAAAAACATCTATAAGACCAGGTTGAGAACAGCCTCGTCGACGGCCTCCTCTACGTCCA    3100
                                                                            SalI
CCCGGTGTGCATAAACCTCTCCTACTTCTTCTTCTTCTTGGCCATGTTTTTGTTTATAAACATCTATAAGACCAGGTGAAGGGGCTGAAGGGGCTC    3200
GACATCTACAAAAAGTTTTTTTGTGGACGGGGGGCTCCATCGTCCTGGGGGCCGTTGGGCTCAGCACGAGCTTAACTGGGCGGCTTTTGGAGCTGGG    3300
ACCAGGTCGAGATCATATCCTTGTCTACTTTGTCTGTGGCCACTCTTCCTCCTCCACTTAAAAAGGCCCGCTCTTCTTCTTCGGCCCTCCGCCTTTTC    3400
TTACTTTTTTTTGGGCTGAGGTTAACTACTTTACCTCGGTCCACTCCTTCGGTTCTAAGGCTGCGGCAGCCCAGACCCTAACTTTTTTTTTTTGGG    3500
CCCTGGCCCTGGCCCTGAGTCCCTCTTAGGGCTCCTTAAGGCCTCAGAGCGCCAACCCTCGACGTACCTGTTTTTTGGGGCCCTTTCATGTTCTT    3600
                                     AflII
CTTGAACTTCTTGGTGGGCCTCGTGTCGTCCTTGACATCAAGAACTTTTTTTCTTACCTCTTCTTTCTACTCGTTCTTTAAAAAGGCCCTTTTAGCCTTTTCG    3700
        XhoI
GCCCTCGCCTTCTGAGACTTTTTTTTTTGGAGAAGCCGCTTGTGCTCCATGCCTTCGTGGTCCTCTAAAGAGGGTCTCTACTACGCAACTTGCGGCGACC    3800
       HindIII  AflII
ACTTTTCCATGATCAGCTCCTTCCTGCTCGACCTTTAAGCTTAAGACCGCCTTCGCCTATAGCGTAATAAATAAGAACAACTTGTCTTTTGC    3900
TAGGGGCCTACGATGGGGTTTGGAGCTACAATACCACGACCCTTAAGCTACGATGGGGTTTGGAGCTAC... 4000
```

(*continued*)

```
                    HindIII
GGCTTTTTTTTCTTCAACCTGTTCATGGTCACCAGAAAGCTTTATTACCTTCCCAGCTTCCTTGCGATGAGTGTGCTCTTCTGGCGCTCCTTCTTAAT   4100
TTGTACTTTTCGGATTTGCACCCTCTGTAAAATTAACTAACCTCTCCCAGAGAAGCGAAAGCTGAAGAAGATTCTCCACTTCCCAACCTTCAACAAGG   4200
                 AflII
CAGCGCCGTCTTCTTCTTCTTCAACACCATCAAGTTGAGTGCGTCACCTTAAGTTCTTGGAGAGGAAGCTAAGGAAGCTCATCGTTAGGCGCAAAAA   4300
                           AflII
CATTACCGTGGGCCAAGGTCTGGATAAACCTTAAGGCCAACTACCCCCTCACCAAGAAATCAAAAAAATTCTACAATGGGAAGGCAAGGGCCGCTCC   4400
            EcoRI    KpnI         HindIII   AflII
TCCGCTGGGTCGTGACCATACGCCCCGCAACAAAGTTTGCAGAATTCGTAGGGTACCCTTCAGTTCCTGACCAAGTTAAGCACAGCCTGGGGTGCTG   4500
CTACCGCTCCGACGTCTACTTCTTTCACGATTTTGAAAGCCACCCCGTGTGGTCAGGCCCCGAAGACGAGAGTTCTTCAACCTAGACAAGAGGAGAACC   4600
TCTTCTAGGCTTCTCGTAGAATTTGAACCACAAAGCTCACCTAAGCCTCCACTTCATCATCCAAGACGTCGAGATGGCTTTCAAAAAAAAGAATAGGATA   4700
AAGATTTCGATCAAATCCTTCCTCTTTTTTATGGTTATTCTAACCCTCCTCTTCAAGAGCGTCGAGATGGCTTTCAAAAAAAAGAATAGGATAA   4800
                   HindIII
GCATCCTTAATGCCCCCTATAAGAACAAGCTTGCACAGCGCCAATATTACATCGAGCCGAACTTTTTTTTTATAAAGATCGAGGGCCTAGCCTCCGACAC   4900
        NarI         BamHI
CCCCGAGGCCCTGCTCGGGGCCCTTGGCAAGCTCAGGGCCCTCGGATCCTTTTTTTTTCTACATGACCAAGCTCCGCATCAGGGTCCGTGCCCTATAACAG   5000
AGGGCTCGAATGGGCGACATAATTTTAAATCTAAAGAAGGACCGACTCATGCGCCTCAGAACTCTTTTTTTATTTTTATTTTTATAGGCGATCTGATTTTTTTA   5100
CTTCAGTACTTGTTTTTTGCTGAGGCCTTTATATCCTCAAGTGCCTTTGAGTGCAAGCACGTGACGAAATTTTTTCTTCTTCACTAACTACAACTTT   5200
                                            ORF3(196) start
TCGCTGCCGCTAGGTGGCGGCAGGTGCGAGGTAATAAGGGCCGTCATGCTAGGCT[ATGCTTTGAGAGCTTCTCAATCTTCTGCATCTTCGGTATCGTTT   5300
TCTACAACAACTGGACCTCTACTTCCTTAACTATACCCTCAGCAAGTATGCTCTAAGCAGCAGACCGAGGCTACAGCGCCATACTCGAAGCTCTAT   5400
TACAAGAACCTCGGGCTGCTGGCTTTCTTTTTTATTCTTTTCTTCCTGCTAGCTTTTTTTTAGCCCTCTTATTTTTTTTTTTAGGGGTTCCAGGGCACCCTAT   5500
                                                                    ORF3(196) end
TTTTTAATAGCGTGCTGTTGAGCAACAAGGGCCTCTACTTCGTGCTTCTTTTTATGTTGCAGCCCTTGGCGCTTCGTCTTCGTTCTTCAAACAACCCTCTTCTGAA   5600
GCAGTTCTTCTTCAACAACGCTAGACTTCCTGCTCATACTCTACTTCATTTCTCTTTTTTTTTAACCCTCTTCTCAACCCCTCATTTTCTTCTCAAACAACCTCTTCGCCTTC   5700
                                                    SstI      ORF4(156)
TTCTTTTACCCTAGAGCTTGTAGCTGTTACAAACTTCTCATTTCTTCGGCTCGTGAATTTTTTTTTTCTTGAAAAGTCTCAGGGTGGAGCTTCAAGCC   5800
TCAAGGCCCAAAAAAAAAAAACCTTCTTTAACGTCCGTTTTTTTT[AACTTTTGGAGCTCTTTTTTTTTTTAGCTCA[ATGTTCCTCCCTCACTCTATCATGACAA   5900
CTTACTTCCCTCTTCGGGACCACGGATTGAGCAACCCTTAACTTCTTGGTGAGCGTTTGCCTTGACAACGACTATCTTTATAACAAAAACAGCCTCTTCTT   6000
```

```
                                                         HindIII
TACCTTTTTATTTTTTTAGCAGCCGTGTTCAGCCTGGCCTCCCCCCCTTCTTCTTCTTCAAAATTGAGGTCTATAAGGCCTCCCCTCTTCGTA   6100
ACCTTCTTTATTCGGTATTTTTTTTTTTTAACTACCTTCGGGCTTCTTCTTCTTTTTTTGTGTTCTTTACGAGCTTCTTTGTACTTCTCGTACC   6200
TTCTCTTCTTCGTCCCCGTCTTCGTCCTTTTTTTTTTTTTTATCTCACGAGTAGCACAACCTCAACTGCTTCTTTCGATAAGTTCTGTAATTAA   6300
       ORF4(156) end       HindIII AflII TRNAPHE
CAGCACGTTCTTGGTTTACTTGTTGGCCAGCTCTTTTTT[AAGGTAGCGGGGCTGAGCCAA[GCTTAAGTAGCTCAGTGGTAGAGCGTTAGACTGAAAA   6400
TRNAPHE                                                                              BamHI   6500
TCTAAAGGTCGTTGGTCAATTCCAATCTTGAGCA]GTCCATAGTGGTTTCCTGCCCTGTTTAAAAGCCGCCACCTTCGTGCCCTTGGGATCCTCT   6600
TGCGAGGAACCCTCGCCAGTATAGATTCGATCAGCTCCTTCAGCTCTTCTTGGAAGTACTTTTTTTTTTATCTGACTCGGCTTCTGTACTTTGAACGTAGCC   6700
CTGATCTCCTTCTTTGATTTTTTATTTTTTTAAGCTGTATCCTTGCAAACCCCCAAAGCTCGACCCAAGATCGAAGTCCCTTCTCATCCCCCAAAGACA   6800
SstI
TCCTCTGAAAAGTGTGAGGCTCAACCTCTGTGCTTCTTTGAGGTGAAGTCGAGGAGCCAGCGGCCCTCGACAGCAAGTATAGGGGCAGAGTTAAGTT   6900
CATGCTAAACCTCGACGAGCTCGGTACGGCCCCATCAGCCAAATCTTTTAAGGCCCCACGCAGCCTTTCGCCCAAGATCGGGCTCCTAGCCCTCCAGG   7000
SalI
CTTCGCCGCCCTTCCTCCAGGAGCACCCTCAAACTATAGGGGAGCACCTTCGCCCAAAAAAGTTCTCAGCGAGGAGCCCGCAGCCTCCTTTTAGGAG   7100
GGGGCTTCTCTAAGACCACCGTGACCTCTTTCGAACCTATAGGCAGGCCCGCTCAGTCGAGCCTTTGAGGCTTCGGCGTGGCCTGGTCCTACCTCA   7200
AflII                                                              P2 PstI P9   7300
AGGTGTCCTATAAAGTTTTATCGAGCGCTTCTTCGCGGCGACCTAGACATCGTCGAGGGCCTCCTCTTCGTCCCCTTCAAGCCCGCTTTTGGATCTCG
XhoI
CAAGGGGCGCCGCCTCGACTTCCTTAAGTTTTTTTTTAGCTGAGTACGTGCGCTCCGAAACCTTCTTAGGGGAGCTGGGCCCTCGCCCGAGACCCCTGCA   7400
                                                                    BamHI   7500
GGGGTTCCAGACCTCGAGGTCGCCGACGAGCGTGAAGCCCAGCCCTTCGAGGACCTAAGGCCCTAAGAAAGATTTTTGAGCAACCCTACTTCTTCA
ACAAGACTTTTTGTTTTTTGTTTTTGTTTTTTGGGACCTCTTTACCACGAGCAACGGGACTTCTGGGGCCTTGACGATGGCCAGGCCTGGATCCCGAAGCC   7600
TACCAAAAGTGGCTTGACAGCATCCTACCATACCAGAGGCCAGGCCGATACTTATGGGCTGAGGAAGAGGGGAACCCCGAAACTTTGTCTTTAACTATG   7700
                                                             SalI
AGACCTTCGAGGCTTGCGAGGTCGATAAGCTCTTCTTTAATAAAAATCTACCCCCCTCCCCCCAGATCGACGTGGTCGACCTCCCCTTCAAAGATATCTA
NarI         P9 PstI P8                              NarI   7800
CGAAAGAGGCGCCCACTCCAGGAGTACTTCCTGCAGATTTATTACGGGGAACAGACGAGGGAGCCAGAGGCGCCTTTCCACAAGAAGAACAAGGACTTT
                                        HindIII              SalI   7900
TTTTTCAAAAAAACCTACATAAACAAGTTTAACGCCTACTTTTTTTAGGGGGGCAAGAAGTTACGTCCCTCAAAGCACTTACGGCGGCTATGTCGACAT   8000
TCCTCCACGGACTGTTGAGCGCGACATAGCCCCTGAGTTTAGGAGCAGCAGGAACATATTTTTCTTTTTCATTTCGAGACCATCAGTAGAGTTAA
```

(*continued*)

```
                    BamHI
CATCAACGTGGTCTTGAACTGGATCCTCGACTTTACGGGGTTTATCTTTTTTTTTTTTAGAACGAAGGCCGTGCCAAAGTTCTTAAAGAAGCAGACGAAGAAG  8100
                            AflII
AAATATGCCGTCGAGCCCTTCTTGGTCAGGCGGCCAAGTTAGATCAAAGTACACCCTTAAGTAACCTTACTATTTATCGAAAAGGAGGAGTTTGCAAAGC    8200
XhoI      SstI                                                                          HindIII
TCGAGGCCAGAGCTCACAACACGATATGCTCCGTTGCTTACGAGTACCAAGACTCGAAGTTTTTGAATACAAGCTTGTGCCCTAAAGAAGATCGTATA       8300
                                                                                P8  PstI P4
CCAGTTTTTGAAGAAGTAACCCAGAGCAT[ATGTTACTAGTACTAGCAATCAAAACATTAGTATTGGGTTTATGTATGTTACCTATCTCTGCAGCAGCTCT   8400
               ATP9 start
AGGCGTAGGTATCCTCTTCGCAGGTTACAATATCGCAGTTCAAGGAATCCTGATGAGGCCGAGACCATCTTCAACGTACTCTCATGGGTTTCCACTC       8500
                                                     ATP9 end          XbaI
GTTGAAACTTTCGTCTTCATGTCGTTCTTTTTGGAGTCATCGTCACTTCATC[TAAAATGTTCTAGAAAAGAACGCTCCTAGGCCCGGCCTACACCCCT      8600
                                                                                                        8700
CCCCCAAATAATGCTTCGTCGATTCTTAGAGTATGCCGGTTCTTTTGATGAGCCACGTTGCTTCCTTTTTTTTTAATAAAACTAGGGCCCTAAAAAACAA   8700
HindIII
AGCTTCTGACTCTTCTTCCTTTCTATAAACCACAGGGAGGTCATCGAACCCCTTGCCGAGTGCTCCTTCCTTCAACAAAAGCTCCCTTCTTGGCCCCG      8800
                                                                                                        8900
CTGTGGTCTTGACGGCCCTGCTCGCCTACCTCTACACTTTTATGTTTGATGATGCTTCTCAAAGTTCTTCCTATTCGAATTTTTGCAGGCCTCCTCTTCTTCCTCGATAGGAAA  9000
CTATCTAAGTGCCAGCATGCTAACCTTTGGAGCATCCCGTCTCAAAGTCTTCCTATTCGAATTTTTCGAATTTTTGCAGGCCTTCCTCTTCCTCGATAGGAAA  9100
                                                                BglII
CTTACGCCCCTCGCCTCCTTCTTTCTAAGGCTTTTTTGTAGGAACTACTCAGGAGAGCAAACTAAAGTTCAAATACCAGGGTCGAGCCCACACTTACAAAGATCTGCAGTACCTCAGAAA  9200
                                                     HindIII AflII
GAGGTAATTTCTAAATAATTTTTTTTGTAGGAACTACTCAGGAGAGCAAACTAAAGTTAAGCTTAAGCTACTTCTTAAAAAACGAGGCCGA             9300
SstI         SstI
GGCCTGGACCTCGCTATTTTTTTTTTTTTTTTTTTAGGACGAACCTAGAGGGCCCCTCCCTTTTTTTTTAGGGGCCCCTACCTTCTTCTTAAAGACCCTCG   9400
CCGAGCTCGGAATCGAGGAGAGTCCCGAGCTCCCATGAGGACGCTGAGGTCGAGGACGACCCTACAGGTTCCCTAGGCTCACGCGAGCCCTTCTGTAG      9500
       XhoI
GAACTTACTTTATTTTTATTTTTTAGGGGAGCTCCCGACCTCGCTCGAGGAGGGCTCGAGGAGGAGGCTCAAGTCTTCAAAAAACTTTTCTTTTGCTCGT    9600
                                                 SalI                                        XhoI
GGGCACAGCTTAGGCTGCTTCTCGCCCATGCATGAACGAAAGGTCGACCCCTCGATAATAGTCCACGACTGGCTCTCGAGTTTCTCGAGGAGCTAGCT       9700
                                                                                                SstI
GGGCTTGAGGTCGGAGACTACGAGTTTCTCGATCCCAAAAAGCCTCGATTTTTTCCTAAAGCAGGCTATCTGAGAAGAGGCCTTCAAGACGAGCTCCTG     9800
                                                                                                        9900
GGCCTCTTCTCACCCGAAGACGGGAGCCGAGACCAAGCTGGGCGTGCTTCATGGGCTCGATGACCTTCGATGGCTCGAGGCTCGAGGACGCCAACCCTCTCCGTCGCCCGA        9900
TCCCGATCCGGAAGAGGAGGCCGGAGACGGGTGCCGACTTTTATAACTTTTCTTTTTGAAGACCCCGCGAGATCAAAAAATCTCGTGCCTAGGTAGTGGC     Sal1 10000
TTATTCTTTTTCGATCAGGCCCTTCTTTTTTTTGTCGAAGAAGAGCCCCTAGCTCGAGAAGAGACGCCCATAGTCGCAACGCGTCCTCCTCGGGCTCCTCGGGCGTCGACG
```

```
                                    XbaI                              XhoI
AGGGCTTTTATGCCGAGGACCTAGACTCGCTGGACCACCAGAGTTTCTAGAGCCCTCCCTCTACGGAGACCTCGAGGAGGCTAGCATTGAGGCCCTTTT   10100
     KpnI   NarI                                  XhoI                                                   10200
TTTTGGTACCTCCGCCCTGGGCGCCTGCGGCCCAGAGCAAGCACTGCTCGAGACCTACTTTGTGGACGAGCTTTCCGAGGCCCCCGACGGCGGGGGAGGGTC
                                                      XhoI                                                10300
GAAGATGAGGAGGTCGAGGAGGGGTTGAAGACCTATCGACGCCCCTCGTCCTCGAAGGAGCCCTTCTGCGAGCCTTATGCGCGGATGAGGAAACCTCCTCATC
                                                                                                          10400
CGACGACGAGGAGGCCCCCCTCTTTGAGAAAAACTTTCTGCCCTTCTTTTTTGTCTTGAAGCGCTTGCGACGTTCTTCTTCTTTAAGAACATCTTTAGTT
       SstI        XhoI  XhoI                                                                             10500
TCTTCTTCTTCTTCTTCCCCCTAGAGCTCTTTTACTTCGTGTACTCGGACCTCGAGGAGGACTTTGAGTCCTCTGAAATGTCCGAGAACTTCTTTTTCCAG
                                                                                                          10600
ACCCTAAGGGGCCTAAAATAGCTCGCGAGCTAAACAAGACCCCCGTCAGCCTGGTCGAGCTATCCAAAGCTACATTTACGAGCCTAGCAAGCAGGGAA
                                                        XbaI       10700
AGCACAAGAGCTTTTTCTTCTTCCTGTTCCTCTCCGGCTTCTTAATCAGAAGCCCTTCTTTTCTTTAACACCCCTGAGTCATTCAGTC
                                                                                                          10800
GAAGGACTTCAGCATCTTGTCCTCAAAGTTAAGGACCTGAGTGCCCTCGCAACATATGTAAAGATTCTTTTGAAAGAATACAGATTAAATTCCATAAGGTCTTCTTGA
                                                                                                          10900
CTCTTCTTGTGCCTCAAAGTTGTCCTTCTTCTTTAACAAGTTGAGGTCGAAGTTCGGCGTCAAGGGCTTTTCCTCGACGTTAGGGGAAAAGTGTCGGTTGT
                     KpnI                                                                                 11000
GGTAATTCTAAGAAGCGCCACGTGCATCAAAAAGGGGTACCTCTCTAAAACAAAGAAAGAATTGAGGTTCTTTTTTATGAAGAACCAAATTAACACG
                                                                                                          11100
ACCACAGGGCGTCCTGGGTGCGTCCTACCTACCTCCTAAGCTACTAGCCTAGTCCCAGGGGCCTAGAGATGGCTTTGTACTTTGAGCAAAAAAAACGTGCTTATGC
                                                                                                          11200
CCCTCGCTTCGTATACCAGCTTATGGGCTTGATTGTTGCCCAGGGGCCTGGAATGCCCAAAAAAAATTAAATTTTTTCTTTCGTCTTTCTTAGCT
                                                                        AflIII  HindIII                   11300
TTCGGTTATTCTCTTTCCAACGAGGTTAAAGTTAAAGAAGTAGACGGGGTGGGAATGCCCAAACACATAATTATTTAGCTTCTTGTTTTATCTTTTTGTTCTTCTCTC
                                                                                                          11400
TAAAGCCCTTACCCTTGAAGTTACCTTCTTGAAGAAGAAGCGGCCAAACAGTATGTGGGTCTATTTGACTCCTTGTGGTTGTGGTGTTGCCTGAAAAGTCTCCTCTCGAGG
                                                                           ORF5(70) end                  11500
GAGGAGGAGAGTATGCCGCCCTCTATGCTTAACGACGTATGAAGTTCACAACCAAAGCTACCGAGGGCGATCTTCTTTATTCTA
 XhoI                                                                                                     11600
CTTCCAGCAACGTAGCGCGCCCTCTATGCTTAACGACGTATGAAGTTCACAACCAAAGCTACCGAGGGCGATCTTCTTTATTCTAGCCATTCAAAGCCTTT
                                                                        ORF5(70) start                   11700
                                                                                                          11800
TTTACTTTAGTGAAGGAGAAGGTGTTGTTATGCTTCTTCTTTTTTATTGAAGTGAAATTCTATGTTTTTTGGACCCCACGTTCTTTACGATGGACATCTTCTTGA
 RPS14 end                                                                                                11900
AGTTCTTTCGAATCTTAGACTTAGTGAAGGTTAGCTTGAACTTACTTTGTTGTGTTAATTGAGCAGTTTATTTGTTTAAAAAGCGCTTCATTTA TCA
                                                                                                          12000
AGATTTTGTTTTGTATTCTGGAGGAGCCCTCGAAGGCGTGCATGCCTGCCAGATTGCTTCGATGATGCTTTAGTCCTGCCT
```

(continued)

```
                                                                    AfIII                                           12100
GTAAGGCAGCACCTATCGACCTGTTAGTAAGCCTGCTCTTGTCTTGTTTTTGATACCTAAGCTGCCATAGGTGGTTCCTTAGCTGCTTGGCTGATTTT
                                                                              RPS14 start 12200
GGTTTTTTTAAGCATTCGAGGATCTTCTTCTTTTTTTCGAAATCTCTATAGCCGATCCATCCTTTTAAAAATTCTATGTTTATTAAGCTGAGCTGATAT
                                                                                                12300
         ORF400 end
CAT]TTA(ACGATCTATCTCTCCAAAAACGATGTCTATGGTTCCTATTAAAGCCGAAAGGTCGGCTAGGAGGTGCCCCTTAGACATCTTAGGAAGCACCTGA
                                                                                                12400

AGGTGGTGATAAGCAGGGGACCTGACCTTGCATCTCTGTAGGGCTTGTTGAGCCATCAGAAACCAAGGTTACGCCGAACTCGCCCTAGGGGACTCGACGG
                                                                                                12500
CCTGGTAGGTCAATTGCTTTGAATCTTAAAGCCCTCGCTTCAGTAGTACTTAAAGTGGGTTATGAGCTTCTCCATCGAAGAGTACTCGTTCTTGTTAGGTTTG
                                                                                                12600
GGAGATGAACTTTTTTTGTTTAGCGCCTTGAGGATGCTTGCTGGGCTACAAGTATTTTTACTTGTTGTGAGCTTAAAGATAGCCTGGCTTATGATGTTG
                                                                                                12700
ATGCTTTCAGACATCTCATTCATTGCGAATCAGAAACCTATCGTAAGAATCTCCAGCCTGCCCAGTATAGCTCCTAAAGTTTAGTAGTAATAGTTTGCAT
                                                                                                12800
AGGTCGAAGGCGTCCATCCTCAGGTCTCTCTTGATACCTGTCGATCGTGCCATCAACCCCCGTCAATCCAAAGTCTAGGCAATCCTGTAGGAGATGCT
                                                                                     XhoI 12900
TCCGATGTTTACGAGCCTTTGCTTCCAAATTTATTGTATGTTAAGATGTTATGCATCTCTGAAAGAGTCGTGAGGCAATTATTGTTAAACTCGAGGATG
                                                                                                13000
TCAGTTAACAGCTTTACGGATAAGAAGTTTAGGTTAACCTCGTTGGGCGTGTAAAAGGCTGCGTGCATGCGGCCTCCGCAAACTCGCTCATAAAACTCCA
                                                                                                13100
TGATTTTTTCGTTCTTCAAAGGGCCCAAAAAAATAGATGACATGTCTAGTGCATGGCAAGCAATTGCGAGCATGTGTTCAAGATCCTAGT
                                                                                                13200
TAGCTCATCGAACATGGTTCGAACGAGCACGAAGTTAGCTGTAGTTTGCTGGTGTTTAACAGGGCTTCGATTGCTAAGCGTAAGCATGCTCCTGCACC
                                                                                                13300
      ClaI          P4 PstI P7
ATCATCGATACGTAATCCAACCTATCAAAGTAGGGCATTGACTGCAGGTAGGGCTTTGTCGTCTCCATTAGCTTTCTGACCCCCTGTGAAGAGCCCGATGT
                                                                          PvuII         ORF400 start 13400
GGATGCCATCTTTTCTACAACCTCCCCGTTTAGCTGAAGGATCAGCCTAAGGACCCGTGAGCGCCGTGGGTGTTGAGGGCCGAAATTTACCGAGATGGC
       NDH2 start  SstI                                                                        13500
CTTTAT]TTTT[TTGAGAGCTCATTTGCTTCATTGTGAGTGGCTTCAGTTTGGAAAATACTTCTATTCAACAAGTTTTTGAATCTCCTTATGATAAA
                                                                                                13600
TCTTATGTTTCGAAACTCGGGGCTATTTCTTCTTCTCTAAACTTGGCCCCTCTATCTCTTAGCCCTGGCCCTCTTTTTTTTTTTTCTGTTTAACGTCAAGGTC
  AfIII                                                                                        13700
GCGGCTCCTTAAGAGCGTTCGCAAATCTACTACTTTAACAACATCTTTTCTTTCTTAAGTTCTTCGTTCTTCTTATATTTTCCTAAACCTCGCAGCCATACCCC
                                                                                                13800
CGCTTCTTGGGTTTTTCTAAAGTTTTAATTTTTTTTTTTTTTATTCTTTAAAACAAACCTAGCCTTCATCCTTATTTCTTAGGATTTTCTTAGGAGATTAACATGGCCAC
                                                                                                13900
GCTCTTCTTTATCTAAGCACCGTAAAGTCCTTTGTAAATAGGAAGCAGGCCTCCGTACTAAACTCCTTAATTTTTATAAGAGCTGAACTTAGCTTC
            NDH2 end                                                                           14000
CTTTATTTTTTAATTTTTTTACTTTTCCTATTCTTGCTTTTTTTTTCTTAGACTCGACCTTCCTAATTTTTTTAATTTTTTTAATTATTTTT]TAGGCATCTT
```

```
                                                                                          14100
RPS12 start
TTCCTTCATGATAGATAATAATTTTTTATTT[ATTTTGTCAACGTTGATCAAAACGATCTAAAGAAGAAGAAGAAACAGCGAAGAAATAGGAGTGCGG
                                                                                          14200
CCCTGGTTTGCTGCCCCCAAAAAGAGGGTAGCGTTCTAAAGCCAAGGATCGTAACCCCCAAGAAGCCAAATTCTGCACGTAGGCCGTAGCCAAGGCCAA
                                                                           P7  PstI P5  14300
ACTCACAAATAAGAAGTTCGTTGTAGCTCATATCCCAGGAACGGCCACAACCTACGCAAACATTCTACCATTTTAGTTCGCGGAGGGGGCTGCAGGGAC
                                                                                          14400
CTTCCCGGTGTAAGGCATACTTGCATTCGGGGTGTTTCGGACTTCTTAGGTGTGCGAGACAAGACTAAGCGCAGTCCATCTATGGTATAAAGAGGCCCC
                                                                                          14500
                              RPS12 end    PSBG start
CCGAGATGGCTAAGCGCGTACGAAGGAAGTTAGGGCCATCTTTGGGCA[ATAATTTGAAGGCTGATTTTTAAAGTTATCTGCTAATAACTAATCTCC
                                                                                          14600
TGGGCTCGGCAAGGGTCTTTTGGCCCCTAACCTTTGGCCTCGCCTGCTGCCCTAGAGATGATGCATGCAACTGTTAGCCGCTACGACTTTGATCGCT
                                                                                          14700
                       NcoI      PvuII
TTGGGGTGATATTTAGAGCAACCCCCCGCCAAGCCGACCTGATCATCGTTGCAGGAACGGTAACGAATAAAATGGCCCCCGCCCTACGAAGGCTCTATGA
                                                                                          14800
CCAGACGGCCGACCCAAAGTGGGTGTTGTCCATGGCAGCTGCCGTAACCTGCCGAGGGGGCTACTACCACTTCTTATGCCGTTGTGAAGGGGTGCGATAAG
                                                                                          14900
                                                                          HindIII        15000
ATCATCCCGTCGATATGCTTTGTCCCAGTGCCCACCCACGGCCGAAGCTCTTTTTTTGGAGTTTTCGCAACTTCAAAAAACCTTGATGAAGACCATAA
      PSBG end     ORF6(178) start                                                        15100
ATGAAAAGAAAGTATTC[TAGT[ATTCGGCCGAGCCTTATTGAACTTGACAGAGTGAAAGGGTGCTTTTTTACGAGGAAAAAGGAAGGCAAAGAAGCCTTA
          KpnI                                                            SmaI           15200
ATTCCAGCGGTGCTTTGCTTTGGAAGCTCCTATGCTGGTACCTACTTGCTGGATAAGCACTCTCGTAGAAAAAACAACCTCTACCTCAGGTCTTTTT
                                                                                          15300
TTTTTTAAGTTTATTATTGCAACCTCTTTAACTTAATCCTGGCTTATTGGGCAGTTGCGGCATTTACCATACCGCCCGGGTCTTTTTTCTCGACCACTTT
                                                                                          15400
TTTTTCTTTAATAACTTTTTAGAGATACCCTCGTTTTTTTCTTAATTTTTGCCTACGCCCGGCCGTAAACTTGACCGGGGGGGCTGTAAACTTTGAGAATGCTGCTGAAGCAGAGCAGCCAGCTAATGCTCGTCGAGCGTCGAAGACTGCGCTG
                                                                                          15500
                      ORF6(178) end              RPL2 start
CCCAGACGACAGGCCGCGGCGTGTAAACTTTGAGCGGGGGGGCTGTAAACTTTGAGCGGGGGGC[TAATTAACTTTAAC[GTGTACGAGTTAAAAAAAGTAAAAAATCG
                                                                                          15600
AACAACAAACAGGTTGGCATCCTTTTTTTCTTCTTAATTTTTGCCTACGCCCGGCC... AAGGCTCCATTCTTTCAACAACTTCAACTATTTCTCATACCGAAAGCACCAGCTCTTAAATTTACATACCAAAGAACCCTAGTAGGAACAATACGG
                                                                                          15700
GCAAGGTCAGCATGCCATGCAGCCAAGAAGTTAAAAAAAACCTACGTTGCCTCGTCAGGGACTCTTATAACTCACTATCCTACTACGGTGCCTGCCTAACTGCCTCATAACCAA
                                                                                          15800
GCTCCTCTTTTGGAAGGAAAAAAAAAAACCCTACGTTGCCTCGTCAGGGACTCTTATAACTCACTATCCTACTACGTGGCCTCGGAACCTTCGTGCCTCGGAACTACGTCATGCTGAGGGTCT
                                                          AflII                            15900
AGGACCCTCAAGACGCTTACCTACCAAGCATGACCATGCTGCATCCTCCAAGCATGACCATGCTGCATCCTCTCTCCCCCTCGAGGGTCCTCTATAGGTTAGCTTTAGCGCGCCGGAACATACTTTAAGATCCTTTATCAGAG
                                                          XhoI                            16000
TAAGGGTTAACGACGTTATTTTTAACTTCTCTCCCCCTCGAGGGTCCTCTATAGGTTAGCTTTAGCGCGCCGGAACATACTTTAAGATCCTTTATCAGAG
```

(*continued*)

```
CTTCTGCAAAACCTTCTACGTAATGACCATCCCCTCGGGACTCATTATACGAGTACCCTCCGAGGGCCTGGCCGTCATGGCCGAAATTCGAATACCCAA   16100
AATAACAAGAGAGTCCTTGGGCTCCGCTGGAGTTAACTTCTTTAACGGAACTAATCCCGCGTAAGGGGGTAGCAATGAACCCCGTCGACCATCCAAACG   16200
                                                                             XbaI  BglII   16300
GGGGGCGTACCAAGACTCCTAAACCTGAGCGCTCCCTCGGGGTGAC[TAGCGAAGATTAAGAAATAAGGAGAGTAGCTTTTGTCTAGATCTGCTTGAAAG
              RPL2 end                                                                        16400
              XhoI
GGCTTTGTTTTTGTAAGAGCACGAAGCGCGCCCTGCTCGAGGGCACTTGGCAGGGGAGAAGGTTTGAGCTTCTTCTTACTTAGAAGCTCCACATCC   16500
                                                                        SstI
TCCCACACTTCTGGGACGTCGGCTTAAAGGTCCACAACGGTAGGTTCCTAAAGAGAGGCGGCCTCGGGGCAGCTCGTAGTCTGTAGAAAAGCTGGAGAGTT   16600
TGCTTTCTCACGAAAACCTTATTTTTTCCCATAAAAAAAAATAGATAGAATAGCTCTTGGGTCAAAGCAGTACTTTCTTACCTGATAGGCAGGGCGGTT   16700
                                                   BglII  ORF7(125) start
TTTCAATTTTGGAACGATGTTTGAGATTCCAAGGAATCTTTTCAAAAGTCTAACGGCGA[ATCTTACTCCTAAAGAACTGCTTAACGATATTTTTTAAAA   16800
ACAAAAACTATGTGAGCTGGCGCTTCTTAGAGGTAAATGACCGCGAGACCTGGCCGAAGACTATGCCCTCGACCTCGATAAGTACGCCCTAGAGGCGTA   16900
CTGGCTCGAAGCCAAGGAGGTGAAGCAGGACGATGCTTCCGTCCTTCCGTCTTGTCTTAAGTACCAGGGCTGGTTTTCTTTTTTATCTACTACTAT   17000
          SstI                                                                ClaI
CTTTTTAAAACTCAGGCAGAGCTCGATGCGGAAGCTAAGATAGACGAGGAGCTTTCGGACGACGAGGTAACATCGATGAGGACCTAGAGGACTTTCAGA   17100
          ORF7(125) end                                   ORF8(241) start      AflII
ACGAGGCCAAGGTGTCCACGAGTCTCGAACTCTTT[]AAAAAAAATATTTTTTAGTTCTTGGATCA[ATTTTAAGGTAAATGATGTCGTGCTTAAGGAT   17200
GTCGTAAGCTACCATAGCTTCTTCTCTAAGCACGGCCCCTGGTGTAAAGGGGCCCACAATTGAAAAGTTTTAAAGCGCTTGAGTACAACGCCCCCCTGG   17300
ATCTTTACCAGGTCGTAGACCTCGAAGACTTCAATCTTTTTTTTTTTAGATTTTTTTTAAAATCTTCCGAAAAATTAAGCATCTTCGATCGCCAAGC   17400
                                                                 NarI
TGCCAATATCGTAACCGCTAACATAATTTGAAGCTACAGGAGTTGGCGCCACTTTAAGGGCCTTCCTTGTCGGGGGCAGCGTACGTGGTCCAACGCTTCT   17500
      BglII
TCTTGCTACAGATCTAATCTTATCCTGAGGGACTATAAAAAAAAATGTTAGGAAGATTTTGGTAAGTACGGCCGGCCCGAGCAGAAAATTTGCTTCC   17600
      XbaI
TCTTGCGAGTACATAAACTATCTCTGAAAGTCCCAATGATTTAGTGAGTGAATGCACTCTCGAAAATGGATCAAATACACTCTAAAAAAAAAAAGGTCGT   17700
CTTCTACCTCGACCTCTACGCAACCTCGAAGGGCCTACTTGCCAACCTTAGGACGCGATGCCAAGGGGGTTACTAAAAAAAAAAAGATGCTCACGGGC   17800
                                                            HindIII  ORF8(241) end
CACGTTGGGTTTGATCAGGGCTTTACTAAGGATCTCACCTAAAGGCTAAGTATGCTGTTTCTAAAAAAGTTAGAAGAAAGCTTAGCCTTAGA[]AAAAAATT   17900
AGCTTTTGGAAACTTCACCTCTAGATTTACAAGAACTTCAGCTAGACACCCCAGTATCCCTAAAGACCCGATCCCGACGAGGTACT   18000
TCTTCGCGCCCTTTTTTCTCAGCACGCCCTAAAGCCCCTAAATTTAAACTTTTTTTTTTTTGATGTACGATGGGCCTATGATGCAAGCATCTTCTTGAGCAGGCTAGAG
```

```
                                                                                              AflII                                18100
GACCTCGACCCGCCCTCTTTACACTCCTTAGTGCCGTGGATGAGCACGAGTACATTGTTAGCCTTAAGCGGCGAGCACTATTTTTTTTTCGACTCTTTG
                   P5 PstI P5X    ORF9(204) start                                                                              18200
GCTTTGAAGCTGCAGAGCGCTTTCCTC[ATCACCCGGGATAGCTCCATGCTTCGGAAGGCCTTGAGAGGCGAGCCTACGATTCATCGAGGACGGCGAG
                                                       SstI                                                                     18300
CCTTGCGCAGGAGGAAGAGGGGGAGGGCTTGATTTTTTTTTAAAAGAGCTCCTTCCTGGCCGCTACTGACTCCTCGGACGCGACGCAGGGGGGGCTTCC
                                                                                                                               18400
TCGGGCGACGAGGGGGAGGGCTTCCTCGGACGACGAGGGGGATGCTTCCTCGGACGACGAGGAGAGGCTTCGGACTGCTGTAAGCTAAAGAGGCCGAGA 18500
AGGAGACCTTGGACGCTGAGGGCGAGGACTCGGACGAGGAGGGCGGCATGCAGGCCGTAGACTGTAAGAAATTTTTAAGCTAAAGAGAAAGTCTACCTGGAG 18600
                         ClaI                                                                                                  18700
AATCCTCTTTTTCATGCGTAAATTTTATTTGGGCTAAAGTTACGAGATCCCACGTCCACCTATTCCTAAAGAAGCTAAAGAAAGTCTACCCTGGAG
                                                         ORF10(105) start    SstI                                             18800
ATTTGCAATTGCGCGGCAAATAGCTCATCGATGCTGCCTCCTCTCTTTTTTAACGTCAGCTATCTAAATTTCTTCATAAGGCAGGGCTATC
          ORF9(204) end                                                                                                        18800
TTTACCAAAACTTCAATCGCATCGTGCAAGGCCCCCAAC[TAAAGTTTGGTTCCTT[CGTGTCTTGCATCTTTTTCTCCGAGCTCTTTCCGATCTTTACC 18900
                                                                   AflII                                                        19000
TTCTTTAAAG1AAAGATAGGCCACTATGCTCGTGTATTGGCACGCAAGCTAGATAGGCACTAGATATTCGCAACCTCCTAACGCTTAAGACGA
                    SalI                                                                                                        19100
AGAACATGAACTTTTAAAGTTTTTAAACAGGTCGAGCCCCGTCGACACGAGGGCCATTGAGATGACTACCTAAGCCTGAGCTTTTTTTTTTTGGAGG
   XhoI                                          ORF10(105) end                                                                19200
CTCGAGCTGGAGCTTAACTCGAATCACGGCCTCAACGTCTTCTTGTACCGACTCATCTCCTTAGGTAGACCC]TAACCCTCTTGGACCCGAAATTTCT
                                                                                                       AflII                   19300
TAGAAAAGACATCTTTTTTTTTGTTAAGTTAGTGGAGTTCAGAAATCGAAGCCCTCTTTTTCTAAAAGAAAAAGAGAATAAAGACCTTTTTTTAGCATAA
                                                                                  CYTB start                                  19400
AGAAAAAACAACAGCTTTCAAAGCACACATTTTGAAGAGAAAAAGCTACCTCTTCTTGAGGTGGCTATCGCTCGCTGGCAGCGTAGATTTAGGT
                                                                        ATCTTAAACATTTTCAACTACTTT                              19500
GAGTGGAATAAAGAAAAAAAACACTAGGCTTCTTGATTACTCCGTGTTCAGGCAGGGGGTTACTTTAACTTAAG[
                                                                                                                               19600
AAGAACCTCAGGGTTAGCTTCCACGAGGTCTTCTCCGATCTTTATGACGATCATAGTTCAGCTCGTATCTGAACCATGCTGGCTT
                                                                           XhoI                                                19700
TCAGCTCCGTGCCCGAGCCCATGCTCATCCTACCGTACGTGACGAAGAGGATATCGAGGATCTCACCGACGACTCTTCTGATTACACGAGAGGGG
   SalI                                                     HindIII                                                            19800
CGTCGACCTCATCTTTCTTCTTACTTCCACCTACTCAGAAAGCTTTACTTAAATGTTTCGACCTGAGACCGAGGCGTCCTGAAAGAGCGGAGTC
                                                                                                                               19900
TTCTCCTTCTTGGTTTTCAGTTGTCGTTTTTTTTTGGCCTTGTCTTCTTGCTACACACCTTAGTGAGATTACTCTTACTATTGCAGCAAACATCTTCC
                                                                                                                               20000
ATACTTCTTCATGTTCAAGGGCAAGGCTTACTGATTCTTGTTTACGGACAAGCAGCTAAACACCGACACCCTCATAAGGTTGGCTTATGCTCACTACGT
GTCTGCCTTCTACCTCAGCTTCCTTGGTTTGCTCATGTATAGACATCCACTATGACTGAAAGAACGAGCCCTTCTACACGGCCTTAGCTCTGGAGATG
```

(continued)

```
CTTTGATGGGACGAGGCCCTCTCAAACGAGCTTACAAACTTCTTTGTCCTGCTCTTCATCACTCTCGCCCTTTTTTTTACTTTTTGAGGAGCCCGAAG   20100
            BglII
CCCTAAGCTACGAGATCTTTATGTGGGGCGACATTGGCCTCTCTCCTACCGATGTTAGTTCTACGGCGTTGCACCCATTGATATTTAGGCCCTTCATGGC  20200
                                                                                                       20300
ATGATTGATTGCTTGCCCTTTCCACAAAACAGGAATTTTTGGCCCTTTGTCTTTTTCTTTTTCGTAACGCTTTACTACCAGCCTCACGGAGTATCAGAT   20300
                                                                                                       20400
CAGAATAGCTATGGGAAAAAAACCCTCACCATCTCCTCTACCGTACTAGCTAAAAAAAATACTGCCACACCCTTTAGCATCTCAATCGACTCAAACCTCT   20500
ACCACCAGATCACCTACTTTTTTTCATCATGTGCTGCCTCTACACACCCTCATTCCTACCCTACGGACGGTTCTTCAACCAGATTGGAGGTAACTGAGG   20600
                        CYTB end
CTTCTTGTCTCTTCCTTACTTTTGCTATCTAGCTTTTACTGA⌐TAAGAAGGCCCTCTTTTCCTTGATTTTTCTTTAGAAAAGCCTACCACGAAG         20700
TTTATTTTTAAAGGTTGCTAGATAATCTAGCAGCTTTTAGGGGGCTTCGGGGTGGGGACGCCTAATCCCCTGTTCGATCTTTATCAATACCTCTTCT     20800
TCTGATCTGGATTTATGTGCGGCTTAACCACGATCTACCTTTTCTTTACCAGCTATTTCACTAAGTTCCTCCGGGCCCTGGCCTGATTTTTAAAAATTT   20900
                                                                                    PvuII  20900
AACTTTACCCGTGTTCGGAGTACTTTCGACCCTTTGTATATCTTCACAATCCTTCGCTGCTTATGCATACTTCACCTCCGACGATTTTGT            21000
TTTGGTACTATTTGGGGGCATTTAACGTTGTATATCTTCCTTCCTTCCTTACGCACTAAGCGGTGATCTTCTCTCTCTCTATTTAAACACTTTTAAGCTCTAAGGGGTCTTCCTTA  21100
TATTGAGTAG⌐ATGTTCCTTCCTTCTTTTTCCTTCTTCCTTCTCTCTTCCTTCTTTTTCCTTCTCTTTACGCACTAAGCGGTGATCTTCTCTCTCCTATTAAACACTTTTAAGCTCTAAGGGGTCTTCCTTA  21200
          NDH5 start
TTAAACACGACCTCCATCGGCCCTCTTTGAGCTTATAGCCTGTCCAACCTTAATTATTTTTATAAAAAATAAACTGATTGCGATTCACCTCTTCAGGT   21300
                                                        ClaI
GGTTCCCCCTTCAGCAGGCTACCTCGTAAACTTTAGCTTTTACATCGATACGGTACGCCTCATCTCCCTCATAAAACGCTTTATTTGCAAGCATGATTATTTTAGTAAAT  21400
GAACTTATACACGTATTCTTATTTTAGGTACGAGCCCCATATCAGCCGCCTCATCTCCCTCTTCCTAATTAATTTTGAGGTGAAAGAGCCCCTACCTTCAAAT  21500
TCGGGAAACCTCGTTGGTTTTTTTTTTGGTTGAGAGCTGATTGGTATAACCTCCTTCTTCCTAATTAATTTTAATTTATGCCAAGCTGCACGACCTTAATTTGAGGC  21600
CAGCCTTTAAGGCCTTCTCCTCCATAATAAGTTAGTGATTCGGCCCGTCCTGGCCGCCGGGCAGCTCCACCCCCCCCCAAATAAATTCCTGAAATCTCATCTCCTTTGCTCTTTGCAGCC  21700
CATCCTCAACGTCTCCCATCTCTATTCTGAGATGAAGCTAGGCTCCACCCCCCAAATAAATTCCTGAAATCTCATCTCCTTTGCTCTTTGCAGCC       21800
TTTGTTAAGTCTGCCCAGTTGGGTTCCATGTATGGCTCCCAGATTCTATGGAAGCTCCCGCCTCGCGCTAATTCACTCAGCCACGCGTCGTGT          21900
P5 PstI P6                          XhoI SstI             HindIII
CCCGAGGTGTCTTCCTCATCATGAGGTTCACCCCATCCTGAGCTCAGCCTACTTTAAGCTGTTACTGCACTCGTAGGAGCCCTAACAGCTCCGCGT       21900
                                                                                                       22000
TGGCGGCCCTCTCGAGTGTTCCAAAACAGACCTAAAAAAAAATACTCGCCTACTCTACTATAAGCCATTGTGGATTCCTAATCTTCTTGTGCAGCTTTGGA  22000
```

```
AATTTTAAGCTCGTCATAGTGTATCTTTCGTGCATGGATTTTTAAGGCTATTCGTTCCTCTGCGTCGGAAACTTAATACGCTTTCAAAGAGCTATC   22100
                    PvuII                                                                          22200
AAGACTTGAGGAGGATGGGATCTTCTTTAAGTATCTTCCAGCTGAATTTTTTTTTAGTCTTCTCATTACTTAACCTCAGCGGGCTCCCCTTCTTTTT   22300
TGGATTCTATTCGAAGACACTCTGTTCGTCATGATTTCAGACGTGCTTATTTTAGAGATGCTATCTTTTGCATGATTTTGCTAAGCTGCATCACAGGCCTT   22400
TTCTATTCTTCAACATACTTTACTATTCTTTTTGATTCCAAAAAGGCAAGGAAGAGCATATACGGTGGCGTCATTAGCGAGTACCTTAGGTCTTACT   22500
ACTATAGCAACACCACGATGGCCTCCAACATTGCTATCTTTTATTGATAGTCTCGTCCTGCCTACCTCATAAACTTCTACCTCCTCAG   22600
CCTCTCAACTGCAACTGATTTTTATCTCGTCTACGTTAAGACCTTCTCTTCACCCTGAGCCGAGGCCGCCCCTCTTAAACTATTCTTTTTTT   22700
TATTGGATCATAGCAATCTTCTTCGTGATCTTGGTGCTCTTTCTTCTTTCTTACTACTCAAAAAAACAACGACAGAAGTAAGTCTAGCAGGGTTTTTTGATTTT   22800
                    NDH5 end                                    ORF11(367) start
TTCTTGGAGGGTTTTTTTT]TAATTTTACGCCCCTTCCTACTTCTTTTTTGTGAGGTTGT[ATTTCCACAGATTTTGTATTTACCCCAACTTTAGCACAG   22900
GCTTCTCTTGTTGAAGTCCGACGTCGTCATCCACATTGCTCAATGGCAGTATTGCTGATGATTTTGGTTTACATACTTGTGATCCCTGTATTTTTTTTT   23000
                                                                                            AflII
CATTAGCAGTCCATCAGGGCTAGAACCTTTAAAATGCGCCCAAAAATCTATACTAGTTTTAGTCTCACGGAAAGTGGGTGACTTTTTAGCTTGCATC   23100
GTGCCCGTTATTGATGCTTAACATTCTCGTAAATTCTAATTTATCTTAAGCTCATGGAGTGGCAAAACGAGTCCACCATCTTTACCAGTTGCATTC   23200
GTGCTAGGCAGTGATACTGGATTTATAAGTTGAGCTGAAGAACATCCTCGACCTGTTGACGGTCCCAAAGAAGAGTTGGTTGAAACAAGTGGCTCATCCA   23300
TACGGGAAACTCAATCGAGGTTGCTGATGACTACTTCTATGCTTAAGGATTAGGGCTCAAAATAGCTGAACCTCAAAGTACTGAAAAAACCTTCGTGAGG   23400
                                                                                            HindIII
AGCTTAAAGAAGTTAAAGTTAGCAATAACATCCCACTTTATTGATGACGTAGTGACTGACTACTTAAGGCTTAAGCCTCCGTCCTAAGCTTCATGCCCTACA   23500
                                                                      XbaI
ACCAGTTTGACGACAAGACACCCTAGGCTTTAACTTAGCTAAAACAAACGACTCTACTTGTGTTGGGGCAAGAAGAGATTGGGGCAAGAGCGACACCTTGTTCAATGCCCTAAAT   23600
GGGCGGCGATTCTTGATTGTGCCAAAGCATCGTCTTCAATCACTTCGTTATACGACATCGATGCCCTCGTGAGCAATGCCTCTTCTTAGAAAAAAACCTAAGTACGTAA   23700
                                                              ClaI
AAGACTATGTTTTAAACAGAAAATTTTTTAAAAACTTCTTATACGGATTCTCTTCTTCAGTTCACTAAGAAGCGGGTGTTTTGAAGAGGCCAATTTTAATTACCAAAGCCTTCTTCCC   23800
         HindIII     ORF11(367) no stop but overlaps                   COII start
CCTAAACATGGAAGCTTTGATAATGAAATCCTCAACGCTCTTCGATGCAACCGTTACC[GTGGAGAAGAAGATGAACACCGACCTCTTCTACTTAACC   23900
                                                                                            AflII
CTGAAGCAAAAAAGATACAAAAGAAGAAAAAAAAAAACATCCCCTTAAGAATAAGGCTAGACAAGAATACTGCCTCCATCAAGTTTACGG   24000
```

(continued)

```
                                                      HindIII           24100
ACAAGCCTTATCTGGTTGCTAACAAAATCATAAAGAGCCTGGAGTTAATGCTACGAGCCTTTACAGGTGCATTAAAAAAAATAAGCTTAGAAGTGAAAA
      XbaI                                                                            24200
TTTTTCAGTTCAGCTTTCTAGAAGACTCTTGAGAACTAAAAAAACTCTAGTGCTCCCTTCCCACGTAAACATCACCCTTATTCAAACTCTTACGACGTG
                                                                                      24300
ATCCATTCTTCTGGTTCATTCCTGCCCTAGGAATAAAGATTGACTGGCTTCCTGGAAGGGCCACACACCATACTTTCTACTGCGATAGCGTAGCCTTTACT
                                                                                      24400
ATGGTCAGTGTGCTGAAAATTTGTGGACGATACCATCACCACATGCCCATTAAGCTCTGCACTCTCCCCTCGAGCACTTCTTAATTGATGACAGCACTT
                   COII end                                                          24500
TGGCCTGCCAAGCTCTTGTTTACAGAATCTAAGAATAGGTTTGAGACGGACTACGGACTTAAAAAATTTGCTG TTAGTTGCAGCC
SSU rRNA 5'  E1 EcoRI E6                                                               24600
AGGCTGCATACTTGGATGTAAGAATTCTTTTTTAGGAGTGTGATCTTAGCTTTGATTTAACGCTAATCAAATGCATTACACACGCAAGTTTTATTTTACA
                                                                                      24700
CAAGAACCTTCTTCTTTGCTTAAAATGCAAAGATCTAGTTTTAAAAATAGCGTATTGGTGCGTAAAATATGTCTTTTATTCGTATTGTACCTTAGATAGG
         BglII                              ClaI AflII                                 24800
CTAGGTTTTCTTCCTTTAAGAAGTTTAATGCCACCCAAAAAAAATTTTTAAAAAATCGATACTTAAGTATTGAACTCTTTTTTTGATCCAATAGGAA
      E6 EcoRI E2                                                                      24900
AAAAGATATAGAGCCGCTTCGAATTCGAAGGGGGGCCACTTGGGCCTCCTGCCGTTAAGCGTGACCCAGCAAATTGATGTTGCGAAGAAAATAAGTGAGGTGCCTGCAGCAATG
                  AflII                                              P6 PstI  25000
CCTTATCTCGGACAGAAGTGGGGAATTTTGGGCAATGCGCTTAAGCGTTAAAGCTGAGAAACGAGTATAATTAAGCTAAGTTTGTTAGAAGCGATGGCCAATACATGCCAGCAGCC
                                                                                      25100
TGGGATATGCCTCTTTGACTTTTTTTAAAACGCAGGCCAGAAACGAGTATAATTAAGCTAAGTTTGTTAGAAGCGATGGCCAATACATGCCAGCAGCC
                                                                                      25200
GCGGTAAATACAATGAGTGAGCTAGCGTTAATCGTCCTTATTGAGCGTACAGAGTGAGTCGGCGGTCATAACAATTAATAAACTATTTTATATAAAAGATAT
                                                                                      25300
TAATAAAAGGACTAGTGATTTTTTTTATGTAAGTGAGCTTGTAAACTAGTGATAAAATACAGGGAGTTTACAGGAACATTCGACGCGGCAAAGCGACTA
                                                                                      25400
ACAATTAAAAAAATGACGCTATATCACGAAAGTATAGGTAATGAAAGGCATTTAAACCTAGTAGTCTATACTGTAAAAGATGAGTATTTAATACTAAGT
                                                                                      25500
TAACACAAAAAATACTCCGCTTGGGTAGTAAAGCCGCAAGTCAGAAACTTAACAGAATTGGCCGGGGGACTTGTTCAAACGGTGGACATGTGGTTAATGCG
                  HindIII                                                            25600
ATAATCCACGTAAAACCTTAGCAGCGTTAAATTTTTCTTTAGTTAAGCTTCATGGCTAAAATTTTTAGAAAAAAAGAGATGGTATTGCATGGCCTGTCG
                                                                                      25700
TCAGTTCCGTGTTTTGAAATTTGAAATTTAGAATTAGAGTTTTATAAACGAACGCAACCCTTGTTTTGTGTGTTAATACAAAATAAATGATTAAGGAATTATTCATTT
                                                                                      25800
TAGCTAGAAGGCTGAAGTCAAAGTCCTTACGGTCGCTGGCTACGCTGGCTACCAGCATGAAGAAGAAAATCGTTAGTAATCCGCAAATAGTATGTCGGTGAAATAT
                                                                XbaI    SSU rRNA discontinuity  25900
AGTTTTAAAAAATTACCACAGTACGAATTGTTTCTGAAACTTGGAAACTTGAAACGCTCAAAGAGTTTAAATCTAGAAGCTGAATTTAG GATGTGA CCCTACCCATTTTTTTTTAAAAAAAC
                                                                                      26000
AAGTCAAGTCCTGCACACACTGCCCATCACGCTCAAAGAGTTTCAAAGAGTTTAAATCTAGAAGCTGAATTTAG GATGTGA CCCTACCCATTTTTTTTTAAAAAAAC
```

```
TCGCCGGTGCACCAAACACCGGTGTTCTAAACAAGCTACTTGTTGTCAGCACAAAGAAGAGTAAAGGCGAAACCATTTGAAATGATCGGAGTGAAGTTGA    26100
        KpnI                                                    SSU rRNA 3'
CACAAGGTACCCGTATGGGAACTTGCCGGTGGAGCAAGGTGAACAGACCAATGCATGCACCCAGAACGGGGCGCG|TTAGAAATGTTTGAGACTTTTTA      26200
AATTTCTTAAACGAATTTGTCGTTGAACATTTTTGTATTTTGATTTATCTTGTTTTTTTAATTTTTTGCTAAATGCCGTCTTGAACG                 26300
AGTTCATGTTTTTTAATTCATTTTTTAACAGAACCAGATCAAGACGCTCAACTATTTTTATTATTTAGATTCTGATCCATTTTTCTAAAAAATTTTT       26400
GTTTAACCTCTTCTTCTAATGGTGATTTTTTTTTAT[ATGAAGTCCCTTTTTATTTTTTGAAAAGGATAAATTTTTGACTGCCGATTTTTTTATTAAAG     26500
                                     ORF12(265) start
AAGATTAGGATCGGTGCCGATAACTTAAAAGACGTCAGCTCCCTGATCTTTTTTTCGTACATATGCACCACAGGGGCGTGGCTGTTCACAACATCTTTT     26600
TATTTTTGTTAACCTACGTTCCTTTCTTTCTTAAGAGAGCGGTCCTTGGCTGCTACTTAAAAAAAAAAGCTCTTCATCGACCATGACCTAAAGAATAAAACTA 26700
TTTTAAGATGAAGCCGTGCCAATGCTCTTTTTTTTAAAGGTGACTTCCTCACCTGCTATTTTTTAGTTATTTTAGTAAGGGCTCTTTAAACCGCCGAGACCAGACCTACTTTTA 26800
TTTTCCTAATAAGGAACCACGTCATAACTAAGACAACGCTCAATGCTATTTCATCGTCTTCTTAAATGAGACAAAATAAAGGACTATTGGGCGTACCTAACGAC 26900
ATGCCTACGTCCTCATTTTTTTAAAAAAAAACAAGGTAACGACTTCATCGTCTTCTTAAATGAGACAAAATAAAGGACTATTGGGCGTACCTAACGAC       27000
ORF13(169) start
CA[ATATGCAAGCCCGAGGCCCAAAACTTAAAACTTATCGACCCCGTAGGGTCCTGATTGAACAGTCAAACTCCTATGCCGACAGGCTGCCGACGAGA        27100
                                                                                                       ORF12(265) end
AGGACTTAAGAAAACCAGCACCTACTCAAAGCTCGATGTCGAGGCTATGCATTCTCTTTTTCTTTTACAACGCCCCATTGAAGAAGCTGAGCGCCTC         27200
CTCTCAGCCTGAGGCTCCATTCCTGCTACCC]TAAAAAAAATTTCAATTTTTTTGCAAAAAAAACTACCTCTATAACAAGGTAAGTTTCTAGGAAT         27300
                 ORF12(265) end
AAGCAGACATATAGACGGGGGTTTATCTCTGCATCTGATTAACCGTACTGACGGTCGTCGGGCCTTTATTTTTATTTCTACCTTATGAGCATGAAGTTA      27400
                                                                                      ORF13(169)  27500
CCTACAACTACCTACTCTTTTTTTATTCTTCCTAGGCCTTTTTTTTTATAAATTTTTATAAGAAAAATAACAAAAAATTTGAGGTGCATACTGATCTCTT      27600
end              tRNAtrp
TGAAAATTTT|TAAGATTCGGGCACTACACTC[AGGGGAGTAGTTCAACGGAGAAAAACTTAGCTTCAACGGAAAAACTGACATCGTGGTTCGACTCCCGCCTCCC  27700
                                ORF14(387) start
TT]GTACTTTTTTTTTGTTTTTTTTTAAAATG[ATAGAGAGACCACAGACTTTCTATTTTTTAATAAGTATTCACGAGAATTTCATACCACTTTTTTGGTA     27800
end                          HindIII
AGTTTTTTTTTTTAAAGAAATAAATACAAAGCTTATTTTTTTTTTTTGAGAGTTTAACCTCTTTAGTTCTTCTTACCTTCTTTTAGTATTTTAICTTATC      27900
GAGGAACCTTGTCCACACAAAAAAACTTTATTACAGCACTCTTACCTCTTAAAGCAACCCTAAAGAACAAAAAAAACTTCTTTATTTTTTTTAGG          28000
ANCTCCTTTTCTTTGTCGTCTTTTATTTTTTTTACCTTCTATCTCTATTCCATATACGCTAGAAAACATTCCGTATCCAAAACTCTTTTTTGTTGAGGTAGCT
```

(continued)

```
GAGGTGCTTTTTATACATCTTTATCTCTGGCTTAATTTTTTGAAAAAAGTATACCTATGAAAGTATACTGAGGCTCTGCAAAGGTTTGAAGAG    28100
GAGTTTTAGCATCTTTTGACTTATCGAGGCGTTTGTTTTTTTGTTTTTCAGCCTTCATCTTCCTGACCTTCAGCTCTGAGGTCGTATACAGCTACGACCCC    28200
CAGGCTTTTTTAAACTCCACCTGATCTCTTTAAGATTTTTTTTTTTTTAAAATGCTCGCCCTCACTTTTTAATTCTATGCTTTAGCATGGTCTCCTCTC    28300
TAAACTTAAGGAAGCGCCTTTAACATGTTTATTTTTAACAACATCACTTTTCGTCATCATTATTTTTTTTAATTGAGTCTGACCAGTATATTAGTATTGT    28400
AAACTACTGTGGCTTCTTTGAATGATCCTTTAATCACAGCGACTTTGCACTCGATTCTGACTTTAGGAAATCCCGAACTGTTAACAGCTATGTGCTCTTA    28500
ATCGGCATTGCAAAGTACCTTCACATCCTCTTTATCGTATTTGTTTGATTTTTTAATTTTGCAAAAAACCTTGAAAACAATGAGTCCAGAGACTACATCG    28600
CCGGTACCTGCACCACAAAATGCCATCATCTTGTACCTCCTAAATTGGTTTGCCATCTATCCTTACATTAAGTACTTCTTCAGAACTTATTACTATAGCAC    28700
TTTTCTTGATTTTTTTTGACTTTAAAAATGAGTCTGCTTTAAACTTGCCAAGGTTTTTGTTTAACTTTGTTTAACTTTTGGAGCAGTGCCTTTTTAAACTAGAAATAA    28800
                                                    ORF14(387) end
TAAAAAAAGTAACTTTGGCTAGTTTTTTACTCGTTAATGTGTTCTTCATGAATTAGATTTTCTAACGGGGGTCCTTATAAAGTACCATGTACTTTTTTT    28900
ATTACCTTGCCGTTGGGTTGAATACCTTCTTTATACGCAGCAGTGCAACAGTGGCTTCGGTGGATTCGTAGTGGTGTTCTCGGTTGTTGCCTAGAGCTA    29000
                                                    ORF15(85) start
GTCTTGGTGACTAGCCTAGACAAACTTCCTTGATTCATCTTCTAGGGATACCAGGCACTTTGTCTCTTTTCACGTCGATTTTCGTAAGTTTAAAACTTCC    29100
                                                    ↓ 5' mRNA 23
TACGAAAGCGGCCCCTGAGCAAACTCTTTTTACAATGCAAGGCTTTGCGTGGCTTCTTTTTAAAAGAAGCCAATTACAATTTTCAGCACTCTTGACTTTAC    29200
TTTATTTTGTGTTAAACCAACTTGCTTACTACATGCGCGGAAAATTTTCAGTTATTGCCCTACGAGAATTGAAATGAAGAAGTTAAGCGAGCCTTTGATAGAT    29300
                                                    COI start
ACCGTTATTTTTTAAAAAGCATGTT[ATACCATAAACCACAAAAGGATTGCATTAAACTATTTTACTTTAGCATGTGGACTGGCCTTTCAGGAGCTGC    29400
                                                    ORF15(85) overlaps
CTTAGCTACCACCATCGATCCGTCTTGAGATGGCCTATCCAGGAAGCCCTTTTTTTAAGGGGGATCGATAAAGTATCTTCAAGTAGCAACTGCTCATGCCCTC    29500
ATCATGGTATTCTTTGTGTTGTCCCTATCTTCTTTGGAGGGTTTGCTAACTTTTTGATACCCTACCACGTAGCTCAAAGATGTTGCATTCCCCAGAC    29600
TAAATAGTATTGGCTTTTGGATCCAACCTCTCGGCTTCTCGGCTTCTTTGGTTGCAAAAATTGCGTTTTGGAAGTACTACGACAAGACCTC    29700
                                                    BamHI
TTTTTTCTTACAGCCCTACAACAAATCTCTTTACAGGGACTTCTTAATTTTTTGACGGGGGAGCTTAGCTTTAACCCCTTTAAAAAAGTTTAGATGAG    29800
TCCTTGTTCCTCTTTTATGGAAGCCCAGAAAAAAGTAACCACACGGAGTATACATCCTTCCTTCTTCTTCAACCCCTCAACCTAAGCTTTTGGACTCTT    29900
                                                    HindIII
TTTTTTATTATAGTGACAACCTATCATCCCTGGCTAACAAGGTTGTTCCTCTAGGAGGAAGCAAAATTTACGTTACAAAAGTGTTCGAATCGAGCTCGTCT    30000
```

```
30100 AACAGCAGGTTGGACCTTTATAAACACCCTTCTCTTCTAATATGAAATATTCTGGTTTTGGGGCTCAAGACGTCCTCTCAGTTGCTGTTGTGCTAGCTGGC
30200 ATTAGTACAACCATATCTTTACTAACTCTCATAACGAGGAGGACTCTAGTTGCTCCCGGACTTAGAAATAGGAGGTGCTAATCCCTTTATAACAATTT
30300 CTCTACTTCTTACACTCCGCCTCCTTGCCATCGTAACCCCAATTTGGGAGCTGCGGTCTTATGTCTCATCGACAGGCATTGACAAACTTCCTTCTT   BamHI
30400 TGATTTTGCTTATGGAGGGATCCTATACTATTCCAACATCTTTTTGATTTTTTGACACCCGAGTCTACATTCTAATAATCCAAGCTTTGGAGTT   HindIII
30500 GCAAATATCGTCTTGCCTTTCTACACCATGAGGAGGATGTCTTCAAAACATCACATGATTGAGCTGTCTGTCTATGTTATGCTTACATGGTTTGTTGTTT
30600 GAGGCCACCACATGTATCTTGTAGGGCTAGATCATAGAGCAGAAATATCTACAGTACCCATCATCATGATTGTCTTCCAGCAACTATTAAGCTTGT   HindIII
30700 TAATTGAACTTTGACGCTAGCTAATGCTGCTATTCATGTAGATCTGGTATTCTTGTTCTTCGTGTTCTTCATACGTCTTCTTCTTCTTCTTCTTCTTCTTAACTGGAGTTTTACT   BglII
30800 GGAATGTGGCTCTCTCACGTAGGATTGAATATTAGCGTTCACGATACCTTCTTTCTATGTCGTGGCACACTTCCACCTTATGCTTGCTGGAGCTGCAATGATGG
30900 GAGCCCTTTTACAGGCCTCTATTATTACTACAACACCTTCTTGATGTTCAGTACTCTAAGATCTTGGCTTCCTTCACCTTGTTTACTACTCTGCAGGTAT   BglII   PstI
31000 TTGAACGACGTCTTCTCCCCATGTTCTTCTTAGGCTTTCTGGGGCTTCCCAGAAGGATTCATGACTTCCACAAGGCAGCCTTCTTCCTAGGTTGGCACGGACTTGCC
31100 TCCTGTGGCCACTTCTTAACCCTTGCCGGGGTCGTGCTTCTTCTTCTTCGGAATTTTGACTCCCACAAGCGGAGAATAAGTCTTCGATACTAGCTAACTTTG
31200 GTATACCAAAAATCGCTAAGAGGGCCCCACCTTTATTTCTTAAGATCAGCTACAATACTATACAAATGAAATTGCTAGCGAGCTTCCAAAGGTCGAGGT
31300 AAGGAAGTTTATCATCGAAAACACTTTTGGAGAGTACGAGTGCGTAAGTTGGTGCCCGTCACTAAG TAAACGACTAATATTGATTTTATCTTTTTATG   COI end
31400 GTTGATGCTTTTTTGCACCTCTTAAAATGCTACTTTATTAGCTTCACTAAGAAAAATTCACTAAGCTTATTGAAAGATAGTCTTCTTTTTGCG   HindIII
31500 TTCTTTTGAACCTCCGGACTTTTTTATATTTCTGTATCTTTTTAACACTAGCCTAGGCACCCCCTTTTGAGTGTTTTAGGTTTTTGATAGCCTCGGTTCTC   ↓ 5' mRNA 24
31600 TTTTAGCTTATAGACACAGCCTCGATTACCTATTTGGTTGACTTAGCTCTGTACCAGGGTTACAAGCTCGCACACCACTCCCTCCGCCTTTTACCTAGAGTC
31700 GAGCATGCTCTTACTTAGGAACAAAGTATTCTTCTTACCCTTTTTAG ATTTTAATCTATTCCATCGTTTGATGCTGGTCGTTACCTGATAATCGCA   NDH1 start
31800 TCCATAACTCTCCTAGAGCGAAAGCTGCTTCTTCCTTGGTTCAGAGGGCGCGTAGGGCCTAACTTGTAGGATACAAAGGTCGTCTTCAGTACCTTGCCGATG
31900 CACTAAAACTATTTTTAAAGGGGGTTGCAATCCCTTCTGGCGCTAATTCTTTTTTTTGTAGCCATGCCCTCTCTTCGCGGAGCGGTTTGCTATACTTT
32000 TTGAATGAATTCCATCTGGGGCCCAAGCTTAAGCATGTTCGATGTGAGTACAGAATATCGTCTATGCCTCCTCTGCCTCCTCTCCATCCTCTTTGGACTCTGCGTC   E2 EcoRI E4   HindIII AflII
```

(continued)

```
ATGCTTACCGGCTACTTTAGTAAAAATAAGTATTCGGTGATGGCTGGCCTACGAGCAGCTATCCTCATGCTAAATCTCGAAATTTCTTGGCATCGTCT   32100
TTTTAAATGTATGCTTTCTCGTCGAATCTTTTTTCTTTTGCAGCCTTTGCTGTTTATCGGTAACTTTTTGACTTATTTTCTTATTTTTTTCCTCCTATC   32200
AAACATCCTCGCTCGTTTCTTTATTGGAGGTAACAGGACCCCCCTTTGACCTTGCTGAGGCCGAGTCCGAACTCGTTACAGGGTACACTACAGAATACGGG   32300
GGCTTCTACTTTGCACTATTTTACCTAGGTGAGTACTTTCACCTCTTCTTCTTTTTCGTCGCTGATCAGTGTAGTTTTTTTGGTTCATGGGAACTCTTAA   32400
                 NDH1 end                           KpnI       BglII                                  32500
AGCCCCTTCTTGTTCTTGCATAACTATACGGT]TAGGCTAGAATAGTGATTTTTTTTTTTTTTGTCTTCACTGGGTACCTTTTTAGATCTATGATTTCCAT
                        HindIII                                              XbaI
GAACGAACTTTTTATCCGAGGTAAGCTTGTTGTCGCAGGTATTGACCTTGGTAAACTTTTTTTTTTTTTTAGCTCTAGAAATGTTTTTACTCTGTGTTTA   32600
TTTTTTTTTAAATATTCTAACCCTTCGGGGTTTTCCTTGCCATTCTTTAATTTTGAATTTTTAACTGGATTCTTCTGAGTCGTTGAGTTTACTGTTTTTT   32700
ATATTCCTGGTATTTACTTTATTTTTCTCAGCAACGGAACCATTGTATTCAATAAGAACTTCTTCTTTTTTTTTTTATTTTTTTTTTTTTATTTTTTCTTTG   32800
TTTTCTTTTCTCAATTTGAACCTTTTTCATCCTTTGAAAAAACTAGCCTAAACTTTTTTTTCTTTTCTTTTTTTTTTTTATTACGGTCTTAATCTTTAATGT   32900
TATGAACGATCTCAACTCTCTTTTTTCTAAATTTTTTATCTTTAACTCCTTCCTTCTTCTTTTTTTTTAATTTTTTTAAAGACTTCAGCCGCGTTCTGGTTTGC   33000
ATTAACCTGTACCTAAACTCTAAAAAAAAATAGCTTGGGAATGGTAGCTTCTTTTTTTTAATTTTTTTAAAGACTTCAGCCGCGTTCTCCTTTA   33100
TGAGAAAGCAAAGCTTAGTCATTCAAAACTTAGGAAGCCCGTGAATCGCCGTAGTCCTCAAAGAGCAGTTTACTCCAAGTTTACAAGAAGAGGAGGAGA   33200
       HindIII
CAAGAAGGAGGAGGGGCCAGAAGGAGGAGGGGCCCAGAAGAGGAGGGGCTCCCTAAGCTAGTCCTTACCGACTAAGA[ATGATCCAGAAAGAAACACAGCTC   33300
                                                                          RPL14 start
AATGTTTGTGATACCAGCTCGTGTTTGGGTGGTCAACACCTTTCATATTTATAGGGGATTTAGGCATCGTATAGGACGCGCCTAGGGGACTATATCAAGGTCT   33400
CCATACGTAGTACAAAGCCAGAGTGCACGATTAAGCGTGGTAAAAAAAAAAAGTATAATAGTTCGACACGCCTTTGGAAGATTGAAGAGGACGGGCTC   33500
TTTTTCTAAGTTAGTAGCAACGTTTGCGTTTATTAAAGAAAAAGAACTGCTCCCCCTAGGGCGGAGAGATCAAAGGCCCTATCTTTATGGTGTAAAAAA   33600
AAAAAATTTGTTGCTTCCTTTCCCGGACGGGTG[TAGCAAAGCTAGACCTATAACCTCTAGCTTGCAAATGTTGAGTAGCTTGAGTAGCTTTGTTGAGTCGCG   33700
                        RPL14 end
AACGAGGCTTGTAAATGTACTAACTGCTAAACTCGGCTTCTCTCGAATCTTTTGATCTCCGCAGAGAGCAACCAAGCGATAAAAAACATTCCTTGTTT   33800
TTTTATAATATAAATAACTTGAAGGTTCAAAACATTGTTAACCTAATTTTTCTTGAAATAGCTTTTTGAGCATTTCATATTTTTCTATGTTCTTCA   33900
CAGTCCTCTTTACGTTCTTATTATTATTTTTTTTTTAAGTCTAAATAATTTTTTTATGCATGTCTGGAGTACCTCAAGCTTGCCTTCCTTCCTTCCTCCATGCT   34000
                                                                          HindIII
```

```
                                                    NDH4 start                                    34100
CGGAACGTGCTCTTCCTAGACTTTTTTTTTACAGTCTAGTAGTACTCGTCCATCATCCACCTCACC[ATGTTTGCAGTTTATCTTGTTTCAGCTTTA
                                                                                                  34200
AAAAACCCGAGGCCTCTAAAGCCGACCTTGATTTCTTTATGCTTACCTTAAGTATATTTAAAGGGCTCTTCTTTTTTAATGCTGTGCTGGCCCTTTT
                                                                                                  34300
CTGTGCACTCTTTACCTTCGATCTTATGTTTTCTGCTAAAAATCTTCTCTATCCAAACGAATACATATGGACTCGGGTGACTTCTTTTTTACAAAAAT
 AfII                                                                                             34400
GGAGCCCTTAAGTTCTCTTTAAATTTATACGGCCTTATCCTAGTCTTTTCTCGCCTCCTTACTGGGTTTGTTGCCATCTCAACAGTAGACAATCTCTATT
                                                                                                  34500
CTGAAGATAAGCTCAAATTTTATTTAATTTTTTTCAATTCTTCCTGCGTCCTCGGCTTTATAAAGTGTAGCGACCTCATTGCTTTTTTTTTTTTTA
                                                                                                  34600
TGAGGTGTTGATGCTCGGCTCAGTCCTCGTTGTTTTCTTTGGAAGCTATTCTAAAAATCAATACATGCAGTTATCTACTTCGTAGCATGGACGAGCTC
 BamHI                                                                                            34700
GGATCCCTCTTCGTCTTATTAGCATGCCTTACATCTACAGCTTGACAAACTCTCACGAACTTTTTGTCATTAAAACTTTTGTCTTCTCAAGACCCAAG
                                                                                                  34800
CAATGACGATTTACTCGCTCCTCCTTCGTGGGCTTGGTATTAAGTTCCCATCTGACCACTTCACTATTGGCTTACCAAAACTCACGTAGAGGCCTCTAC
                                                                                                  34900
TGGCTTTTCTATATACTTGAGTGGCTTTTAGTAAAAACAGCTCTCTTGGATTTATAGGTTAACAAACTTAATTCAAGTAGAGCTGGATACAACTTTC
                                                                                                  35000
TTTTTAGCAGTTCTTGTTGCAGGCGGTTATTGACTCCTCCCTCAACATGTGGAGCCAGACCGACCTTAAAAAACTTGTAGCTTACTGTTACGATCCAAGAA
                                                                                                  35100
TGAACCTCATTGCCATCTTCTTTCTTTTAAAAGGGATTCCAGCCTCATCGGCCTCGCCTACGGCTTCCTTTTTACCATCATGCAGCGCGCTCATGTCTACGCTCATGTT
                                                                                                  35200
TTTCCTAGTTGAGTGCATCTACTCAAGGTACAAAATCTAGGTCCACTCTAGTTGTTAACGGTGTTTTCTTCTCGTTCAACAACCTCGCATTAGCAATTATT
 E4  EcoRI E5                                                                                     35300
TTTATGGTTCTTTTTTTTCTGGAATTCTAGGAACCCTAAAGTTTGTTGTTGAATTTTTTGTCTTCAACCTTACCCTTCGTTCGTGACCTATAGGCG
                                                                                                  35400
TTATTTTTGTGGTTGTTGTGAGCGGCTATCGGCTTGATTGGCTTTTCTAAAAATTGATTTAATGCTATCTTTGTGCACCAAGCAAAGACGTAGGCCCCGA
                                                  NDH4 end                                        35500
CGCCCTCGACCTCTCCAAAGAAGGAAGGAACTTTACATTATCTTTTTTGTGCTTCGCAGGCTTGATCTTCTTTCTTTCTTAATCTTCTTAATGAT T AATCTTAG
                                                                                                  35600
GGGCTTTATATTATCTTTTTGTGCTTCGTAGGCTGGGCTTCTTAGCCCTCTCTTAATAATTAATTTAGTATGCTTTCAATCCTCTTTTTATTTTTAG
                                                                                                  35700
AGTTTTGAATTTCTTCAAGAGTGGTTTTTATATAGCTACACTACATCATAAAGAAGGTTACTGAAGCTATTGTAAAGAATTATTATTTTTCTTCATTCTTC
                                                                   ORF16(189) start              35800
TTTGCCGGAAGTTTTTAATTGAGTTTTAACTAAAAAAAGCGTAGACAACGTAATGATAAAACTAAACTCCTTGCACTTGACGAGATTTAACTTAATA
                                                                                            35900
CCTTCTTCTCTCTTTGTTGTTTTTCTTGTATTTTTTTTTTTTTTTTTTAGAGATGCTTTCTTGTTTTTTTAATTTTTAGGACGCCTA[ATTCTTTTTTG
 HindIII                                                                                          36000
GAAGGTAGTTACCTGGGACACCTTAGTATTTTTACCCTTGTATGAGACTTCTCCAATATGACTTATGTCGCATTGCGCAAAGCTTCTTTCTTTGCATTCGAGATG
```

(continued)

```
SstI                      E5   EcoRI E3
AGCTCAAGGGTTCCGAAGTCTCTACATACTTCGCAAGCAACGAATTCTTCGTTCTACATGAAATTTCTCCAAAGCATCTTATTCTCTTTTAGTTTTCG   36100
SstI
AGCTCTTTTTAGCAATCTTATGGACCGCCTTTTTAAACAATATCGTGGTTTTTTTTGAGTTGATGAGCCTACTCTTAGGAGGTACATTAGCTTCATGGA   36200
                                                                                    HindIII            36300
CCCACCCTCTTTTCTCTATCTACCACCCCGAGGCCGTTTTTATTCAAAAGAGCCTTCTTACCACAAATCATCTAAGTTATTTTGCAAATTTTAAGCTTAAC   36400
GTAGCTGTACTCTTGGATAAGAGCAATGCTATAGAGCCCCTAAACTATATGCTACAGGTCAGCTTCTACTATACTTCATTGTTCTTTTCTTTTCAATTCTTTT   36500
           ClaI       ORF16(189) end
TTTTCTTCTTTTCTAAAAACAACGAGGAGTACCAAATCGATGTTGAGCACTCTGTAATTAACCTCTCCTCCTGAGGCTGAAAAAGAAATCTTTTCGGTA   36500
  ORF17(221) start                                                                    SstI             36600
GAC[ATGCCCTAATCTTTTATTCTTTTATTCTTCGTTGTTATTTTTTTTTTTTATTGTTTTGATCCCCTTCAACCTCTTATTTGATTTGGTCTTTTTTTTTGTTTTGTATACTTGCGTGG   36700
TGTTCTTTGTGCCTCTCTTGTTTATTTTTTATTTTTTATTGTTTTGATCCCCTTCAACCTCTTATTTGATTTGGTCTTTTTTTTTTGTTTTGTATACTTGCGTGG   36800
TTCTTCAAACACAACTAGCCTATTCTTTGAGTGCGTTTATGAACTACATCGGGGTCTTAGCATTCTTCACCAGACTTATTGTTCAGTTTCTTAGGCTTGT   36900
TCTAATGTTGTTGTTTACTGCATGATGCACGATACCGTCATGTTACAAAACTATTCTCAAAAAATTCTTGGTTGGAGATAGCTTTGAGAGGAGTTA   37000
ATGAGCGGTTCAGCCAACGGAACCTCGATCTTCTTTTCTTCTTTACTATTGTTTTTGATGTTCTTCTTCTCTATACTTTTTTGTTCTACTATCTTTTTTT   37100
TTGTAGTTACTGTACAGTTTCAGCATTCTTTACTATTGTTTTTGATGTTCTTCTATACTTTTTTGTTCTACGAGAAGTATGAGCCACCACTT   37200
CGACAATATAACAAAGATGCATAAAAAGCTCATTGAAGAGCTTAAGTCCTTAAAGAAGGACTCCTTTTAATTTGATTTTTTTT[TTTTGCAGTCTAAC   37300
   AflII    ORF17(221) end                                                       HindIII               37300
AAAGCACTAGACGGATGCCTAAAAATCTTGGTTGAGGGCGTAAATCTAAATACGATATTTGTATGTAAGCTTCTTCTTTTGATATATAGAATTGCTTAGGAT   37400
   5' 5.8S rRNA
GTACAAGGGCCCTGAAAATTTATGAAGCGAAACATCGTAGTAATAAATCTAAAACATAATCAACTGAGATGTAAAGTAACGGTGAGTGAAACAAAG   37400
                                                 5.8S rRNA 3'  HindIII                                 37500
TAGCTCAAAAATTAGAAAAGAGAGGGCTGAATTACTTCTGAAAAGTAAGCTACAGTGGGTGATAGCCCCGTAAGCCTTT]TGATTTAATTATTTTTTT[TGTGC   37500
  LSU rRNA 5'
TGCTAGGGTGGCTCTCAACCTGGGTTCATTCCCCAGCTAATAAAGTTCATTCTTGTAGGCGGCTCCGAGTATATATTATTAACAAAATAATATAGAAATG   37600
                                                                                                       37700
TTTTTCAAAAAAAAAGCTAATAAAACCAAATTTTGATAGAAAATAAGTACCGTGAGGGAAGGTGAAAAGAAATTGTAGAAGTGTTAAAAGATTCTGA   37700
AATCTAGTGCAGTGAAACAGTTAAAGCGTGTTGTTTTAACGTACCTTTTGTATAATGGCCAACTAGTTATAAAATTAGCGAGCTTATCAAATCGCGTA   37800
ATGAAAATGTTATAGTTAATTTATAAGACCCGGAAGTCAAGTGATCTAATCATCGGCTAGGTAGAGAGATCGAACCCATAAATGTTGCAAAATTTCGGGAGA   37900
AGCTGATTAGGGGTAGGGGTGAAAAGGCTAATCAAACTTCGACGATAGCTGGTTTTTCCGCGAAATCTATCTACGTAGAGTATTTTTTTCTTTTGTGCGGGGTAGT   38000
```

AflII *BglII*

GTAATCTTTTCTTAAGRAGATCTGATCTTTTTTCGTGAAGAATCTTCCAATACGCATTAAAATAATAAATAAAACGGACTGAGAGTGCAAAGATTCTTGG 38100

TCGAGAGGGAAACAGCCCAGACCGTACGATAAAGTCATAAACAATGCGAAGTAAAAGAATTGTTTTTAAAAAAATATCGGGAGGTAGGCTTAGAATCAG 38200

CCAGCCTTTAAAGAAAGCGTAACAGCGTAACAGCTCACCAATAGCTACGATTATTTGAACAAACATCGTATTAAAAATAATTTTTAAATTAAAGCGCTAAGCATTGTA 38300

CTGAAGCGGCGGGTAACTCGGTAGCGAAACGTTTGTTAGGTTGTTGAAGGTTTATTGAGAAATAGGCTGGAGATATCAAAAATTGATAATGTTGGCATGAG 38400

TAATAGACAAAATGTTCAAATCATTTTGTTTGTTGATAAGTTAGGGTTGCTTGTTTGTTTGATCATCTTACAAAGTGTGATCCGGTCTCTAATTTTTAAATAG 38500

SstI

ATATTTTTAAAGATGAGCTCTGTATTTTTTCTGGCCTCCTTTGTGTTCTTTTTAATTTTTTAGAAGAGTGCGTGGAGGCGGACTGGAAACATTTACA 38600

TGGTGTACCGTACTAACACTAACGCAAGTACTTTAGTCGAGCAGATGACGACAAAAGAGCTAATGATATTGAAGGAACTCGGCAAAATTACTTTGTAACT 38700

TCGGGATAAAAAGTGCGTCAGCAAACAAAAAAATGGGGGTAGCGACTGTTACTAAAAACATAAGATTTGCAAAATTTAATTATGATGTATAAAATCT 38800

GACTCCTGCCCGGTGTTGTCATGCAAAGTTTAACGGTTAGCGCCTGTGTTGATAACAGCAGCAATAAACGCGGGCCATAACTCTGATGGTCCTAAGGTAGC 38900

AAAATCCCTTGACGGGTAAGTTCCGTCCTGCCACGAATGGAGTAACGACTGCCCTACTGTCTCCAATATCAGCTCTATGAAATTGAATTGCTGTGAAGAT 39000

GCAGCTTTTACAACTAGACGGAAAGACCCTATGCCACCTTTACTGATGCTAGGAGAACTAAAGGGATATACTGGAGATAAATTAAGGTAGGAGTTAAATCGC 39100

AATTGAAAAAACTACTTCTCTTTTACGTCTCTTTCATAAAAAATGACTAGGTTATTTTACTTTTCTGGCTGTTAGTTAACTGGGGCGGTTGCCTCCTAAAA 39200

AGTAACGGAGGTGAGCATAAAGTTACGCCTTGTGAGGAATTTTCTTTGCAAAATGAGTTAATAAAACTGCGTAATTTGATTAAATTACAAACTAGTAAT 39300

TTAGGGGCTAATTGCCTGCTATAATGATCCGGTGTTTTTTTTTTGAATAAGAACATGCTCAACGAATAAAAGGTACGCTAGGGATAACAGGCTTATAAATT 39400

CTGAGAGTTCCTATTAAAGAATTTGTTTGGCACCTCGATGTCGGCTCATCACATCTTGGTGGTGCAAGGGTTTGGCTGTTCGCCAATTAA 39500

AGTGGTACGTGAGCTGGGTTTAAAACGTTGTGAGACAGTTTGGTCCCTATCTGTTGTAGAATTAAGAAACGAGGCTGAGATCGACGCCAGTACGAGAGGA 39600

HindIII

CCGACTAGAACTAGCCTCTGGTTCGGGATCTACTTTAAATAAAGTATAGATGCTATGCTAAGCTGGATAAGCTAATTTTTAATTAAAATTTTTAGCT 39700

TAAAGAATTAGAAGCTTTTTAGCAACTCGTTTCTAATAGTATTTTAGAAGACTACTAAGTTAATTGAAGGTAGCTGTTATAGGGCCCTTAACGGCGTA 39800

LSU rRNA 3' tRNA^{tyr}

CCAGCTCCCCTTAAAAAAAAATACTAAGACTCTTATTATTATTATA|GAAGTAATGGCTGAGTGGTTAAAGCGCAGACTGTAAATCTGTTGGTAGTACCGT 39900

CGTTGGTTCGAAATCCAACTTACTT|CATACTAAGAAATTAGATTAACTTTGTAAACAATAAAATTACAAAAAATTTTTTTTATAAAAAAAATTTAGAACA 40000

(*continued*)

```
                              HindIII
TCTTGTTTTTTTTTAGGAACTAAATTAGCTCCTGATTCTTAAAGCTTAGAACGCGATTGTTAAAAATCTTTAATAAGCCGATGTCTATCTTTAAT  40100
TTTTTTTATAAAAAGATTTGTTCTTTTTTTAATTTTCTTTTTTAAAAACATAAAAAAATCAAAACTCCTTTTATTGCAGATAAGAAAATATACTATA  40200
AGTACGAGTACGTTAAGTACTTAACTACCTGTTGATAGAGAAGATTTTTATAAAAATATAGATTGTATAGGCTATATATTCGCGAACGTATGAACAT  40300
ATTTATTCAGCATCCAGAGCTTTTTTTTTTTTTTTTTTACTAAAAACAAAACCTTGTTAAATCAATTAAACACCATGCCCGTATGCACAAATATTTTTA  40400
TTTTCAGATAATAACATGTCCTTTCTAATTTTTTTTTTAACGTACCGCTGATCATTAGTTATGAATTC
                                    E3  EcoRI 40469          continues ~200 bp
```

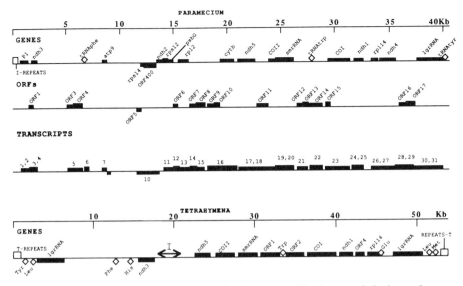

FIG.6 Organization of *P. aurelia* and *T. pyriformis* genomes. Blocks or symbols shown above the line are transcribed left to right and below the line right to left on the opposite strand. I, initiation; T, termination (*T. pyriformis* map kindly provided by Y. Suyama).

are absent in *Paramecium*, which effectively pushes the ndh3 gene closer to the 5' end. The direction of transcription for the ndh3 genes in the two ciliates is also different but this is most likely due to their orientation relative to the replication origin. In *Tetrahymena*, the location of this origin has not been precisely determined but based on the transcription map it is presumably positioned between the ndh3 and the ndh5 genes. The majority of *P. aurelia* genes are transcribed from the same strand but two are not (rps 14 and ORF400).

The most significant features of the *P. aurelia* genome involve tRNA genes, the presence of ribosomal protein genes, and the genes previously identified as chloroplast genes. Suyama (1986) reported that based mainly on hybridization data, *T. pyriformis* mitochondrial DNA encoded at most 10 tRNAs. This assertion has been borne out. As can be seen in Fig. 6, so far only eight tRNA genes have been identified (Suyama, 1985; 1986; Suyama *et al.*, 1987; Suyama and Jenney, 1989; Morin and Cech, 1986; Heinonen *et al.*, 1987). As we will discuss later in connection with the structure of the large ribosomal RNA gene, one of these, tRNA[Met], has an unusual structure. With *P. aurelia*, the number of tRNA genes is even more sparse. We have been able to identify only three tRNAs whose proposed structures satisfy the requirements of tRNA molecules (see later, tRNA^{Phe}, tRNA^{Trp}, and tRNA^{Tyr}) and all are encoded on the same strand as almost all the other genes. Their counterparts also present on the *T. pyriformis* mitochondrial genome have greater than 77% sequence homology, much greater than the 64% homology between the tRNA^{Trp} genes for *P. aurelia* and *P. anserina* (Cummings *et al.*, 1990). The tRNA^{Tyr} gene is in the identical position relative to the large subunit ribosomal RNA gene for *P. tetraurelia, P. primaurelia,* and *T. pyriformis.* Heinonen *et al.* (1987) sequenced the actual tRNAs for Leu and Met and found complete agreement with the deduced sequences. For all these tRNAs, the CCA 3′ terminal sequence must be added post-transcriptionally, as with all other mitochondrial tRNAs. Thus far, there is no direct evidence for the import of the missing mitochondrial tRNAs, but such must be the case.

Quite unexpectedly, several genes encoding presumed ribosomal proteins have been detected (Pritchard *et al.*, 1989; Suyama and Jenney, 1989). Two small ribosomal proteins (rps 12 and 14) and two for the large subunit (rpl 2 and 14) were found in *P. aurelia* and rpl 14 in *T. pyriformis.* The rpl 14 proteins common to both ciliates possessed 42% identical amino acid residues and were of identical size (119 aa). In some respects we should not have been surprised by the presence of ribosomal protein genes in the ciliate mitochondrial genomes. Several years ago, Beale *et al.* (1972) obtained evidence that erythromycin resistance in *P. aurelia* is associated with an alteration in a single mitochondrial ribosomal protein, and later Tait *et al.* (1976), using "hybrids" containing species 7 nuclei and species 1 mitochondria (obtained by microinjection), showed that both nuclear and mitochondrial genomes had to be involved in the expression of ribosomal proteins. A. Tait (personal communication), based on 2D gel electrophoresis, concluded that as many as eight ribosomal proteins were encoded on the mitochondrial genome. For *Tetrahymena*, Millis and Suyama (1982a,b) suggested that all mitochondrial ribosomal proteins less than 25 kDa in size are of mitochondrial origin. Whether these mitochondrially encoded ribosomal proteins contribute to the failure to dissociate *P. aurelia* and *T. pyriformis* mitochondrial ribosomes into two separable particles (Tait and Knowles, 1977; Chi and Suyama, 1970) is not known. It may be that ciliate mitochondrial ribosomes form a class or type different from those of bacteria, fungi, and higher animals (see Borst, 1980, for a review). The genes coding for ribosomal proteins rps 12 and 14 and rpl 2, although absent in animal, fungal, or other protozoan mitochondrial genomes, have been found in chloroplast genomes (Montandon and Stutz, 1984) and as mentioned earlier, rps 14 was identified in a plant mitochondrial genome (Wahleithner and Wolstenholme, 1988).

Two other genes associated with chloroplasts were also identified in the *P. aurelia* mitochondrial genome. In fact, these two genes, specified as ORF400 and psb G, gave the highest similarity scores recorded for

Paramecium (see later). In an alignment with the *Marchantia* chloroplast psb G product, almost 50% of the *Paramecium*-encoded amino acids were identical (Pritchard *et al.*, 1989). There is some controversy about the identity of this psb G gene product. Steinmuller *et al.* (1989) isolated membranes from maize plastids (as well as from the cyanobacterium *Synechocystis*) and determined that a particular protein, psb G, was part of photosystem 2 (PS2). Nixon *et al.* (1989), using similar procedures, concluded that membranes are easily contaminated and that the so-called psb G product was actually a subunit of the NADH–ubiquinone oxidoreductase complex. This gene product was never reported as part of a mitochondrial genome until it was found in *P. aurelia*. Similarly, ORF400, also associated only with chloroplasts, has recently been identified with an NADH complex subunit in the bovine nuclear genome (Fearnley *et al.*, 1989). Marked differences appear to exist in how the ciliate mitochondrial chain carries out oxidative phosphorylation. Their respiration is partially insensitive to cyanide and they have in addition to a standard cytochromic respiratory chain two alternative pathways to molecular oxygen: a cyanide-resistant oxidase like many higher plants and microorganisms (Lloyd and Edwards, 1978) and an auto oxidizable *b*-type cytochrome (Doussiere *et al.*, 1979). It is possible that the two chloroplast-type NADH subunits, ORF400 and the so-called psb G gene products, are involved in these features unique to the ciliates. Thus far, we have identified ndh1, -2, -3, -4, and -5 in the *P. aurelia* genome but not ndh4L and -6; these last two may be either nuclearly encoded or as yet unidentified. As intimated above (and later), the identified *P. aurelia* mitochondrial genes are quite divergent and are difficult to identify by computer comparisons with known genes. But the presence of chloroplast-type ndh genes may indicate that *P. aurelia* mitochondria have an unusual NADH complex.

Other genes also appear to be absent from the *P. aurelia* mitochondrial genome. We have yet to identify convincingly cytochrome oxidase C subunit III (COIII) and ATPase 6 and 8 of the Fo moiety of the ATPase complex. Thus, the *Paramecium* genome appears to be also unique in encoding atp9, which is not found in animal mitochondrial genomes, and not encoding atp6 or -8, which are found in most fungal and animal mito-chondrial genomes. It is also somewhat unique in that, although its genome is 2.5 times as large as animal mitochondrial genomes, like them no introns have been detected. In *P. anserina*, the huge size of the mitochondrial genome (100 kbp) is due primarily to the fact that 60 kbp are taken up by introns (Cummings *et al.*, 1990). Since this is not the case in *P. aurelia* it may be that there are other unusual genes not present in nonciliate organisms. Certainly, the transcript map of *P. aurelia* (Fig. 6) almost completely occupies the entire genome.

B. Divergence of Ciliate Genomes

The ciliate mitochondrial gene products are among the most divergent known (Pritchard *et al.*, 1990a). When amino acid sequence comparisons between predicted *P. aurelia* mitochondrial gene products were quantitated by means of either identical residues or similarity scores, the homologous proteins from a group of organisms as diverse as other protozoans (with the exception of *Tetrahymena*), vertebrates, fungi, plants, and prokaryotes had about the same (low) score. The ciliate mitochondrial gene products appear to be equally divergent from proteins representing a number of different kingdoms and organelles. Some of these comparisons, taken from Pritchard *et al.* (1986, 1989, 1990a; Mahalingam *et al.*, 1986) are presented in Table I.

Several issues need to be addressed in considering the data presented in Table I. First, not all the gene products are as divergent as others, even between proteins of similar function. For example, the ribosomal proteins appear to be a divergent group themselves. The rps 12 gene shows rather high homology (47%) to both chloroplasts and *Escherichia coli*, whereas rps 14 is much more homologous to chloroplasts than to *E. coli*. The rpl 14 gene is more homologous to *E. coli* and *Tetrahymena* than to chloroplasts and although some identities are shared, some are not. What this says about the origin of these proteins is unclear. Both of the so-called chloroplast genes (ORF400 and psb G, each now thought to be part of the NADH subunit complex) display quite good homology to their counterparts in chloroplasts. This common occurrence in both chloroplasts and mitochondria of components of the respiratory chain NADH dehydrogenase suggests that chloroplasts and mitochondria arose from a common ancestor. Alternatively, it is possible that the chloroplast ndh products are an example of transposition from the mitochondrial genome to the chloroplast, or the converse. Neither *P. aurelia* nor *T. pyriformis* exhibit chloroplast or chloroplast-like structures now but it is possible that their ancestral progenitors could have. It is not at all clear why *P. aurelia* and *Tetrahymena* have these genes whereas mitochondria from other organisms do not. With the absence of introns in these genomes, the presence of such genes does provide a partial explanation for the size of these ciliate genomes. Again, it may be that as yet unidentified open reading frames also represent genes unique to ciliate mitochondrial DNA.

Other genes in Table 1 illustrate consistently the low homology of ciliate genes to those from other organisms. The ndh subunit complex shows only 16 to 35% identity. With only two exceptions for all the genes, the size of the ciliate genes closely resembles the homologous gene size from all the organisms studied. For the psb G gene, the *P. aurelia* gene appears to be truncated at both ends. These terminal regions are the least conserved

TABLE I

Divergence of Ciliate Genes

Gene	Organism	% Identity	Residues
rps 14	*P. aur*		102
	Marc cp[a]	35	100
	E. coli[b]	20	99
rps 12	*P. aur*		139
	Marc cp[a]	47	123
	E. coli[b]	47	123
rpl 2	*P. aur*		259
	Marc cp[a]	21	277
	E. coli[b]	22	272
rpl 14	*P. aur*		119
	T. pyr[c]	42	119
	Marc cp[a]	28	122
	E. coli[b]	36	123
ORF400	*P. aur*		400
	Marc cp[a]	34	392
	Tob cp[d]	35	393
psb G	*P. aur*		156
	Marc cp[a]	48	243
	Tob cp[d]	45	247
	Maize cp[e]	45	248
ndh1	*P. aur*		261
	T. pyr[f]	65	284
	Human[g]	32	318
	N. crassa[h]	35	371
ndh2	*P. aur*		193
	Marc cp[a]	20	501
	Human[g]	16	347
	N. crassa[i]	17	583
ndh3	*P. aur*		120
	Marc cp[a]	25	120
	P. ans[j]	24	130
	Human[g]	21	115
ndh4	*P. aur*		474
	Marc cp[a]	23	499
	A. nid[k]	22	488
	Human[g]	23	459
ndh5	*P. aur*		570
	Marc cp[a]	27	693
	N. crassa[l]	28	715

(*continued*)

TABLE I (*continued*)

Gene	Organism	% Identity	Residues
cyt *b*	*P. aur*		391
	N. crassa[m]	20	385
	Human[g]	24	380
	Maize[n]	21	388
ATP9	*P. aur*		75
	N. crassa Nuc[o]	28	81
	S. cerevis[p]	32	76
	Maize[q]	25	74
COI	*P. aur*		645
	T. pyr[r]	58	698
	S. cerevis[s]	29	512
	Human[g]	27	513
COII	*P. aur*		205
	T. pyr[t]	57	205
	S. cerevis[u]	29	251
	Human[g]	35	201

Note: [a]Ohyama *et al.*, 1986; [b]Post and Nomura, 1980; Kimura *et al.*, 1982; Ceretti *et al.*, 1983; [c]Suyama and Jenney, 1989; [d]Shinozaki *et al.*, 1986; [e]Steinmuller *et al.*, 1989; [f]Y. Suyama, personal communication; [g]Anderson *et al.*, 1981; [h]Burger and Werner, 1985; [i]De Vries *et al.*, 1986; [j]Cummings and Domenico, 1988; [k]Lazarus and Kuntzel, 1984; [l]Nelson and Macino, 1987; [m]Citterich *et al.*, 1983; [n]Dawson *et al.*, 1984; [o]Sebald *et al.*, 1979; [p]Macino and Tzagoloff, 1979; [q]Dewey *et al.*, 1985; [r]Ziaie and Suyama, 1987; [s]Bonitz *et al.*, 1980; [t]Mahalingam *et al.*, 1986; [u]Coruzzi and Tzagoloff, 1979.

regions in other organisms (Steinmuller *et al.*, 1989) and may represent nonessential components of the subunit. The *Paramecium* ndh2 gene is the smallest of this subunit studied to date. The region corresponding to the amino terminus appears to be absent from the *Paramecium* sequence. Using the N-terminal region from any of the other organisms did not reveal any significant homology in the rest of the *Paramecium* genome. We have emphasized that no introns have been detected in the *Paramecium* mitochondrial genome but with the poor homology detected for the genes, it is difficult to be certain. Later we will see definitive evidence that at least at the normal locus of the large subunit ribosomal gene, no intron is present in the ciliate gene. For the COI gene in both *P. aurelia* and

Tetrahymena, the DNA sequence and its putative protein sequence are larger than those from other organisms. The increase in size arises at two positions. As can be noted in Table I, the two ciliate genes are 58% homologous. Both contain an additional 50–57 aa residues at the N-terminus, which is only 20% homologous between the two ciliates. The second "insert" in the ciliate COI gene occurs internally. There is a 108 aa region not present in nonciliate organisms. Hydropathy plots indicate that this region is flanked on both sides by regions that are homologous to genes from other organisms and the open reading frame is not disturbed. For the few genes we have been able to compare, the NDI, COI, and COII genes exhibit the greatest amino acid similarity between the two ciliates. With such poor homology between *P. aurelia* mitochondrial genes and genes identified in other organisms, it is reasonable to ask how we can be certain of the assignments specified. In most cases, hydropathy plots are quite similar, as are the size of the homologous genes. For cyt *b*, the homology was particularly poor. Yet, there are four invariant histidine residues that have been identified which are believed to bind two heme groups (Howell and Gilbert, 1988) and all of these are present at the same position in the *Paramecium* gene product (Pritchard *et al.*, 1990a).

C. Comparison with Other Mitochondrial Genomes

As we have indicated before (Pritchard *et al.*, 1986; 1990b), the gene organization of *P. aurelia* (and as we have noted here, *T. pyriformis*) is unlike any other known. In Fig. 7, the *P. aurelia* gene organization can be compared with those of several other organisms. Except for *P. aurelia*, all of these organisms have circular genomes and we have aligned them relative to the locus of the cyt *b* gene. The initiation region (I) of the *P. aurelia* genome is indicated. It is difficult to say whether any of these variations in gene organization have evolutionary significance. For the human map, a representative of the highly compacted genome, the ribosomal RNA genes are quite close and in the small–large orientation, whereas for *P. aurelia* and the fungi these genes are well separated and sometimes in the large–small orientation. But not even all compacted genomes show a close separation of these genes. In the sea urchin *P. lividus* these two genes are separated by the ND1 and ND2 (ndh) genes (Cantatore *et al.*, 1987). For chloroplast genomes, which are uniformly large ranging in size from 130 to 200 kbp, the gene order of the ribosomal RNA genes is invariant in six quite different organisms (Palmer, 1985). It has been argued that the evolutionary reduction in size and gene organization of chloroplasts took place during a relatively brief period shortly after their endosymbiotic origin (Gray and Doolittle, 1982). The apparent wide

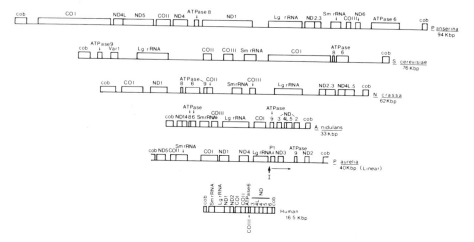

FIG. 7   Gene organization comparison with fungal, human and *Paramecium* genomes (updated versions adopted from Dujon, 1983). NADH genes are labeled ND rather than ndh. To facilitate the comparison, cob (cytochrome *b*) began and ended each genome. Note the absence of the ND complex in *S. cerevisiae*.

differences in genome organization could argue for a polyphyletic origin of mitochondria. In *S. cerevisiae*, Clark-Walker (1989) studied a revertant from a high-frequency petite strain and found its mitochondrial genome to be rearranged. Mapping and sequencing analyses showed that the rearrangement occured by means of intramolecular recombinations involving short GC-rich repeats. As can be seen in Fig. 5, the mitochondrial DNA sequence of *P. tetraurelia* has several short GC-rich regions and we have not been able to rule out intramolecular recombination. The most remarkable feature about this gene organization comparison is that involving *P. aurelia* and *T. pyriformis*. As we noted earlier, the known genes of *T. pyriformis* are in the same orientation as in *P. aurelia*. Whatever caused the gene organization in these two ciliates must have occurred before their divergence and then was maintained. The differences in gene order observed in the four ascomycetes in Fig. 7 may be due to the fact that all these fungi readily undergo intramolecular recombination.

D. Phylogenetic Tree Analysis

As discussed in the introduction, evolutionary relationships are best studied by comparing macromolecular sequences. In particular, Fox *et al.* (1980) proposed that the ribosomal RNAs, which are ubiquitous and functionally related, provide excellent molecular chronometers. The small subunit rRNAs are of an optimal size since they are large enough (approximately 1600 bases) to provide a greater number of independently variable nucleotide positions and yet are small enough to lend themselves to analysis. The use of 16s rRNA from *E. coli* can serve as a model for comparison with all other small subunit rRNAs (Gutell *et al.*, 1985). The secondary structure diagram for the small subunit ribosomal RNA from *P. aurelia* is presented in Fig. 8. Although the *E. coli* 16s rRNA model is not presented, the similarity to the *P. aurelia* mitochondrial structure is apparent. As we will discuss in more detail in the next section, the secondary structure for the *P. aurelia* 16s rRNA is quite similar to its counterpart in *T. pyriformis* (Schnare *et al.*, 1986). The complete sequence can be divided into nonconserved or variable regions, conserved or "universal" elements, and semiconserved sequences. The universal sequences are almost identical in all prokaryotic and eukaryotic small subunit ribosomal RNAs and although they are essential for the construction of the secondary structure, they do not contribute much to the understanding of sequence divergence. The semiconserved sequences are necessary for determining distant relationships and the nonconserved for resolving close phylogenetic relationships. By juxtaposing those sequences that form evolutionarily conserved secondary structures, Sogin and Elwood (1986), constructed a phylogenetic tree for the nuclear 16s rRNA from *P. tetraurelia* and *T. thermophila*. In this tree, as mentioned earlier, the ciliates branched as a cohesive phylogenetic grouping with *P. tetraurelia* and *T. thermophila* closely paired, as were the fungi. The methodology of Sogin and Elwood (1986), with the advice of M. Sogin, was used for the construction of the mitochondrial phylogenetic tree shown in Fig. 9. As has already been noted (Woese, 1987), there are two primary prokaryotic lineages, the archebacteria and the eubacteria, and chloroplast 16s rRNAs share a relatively recent ancestry with the blue-green algae, *Anacystis nidulans*. Yang *et al.* (1985) first found that the closest relative of the mitochondrial small subunit rRNA sequence was the purple bacterium, *A. tumefaciens* and this is further substantiated here. The plant mitochondrial sequence is shown to be closer to *A. tumefaciens* than are the other mitochondrial sequences and this is due to the more rapid rate of evolutionary change, a "fast-clock" in fungal, animal, and now ciliate mitochondria (Brown *et al.*, 1979). The fungal mitochondrial DNAs form a separate group, as do the two ciliates. Not surprisingly, *P. primaurelia* and *P. tetraurelia* are indistinguishable and *T. thermophila* appears to have a faster clock than either of the *Paramecium* RNAs. Not included in the tree is the human 12s RNA. Because of the distortions in the matrix distances for mitochondrial genes caused by the fast-clock, the matrix distance for mammalian 12s RNA is even more distant. It is also clear from this tree that although *P. aurelia* and *T,*

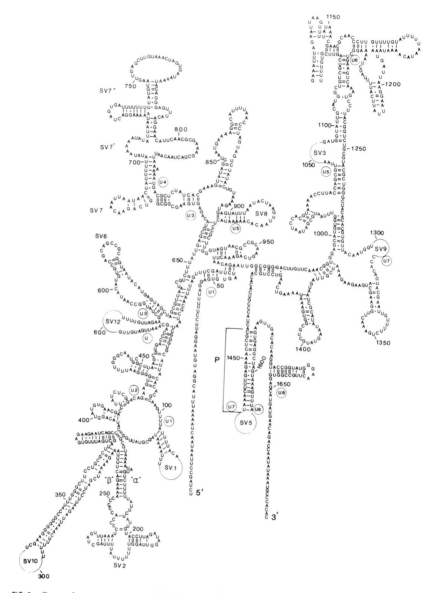

FIG. 8 Secondary structure model of the small subunit ribosomal RNA from *P. primaurelia*. *P. tetraurelia* is 94% identical in sequence. SV signifies small subunit variable and U, universal or conserved sequences (nomenclature introduced by Schnare *et al.*, 1986). P signifies the so-called penultimate helix in the structure (see Seilhamer *et al.*, 1984b).

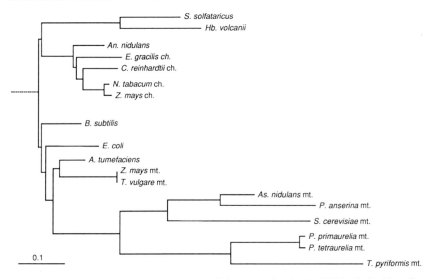

FIG. 9 Phylogenetic tree diagram constructed from small subunit rRNA similarities. See Cummings *et al.* (1989b) for detailed organismal names and references. Tetrahymena sequence analysis taken from Schnare *et al.* (1986). The 0.1 marker depicts a 10/100 bp evolutionary change.

thermophila are in the same grouping, *Tetrahymena* appears to have evolved more rapidly. This agrees with the degree of identities noted for the rpl 14, ndh1, COI, and COII genes given earlier, which were greater than between other organisms but were also divergent.

IV. Structure and Organization of Ribosomal RNA Genes

The structure of ribosomal RNA genes displays unusual variation from organism to organism as well as in organelles. Prokaryotic ribosomes lack the 5.8s RNA found in eukaryotic ribosomes (Pene *et al.*, 1968) and *C. reinhardtii* chloroplast ribosomes contain unique 3s and 7s rRNAs (Rochaix and Darlix, 1982). It has been proposed that in prokaryotes, the 5' end of 23s rRNA is the structural analog of the 5.8s eukaryotic rRNA (Nazar, 1980, 1982). Spacer sequences separate the 5.8s rRNA, as well as the *C. reinhartii* chloroplast 3s and 7s rRNAs from the equivalent 23s rRNA and these sequences are absent in the prokaryotic rRNA. No discontinuities in structure have been noted in the 16s rRNA from either prokaryotic or eukaryotic organisms.

As discussed earlier, the structure of mitochondrial ribosomes from both *Tetrahymena* and *Paramecium* is unusual. Although each 55s subunit of the 80s complex is morphologically distinct, they are not separable (Tait and Knowles, 1977; Stevens *et al.*, 1976). This could be due to the fact that these ciliate ribosomes contain unique mitochondrially encoded proteins but it could also be a reflection of unusual ribosomal RNAs. Some 9 years ago, we initiated studies on mitochondrial rRNA genes from *P. primaurelia* and *P. tetraurelia* and found that the size of the rRNAs was smaller than expected [20s and 13s for the large (LSU) and small (SSU) RNAs] and that two small RNAs of about 300 and 200 bases were present in equimolar amounts (Seilhamer and Cummings, 1981). A schematic diagram of these RNAs is presented in Fig. 10. The identical pattern was also found for the ribosomal RNA from *T. pyriformis* mitochondria (Schnare *et al.*, 1986). As we will see the two lower-molecular-weight RNA, labeled α LSU and α SSU, represent distinct components of each ribosomal RNA.

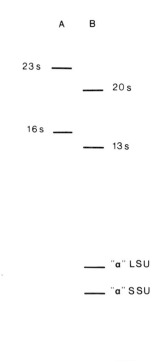

FIG. 10   Idealized RNA gels. See text for description of labels.

A. Discontinuities in the Small Ribosomal RNA

Mitochondrial DNA clones from *P. primaurelia* and *P. tetraurelia* identified as containing the small subunit ribosomal RNA gene were sequenced
and their sequences compared with each other as well as that from *E.
coli* (Seilhamer *et al.*, 1984b). The sequence from *P. tetraurelia* is given in
Fig. 5, lines 24500 to 26200 and the potential secondary structure for *P.
primaurelia* SSU is shown in Fig. 8. On the basis of its alignment with the
E. coli sequences, we deduced that the *P. aurelia* SSU gene was 1680 bp
in length. This contrasted sharply with the size of the rRNA determined
on gels (1400 bases, see Fig. 10). In comparing the secondary structure
with that from *E. coli* SSU (not shown here, but see Cummings *et al.*,
1989b), for a comparison of *E. coli* with *Podospora anserina* SSU), we
noted that the region marked SV5 at the 3′ end was quite different. In *E.
coli*, this region is highly structured, whereas in *P. aurelia* it is devoid of
possible helices. S1 nuclease analyses of the 3′ end showed that although
much of the DNA probe was fully protected, a major protected site was
located precisely at the SV5 region. A minor protected site was at the SV9
region. S1 analyses at the 5′ end showed that the vast majority of the DNA
probes were fully protected. Even though we were not able to demonstrate
a precursor relationship with the 13s rRNA we proposed that the SSU
rRNA was discontinuous, consisting of the main RNA and the SV5 "insert." We were concerned, however, with the poorly defined S1 analyses
at the 5′ end, but because it aligned so well with its *E. coli* counterpart we
attributed this to uncontrolled technical difficulties.

   The location of the discontinuity appears to have been resolved when
Schnare *et al.* (1986) isolated and sequenced the 13s SSU RNA from *T.
pyriformis*. Quite unexpectedly, the RNA sequence lacked completely the
entire U1 region at the 5′ end of the secondary structure model (see Fig.
8). This region is present within the 5′ terminal region of all SSU rRNAs
examined to date. Determination of the DNA sequence showed that as
with the *P. aurelia* gene, this U1 region is indeed present. Northern
hybridization experiments demonstrated that the DNA probe hybridized
primarily to the 13s rRNA but a faint hybridization was found to one of
the small RNAs depicted as α SSU in Fig. 10. Seilhamer *et al.* (1984b)
failed to detect any hybridization to *Tetrahymena* RNA. Schnare *et al.*
(1986) isolated this RNA and its sequence was identical to the missing U1
region found in the DNA. The 3′ end of the *T. pyriformis* and *P. aurelia*
SSU DNAs were 89% identical. It is clear that in both *P. aurelia* and *T.
pyriformis*, the SSU mitochondrial RNA is discontinuous. In *T. pyriformis*
it consists of two parts, α and β, 203 and about 1400 bases, respectively,
and it is most likely that the discontinuity in *P. aurelia* is the same. But
there are differences between the two. For example, the sequence of the

5' end, including U1, is not identical (Fig. 11). U1 itself is 63% homologous but the 3' end of α and the 5' end of β show less than 50% homology between the two ciliates. This is reflected in the helix "joining" the 3' end of α and the 5' end of β in the proposed secondary structure (Fig. 8) where the helix ends well above the SV2 region, being interrupted by C-rich loops. In *Tetrahymena* this stem continues down to SV2, without a C-rich region (Schnare *et al.*, 1986). The SV2 region itself is structurally different as well; in *Paramecium* SV2 is quite similar to its *E. coli* counterpart and to that in *P. anserina*, whereas in *Tetrahymena* it is a highly AT-rich unstructured sequence. Although our results on the S1 analyses of the 3' end of *P. aurelia* SSU did indeed show full protection to the expected 3' end, a major protected site was the SV5 region as well as another site in the SV9 region. It may be that further surprises are in store when these molecules are reexamined. The failure to detect a precursor relationship with the α and β components in both *P. aurelia* and *T. pyriformis* suggests that the two components are not cotranscribed. The assembly of these molecules into the ribosomal particle may involve other unique characteristics.

B. Discontinuities in the Large Ribosomal RNA

Northern blots of *Paramecium* mitochondrial RNA revealed three RNAs that hybridized to appropriate large subunit rDNA clones: the abundant 20s species (see Fig. 10), an equally intense 300-base species (the so-called α LSU in Fig. 10), and a significant amount of a 23s molecule (Seilhamer *et al.*, 1984a). This indicated that the 20s LSU rRNA was part of a larger 23s precursor molecule that gave rise to both the 20s molecule and a 300-base element. The 3' end of the large subunit rRNA gene complex terminated one base from the start of a tRNA^{Tyr} gene (see line 39900 in Fig. 5), the last known gene in the genome. DNA sequence analysis showed that this "5.8s"-like element was separated from the 5' end of the 20s component by a 20-base T-rich region (see line 37500 in Fig. 5) with the entire gene spanning 2654 bp. The 5' end of the 5.8s element is preceded by a 15-bp poly(T) tract and the 3' end of the 20s component terminates randomly within a 20-base AT-rich segment immediately abutting the rRNA^{Tyr} gene. The entire rRNA conformed to a secondary structure model quite similar to that from *E. coli*, with the 5.8s-like molecule constituting the 5' end. Curiously, although the 5' end of the 20s component formed an integral part of the secondary structure model, the first few bases (lines 37500 and 37501 in Fig. 5) could be folded separately into a tRNA^{Met}-like molecule (called structure A in Seilhamer and Cummings, 1981). Because base-pairing in both the amino-acyl arm and the D arm was so poor (see

P.a. A A A A U U U A C G G A U G U A A G G A U U C C U
T.p. A A A A U U a g g u u g U a a u c a a A a a a u U
    ↑5'-α

               ↓U1
P.a. U U U U A G G A G U G U G A U C U U A G C U U U G
T.p. U a U U g a G u G U G a u c a u g g c u c g g - G

P.a. A U U U A A C G C U A A U C G G G U G C A U U A C
T.p. A U U - A A C G C U A A U u a G a c G C c U a A C

             U1 ↓   ↓SV1
P.a A C A C G C A A G U U U U A U U U U A C A C A A G
T.p. A C A u G C g A G U U U U A U a U a u - A u u A G

P.a. A A C C U U C U U U U U U U G C U U A A A A U G C A
T.p. A u a a c a a U U a a a U - C U a A A u g U a u A

                   SV1 ↓
P.a. A A G A U C A A G U U U U A A A A U A G C G U A U
T.p. g c G - U C A A G g U g a g u A u a A u a c a u a

P.a. U G G U G C G U A A A A U A U G U C U U U U U A U
T.p. a a u a u g c c u u u A a g c c U a U U U a a A U

                ↓SV2
P.a. U A U U C C U A U U G U A C C U U A G A U A G G U
T.p. a A U a g u U A a U u U u a a a a c a A U A a c a

P.a. U A G G U U U U U U U A U U U U A G C U A G - U U
T.p. a A u G c U a U a a a A a a c a A u u a u a a a U
                3'-α↑

      SV2 ↓
P.a. U A A A U C C C C A C C C C C A A A A G A - C U U
T.p. U A u A a u a a a c a a u C a u A u A a u u C U U

P.a. U U A A A - - - G A A A A U C G A U C C U U A A G
T.p. U a A A A c c a a A A A A a u u A U u a g c - A G
                ↑5'-β

P.a. A U U A A A C U U U U U U U U U G A U C C A A U A
T.p. c U U A A A g a U U a a U U U g u A U a a u A U u

FIG. 11 Comparison of the U1 sequence region in the small subunit ribosomal RNA sequence from *P. aurelia* and *T. pyriformis* (Seilhamer *et al.*, 1984b; Schnare *et al.*, 1986). Uppercase shows identical nucleotides. See Figs. 8 and 13 for U1, SV1, and 2 and α and β subunits.

Fig. 14) we could not be certain of their function. A similar tRNA^{Ile}-like structure was also discernible in the middle of the 20s sequence (38351–38435 in Fig. 5). We considered the possibility that these psuedo-tRNA genes represented introns or were themselves interrupted by introns, but we had no evidence to support this conjecture. In fact, the one sequence position where an intron has been found in *S. cerevisiae, P. anserina, N. crassa,* and *A. nidulans* (Cummings *et al.*, 1989a) LSU rRNA is continuous in *P. aurelia* and *T. pyriformis* (M.W. Gray, personal communication), located at 39376 in Fig. 5: CTAGGGAT/AACAGGCT. This sequence is remarkably conserved in many organisms including nuclear and chloroplast genes (Wallace, 1982) (Fig. 12) and its continuity in the ciliates supports our failure to detect introns. S1 nuclease determinations clearly showed the location of the 5.8s and 20s rRNA termini. The only additional protected molecule was one that included the tRNA^{Tyr} gene abutting the 3' end of the 20s component. This suggested that the rRNA and tRNA may be derived from the same nascent transcript.

This analysis of the LSU rRNA structure from *P. aurelia* was confirmed in large part, but with some totally unexpected findings, when Heinonen *et al.* (1987) examined this gene in *T. pyriformis* mitochondria. As with the SSU rRNA, Heinonen *et al.* (1987) first sequenced and analyzed the RNA molecule itself and found the same results as reported previously with *P. aurelia*. The LSU rRNA had two components, a 5.8s element and the major 20s molecule. The 5.8s α LSU (see Fig. 10) displayed quite good homology to the same region in *P. aurelia, C. reinhardtii* chloroplasts, *E. coli,* and mouse nuclear rRNA and left little doubt that this was indeed the equivalent molecule. In ciliates, then, both the SSU and the LSU mitochondrial rRNA molecules have discontinuities at their 5' ends. Unexpectedly, the 5' terminus of the 20s rRNA from *T. pyriformis* commenced just downstream of the position occupied by the so-called *P. aurelia*

↓

E. coli	5'	AAGGUACUCCGGGGAUAACAGGCUGAUACCGC 3'
Z. mays Chl.		AAGUUACUCUAGGGAUAACAGGCUCAUCUUCC
Human Mt.		AAGUUACCCUAGGGAUAACAGCGCAAUCCUAU
S. cerevisiae Mt.		AAGUUACGCUAGGGAUAACAGGGUAAUAUAAC
P. anserina Mt.		AAGCUACGCUAGGGAUAACAGGCUAAUUUGCG
P. primaurelia Mt.		AAGGUACGCUAGGGAUAACAGGCUUAUAAAUU
P. tetraurelia Mt.		AAGGUACGCUAGGGAUAACAGGCUUAUAAAUU
T. pyriformis Mt.		AAGGUACGCUAGGGAUAACAGGCUUAUGAGUU

FIG. 12   Intron locus of LSU rRNAs. Arrow depicts the position of the intron in *Z. mays* chloroplasts, *S. cerevisiae,* and *P. anserina* mitochondria but the continuous sequence for *E. coli,* human, *Paramecium,* and *Tetrahymena* mitochondria (see Cummings *et al.*, 1989b, and Wallace, 1982, for complete set of references).

structure A discussed above. Remarkably, when Heinonen *et al.* (1987) isolated and sequenced DNA clones of the LSU rRNA genes, at either end of the linear molecule, the order of the two major components was reversed! α LSU was not encoded at the 5' flank of the β LSU gene (see Fig. 13) as it is in *P. aurelia*, but was located at the 3' end with the two genes being separated by a tRNA^{Leu} gene. This was the first demonstration of a discontinuous rRNA resulting from or coinciding with a rearrangement of its coding sequences, and the same rearrangement occurred in both copies of the inverted repeats. As with *P. aurelia*, the 3' end of α LSU is

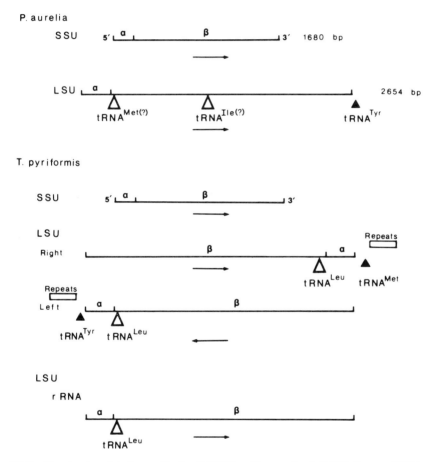

FIG. 13   Structural models for small subunit (SSU) and large subunit (LSU) ribosomal RNAs and their genes in *P. aurelia* and *T. Pyriformis* (Seilhamer *et al.*, 1984a,b; Schnare *et al.*, 1986; Heinonen *et al.*, 1987). Solid triangles are tRNA genes outside the structures and open triangles are presumed tRNA genes within the structures.

quite close to tRNA^{Tyr} on one terminal repeat and to tRNA^{Met} on the other. The structure of the LSU gene complex (and SSU) for both *T. pyriformis* and *P. aurelia* is shown in Fig. 13. Although our S1 analysis clearly showed that the 5' end of the 20s component in *P. aurelia* was at the 5' end of structure A (see Fig. 5 in Seilhamer *et al.*, 1984a). Heinonen *et al.* (1987) argued that the probe we used was just within the 3' end of structure A and hence could not detect the "true" 5' end of the β subunit. Since repeated attempts to detect tRNA transcripts from this region in the gene complex were not successful, we were not aware that the 5' end detected by the S1 analysis could be artifactual. Because of the similarities between the LSU genes in other respects, Heinonen *et al.* (1987) proposed that the structure of both genes was the same, i.e., 5'-(α LSU)–(tRNA^{Met} or structure A)–(β LSU)–(tRNA^{Tyr}) in *P. aurelia* and 5'–(β LSU)–(tRNA-^{Leu})–(α LSU)–(tRNA^{Met}) in the right-hand copy of *T. pyriformis*. For *T. pyriformis*, the tRNA^{Leu} gene appeared to be quite normal for both the amino-acyl arm and the D arm, whereas for *P. aurelia*, structure A is aberrant. There are many examples of aberrant-appearing tRNAs (Dirheimer and Martin, 1990), so this may not be a serious argument. Clearly the only way to resolve this question is to sequence the 5' end of the *P. aurelia* 20s rRNA. The functionality of the putative tRNA^{Met} in *P. aurelia* should also be examined. It may be that this small gene serves a dual function, as both tRNA and a portion of LSU rRNA. Examination of the *E. coli* sequence in this region shows quite good homology (Seilhamer *et al.*, 1984a). In this regard, the role of the second putative tRNA within the 20s rRNA component of *P. aurelia* (structure B) might also be considered a possible rare cleavage site. The substantial amount of unprocessed 23s precursor rRNA present in both ciliates suggests that cleavage is either relatively slow or does not always occur. In any case, the structure of the mitochondrial SSU and LSU gene from both these ciliates is quite unusual and further study may reveal unexpected results on ribosome assembly and function. It would be useful for similar studies in other ciliates to be done.

V. Mitochondrial Genetic Code

A. The tRNA Problem

The paucity of tRNA genes in the mitochondrial genomes of *P. aurelia* and *T. pyriformis* is interesting in three respects. First, there are only two other mitochondrial genomes that have so few tRNA genes: the kinetoplastid protozoa (trypanosomes), which have none (Simpson, 1987), and *C. reinhardtii*, where only three have thus far been found (Boer and

Gray, 1988). This raises important questions not only about the possible mechanism(s) for the import of required tRNAs from the cytoplasm into the mitochondrion (see Suyama, 1982; Benne and Sloof, 1987), but also about the evolutionary origin of mitochondria (see later). Second, it has been known for quite some time that mitochondrial genomes code for fewer than the 32 tRNAs required by the wobble hypothesis (Crick, 1966). Human and mouse genomes code for only 22 (Anderson *et al.*, 1981; Bibb *et al.*, 1981), *S. cerevisiae* for 24 (Bonitz *et al.*, 1980), and *P. anserina* for 25 (Cummings *et al.*, 1990). This is accomplished by the establishment of 8 four-codon families and 15 two-codon families where any of 4 bases can be in the third position; i.e., the two out of three model of codon usage (Lagerkvist, 1981). For the four-codon families, in both mammalian and fungal tRNAs, this reduction in number correlates with the presence of uridine in the 5' wobble position of the tRNA anticodon. An unmodified U at the first position of the anticodon functions to recognize all four codons, but a modified U (methylated) cannot recognize U or C at the wobble position of the codon. Remarkably, none of the three (or five) *Paramecium* tRNA genes (Fig. 14), the seven *Tetrahymena* sequences, or the three known *C. reinhartii* mitochondrial tRNAs are members of four-codon families. Suyama (1986) has speculated that the development of the U/N wobble occurred in ciliate mitochondrial genomes (and now also perhaps unicellular green algae!) after the branching of the ciliates from their primitive eukaryotic ancestor but before the branching that led to fungi, mammals, etc. Finally, there is the problem of the "modified" tRNA genes within the large subunit rRNA gene. As discussed in Section IV,B, the two subunits of the LSU gene, α and β, are separated in the *Tetrahymena* genome by a tRNA^{Leu} gene. At the same gene locus in *P. aurelia*, there is a sequence whose presumed secondary structure resembles tRNA^{Met} (Fig. 14). Seilhamer and Cummings (1981) argued that because the D arm of this putative tRNA was so unusual it was unlikely that it was a functional tRNA. The results of Heinonen *et al.* (1987) with *T. pyriformis* forces a reexamination. One difficulty with assuming that either sequence is a tRNA gene is that the product itself has not been sequenced nor has a charged tRNA been isolated. If this sequence in *P. aurelia* truly is tRNA^{Met}, then what does this say about the other sequence within the LSU β subunit whose putative secondary structure resembles tRNA^{Ile} even more closely than did the tRNA^{Met} (Fig. 14)? Is there another possible discontinuity in the LSU β subunit? This seems unlikely since the size of the 20s rRNA would decrease by about 800 bases. This structure could serve some function in the mitoribosome, perhaps an unusual interaction with the mitoribosomal proteins.

A separate but important issue of tRNA genes in the ciliates must also be addressed. For both *Paramecium* (Caron and Meyer, 1985; Preer *et al.*, 1985) and *Tetrahymena* (Horowitz and Gorovsky, 1985), nuclear UAA

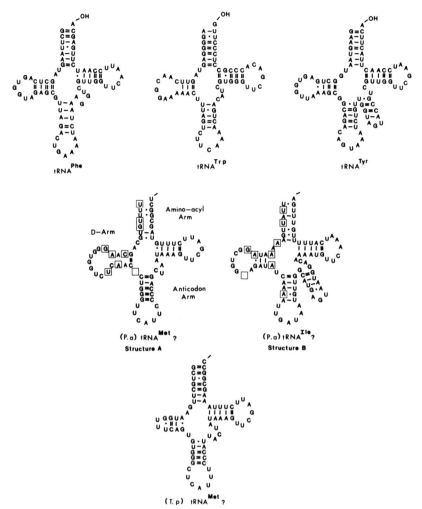

FIG. 14 Structural models of tRNA genes from *P. aurelia*. Heinonen *et al.* (1987) realigned the amino-acyl arm of structure A (tRNAMet) yielding better base-pairing but this distorted the D arm even more than in this representation.

and UAG codons specify glutamine rather than a stop. Only UGA functions as a stop codon. This is precisely opposite this codon usage in the ciliate mitochondria where, like other mitochondria, UGA codes for tryptophan and UAA and UAG are stop codons. Grivell (1986) considered the possibility that the ciliates nuclear genetic code is a relic of a primitive genetic code, one which preceded the emergence of the universal code.

Hanyu *et al.* (1986) clarified this whole problem when they found that *Tetrahymena* cytoplasm had three different tRNAs that could attach glutamine. The first has the anticodon UmUG (Umethylated), which could decode CAA and CAG, but the other two had anticodons UmUA and CUA, which recognize UAA and UAG. All three tRNAs were quite closely related in their sequence (greater than 81% homologous) and Hanyu *et al.* (1986) concluded that the two unusual tRNAs^{Gln} could have evolved from the normal one early in ciliate evolution by gene duplication and then diverged. They proposed further that UAA and UAG were probably stop codons in the ancient ciliate but that they were seldom used for whatever reason, possibly due to the existence of weak suppressor tRNAs. In a sense, their proposal is consistent with the molecular phylogenetic data discussed earlier, which showed that the ciliates branched off very early from their primitive eukaryotic ancestor. A difficulty with this view is that *Dictyostelium* and trypanosomes diverged even earlier and these organisms do not have aberrant tRNAs^{Gln}. The ciliates do represent striking deviations from the so-called universal code. This places unusual restraints on nuclear–mitochondrial interactions. The only stop codon in the cytoplasm is the preferred tryptophan codon in the mitochondrion so that nuclearly encoded mitochondrial enzymes must have distinctive properties. In addition, the mechanisms for transporting tRNAs into the mitochondrion must be quite selective to allow only the import of the "normal" tRNA^{Gln}.

B. Codon Usage

For the past few years we have been reporting the codon usage for the genes encoded in the *Paramecium* mitochondrial genome and will not repeat all the data here. As we have pointed out before (Pritchard *et al.*, 1986, 1989, 1990b; Mahalingam *et al.*, 1986), codon usage in *Paramecium* is not that unusual and does not reflect the many unique features of this genome. Like other mitochondrial genomes, UGA codes for tryptophan, whereas in chloroplasts it is a stop codon. The genes coding for ribosomal proteins, ORF400 and psb G, which are unique to the ciliate mitochondrial genome but common to chloroplasts, yield a mixed result for the use of UGA. The rps 14 and ORF400 genes both display UGA in their translation product but the other genes do not. In human mitochondria the AGA and AGG triplets specify stops (Anderson *et al.*, 1981), but in *Paramecium* they have the standard arginine specification. If one uses the so-called "usual" mitochondrial genes in *Paramecium*—ndh1, ndh4, COI, and COII—as a basis for a codon table and compare this usage with the "unusual" gene products, no clear correlation emerges (Pritchard *et al.*,

1989). Usual or unusual genes have similar codon usage. It is also not clear why the ciliate genes are so divergent. Since most of the tRNA genes are absent from the mitochondrial genome, the nuclear-encoded tRNAs must be utilized. Therefore we should not be surprised to find that except for UGA, the universal genetic code is adhered to. But, as discussed earlier the ciliate nuclear genome encodes two unusual tRNA^{Gln}, which serve as stop codons and this is not reflected in the mitochondrial codon usage. Another area where the *Paramecium* mitochondrial codon is variant is in the initiation codon. For most genes (ATP9, ndh3, ndh4, ndh5, ORF1, and rpl 14), the usual ATG is that codon. But other genes have unexpected initiation codons: GTG for COII and rpl 2, ATT for ndh1 and rps 12, TTG for ndh2, and ATC for cyt *b*. Other genes have initiation codons not too different from the norm: ATA. Similar unusual initiation codons have been reported in the *Tetrahymena* mitochondrial genome (Ziaie and Suyama, 1987).

As shown in Table I, the closest relatives of the divergent *Paramecium* genes were those homologous genes in *Tetrahymena*. A comparison of three "normal" genes, ndh1(ND1), COI, and COII, illustrates best the important features of the ciliate codon usage (Table II). In some instances, the codon usage for these closely related ciliates shows marked differences. For most families, A is preferred over C in the wobble position for *Tetrahymena* but C is dominant over A for *Paramecium*. The difference in AT content in the two ciliates (75% versus 59%) is not sufficient to account for this disparity. An obvious difference in the codon usage is the rare use of the entire CTN family in *Tetrahymena* and the normal frequency in *Paramecium*. The tRNA^{Leu} encoded in the *Tetrahymena* genome (but not *Paramecium*) has the UUA anticodon, so this cannot be the explanation. The two ciliates are similar in the rare use of the CGN four-codon family for arginine. They do differ in the preferred use of AGA in *Tetrahymena* and the use of AGG for the arginine codon in *Paramecium*. This disparity could be related to the preferred use of A in the wobble position for *Tetrahymena*. It should also be emphasized that even though nuclear-encoded tRNAs must be utilized in both ciliates, the base composition of both nuclear genomes is the same, 75% AT. Finally, it is curious that in *Paramecium*, the codon usage for tryptophan is evenly divided between TGA and TGG, whereas in *Tetrahymena*, only TGA is utilized. Both ciliate genomes have a highly homologous tRNA^{Trp} gene, so this difference is also inexplicable except on the usual A wobble position in *Tetrahymena* codons. That this is not reflected in the wobble position of the anticodons (C or G in four of the five *Paramecium* tRNAs and C or G in four of the seven unique *Tetrahymena* tRNAs) is another indication of our lack of understanding of the basis for codon usage.

TABLE II
Codon Usage

aa	Codon	COI[a] P.a.	COI[a] T.p.	COII[b] P.a.	COII[b] T.p.	NDI[c] P.a.	NDI[c] T.p.	Codon	aa	COI P.a.	COI T.p.	COII P.a.	COII T.p.	NDI P.a.	NDI T.p.	Codon	aa	COI P.a.	COI T.p.	COII P.a.	COII T.p.	NDI P.a.	NDI T.p.
F	TTT	40	52	8	8	23	30	ACT	T	11	18	4	7	3	5	TGT	C	4	2	2	3	0	5
F	TTC	36	10	4	2	11	2	ACC	T	13	1	1	0	3	1	TGC	C	2	0	5	1	4	0
L	TTA	9	55	2	16	7	29	ACA	T	10	20	2	7	2	3	TGA	W	7	15	2	5	2	6
L	TTG	12	0	3	0	7	1	ACG	T	8	0	3	0	1	0	TGG	W	7	0	2	0	2	0
L	CTT	25	0	5	1	7	1	GCT	A	23	25	3	4	5	11	CGT	R	1	0	0	0	1	0
L	CTC	11	0	5	0	13	0	GCC	A	8	1	3	2	5	1	CGC	R	1	0	0	0	1	0
L	CTA	14	4	3	3	8	4	GCA	A	10	10	1	0	6	3	CGA	R	1	0	1	0	2	0
L	CTG	2	0	3	0	4	0	GCG	A	2	1	0	0	2	0	CGG	R	0	0	0	0	0	0
I	ATT	18	30	17	4	4	8	TAT	Y	15	27	2	8	6	19	AGT	S	5	7	1	0	1	8
I	ATC	15	2	6	0	11	1	TAC	Y	19	11	6	5	9	1	AGC	S	10	3	4	0	2	0
I	ATA	12	20	3	11	2	24	TAA	*	1	1	0	1	0	1	AGA	R	5	31	5	12	0	7
M	ATG	19	35	1	3	7	12	TAG	*	0	0	1	0	1	0	AGG	R	16	1	4	0	2	0
V	GTT	18	18	3	6	12	11	CAT	H	7	21	3	6	1	1	GGT	G	7	35	1	10	3	13
V	GTC	10	1	0	1	5	1	CAC	H	14	5	6	4	1	1	GGC	G	11	0	2	0	6	1
V	GTA	10	25	2	4	5	12	CAA	Q	5	11	0	6	0	2	GGA	G	16	5	4	0	3	2
V	GTG	6	1	2	0	2	0	CAG	Q	2	1	3	0	3	1	GGG	G	8	1	0	0	4	2
S	TCT	17	18	4	4	5	5	AAT	N	13	13	6	9	6	10								
S	TCC	7	3	2	0	7	0	AAC	N	13	5	3	2	4	1								
S	TCA	5	11	2	5	3	8	AAA	K	12	29	7	18	3	5								
S	TCG	3	3	0	0	2	0	AAG	K	18	1	9	0	4	1								
P	CCT	6	13	4	5	2	6	GAT	D	9	13	1	5	2	3								
P	CCC	12	0	4	0	3	0	GAC	D	8	3	6	2	1	0								
P	CCA	7	12	0	4	1	2	GAA	E	2	11	3	8	6	13								
P	CCG	0	1	0	0	0	0	GAG	E	10	0	3	0	6	1								

[a] Ziaie and Suyama, 1987; Pritchard et al., 1986.
[b] Mahalingam et al., 1986.
[c] Prichard et al., 1990a; Y. Suyama, personal communication.

VI. Perspectives

From what we know about the mitochondrial genomes of the ciliates, it is difficult to formulate comprehensive arguments about mitochondrial evolution. For some time it has been argued that both mitochondria and chloroplasts arose as endosymbionts (Gray and Doolittle, 1982; Wallace, 1982). Based on phylogenetic tree construction from small subunit ribosomal RNA sequence comparisons, this is certainly borne out. The chloroplasts appear to share a close ancestry with the blue-green algae, *A. nidulans* (Fig. 9) (Woese, 1987), and mitochondria are closer to the purple bacterium, *A. tumefaciens* (Fig. 9) (Yang *et al.*, 1985). The relationship between mitochondria from different organisms is confounded by an apparent fast evolutionary clock where the plant mitochondria appear closer to their purple bacterial ancestor but fungal and ciliate mitochondria are more distant (with mammalian off the scale with an even faster clock). Within the ciliates, *Tetrahymena* appears to be evolving faster than *Paramecium*. Because of the existence in all mitochondria of the same genes (COI, -II, -III, etc.) it has been argued that only one endosymbionic event took place and that all mitochondria originated from the same ancient ancestor. This is basically a negative argument since it is so difficult to imagine how a presumably large ancestral precursor genome is reduced to the variety of sizes we see today, each containing the same genes. In some respects, von Heijne (1986) weakens this argument in his considerations on why mitochondria have a genome at all. He concludes that those genes are retained whose gene products could not be exported (and hence imported after transfer of the gene to the nucleus). This removes at least partially the general argument for a monophyletic origin of mitochondria. In the ciliates the transfer of mitochondrial genes to the nucleus is complicated by the fact that in the nucleus, TAA and TAG are not stop codons.

   In arguing for a polyphyletic origin of mitochondria, i.e., several independent endosymbionic events, the diverse properties of the ciliate genome can be used as a springboard. The ciliate mitochondrial genome resembles chloroplast genomes in three respects: the use of two-codon family tRNAs, certain ribosomal protein genes, and the common occurrence of a particular NADH subunit complex. Excluding a polyphyletic origin these properties have at least two explanations. First, it could be suggested that since at least one species of *Paramecium (bursaria)* is known to harbor chlorella algae (Kudo, 1971), these genes were transferred from a chloroplast genome to the mitochondrial genome. Such transfer is known to exist in maize (Lonsdale *et al.*, 1983). Or, these genes could have arisen from a progenitor that was common to both mitochondria and chloroplasts. Both these possibilities cannot be easily reconciled with the use of the TGA codon or with the differences in conserved sequences in the so-called chloroplast genes. Why also are these genes not found in other mitochondrial genomes? Benne and Sloof (1987) in their summary of the evolution of the mitochondrial protein synthetic machinery point out that the diversity of the various mitochondrial translation systems is extremely large. The uniqueness of the morphology of the ciliate mitoribosome is only one example. Probably, the two most striking features of the ciliate genome are the structure of the ribosomal RNA genes and the genome itself. For both RNA genes, the rRNA is discontinuous unlike any other mitochondria, chloroplast, or prokaryote genes. In the large subunit ribosomal RNA, the discontinuity involves a tRNA. Despite these discontinuities, the secondary structure models for both SSU and LSU RNAs closely resemble the prokaryotic model. The linearity of the ciliate genome is also quite uncommon. In fact as we have pointed out the replication mechanism more closely resembles viral models than either other mitochondrial or prokaryotic schemes (Pritchard *et al.*, 1983). The presence of inverted repeats of the large subunit rRNA gene is not unique to *Tetrahymena*. Most chloroplast genomes have one to several such repeats (Palmer, 1985) and the fungus *Achyla ambisexualis* has two copies (Hudspeth *et al.*, 1983). But neither of these sets of repeats is so intimately involved with the replication of the entire genome as they are in *Tetrahymena*. In some respects, the absence of introns in the ciliate genome could be taken as favoring a monophyletic origin of mitochondria since molecular phylogenetic data place the branching of ciliates quite early in the evolutionary tree. Shih *et al.* (1988) have proposed that the existence of introns predated the divergence of eukaryotes and prokaryotes. But both *D. discoideum* and *Euglena* preceded the ciliates on the same tree and are not known to lack introns in their mitochondrial genomes. It could be argued that the lack of recombination between ciliate mitochondria is the basis for the lack of introns but this in itself illustrates a marked difference between ciliate and fungal mitochondria. Taken together, the many unusual properties of the ciliate mitochondria genome strongly suggest, but do not prove, the polyphyletic origin of mitochondria. On the basis of cytochrome *c* data, Dayhoff and Schwartz (1981) proposed at least three independent origins, one involving *Tetrahymena*. That view is supported here.

There are many unanswered questions concerning the characterization of ciliate mitochondrial genomes. First, the *Tetrahymena* genome sequence must be completed so that a more detailed comparison can be made with the *Paramecium* genome. Since both sets of genes are quite diverse, some sense might be made by such a comparison. It would also be useful to complete the sequence of *P. primaurelia* (only half is known) so that the basis for the poor homology between *P. primaurelia* and *P.*

tetraurelia can be better understood. Thus far, the sequence data available do not support the hybridization data. The same may be true of the poor homology shown between *Tetrahymena* species. With regard to the replication mechanisms of both ciliate genomes, it is essential to sequence the termination region in *Paramecium* and the initiation region in *Tetrahymena*. Only with these sequences in hand can we critically examine the similarities in the replication mechanisms. Finally, the apparent diversity of ciliate genes needs to be studied at the RNA level. With their closeness to trypanosomes in the phylogenetic tree, do ciliate mitochondria exhibit RNA editing? Such a study would not only clarify the diversity of ciliate mitochondrial genes but would also directly address the possibility of detecting unusual "inserts" or introns in these genes. There are still surprises ahead.

Acknowledgments

I would like to recognize first Geoffrey Beale, who in his own special way introduced me to and encouraged my research in *P. aurelia*. The colleagues I met in his laboratory, particularly Andrew Tait and Jonathan Knowles, were influential in the development of the work. Although it is difficult to acknowledge everyone in my laboratory, Judy Goddard played a central role in establishing the replication scheme and Rich Maki in starting the restriction enzyme studies; Jeff Seilhamer set a lot of things in motion; and Art Pritchard provided all the initiation region sequences and most of the genome. Ken McNally performed some of the computer analyses and was responsible for constructing the phylogenetic tree. The manuscript could not have been completed without the help of Margaret Silliker and Virginya Tipton. Finally, from beginning to end Wanda Collins was always there with clean glassware, fresh Scottish grass for *Paramecium* growth, and a smile.

The research presented here was supported mostly by NIH Grant GM 21948 and in the last years by NSF Grant DMB-86-05319.

References

Adoutte, A., and Beisson, J. (1970). *Mol. Gen. Genet.* **108**, 70–77.

Adoutte, A., Knowles, J. K., and Sainsard-Chanet, A. (1979). *Genetics* **93**, 797–831.

Anderson, S., Bankier, A. T., Barrell, G., de Bruijn, M. H. L., Coulson, A. R., Drouin, J., Eperon, I. C., Nierlich, D. P., Roe, B. A., Sanger, F., Schrier, P. H., Smith, A. J. H., Staden, R., and Young, I. G. (1981). *Nature (London)* **290**, 457–464.

Arnberg, A. C., Goldbach, R. W., Van Bruggen, E. F. J., and Borst, P. (1977). *Biochim. Biophys. Acta* **477**, 51–69.

Arnberg, A. C., Van Bruggen, E. F. J., Borst, P., Clegg, R. A., Schutgens, R. B. H., Weijers, P. J. and Goldbach, R. W. (1975). *Biochim. Biophys. Acta* **383**, 359–369.

Arnberg, A. C., Van Bruggen, E. F. J., Clegg, R. A., Upholt, W. B., and Borst, P. (1974). *Biochim. Biophys. Acta* **361**, 266–276.

Baroudy, B. M., Venkatesan, S., and Moss, B. (1982). *Cell* **28**, 315–324.

Beale, G. H. (1969). *Genet. Res.* **14**, 341, 342.

Beale, G. H., and Knowles, J. K. C. (1976). *Mol. Gen. Genet.* **143**, 197–201.
Beale, G. H., Knowles, J. K. C., and Tait, A. (1972). *Nature (London)* **235**, 396–397.
Beale, G. H., and Tait, A. (1981). *Int. Rev. Cytol.* **71**, 19–40.
Benne, R., and Sloof, P. (1987). *Biosystems* **21**, 51–68.
Bibb, M. J., Van Etten, R. A., Wright, C. T., Walberg, M. W., and Clayton, D. A. (1981). *Cell (Cambridge, Mass.)* **26**, 167–180.
Blackburn, E. H., and Gall, J. G. (1978). *J. Mol. Biol.* **120**, 33–53.
Bland, M. M., Levings, C. S., III, and Matzinger, D. F. (1986). *Mol. Gen. Genet.* **204**, 8–16.
Boer, P. H., and Gray, M. W. (1988). *Curr. Genet.* **14**, 583–590.
Bonitz, S. G., Berlani, R., Coruzzi, G., Li, M., Macino, G., Nobrega, F. G., Nobrega, M. P., Thalenfeld, B. E., and Tzagoloff, A. (1980). *Proc. Natl. Acad. Sci. U.S.A.* **77**, 3167–3170.
Borst, P. (1980). *In* "Biochemistry and Physiology of Protozoa" (M. Levandowsky and S. H. Hutner, eds.), pp. 341–364. Academic Press, New York.
Brown, W. M., George, M. J., and Wilson, A. C. (1979). *Proc. Natl. Acad. Sci. U.S.A.* **76**, 1967–1971.
Burger, G., and Werner, S. (1985). *J. Mol. Biol.* **186**, 231–242.
Burke, J. M., and RajBhandary, U. L. (1982). *Cell (Cambridge, Mass.)* **31**, 509–520.
Cantatore, P., Roberti, M., Morisco, P., Rainaldi, G., Gadaleta, M. N., and Saccone, C. (1987). *Gene* **53**, 41–54.
Caron, F., and Meyer, E. (1985). *Nature (London)* **318**, 185–188.
Cavalier-Smith, T. (1974). *Nature (London)* **250**, 467–470.
Cerretti, D. P., Dean, D., Davis, G. R., Bedwell, D. M., and Nomura, M. (1983). *Nucleic Acids Res.* **11**, 2599–2616.
Chi, J. C. H., and Suyama, Y. (1970). *J. Mol. Biol.* **53**, 531–556.
Citterich, M. H., Morelli, G., and Macino, G. (1983). *EMBO J.* **2**, 1235–1242.
Clark-Walker, G. D. (1989). *Proc. Natl. Acad. Sci. U.S.A.* **86**, 8847–8851.
Clary, D. O., and Wolstenholme, D. R. (1984). *In* "Oxford Surveys on Eukaryotic Genes" (N. MacLean, ed.), Vol 1, pp. 1–35. Oxford Univ. Press, Oxford.
Corden, J., Wasgly, K. B., Buchwalder, A, Sassone-Corsi, P, Kedinger, C., and Chambon, P. (1980). *Science* **209**, 1406–1414.
Corliss, J. O. (1979). "The Ciliated Protozoa: Characterization, Classification, and Guide to Literature." 2nd ed. Pergamon, New York.
Coruzzi, G., and Tzagoloff, A. (1979). *J. Biol. Chem.* **254**, 9324–9330.
Crick, F. H. C. (1966). *J. Mol. Biol.* **19**, 548–555.
Cummings, D. J. (1977). *J. Mol. Biol.* **117**, 273–277.
Cummings, D. J. (1980). *Mol. Gen. Genet.* **180**, 77–84.
Cummings, D. J., and Domenico, J. M. (1988). *J. Mol. Biol.* **204**, 815–839.
Cummings, D. J., Domenico, J. M., and Nelson, J. (1989a). *J. Mol. Evol.* **28**, 242–255.
Cummings, D. J., Domenico, J. M., Nelson, J., and Sogin, M. C. (1989b). *J. Mol. Evol.* **28**, 232–241.
Cummings, D. J., and Laping, J. L. (1981). *Mol. Cell. Biol.* **1**, 972–982.
Cummings, D. J., McNally, K. L., Domenico, J. M., and Matsuura, E. T. (1990). *Curr. Genet.* **17**, 375–402.
Cummings, D. J., Pritchard, A. E., and Maki, R. A. (1979). *In* "ICN-UCLA Symposium on Extrachromosomal DNA" (D. J. Cummings, P. Borst, I. B. David, S. M. Weissman, and C. F. Fox, eds.), pp. 35–51. Academic Press, New York.
Dawson, A. J., Jones, V. P., and Leaver, C. J. (1984). *EMBO J.* **3**, 2107–2113.
Dayhoff, M. O., and Schwartz, R. M. (1981). *Ann. N.Y. Acad. Sci.* **361**, 92–103.
DeVries, H., Alzner-DeWeerd, B., Breitenberger, C. A., Chang, D. D., de Jonge, J. C., and RajBhandary, U. L. (1986). *EMBO J.* **5**, 779–785.

Dewey, R. E., Schuster, A. M., Levings, C. S., III, and Timothy, D. H. (1985). *Proc. Natl. Acad. Sci. U.S.A.* **82,** 1015–1019.

Dirheimer, G., and Martin, R. P. (1990). *In* "Chromotography and Modification of Nucleosides, Part B (C. W. Gehrke and K. C. T. Kuo, eds.), pp. 197–264.

Doussiere, J., Sainsard-Chanet, A., and Vignais, P. V. (1979). *Biochim. Biophys. Acta* **548,** 236–252.

Dujon, B. (1983). *In* "Mitochondria 1983" (R. J. Schweyen, K. Wolf, and F. Kaudewitz, eds.) pp. 1–24. de Gruyter, Berlin.

Fearnley, I. M., Runswick, M. J., and Walker, J. E. (1989). *EMBO J.* **8,** 665–672.

Flavell, R. A., and Jones, I. G. (1971). *Biochim. Biophys. Acta* **232,** 255–260.

Fox, G. E., Stackebrandt, E., Hespell, R. B., Gibson, J., Maniloff, J., Dyer, T. A., Wolfe, R. S., Balch, W. E., Tanner, R. S., Magrum, L. T., Zablen, L. B., Blakemore R., Gupta, R., Bonen, L., Lewis, B. J., Stahe, D. A., Luehrsen, K. R., Chen, K. N., and Woese, C. R. (1980). *Science* **209,** 457–463.

Gefter, M. L. (1975). *Annu. Rev. Biochem.* **44,** 45–78.

Goddard, J. M., and Cummings, D. J. (1975) *J. Mol. Biol.* **97,** 593–609.

Goddard, J. M., and Cummings, D. J. (1977). *J. Mol. Biol.* **109,** 327–344.

Goldbach, R. W., Arnberg, A. C., VanBruggen, E. F. J., Defize, J., and Borst, P. (1977). *Biochim. Biophys. Acta* **477,** 37–50.

Goldbach, R. W., Bollen-DeBoer, J. E., Van Bruggen, E. F. J., and Borst, P. (1979). *Biochim. Biophys. Acta* **562,** 400–417.

Goldbach, R. W., Borst, P., Arnberg, A. C., and Van Bruggen, E. F. J. (1976). *In* "The Genetic Function of Mitochondrial DNA" (C. Saccone and A. M. Kroon, eds.), pp. 137–142. North-Holland, Amsterdam.

Gray, M. W., and Doolittle, W. F. (1982). *Microbiol. Rev.* **46,** 1–42.

Grivell, L. A. (1986). *Nature (London)* **324,** 109–110.

Gutell, R. R., Weiser, B., Woese, C. R., and Noller, H. F. (1985). *Prog. Nucleic Acids Res. Mol. Biol.* **32,** 155–216.

Hanyu, N., Kuchino, Y., Nishimura, S. and Beier, H. (1986). *EMBO J.* **5,** 1307–1311.

Heinonen, T. Y. K., Schnare, M. W., Young, P. G., and Gray, M. W. (1987). *J. Biol. Chem.* **262,** 2879–2887.

Heumann, J. M. (1976). *Nucleic Acids Res.* **3,** 3167–3171.

Hiratsuka, J., Shimada, H., Whittier, R., Ishibaski, T., Sakamoto, M., Mori, M., Kondo, C., Honji, Y, Sun, C-R., Meng, B-Y., Li, Y-Q., Kanno, A., Nishizawa, Y., Hirai, A., Shinozaki, K., and Siguira, M. (1989). *Mol. Gen. Genet.* **217,** 185–194.

Hoffman, H. P., and Avers, C. J. (1973). *Science* **181,** 749–750.

Horowitz, S., and Gorovsky, M. A. (1985). *Proc. Natl. Acad. Sci. U.S.A.* **82,** 2452–2455.

Howell, W., and Gilbert, K. (1988). *J. Mol. Biol.* **203,** 607–618.

Hudspeth, M. E. S., Ainley, W. M., Shumard, D. S., Butow, R. A., and Grossman, L. I. (1982). *Cell (Cambridge, Mass.)* **30,** 617–626.

Hudspeth, M. E. S., Shumard, D. S., Bradford, C. J. R., and Grossman, L. I. (1983). *Proc. Natl. Acad. Sci. U.S.A.* **80,** 142–146.

Jacobs, H. T., Elliot, D. J., Veerabhadracharya, B. M., and Farquharson, A. (1988). *J. Mol. Biol.* **202,** 185–217.

Kan, N. C., and Gall, J. G. (1981). *J. Mol. Biol.* **153,** 1151–1155.

Kimura, M., Mende, L., and Wittmann-Liebold, B. (1982). *FEBS Lett.* **149,** 304–312.

Kiss, G. B., and Pearlman, R. E. (1981). *Gene* **13,** 281–287.

Knowles, J. K. C. (1974). *Exp. Cell Res.* **88,** 79–87.

Knowles, J. K. C., and Tait, A. (1972). *Mol. Gen. Genet.* **117,** 53–59.

Kudo, R. R. (1971). "Protozoology." Thomas, Springfield, Ill.

Lagerkvist, U. (1981). *Cell* **23,** 305–306.

LaPolla, R. J., and Lambowitz, A. M. (1981). *J. Biol. Chem.* **256**, 7065–7067.

Lazarus, C. M., and Kuntzel, H. (1984). Submitted to the NBRF-PIR Protein Sequence Database.

Lazdins, I., and Cummings, D. J. (1984). *Curr. Genet.* **8**, 483–487.

Lloyd, D., and Edwards, S. W. (1978). *In* "Functions of Alternative Terminal Oxidase," Vol. 49, pp. 1–10. 11th FEBS Meeting, Copenhagen.

Lonsdale, D. M., Hodge, T. P., Howe, C. J., and Stern, D. B. (1983). *Cell (Cambridge, Mass.)* **34**, 1007–1014.

Macino, G., and Tzagoloff, A. (1979). *J. Biol. Chem.* **254**, 4617–4623.

Mahalingam, R., Seilhamer, J. J., Pritchard, A. E., and Cummings, D. J. (1986). *Gene* **49**, 129–138.

Maki, R. A., and Cummings, D. J. (1977). *Plasmid* **1**, 106–114.

Marechal-Drouard, L., Weil, J-H., and Giullemant, P. (1988). *Nucleic Acids Res.* **16**, 4777–4788.

Middleton, P. G., and Jones, I. G. (1987). *Nucleic Acids Res.* **15**, 855.

Millis, A. L., and Suyama, Y. (1982a). *J. Cell Biol.* **73**, 139–148.

Millis, A. L., and Suyama, Y. (1982b). *J. Biol. Chem.* **243**, 4063–4073.

Montandon, P. E., and Stutz, E. (1984). *Nucleic Acids Res.* **12**, 2851–2859.

Morin, G. B., and Cech, T. R. (1986). *Cell (Cambridge, Mass.)* **46**, 873–883.

Morin, G. B., and Cech, T. R. (1988). *Mol. Cell Biol.* **8**, 4450–4458.

Nanney, D. L., and McCoy, J. W. (1976). *Trans. Am. Microsc. Soc.* **95**, 664–682.

Nazar, R. N. (1980). *FEBS Lett.* **119**, 212–214.

Nazar, R. N. (1982). *FEBS Lett.* **143**, 161–162.

Nelson, M., and Macino, G. (1987). *Mol. Gen. Genet.* **206**, 307–317.

Nixon, P. J., Gounaris, K., Coombes, S. A., Hunter, C. N., Dyer, T. A., and Barber, J. (1989). *J. Biol. Chem.* **264**, 14,129–14,135.

Ohyama, K., Fukuzawa, H., Kohchi, T., Shirai, H., Sano, T., Sano, S., Umesono, K., Shiki, Y., Takeuchi, M., Chang, Z., Aota, S., Inokuchi, H., and Ozaki, H. (1986). *Nature (London)* **322**, 572–574.

Palmer, J. D. (1985). *Annu. Rev. Genet.* **19**, 325–354.

Pene, J. J., Knight, E., Jr., and Darnell, J. E., Jr. (1968). *J. Mol. Biol.* **33**, 609–623.

Post, L. E., and Nomura, M. (1980). *J. Biol. Chem.* **225**, 4660–4666.

Preer, J. R., Jr., Preer, L. B., Rudman, B. M., and Barnett, A. J. (1985). *Nature (London)* **314**, 188–190.

Pritchard, A. E., and Cummings, D. J. (1981). *Proc. Natl. Acad. Sci. U.S.A.* **78**, 7341–7345.

Pritchard, A. E., and Cummings, D. J. (1984). *Curr. Genet.* **8**, 477–482.

Pritchard, A. E., Herron, L. M., and Cummings, D. J. (1980). *Gene* **11**, 43–52.

Pritchard, A. E., Laping, J. L., Seilhamer, J. J., and Cummings, D. J. (1983). *J. Mol. Biol.* **165**, 1–15.

Pritchard, A. E., Seilhamer, J. J., and Cummings, D. J. (1986). *Gene* **44**, 243–253.

Pritchard, A. E., Sable, C. L., Venuti, S. E., and Cummings, D. J. (1990a). *Nucleic Acids Res.* **18**, 163–171.

Pritchard, A. E., Seilhamer, J. J., Mahalingam, R., Sable, C. L., Venti, S. E., and Cummings, D. J. (1990b). *Nucleic Acids Res.* **18**, 173–180.

Pritchard, A. E., Venuti, S. E., Ghalambor, M. A., Sable, C. L., and Cummings, D. J. (1989). *Gene* **78**, 121–134.

Rhode, S. L. (1982). *J. Virol.* **42**, 1118–1122.

Rhode, S. L., and Klaasen, B. (1982). *J. Virol.* **41**, 990–999.

Rochaix, J-D., and Darlix, J-L. (1982). *J. Mol. Biol.* **159**, 383–395.

Roe, B. A., Ma, D-P., Wilson, R. K., and Wong, J. F. H. (1985). *J. Biol. Chem.* **260**, 9759–9774.

64 DONALD J. CUMMINGS

Sainsard-Chanet, A., and Cummings, D. (1988). In "Paramecium" (H-D. Gortz, ed.), pp. 167–184. Springer-Verlag, Berlin/Heidelberg.

Schnare, M. N., Heinonen, T. Y. K., Young, P. G., and Gray, M. W. (1986). J. Biol. Chem. 261, 5187–5193.

Sebald, W., Hoppe, J., and Wachter, E. (1979). In "Function and Molecular Aspects of Biomembrane Transport" (E. Quagliariello et al., eds.), pp. 63–74. Elsevier/North-Holland, Amsterdam

Seilhamer, J. J., and Cummings, D. J. (1981). Nucleic Acids Res. 9, 6391–6406.

Seilhamer, J. J., Gutell, R. R., and Cummings, D. J. (1984a). J. Biol Chem. 259, 5173–5184.

Seilhamer, J. J., Olsen, G. J., and Cummings, D. J. (1984b). J. Biol. Chem. 259, 5167–5172.

Shih, M-C., Heinrich, P., and Goodman, H. M. (1988). Science 242, 1164–1166.

Shinozaki, K., Ohme, M., Tanaka, M., Wakasugi, T., Hayashida, W., Matsubayashi, T., Zaita, N., Chunwongse, J., Obokata, J., Yamaguchi-Shinozaki, K., Ohto, C., Torazawa, K., Meng, B. Y., Sugita, M., Deno, H., Kamogashura, T., Yamada, K., Kusuda, J., Takaiwa, F., Kato, A., Tohdoh, N., Shimada, H., and Siguira, M. (1986). EMBO J. 5, 2043–2049.

Simpson, L. (1987). Annu. Rev. Micro. 41, 363–382.

Sogin, M. L., and Elwood, H. J. (1986). J. Mol. Evol. 23, 53–60.

Sonneborn, T. M. (1975). Trans. Am. Microsc. Soc. 94, 155–178.

Steinmuller, K., Ley, A. C., Steinmetz, A. A., Sayre, R. T., and Bogorad, L. (1989). Mol. Gen. Genet. 216, 60–69.

Stevens, B., Curgy, J. J., Ledoigt, G., and André, J. (1976). In "Genetics and Biogenesis of Chloroplasts and Mitochondria (T. Bucher, W. Neupert, W. Sebald, and S. Werner, eds.), pp. 731–740. Elsevier/North-Holland, Amsterdam.

Suyama, Y. (1982). In "Mitochondrial Genes" (P. Slonimski, P. Borst, and G. Attardi, eds.), pp. 449–455. Cold Spring Harbor Laboratory, Cold Spring Harbor, NY.

Suyama, Y. (1985). Nucleic Acids Res. 13, 3273–3284.

Suyama, Y. (1986). Curr. Genet. 10, 411–420.

Suyama, Y., and Jenny, F. (1989). Nucleic Acids Res. 17, 803.

Suyama, Y., Jenney, F., and Okawa, N. (1987). Curr. Genet. 11, 327–330.

Suyama, Y., and Miura, K. (1968). Proc. Natl. Acad. Sci. U.S.A. 60, 235–242.

Suyama, Y., and Preer, J. R., Jr. (1965). Genet. Princeton 52, 1051–1058.

Tait, A., and Knowles, J. K. C. (1977). J. Cell Biol. 73, 139–148.

Tait, A., Knowles, J. K. C., Hardy, J. C., and Lipps, H. (1976). In "Genetics and Biogenesis of Chloroplasts and Mitochondria" (T. Bucher, W. Neupert, W. Sebald, and S. Werner, eds.), pp. 569–572. Elsevier/North-Holland, Amsterdam.

Upholt, W. B., and Borst, P. (1974). J. Cell Biol. 61, 383–397.

von Heijne, G. (1986). FEBS Lett. 198, 1–4.

Wahleithner, J. A., and Wolstenholme, D. R. (1988). Nucleic Acids Res. 16, 6897–6913.

Wallace, D. C. (1982). Microbiol. Rev. 46, 208–240.

Watson, J. D. (1972). Nature New Biol. 239, 197–201.

Wesolowski, M., and Fukuhara, H. (1979). Mol. Cell Biol. 1, 387–393.

Woese, C. R. (1987). Microbiol. Rev. 47, 621–669.

Yang, D., Oyaizu, Y., Oyaizu, H., Olsen, G. J., and Woese, C. R. (1985). Proc. Natl. Acad. Sci. U.S.A. 82, 4443–4447.

Ziaie, Z., and Suyama, Y. (1987). Curr. Genet. 12, 357–368.

Mitochondrial DNA of Kinetoplastids

Kenneth Stuart and Jean E. Feagin

Seattle Biomedical Research Institute, 4 Nickerson Street, Seattle, Washington 98109

I. Introduction

Kinetoplast DNA (kDNA) is the unusual mitochondrial DNA characteristic of members of the Protozoan order Kinetoplastida. This DNA has two distinctive general features: it comprises two classes of unrelated circular molecules, maxicircles and minicircles, and in each cell these molecules are catenated into a single massive network. Each network contains 40–50 maxicircles and 5000–10,000 minicircles. The presence of minicircles reflects the remarkable ability of these organisms to revise the nucleotide sequence of mitochondrial transcripts by RNA editing. The significance of the network organization is not known but it may ensure the proximity of maxicircle and minicircle transcripts, which appear to interact during RNA editing. Alternatively, the network organization may be important to kDNA segregation at cell division.

A. Subcellular Localization and Organization

About 7% of cellular DNA in kinetoplastids is kDNA. This abundance and its concentration at a single subcellular site, as a result of its network organization, led to its detection as a stained granule by 19th-century cytologists (Hertwig, 1893). These workers accorded it organelle status and gave it a variety of names that reflected their views of its function, including a second nucleus. The name kinetoplast, which it has retained, reflects its association with the flagellum. The kDNA is located within an enlarged region of the mitochondrion, also called the kinetoplast, opposite the flagellar basal body. This relationship occurs in all kinetoplastid species, although their morphologies differ, and is maintained during the life cycle even when morphological changes shift basal body location (Vickerman, 1985). This positional relationship and the presence of fibrils between the kinetoplast and the basal body implies a functional relationship such as a role for the basal body in segregating duplicated kDNA at cell division like the structurally similar centriole (Robinson and Gull, 1991). Thus, the kinetoplast is not an organelle but a region of the mitochondrion. Some functions, such as kDNA replication, transcription, and RNA editing, may be restricted to this region and thus it might be considered an organelloid, analogous to the nucleoid of bacteria. Several excellent reviews of kDNA have been presented (Englund 1981; Stuart, 1983a; Benne, 1985; Simpson, 1986, 1987; Ryan *et al.*, 1988).

The kDNA network has a highly ordered structure. The network is formed by catenation of the maxicircles and minicircles since topoisomerase II and restriction enzymes can decatenate or release, respectively, molecules from the network (Borst *et al.*, 1987). Several minicircles are catenated together; clusters of catenated minicircles are released from kDNA by partial digestion. The clusters are catenated together and maxicircles are catenated into the network to form a basket-shaped structure, at least in some species (Barker, 1989). Occasional maxicircles are observed at the edge of spread networks and maxicircles are clustered at the juncture of bilobed kDNA networks that are almost completely replicated. The network is packed within the mitochondrion with the DNA fibrils oriented parallel to one another and has a diameter which approximates that of one minicircle. This implies that the minicircles are stacked in a regular array. All minicircles have a "bend" that affects their electrophoretic mobility in acrylamide gels and maps to periodic clusters of A's that appear to affect DNA structure (Englund *et al.*, 1982). Bends characteristically are specific protein binding sites. It has been suggested that the bends and associated proteins bind the minicircles to the mitochondrial membrane, thus orienting the DNA (Silver *et al.*, 1986).

B. Maxicircle and Minicircle Diversity

The sizes of both the maxicircle and the minicircle vary among species, as summarized in Table I. The maxicircles range in size from less than 20 kb in some African trypanosomes to 35 kb in *Leishmania*. The size differences do not appear to be due to differences in gene content since maxicircles from the different species examined have the same genes in the same order (Simpson *et al.*, 1987; Sloof *et al.*, 1987; van der Spek *et al.*, 1989). Some variation in gene size occurs among species, reflecting differences in the extent of editing among species, as discussed below, but this is minor compared to the extent of variation. Rather, the size differences reflect variation in the size and presumably repeat number of simple

TABLE I
kDNA of Different Species[a]

Species	Maxicircle size (kb)	Minicircle size (kb)	Minicircle classes
Phytomonas davidii	39	1.1	10
Crithidia fasciculata	38	2.5	20
Herpetomonas ingenoplastis	36	23 + 17	Multiple
Trypanosoma cruzi	33	1.4	20
Herpetomonas muscarum	32	0.9	Multiple
Leishmania tarentolae	31	0.9	3
Trypanosoma mega	26	2.3	70
Trypanosoma brucei	22	1.0	300–400
T. equiperdum[b]	14–24	1.0	1
T. evansi[b]	—	1.0	1
Bodo caudatus	19	10 + 12	2

[a] Ordered according to maxicirle size.
[b] Considered here as a variant of *T. brucei*.

repeated sequences in one region, termed the variable region (Simpson *et al.*, 1987).

Minicircle sizes range from 645 bp in *Trypanosoma lewisi* to 2500 bp in *Crithidia* (Stuart, 1983a; Simpson, 1987). There is also substantial variation in minicircle sequence organization, as discussed below. In addition, the sequence complexity of minicircles varies among species by about two orders of magnitude. African trypanosomes have the greatest sequence complexity as determined from renaturation kinetic analyses. The total complexity of *T. brucei* kDNA is approximately 350 kb, which is primarily due to minicircle complexity (Stuart and Gelvin, 1980). However, these minicircles have a conserved ~120-bp sequence and three conserved pairs of 18-bp inverted repeats (Jasmer and Stuart, 1986a) and renaturation kinetic analyses do not effectively measure sequence microheterogeneity. Thus, *T. brucei* kDNA probably contains over 400 different minicircles. In addition, individual minicircle sequences differ in abundance in kDNA. Renaturation kinetic studies indicate that kDNA contains about 60 molecules of one minicircle sequence but about 500 copies of another (Stuart and Gelvin, 1980). The distribution of specific minicircle sequences among molecules is not known since minicircles may be capable of recombination. *Leishmania tarentolae* has less total complexity than *T. brucei* but also has an unequal content of various minicircle sequences, with 5 minicircles composing half of the total (Wesley and Simpson, 1973). Minicircle complexity has been measured in several strains of *Crithidia* and varies between 13 common minicircles (Borst and Weijers, 1976) and one predominant minicircle (Birkenmeyer *et al.*, 1985). *Trypanosoma equiperdum* and *T. evansi* also have a single minicircle as the predominant, and perhaps sole, class. *Bodo caudatus* is the most deviant kinetoplastid since it lacks the kDNA network and minicircles but retains two classes of kDNA molecule (Hajduk *et al.*, 1986). Minicircle sequences have been used as the basis for species identification in several cases (Sturm *et al.*, 1989; Laskay *et al.*, 1991).

Minicircle diversity is correlated with the extent of RNA editing in these species as far as has been examined; *T. brucei* has both the greatest diversity and the most editing. This is reasonable since small RNAs that direct editing are encoded in minicircles (Koslowsky *et al.*, 1990, 1991; Bhat *et al.*, 1990; Sturm and Simpson, 1990; Pollard *et al.*, 1990). All kinetoplastids thus far examined exhibit RNA editing but whether this type of editing is restricted to the Kinetoplastida is unknown.

C. Loss and Alteration of kDNA

There are several instances of naturally occurring, spontaneous, and induced kDNA alterations or loss. kDNA is lost at high frequency in kinetoplastids (Hoare, 1954); this is lethal in most species. It is conditionally lethal in African trypanosomes that do not utilize mitochondrial respiration when growing in the mammalian host. They can tolerate kDNA alteration and even total loss and grow in the mammalian host but cannot grow in the insect host where mitochondrial respiration is required (Vickerman, 1971; Stuart, 1983b). This resembles the ability of yeast to ferment anaerobically with one important difference: production of the respiratory system is developmentally controlled in African trypanosomes, requiring unidirectional progression through the life cycle, but is a freely reversible physiological switch in yeast. Among *brucei*-group African trypanosomes, three species exhibit natural kDNA alterations. *Trypanosoma equiperdum*, *T. equinum*, and *T. evansi* are morphologically indistinguishable from *T. brucei* but lack insect stages, being transmitted venereally or mechanically by biting flies (Woo, 1977). Some stocks of *T. equiperdum* have a single class of microheterogeneous minicircles in addition to the maxicircle, whereas others have maxicircle alterations that appear to be deletions but have not been characterized in detail (Frasch *et al.*, 1980). Stocks of *T. evansi* lack maxicircles and retain a single class of microheterogeneous minicircles; some stocks lack kDNA altogether (Borst *et al.*, 1987), as does *T. equinum* (Borst *et al.*, 1980). Intercalating agents such as ethidium bromide or acridines induce partial or total kDNA loss in African trypano-

somes and other kinetoplastids (Stuart, 1971; Hajduk, 1989). Mutants of
T. brucei with such alterations can be grown in mammalian hosts but
mutants of species unable to grow without the mitochondrial respiratory
system are nonviable. No mutants, natural or laboratory-induced, that
retain the maxicircle but have lost all minicircles have been found.

D. kDNA Replication

Replication of such a complex network would appear to be a forbidding
prospect. However, several studies (Ryan *et al.*, 1988) indicate that minicircles decatenate from the network and replicate by a conventional mechanism that entails the formation of theta structure intermediates. The
replicated minicircles then reattach to the network. The detachment and
reattachment appears to entail topoisomerase II activity and this enzyme
has been purified from kinetoplastids (Riou *et al.*, 1982). Maxicircle replication appears to entail a rolling circle mechanism although there has
been only one study on this process (Hajduk *et al.*, 1984). Cytological
incorporation studies into DNA in *L. tarentolae* indicate that replicated
DNA is added at the periphery of the kDNA network and is later redistrib-
uted, indicating some topological control of the minicircle reattachment
(Simpson and Simpson, 1974). Replication of the kDNA network has been
reviewed (Ryan *et al.*, 1988).

II. Maxicircle Genes

The nucleotide sequences of the *T. brucei* (Eperon *et al.*, 1983; Benne *et
al.*, 1983; Hensgens *et al.*, 1984; Feagin *et al.*, 1985; Payne *et al.*, 1985;
Jasmer *et al.*, 1987; Simpson *et al.*, 1987; de Vries *et al.*, 1988) and *L.
tarentolae* (de la Cruz *et al.*, 1984, 1985a,b; Muhich *et al.*, 1985) maxicircles and substantial portions of the *C. fasciculata* (Sloof *et al.*, 1985, 1987;
van der Spek *et al.*, 1989) maxicircle have been determined. Several
maxicircle genes have been identified by their nucleotide sequence homology with mitochondrial genes, and occasionally nuclear or chloroplast
genes, of other organisms. Not surprisingly, the genes identified (Table II)
are those commonly encoded in mitochondrial genomes. The genes are
encoded in both DNA strands of the maxicircle and are quite tightly packed
in 15–17 kb of sequence, depending on species (Fig. 1). The remaining
sequence is called the variable region and does not appear to contain
protein coding genes (Muhich *et al.*, 1985; de Vries *et al.*, 1988). The
maxicircle encodes large and small subunit ribosomal RNAs of about 1150

TABLE II
Summary of RNA Editing

Gene	*Trypanosoma brucei*		*Leishmania tarentolae*		*Crithidia fasciculata*	
	Edited mRNA(nt)	Uridines (+/−)	Edited mRNA(nt)	Uridines (+/−)	Edited mRNA(nt)	Uridines (+/−)
ND8	562	(259/46)				
2	649	(345/20)				
ND7	1238	(553/89)	1199	(25/0)	1198	(27/0)
COIII	969	(547/41)	904	(29/15)	912	(32/2)
CYb	1151	(34/0)	1150	(39/0)		(39/0)
A6	811	(447/28)	764	(106/5)		
3	"274"	(30/3)				
COII	663	(4/0)	670	(4/0)	643	(4/0)
MURF2	1111	(26/4)	1099	(28/4)	1117	(28/0)
4	"284"	(168/25)				
5	"450"	(211/15)				
6	325	(132/28)	330	(117/32)		

Note: Genes are ordered according to arrangement in the maxicircle irrespective of strand location and abbreviations are as in the legend to Fig. 1. (+/−) indicates uridines added (+) and removed (−) by editing and mRNA sizes in (" ") are estimated.

and 610 nt (Eperon *et al.*, 1983; Sloof *et al.*, 1985; de la Cruz et al., 1985a,b). These RNAs have been isolated from purified mitochondria but not purified ribosomes or ribosomal subunits. Nevertheless, the 12S and 9S rRNA sequences can be arranged into predicted secondary structures that share characteristics with the large and small rRNAs of *Escherichia coli*. It has not been determined whether the mitochondrial ribosomes contain other RNAs but their low sedimentation coefficient (Hanas *et al.*, 1975) and the presence of only two RNAs in mitochondrial ribosomes of most other organisms (Wallace, 1982) suggest that they do not.

A. Protein Coding Genes

The maxicircle encodes proteins of the oxidative phosphorylation system. These include mitochondrial NADH dehydrogenase (ND) subunits 1, 4, and 5, which are part of respiratory complex I (de la Cruz *et al.*, 1984; Hensgens *et al.*, 1984; Payne *et al.*, 1985; Jasmer *et al.*, 1987; Simpson *et al.*, 1987; van der Spek *et al.*, 1989). It encodes two additional components

T. brucei

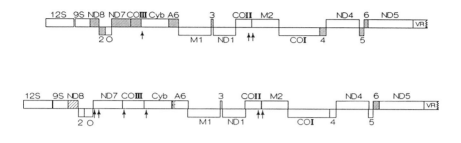

L. tarentolae

FIG. 1   Diagrams of the maxicircles of *Trypanosoma brucei* and *Leishmania tarentolae*. The linearized maxicircle maps are broken at the variable region (VR) that varies in size and sequence among species and stocks. The strand location of each gene is indicated. Shaded areas indicate extensive editing of mRNAs, light shading of ND8 of *L. tarentolae* indicates some editing of mRNA, and arrows indicate sites of limited editing in mRNA. Abbreviations: 12S and 9s, rRNA; ND4, −5, −7, and −8, NADH dehydrogenase subunits 4, 5, 7, and 8; COI, II, and III, cytochrome oxidase subunits I, II, and III; Cyb, apocytochrome *b*; A6, ATPase (ATP synthase) subunit 6; M1 and −2, maxicircle unidentified reading frames 1 and 2; 2–6, unidentified C-rich template or G-rich coding strand genes (6 may be S12); O, unidentified open reading frame.

of complex I, ND7 (Shaw *et al.*, 1988; van der Spek *et al.*, 1988; Koslowsky *et al.*, 1990) and ND8 (Souza *et al.*, 1992), that are encoded in the bovine nucleus and in chloroplast DNA in plants. The maxicircle also encodes apocytochrome *b* (CYb), a component of complex III (Benne *et al.*, 1983; de la Cruz *et al.*, 1984; Sloof *et al.*, 1987), and subunits I, II, and III of cytochrome oxidase (CO), which are part of complex IV (de la Cruz *et al.*, 1984; Hensgens *et al.*, 1984; Payne *et al.*, 1985; Sloof *et al.*, 1987; Feagin *et al.*, 1988a). A protein that resembles ATP synthase subunit 6 (A6) is also found in maxicircle genomes (Bhat *et al.*, 1990, 1991). This gene occurs in mt DNAs of other organisms but there is little evolutionary conservation of the predicted amino acid sequence. Only a few amino acids near the C-terminus, thought to play a role in proton translocation, are conserved in all species (Cain and Simoni, 1988, 1989). This low conservation reduces the confidence in the identification of the A6 gene. Biochemical or functional studies may resolve this uncertainty. Similarly, the CR6 gene encodes extensively edited mRNA (Read *et al.*, 1992b) and may encode S12 ribosomal protein (Maslov *et al.*, 1992). Mitochondrial genomes also often encode additional NADH dehydrogenase and ATPase subunits and mt tRNAs (Wallace, 1982). Although these genes have not been identified in maxicircle sequences, candidate unidentified genes are present, as described below. Some of these may encode the above-noted components but their editing and/or divergence may have prevented their recognition. Evidence has been presented that some mitochondrial tRNAs are encoded in the nucleus in *T. brucei* (Hancock and Hajduk, 1990) and *L. tarentolae* (Simpson *et al.*, 1989), as is the case for other lower eukaryotes, especially *Tetrahymena* (Suyama, 1982). However, the possibility that diverged or edited tRNAs are encoded in kDNA cannot yet be excluded.

The maxicircle encodes additional genes that probably specify proteins but which have unknown functions. One set of two genes has sequences that are conserved among maxicircles of kinetoplastid species but have no detected homology to other mitochondrial genes or predicted proteins. These maxicircle unidentified reading frame (MURF) 1 and 2 genes have large open reading frames, with predicted proteins of 54.3 and 44.5 kDa, respectively, and codon usage similar to that of the other maxicircle protein coding genes (Simpson *et al.*, 1987; van der Spek *et al.*, 1989). A third possible gene, ORF8 in *T. brucei*, has apparent homology to ORF10 in *L. tarentolae* but the significance of this is uncertain due to AT richness of the genes (Simpson *et al.*, 1987). The MURF1, MURF2, and ORF8/10 proteins may be homologous to proteins that are encoded in the nucleus (or chloroplast) of other organisms, such as ND7 and ND8, or they may be specific to the kinetoplastid mitochondrion. The other set of six genes are characterized by G versus C strand bias. Their locations but not their sequences are conserved among maxicircles of different kinetoplastid species (Simpson *et al.*, 1987). Their transcripts invariably correspond to the G-biased strand (Jasmer *et al.*, 1987) and these C-rich template (CR) genes are rather small, 105–368 bp (Simpson *et al.*, 1987). Analyses in progress indicate that these genes encode small hydrophobic proteins. The predicted CR1 protein appears to be a nonheme iron sulfur protein and is homologous to a bovine nuclearly encoded subunit of NADH dehydrogenase; it has been designated ND8 (Souza *et al.*, 1992). CR6 may encode the S12 ribosomal protein (Maslov *et al.*, 1992) but this needs to be confirmed. The potential identity of the other CR proteins has not been determined but they also may be hydrophobic components of the respiratory complexes.

B. gRNA Genes

The *L. tarentolae* maxicircle also includes guide RNA (gRNA) genes (Blum *et al.*, 1990). The gRNAs are small transcripts that are complementary to edited RNA, providing G–U base-pairing is allowed, and appear to specify edited mRNA sequences, as described below (Sections III,C and III,D). The gRNAs are encoded in the flanking regions of the protein genes, in some instances on the strand opposite to protein coding genes, and the COII gRNA sequence may be located within the 3' end sequence of the COII gene. gRNA sequences have been found in identical positions in the *C. fasciculata* maxicircle, although in two instances, the resulting alignment predicts an A–C match between the gRNA and the mRNA (van der Spek *et al.*, 1991). With the possible exception of the COII gRNA, which may be within the 3' region of the COII transcript, as in *L. tarentolae*, no gRNA genes have been found in the maxicircle of *T. brucei;* rather, gRNA genes for this species are in minicircles (Koslowsky *et al.*, 1990, 1991; Bhat *et al.*, 1990; Pollard *et al.*, 1990; Souza *et al.*, 1992; Read *et al.*, 1992b).

C. Variable Region

The variable region (VR) of the maxicircle between the ND5 and the 12S rRNA genes varies in size and sequence among species and stocks within a species. It has been completely sequenced in *T. brucei* (K. Stuart and colleagues, unpublished results). The portion of the VR that is nearer the ND5 gene contains several AT-rich, repeated sequences that occur in superrepeats (Muhich *et al.*, 1985; Simpson *et al.*, 1987; Stuart *et al.*, 1987; de Vries *et al.*, 1988; K. Stuart and colleagues, unpublished results). The simple sequence nature of the repeats suggests that this portion of the VR does not have a protein coding function. The repeats are present in various stocks of *T. brucei* but the size of this region varies among stocks, suggesting that VR size differences among stocks are due to differences in repeat number (Stuart, 1991). The *T. brucei* repeats are not homologous to those of *L. tarentolae* (Simpson *et al.*, 1987) and these regions do not cross-hybridize between these species (Muhich *et al.*, 1983). The VR sequence near the *T. brucei* 12S rRNA gene is nonrepetitive and less AT-rich; it makes up a smaller proportion of the VR (Simpson *et al.*, 1987; de Vries *et al.*, 1988). A single small transcript has been mapped to this region but has not been further investigated (Rohrer *et al.*, 1987). The lack of other abundant transcripts in this VR region suggests that it has no protein coding function. The possibility that the VR may contain gRNA genes has not been fully explored. The VR may contain the maxicircle replication origin. The replication origin has an analogous location near the rRNA genes in other mitochondrial genomes (Wallace, 1982). The similarity is especially evident in *Drosophila* mitochondrial DNA where the replication origin is found in AT-rich sequences that vary in size among species and are adjacent to the rRNAs (Wolstenholme and Clary, 1985).

D. Maxicircle Gene Order

Gene order is conserved between the maxicircles of *T. burcei* and *L. tarentolae* (Simpson *et al.*, 1987) (Fig. 1) and with the characterized regions of *C. fasciculata* (Sloof *et al.*, 1985, 1987; van der Spek *et al.*, 1989), except for the gRNA genes noted above. The maxicircle of *T. cruzi* appears to be colinear with that of *T. brucei*, based on cross-hybridization studies (Affranchino *et al.*, 1986). Maxicircle sequence homology is higher between *L. tarentolae* and *C. fasciculata* than between these species and *T. brucei*. The *T. brucei* maxicircle genes that encode RNA sequences that are not extensively edited show substantial homology to those of *L. tarentolae* and *C. fasciculata* (Simpson *et al.*, 1987; Sloof *et al.*, 1987; van der Spek *et al.*, 1989). However, the *T. brucei* sequences that encode more extensively edited RNAs understandably do not show homology to the corresponding sequences of *L. tarentolae* or *C. fasciculata* (Simpson *et al.*, 1987; Feagin *et al.*, 1988a; Koslowsky *et al.*, 1990; Bhat *et al.*, 1990). This is especially evident in the regions between the 9S rRNA and CYb genes and 3' to the CYb gene. There is virtually no significant homology between the maxicircle sequences of the different species in these regions but the edited transcripts are highly homologous (Feagin *et al.*, 1988a; Koslowsky *et al.*, 1990; Bhat *et al.*, 1990). As noted, the CR1–6 genes are conserved among species in location, G versus C strand bias, and transcribed strand. All CR region genes examined to date are extensively edited (Souza *et al.*, 1992; Read *et al.*, 1992b; L. Simpson, personal communication; K. Stuart and colleagues, unpublished results); comparisons of the edited RNAs to demonstrate their functional homology have not yet been possible except for CR1 (ND8) and CR6 (S12), but this is anticipated.

III. Maxicircle Transcripts

Gene-specific probes were used initially to identify rRNAs, mRNAs, and larger low-abundance possible precursor transcripts in steady-state RNA and to map these to the maxicircles of *T. brucei* (Benne *et al.*, 1983; Feagin and Stuart, 1985; Feagin *et al.*, 1985; Jasmer *et al.*, 1985; Michelotti and Hajduk, 1987), *L. tarentolae* (Simpson *et al.*, 1985), and *C. fasciculata* (Sloof *et al.*, 1987). Subsequent studies (described below) have analyzed these transcripts in greater detail. Less specific probes were used to examine maxicircle transcripts in *T. cruzi* and these appear similar to those of *T. brucei* (Affranchino *et al.*, 1986). In general, the most abundant transcripts in steady-state RNA map to the genes described above which are located on both maxicircle DNA strands (Fig. 1). Lower abundance transcripts with sizes and hybridization patterns consistent with their being precursors of processing by cleavage have been observed (Feagin and Stuart, 1985; Feagin *et al.*, 1985; Jasmer *et al.*, 1985). However, kinetic studies have not been performed to determine whether each DNA strand is a single transcription unit or if there are multiple promoters on each strand. Maxicircle genes are closely packed and overlap in some cases, as discussed in Section III,B. This suggests that each gene does not have its own promoter but further study is needed to resolve the pattern of maxicircle transcription. Additional low-abundance, heterogeneous transcripts map to the variable region but the significance of these transcripts is unknown and their analysis is complicated by the repeat nature of this region (Stuart *et al.*, 1985; de Vries *et al.*, 1988).

A. Editing of Maxicircle Transcripts

The nucleotide sequence data from the *T. brucei, L. tarentolae,* and *C. fasciculata* maxicircles include several apparent anomalies. The COII gene in all three species contains a frameshift as do the ND7 genes of *L. tarentolae* and *C. fasciculata,* and the MURF2 gene of *L. tarentolae* (Hensgens *et al.*, 1984; de la Cruz *et al.*, 1984, Payne *et al.*, 1985; Sloof *et al.*, 1987). The CYb (except for *T. brucei*) and MURF2 genes, among others, lack encoded AUG initiations codons near the 5' end of their ORFs (Hensgens *et al.*, 1984; Payne *et al.*, 1985; Sloof *et al.*, 1987; van der Spek *et al.*, 1989). Most dramatically, the COIII, ND7, and A6 genes of *T. brucei* lack continuous ORFs and their RNAs are heterogeneous in size and larger than expected from ORF sizes; in fact, these genes cannot be identified on the basis of DNA sequence homology (Feagin *et al.*, 1988a; Koslowsky *et al.*, 1990; Bhat *et al.*, 1990). This is also now known to be the case for the CRl–6 genes (Souza *et al.*, 1992; Read *et al.*, 1992b). These gene sequence peculiarities reflect the existence of the remarkable phenomenon of RNA editing, which adds and deletes uridines at specific sites in specific transcripts from kinetoplastid maxicircle genes.

Editing can be quite restricted or very extensive and although the same transcripts are edited in different species, editing is generally more extensive in *T. brucei* (Table II). The frameshift is eliminated from COII transcripts by addition of four uridines at three sites in *T. brucei* (Benne *et al.*, 1986), *L. tarentolae* (Shaw *et al.*, 1989), and *C. fasciculata* (Benne *et al.*,

1986). The CYb (Feagin *et al.*, 1987, 1988b) and MURF2 (Feagin and Stuart, 1988; Shaw *et al.*, 1988) transcripts in these species are edited near the 5′ end by the addition of tens of uridines; additionally, several uridines are deleted from *T. brucei* MURF2 transcripts. The CYb and MURF2 editing creates AUG initiation codons that are in frame with the ORFs. The *L. tarentolae* and *C. fasciculata* COIII transcripts are edited in a restricted region near the 5′ end by the addition of tens and removal of a few uridines, creating initiation codons (Shaw *et al.*, 1988), but *T. brucei* COIII transcripts are extensively edited by the addition of over 500 uridines and removal of tens of uridines (Feagin *et al.*, 1988a; unpublished results). A substantial portion of the *L. tarentolae* A6 transcript is edited but again the editing of the corresponding transcript in *T. brucei* is more extensive (Bhat *et al.*, 1991). ND7 transcripts are edited in two domains that are separated by a short unedited region (Shaw *et al.*, 1988; van der Spek *et al.*, 1988; Koslowsky *et al.*, 1990); both domains are more extensively edited in *T. brucei*. CR gene transcripts are also extensively edited (Table II). Editing frequently creates initiation and termination codons and the consequence of editing is the creation of a mRNA that predicts translation into a protein that is usually homologous to a known mitochondrial protein (reviewed in Stuart, 1989a,b; Benne, 1989; Simpson and Shaw, 1989; Simpson, 1990; Benne, 1990; Feagin, 1990). Thus, editing results in the production of a functional mature mRNA and is a critical step in expression of many maxicircle genes. The ND7 transcripts of *L. tarentolae* (Shaw *et al.*, 1988) and *C. fasciculata* (van der Spek *et al.*, 1988) are edited but no inframe initiation codon results. These mRNAs may use alternate initiation codons, the mature mRNA may not have been identified, or these mRNAs may be nonfunctional in these species. Antibody prepared against a peptide from the C-terminus of the *L. tarentolae* COII protein detects a protein of the size expected from the edited COII transcript on Western blots (Shaw *et al.*, 1989), supporting the translation of edited transcripts.

Although many maxicircle transcripts are edited, a number appear not to be edited (Shaw *et al.*, 1988; K. Stuart and colleagues, unpublished results). RNA sequencing of the 5′ ends of many maxicircle RNAs has shown that several contain encoded AUGs and are not edited. Additionally, a few genes that lack 5′ AUGs also appear to lack 5′ editing. With the exception of the extensively edited genes and those containing identified frameshifts, maxicircle genes have continuous open reading frames that are well conserved between the kinetoplastids, suggesting a lack of internal editing (Simpson *et al.*, 1987). The *T. brucei* CYb (Feagin *et al.*, 1987; unpublished results), *L. tarentolae* A6 (Bhat *et al.*, 1991), and *C. fasciculata* ND7 (van der Spek *et al.*, 1988) mRNAs have been sequenced in their entirety and are edited only in restricted regions, confirming a lack of further internal editing.

Extensive editing is associated with GC-rich sequence (Feagin *et al.*, 1988a; Koslowsky *et al.*, 1990; Bhat *et al.*, 1990) and an internal portion of the COI gene with this characteristic has been investigated and found to be unedited (K. Stuart and colleagues, unpublished results). These studies confirm that only certain regions are edited. In addition, approximately 30 and 15 nt are not edited at the 5' and 3' termini, respectively, of any RNAs (Stuart, 1991). The mature mRNA sequences thus can entail no editing, editing in restricted regions, or editing throughout except at the termini. Some transcripts are potentially translatable in both their edited and their unedited forms and thus there is a possibility that both versions are functional in these cases, as discussed in Section IV,D.

B. Transcript Termini and Overlapping Genes

The 5' ends of most maxicircle transcripts have been determined and all edited transcripts have a short 5' (28–57 nt) untranslated sequence (Stuart, 1991). For edited transcripts, the 5' 30 nt are not edited, perhaps reflecting requirements for editing, and there is minor editing in the 5' untranslated region that sometimes varies among cDNAs from the same gene (Bhat *et al.*, 1990; Koslowsky *et al.*, 1990; Read *et al.*, 1992b). The variation in 5' editing may reflect incomplete editing in these cDNAs or less stringent editing outside the open reading frame. The poly(A) addition sites have been examined for numerous maxicircle transcripts. There appears to be a preferentially used poly(A) addition site for each mRNA, based on cDNA analysis, but multiple sites can be used (van der Spek *et al.*, 1990; Bhat *et al.*, 1991). The poly(A) addition sites tend to be rich in adenines and uridines. Since there are often uridines as well as adenines present in the poly(A) tail (van der Spek *et al.*, 1988, 1990; Feagin *et al.*, 1988a; Campbell *et al.*, 1989; Bhat *et al.*, 1990; Read *et al.*, 1992b), the exact poly(A) addition site is sometimes ambiguous. The uridines may be added to the poly(A) tail by poly(A) polymerase or alternatively by the mitochondrial terminal uridylyltransferase (Bakalara *et al.*, 1989; Harris *et al.*, 1990) which may have a role in editing. The sites of uridine addition in the poly(A) tail vary among the cDNAs (van der Spek *et al.*, 1988, 1990; Feagin *et al.*, 1988a; Bhat *et al.*, 1990), suggesting that uridine addition in the tail is not directed by gRNAs. Interestingly, unedited transcripts of some genes have poly(A) tails that have added uridines (van der Spek *et al.*, 1990) and edited transcripts of other genes have poly(A) tails that lack added uridines (Koslowsky *et al.*, 1990). Thus, editing, polyadenylation, and addition of uridines in poly(A) tails appear to occur independently. The 12S and 9S rRNAs have 2–17 and 11 uridines, respectively, that are added post-transcriptionally, to their 3' ends by an uncharacterized mechanism (Adler *et al.*, 1991). In general, the 3' untranslated regions of the mRNAs are larger (8–82 nt) than the 5' untranslated regions and can also be edited. The most 3' 15 nt are not edited in any RNA, suggesting that this may be a requirement for editing, as discussed in Section III,C.

Detailed analyses of adjacent mRNAs have revealed the extent of close packing of the maxicircle genes. The 5' end of the *T. brucei* MURF2 transcript maps 10 nt into the open reading frame of the upstream COII gene (Feagin and Stuart, 1988). Since the 3' end of the COII transcript has not been determined, the full extent of gene overlap is not yet known. Similarly, the 5' end of the COIII transcript maps 28 nt into the 3' end of the ND7 gene but not into the open reading frame (Koslowsky *et al.*, 1990; unpublished results). Intriguingly, a single added uridine is added to the ND7 RNA by editing (Koslowsky *et al.*, 1990) but the same sequence is not edited in the 5' end of the COIII transcript (K. Stuart and colleagues, unpublished results). The ends of the COIII and CYb, COI and CR4, and CR6 and ND5 transcripts also overlap, by 5 (Feagin *et al.*, 1987, 1988a), 41, and 37 nt (Read *et al.*, 1992b), respectively. In each of these cases, one member of the pair is edited in the overlap but the other is not. A relatively abundant potential precursor for COII and MURF2 has been detected in Northern blots; it is large enough to encode both transcripts and maps to both genes (Feagin and Stuart, 1985). The other pairings listed above have less abundant potential precursors (Feagin *et al.*, 1985; Jasmer *et al.*, 1985). A cDNA for CR6 also contained ND5 sequence, verifying the existence of a precursor. Somewhat unexpectedly, it is edited, indicating that editing can precede cleavage (Read *et al.*, 1992b). However, in all the above pairs, neither potential precursor could produce both mRNAs by cleavage (and editing) alone since a single gene sequence is shared by both mRNAs in both cases. If the larger molecules are precursors, then each molecule must lead to one or the other mRNA or a mechanism exists to ensure that the same sequence is retained on both mRNAs.

Another possibility is that the low-abundance transcripts are not precursors but are transcribed from different promoters. Maxicircle promoters have not been identified. Small mitochondrial genomes generally have few promoters and some transcribe each DNA strand from a single promoter (Wallace, 1982). The large low-abundance maxicircle transcripts are consistent with this pattern (Feagin and Stuart, 1985; Feagin *et al.*, 1985; Jasmer *et al.*, 1985). However, maxicircle rRNAs and other maxicircle transcripts can be radiolabeled using guanylyltransferase, suggesting that they are primary transcripts, which in turn implies multiple promoters (Simpson *et al.*, 1985). gRNAs can also be labeled using guanylyltransferase (Blum and Simpson, 1990; Pollard *et al.*, 1990; Pollard and Hajduk, 1991) and gRNA/mRNA chimeras have been observed (Blum *et al.*, 1991; Koslowsky *et al.*, 1991; Read *et al.*, 1992a); these may explain the labeling of mRNA-sized transcripts. The identity of the labeled RNA is thus uncertain and the number and location of maxicircle promoters remain enigmatic.

C. gRNAs

The maxicircles of *L. tarentolae* and *C. fasciculata* encode gRNAs (Blum *et al.*, 1990; van der Spek *et al.*, 1991). Although the *T. brucei* maxicircle may also encode gRNAs, all gRNAs identified thus far, except for a possible COII gRNA (Section II,B), are encoded by minicircles in this species (Koslowsky *et al.*, 1990, 1991; Bhat *et al.*, 1990; Pollard *et al.*, 1990). *Leishmania* minicircles also encode gRNAs (Sturm and Simpson, 1990, 1991). Regardless of the gene location, gRNA transcripts share some general characteristics. They are small, usually less than 60 nt (Blum *et al.*, 1990; Sturm and Simpson, 1990; Pollard *et al.*, 1990; Pollard and Hajduk, 1991; Koslowsky *et al.*, 1990; Bhat *et al.*, 1990), and have a 3' poly(U) tail that is added post-transcriptionally (Blum and Simpson, 1990; Pollard and Hajduk; 1991). The tail varies in length but averages 15 residues. As noted in Section III,B, these gRNAs may be primary transcripts (Blum and Simpson, 1990; Pollard *et al.*, 1990; Pollard and Hajduk, 1991). The 5' sequences of gRNAs are very similar, with a RYAYA motif commonly as the most 5' nt (Pollard *et al.*, 1990; Pollard and Hajduk, 1991) although not all gRNAs have this 5' sequence. All but about eight 5' end and a few 3' end nucleotides of the gRNAs can form a perfect duplex with corresponding edited mRNAs. Much of the duplex entails substantial G:U, in addition to Watson–Crick, base-pairing but the complementarity at the 5' end of the gRNA employs predominantly Watson–Crick base-pairing (Blum *et al.*, 1990; Koslowsky *et al.*, 1990; Bhat *et al.*, 1990; Pollard *et al.*, 1990). This is called the anchor sequence and may provide the initial recognition and binding between the gRNA and mRNA for editing. Only a few gRNAs for extensively edited mRNAs can form an anchor duplex with unedited transcripts. The rest require an edited sequence that is created by another gRNA. This apparent requirement may impose a progression in the sequence in which gRNAs specify the edited sequences.

D. Editing Mechanism

It has recently been proposed that editing may involve transesterification reactions similar to those occurring in splicing (Cech, 1991; Blum *et al.*, 1991). Following anchor duplex formation, the 3' oligo(U) tail may participate in a nucleophilic attack on the mRNA, resulting in production of a chimeric molecule consisting of the gRNA and the 3' portion of the mRNA. The predicted 5' truncated molecules have been observed (Read *et al.*, 1992a), as have chimeras formed *in vitro* (Koslowsky *et al.*, 1992a; Harris and Hajduk, 1992). An additional transesterification would then allow reattachment of the 5' end of the mRNA to its 3' end, leaving behind uridines from the gRNA tail (or removing them from the mRNA) as directed by the gRNA sequence. The chimeric molecules predicted by this mechanism have been observed in *L. tarentolae* (Blum *et al.*, 1991) and *T. brucei* (Koslowsky *et al.*, 1991; Read *et al.*, 1992a), providing support for the proposed mechanism.

IV. Developmental Control in *T. brucei*

The production of the mitochondrial respiratory system is regulated during the life cycle of *T. brucei*. Procyclic (insect gut) stages rely on aerobic respiration for energy production and have a functional mitochondrial respiratory system. Mammalian bloodstream stages obtain energy by glycolysis and lack cytochromes and Krebs-cycle enzymes. The bloodstream stages can be further divided into slender and stumpy forms; the former is the first to appear after infection and the latter appear later and may be the forms capable of infecting the insect host (Vickerman, 1985).

A. mRNA Abundance

The abundance, editing, and polyadenylation of specific mitochondrial transcripts are regulated during the life cycle of *T. brucei*. For example, CYb, COI, and COII transcripts are more abundant in procyclic forms than in slender bloodstream forms (Feagin and Stuart, 1985; Feagin *et al.*, 1985; Jasmer *et al.*, 1985; Michelotti and Hajduk, 1987). In contrast, ND5 transcripts are more abundant in slender bloodstream forms (Jasmer *et al.*, 1985). The MURF2 transcripts are similar in abundance in both life cycle stages (Feagin and Stuart, 1985). This pattern is somewhat altered in naturally occurring and DFMO-induced stumpy bloodstream forms and varies somewhat on growth of the organisms and among strains (Feagin *et al.*, 1986; Michelotti and Hajduk, 1987). Thus, in general, transcripts for the cytochromes are more abundant in the procyclic forms that rely on oxidative phosphorylation, whereas transcripts for complex I components are more abundant in bloodstream forms that rely on glycolysis. The role of complex I in bloodstream forms is uncertain but it may be involved in transferring electrons to an alternate oxidase (Koslowsky *et al.*, 1990).

Thus, physiological and developmental factors may alter the pattern tran- script abundance. Additional mechanisms, discussed below, also serve to regulate transcript function during the life cycle and perhaps serve to control metabolic processes.

B. Polyadenylation

Two size classes of transcripts are observed for most genes. Treatment of RNA with RNAse H in the presence of specific oligonucleotides, including oligo(dT), indicates that the small transcripts have a short 20-residue poly(A) tail, whereas the larger transcripts have poly(A) tails that are about 120 nt larger (Bhat *et al.*, 1991; 1992). Both sizes of poly(A) tail can also contain noncoded uridines (Bhat *et al.*, 1991). COI, COII, and CYb transcripts with larger poly(A) tails are more abundant in procyclic forms than in bloodstream forms. However, the reverse pattern is seen for CR1 transcripts that have longer poly(A) tails in bloodstream forms than in procyclic forms. This pattern is reminiscent of the pattern of transcript abundance. The transcripts that are more abundant in one stage than the other tend to have longer poly(A) tails in that stage (Bhat *et al.*, 1992). It is not possible to determine whether long poly(A) tails are responsible for greater transcript abundance, by slowing the rate of degradation, or whether the greater abundance is responsible for the longer poly(A) tails, by allowing time for greater accumulation of the tail.

C. Editing

RNA editing is also developmentally regulated in *T. brucei*. The CYb and COII transcripts are preferentially edited in procyclic forms compared to bloodstream forms (Feagin *et al.*, 1987; Feagin and Stuart, 1988); CR1 and CR6 transcripts are preferentially edited in bloodstream forms (Souza *et al.*, 1992; Read *et al.*, 1992b); and A6, MURF2, and COIII transcripts are edited in both life cycle stages (Bhat *et al.*, 1990; Feagin *et al.*, 1988a; Feagin and Stuart, 1988). The 5′ domain of ND7 transcripts is edited in both bloodstream and procyclic forms but the 3′ domain is preferentially edited in bloodstream forms (Koslowsky *et al.*, 1990). As with abundance differences, the proportion of edited transcripts varies with growth of the parasite, between slender and stumpy bloodstream forms and among strains. However, some mechanism regulates editing in a transcript- specific fashion. This regulation is not at the level of absolute gRNA presence since gRNAs that specify edited sequence are present in life cycle stages in which the corresponding mRNA is not edited (Koslowsky *et al.*, 1992b), but may be affected by the relative proportion of specific gRNAs to the corresponding mRNA.

D. Potential for Alternate Proteins

In many edited transcripts, the initiation codon is created by editing, providing a mechanism to avoid translation of unedited or partially edited transcripts. However, two edited *T. brucei* transcripts can potentially be translated in their unedited versions as well. The frameshift editing in COII occurs near the 3' end of the transcript (Benne *et al.* 1986; Shaw *et al.*, 1989). As the initiation AUG is not affected by editing, a carboxy-terminal truncated protein may be made from unedited COII mRNA. This would be most abundant in bloodstream forms, since COII editing is developmentally regulated. Similarly, *T. brucei* CYb transcripts have an encoded AUG 20 amino acids downstream of the created one; this may initiate protein synthesis from unedited message (Feagin *et al.*, 1987; Feagin and Stuart, 1988). An amino-terminus truncated protein would result, again probably most abundant in bloodstream forms because of the developmental regulation of CYb editing. In a third example, ND7 transcripts are constitutively edited by the 5' domain, but the 3' domain is incompletely edited in procyclic forms (Koslowsky *et al.*, 1990). An amino-terminal peptide could thus be made in procyclic forms.

V. Minicircles

Both minicircle size and sequence organization varies among species. All minicircles have sequence that is conserved among different minicircles within different stocks of the same species (Simpson, 1987). This sequence is about 120 nt in the 1000-bp *T. brucei*, 850-bp *L. tarentolae*, and 1400-bp *T. cruzi* minicircles, respectively. The conservation is at about the 90% level but some additions and substitutions occur between conserved regions of minicircles from the same stock. A 13-nt sequence is contained within the conserved sequence in all kinetoplastid species examined. This conserved sequence occurs as a single-stranded region (or gap) in replicating minicircles and a small RNA that is homologous to this region has been detected (Rohrer *et al.*, 1987). These data suggest that the conserved sequence is required for DNA replication and the RNA may be the primer that initiates replication. The 120-bp conserved sequence contains numerous termination codons and nucleotide additions and deletions. Additionally, it lacks AUGs at the beginning of its short open reading frames and lacks abundant transcripts. These points suggest that it does not encode a protein(s).

All minicircles examined also have a region that results in lower mobility than expected on acrylamide gels as a function of the acrylamide concentration (Englund *et al.*, 1982; Jasmer and Stuart, 1986b). This is the characteristic of a "bend" or "kink" in the DNA and appears to be associated with regularly spaced runs of adenines. Although this feature is conserved in all minicircles, its position is not conserved; it is found at the edge of the conserved region in *T. brucei* (Ntambi *et al.*, 1984; Jasmer and Stuart, 1986b) and *L. tarentolae* (Kidane *et al.*, 1984) but is located at a different position in *Crithidia* (Ray *et al.*, 1986). Bends in DNA are associated with DNA binding proteins in other systems (Englund *et al.*, 1982). It appears unlikely that the minicircle bend is associated with DNA replication (i.e., DNA polymerase binding site) since its position relative to the putative origin of replication in the conserved region varies among species. However, it might bind a protein that has a role in orienting the minicircles in the highly structured arrangement in which they occur in the kinetoplast (Silver *et al.*, 1986).

Minicircles vary in their overall organization among species. The minicircles of *L. tarentolae* have one conserved sequence and one bend and the remaining sequence varies among minicircles (Kidane *et al.*, 1984). The minicircles of *T. brucei* have, in addition to the conserved sequence and bend, an organization that is defined by pairs of inverted 18-bp repeats (Jasmer and Stuart, 1986a). There are three pairs, or four in *T. equiperdum*, of inverted repeats that flank about 100-bp units of sequence. All gRNAs that are encoded in minicircles in African trypanosomes have their coding sequence in the middle of the 100-bp sequence (Koslowsky *et al.*, 1990; 1991; Bhat *et al.*, 1990; Pollard *et al.*, 1990; Pollard and Hajduk, 1991). Thus, African trypanosome minicircles contain three or four gRNA coding cassettes in addition to the conserved sequence. Neither the 18-bp repeats nor their analogs occur in minicircles of *L. tarentolae* (Jasmer and Stuart, 1986a); these appear to encode single gRNAs 150 bp from the conserved region (Sturm and Simpson, 1990; 1991). They also do not occur in *T. cruzi*. However, the entire *T. cruzi* minicircle has a repeat organization that comprises four repeats of the conserved sequence and an associated variable sequence (Riou and Yot, 1977). This variable sequence may encode gRNAs in *T. cruzi* and thus each minicircle may encode four gRNAs. The significance of the different minicircle sequence organizations and the 18-bp sequences is not known.

VI. Minicircle Transcripts

Two classes of minicircle transcripts have been described. One class is encoded near or within the conserved sequence (Fouts and Wolstenholme, 1979; Rohrer *et al.*, 1987). These transcripts are heterogeneous in size and low in abundance. Their function or significance is uncertain. The second class of transcripts are the gRNAs that direct editing (Blum *et al.*, 1990; Sturm and Simpson, 1990, 1991; Koslowsky *et al.*, 1990, 1991; Bhat *et al.*, 1990; Pollard *et al.*, 1990; Pollard and Hajduk, 1991). The gRNAs of *T. brucei* and *T. equiperdum* are all transcribed from the same strand relative to the conserved sequence and from the opposite strand to that from which the conserved sequence transcripts are derived (Pollard and Hajduk, 1991; Koslowsky *et al.*, 1991). The ability to label gRNAs with guanylyltransferase suggests that they are primary transcripts (Blum and Simpson, 1990; Pollard *et al.*, 1990; Pollard and Hajduk, 1991). However, additional studies are needed to confirm this hypothesis.

VII. Conclusion

Kinetoplast DNA maxicircles are the genetic equivalent of mitochondrial genomes but, until recently, minicircle function was undetermined. The finding that minicircles encode gRNAs now makes their presence, abundance, and heterogeneity comprehensible. Catenation of maxicircles and minicircles into a network may function in kDNA segregation during cell division and may also promote physical proximity for gRNAs and pre-mRNAs, enhancing RNA editing. A variety of maxicircle gene characteristics, including frameshifts, G versus C strand bias, and the absence of AUG initiation codons, reflect the phenomenon of RNA editing. The remodeling of RNA structure permitted by editing, combined with processing of transcripts from overlapping genes and developmental regulation of transcript abundance, polyadenylation, and editing, provides the kinetoplastid protozoa with a variety of means to modulate gene expression. The study of kDNA has elucidated its genetic content and has provided a rationale for its organization and insight into the mechanisms that regulate its expression.

Acknowledgments

We acknowledge G. J. Bhat, R. Corell, U. Göringer, D. Koslowsky, A. Morales, P. Myler, A. Perrollaz, L. Read, G. Riley, H.-H. Shu, B. Smiley, A. Souza, and K. Wilson for their enthusiasm, helpful discussions, and numerous contributions to the study of RNA editing. Our studies of RNA editing were supported by NIH Grants AI14102 and GM42188, and K. S. is a Burroughs Wellcome Scholar in Molecular Parasitology.

References

Adler, B. K., Harris, M. E., Bertrand, K. I., and Hajduk, S. L. (1991). *Mol. Cell. Biol.* **11,** 5878–5884.
Affranchino, J. L., Sanchez, D. O., Engel, J. C., Frasch, A. C. C., and Stoppani, A. O. M. (1986). *J. Protozool.* **33(4),** 503–507.
Bakalara, N., Simpson, A. M., and Simpson, L. (1989). *J. Biol. Chem.* **264,** 18679–18686.
Barker, D. C. (1989). *Micron* **11,** 21–62.
Benne, R. (1985). *Trends Genet.* **1,** 117–121.
Benne, R. (1989). *Biochim. Biophys. Acta* 1007, 131–139.
Benne, R. (1990). *Trends Genet.* **6,** 177–181.
Benne, R., de Vries, B. F., van den Burg, J., and Klaver, B. (1983). *Nucleic Acids Res.* **11,** 6925–6941.
Benne, R., van den Burg, J., Brakenhoff, J. P., Sloof, P., Van Boom, J. H., and Tromp, M. C. (1986). *Cell (Cambridge, Mass.)* **46,** 819–826.
Bhat, G. J., Koslowsky, D. J., Feagin, J. E., Smiley, B. L., and Stuart, K. (1990). *Cell (Cambridge, Mass.)* **61,** 885–894.
Bhat, G. J., Myler, P. J., and Stuart, K. (1991). *Mol. Biochem. Parasitol.* **48,** 139–150.
Bhat, G. J., Souza, A. E., Feagin, J. E., and Stuart, K. (1992). *Mol. Biochem. Parasitol.,* **52,** 231–240.
Birkenmeyer, L., Sugisaki, H., and Ray, D. S. (1985). *Nucleic Acids Res.* **13,** 7107–7118.
Blum, B., Bakalara, N., and Simpson, L. (1990). *Cell (Cambridge, Mass.)* **60,** 189–198.
Blum, B., and Simpson, L. (1990). *Cell (Cambridge, Mass.)* **62,** 391–397.
Blum, B., Sturm, N. R., Simpson, A. M., and Simpson, L. (1991). *Cell (Cambridge, Mass.)* **65,** 543–550.
Borst, P., Fase-Fowler, F., and Gibson, W. C. (1987). *Mol. Biochem. Parasitol.* **23,** 31–38.
Borst, P., Hoeijmakers, J. H. J., Frasch, A. C. C., Snijders, A., Janssen, J. W. G., and Fase-Fowler, F. (1980). *In* "The Organization and Expression of the Mitochondrial Genome" (C. Saccone and A. M. Kroon, eds.), pp. 7–19. Elsevier/North-Holland Biomedical Press, Amsterdam.
Borst, P., and Weijers, P. J. (1976). *Eur. J. Biochem.* **64,** 141–151.
Cain, B. D., and Simoni, R. D. (1988). *J. Biol. Chem.* **263,** 6606–6612.
Cain, B. D., and Simoni, R. D. (1989). *J. Biol. Chem.* **264,** 3292–3300.
Campbell, D. A., Spithill, T. W., Samaras, N., Simpson, A. M., and Simpson, L. (1989). *Mol. Biochem. Parasitol.* **36,** 197–200.
Cech, T. R. (1991). *Cell. (Cambridge, Mass.)* **64,** 667–669.
de la Cruz, V. F., Neckelmann, N., and Simpson, L. (1984). *J. Biol. Chem.* **259,** 15136–15147.
de la Cruz, V. F., Lake, J. A., Simpson, A. M., and Simpson, L. (1985a). *Proc. Natl. Acad. Sci. U.S.A.* **82,** 1401–1405.
de la Cruz, V. F., Simpson, A. M., Lake, J., and Simpson, L. (1985b). *Nucleic Acids Res.* **13,** 2337–2356.
de Vries, B. F., Mulder, E., Brakenhoff, J. P., Sloof, P., and Benne, R. (1988). *Mol. Biochem. Parasitol.* **27,** 71–82.
Englund, P. (1981). *In* "Biochemistry and Physiology of Protozoa" (M. Levandowsky and S. Hutner, eds.), pp. 333–381. Academic Press, New York.

Englund, P. T., Marini, J. C., Levene, S. D., and Crothers, D. M. (1982). *Proc. Natl. Acad. Sci. U.S.A.* **79,** 7664–7668.

Eperon, I. C., Janssen, J. W., Hoeijmakers, J. H., and Borst, P. (1983). *Nucleic Acids Res.* **11,** 105–125.

Feagin, J. E. (1990). *J. Biol. Chem.* **265,** 19373–19376.

Feagin, J. E., and Stuart, K. (1985). *Proc. Natl. Acad. Sci. U.S.A.* **82,** 3380–3384.

Feagin, J. E., and Stuart, K. (1988). *Mol. Cell. Biol.* **8,** 1259–1265.

Feagin, J. E., Jasmer, D. P., and Stuart, K. (1985). *Nucleic Acids Res.* **13,** 4577–4596.

Feagin, J. E., Jasmer, D. P., and Stuart, K. (1986). *Mol. Biochem. Parasitol.* **20,** 207–214.

Feagin, J. E., Jasmer, D. P., and Stuart, K. (1987). *Cell (Cambridge, Mass.)* **49,** 337–345.

Feagin, J. E., Shaw, J. M., Simpson, L., and Stuart, K. (1988a). *Proc. Natl. Acad. Sci. U.S.A.* **85,** 539–543.

Feagin, J. E., Abraham, J. M., and Stuart, K. (1988b). *Cell (Cambridge, Mass.)* **53,** 413–422.

Fouts, D. L., and Wolstenholme, D. R. (1979). *Nucleic Acids Res.* **6,** 3785–3804.

Frasch, A. C. C., Borst, P., Hajduk, S. L., Hoeijmakers, J. H. J., Brunel, F., and Davison, J. (1980). *Biochim. Biophys. Acta* **607,** 397–401.

Hajduk, S. (1989). *In* "Progress in Molecular and Subcellular Biology," pp. 158–200.

Hajduk, S. L., Klein, V. A., and Englund, P. T. (1984). *Cell (Cambridge, Mass.)* **36,** 483–492.

Hajduk, S. L., Siqueira, A. M., and Vickerman, K. (1986). *Mol. Cell. Biol.* **6,** 4372–4378.

Hanas, J., Linden, G., and Stuart, K. (1975). *J. Cell Biol.* **65,** 103–111.

Hancock, K., and Hajduk, S. L. (1990). *J. Biol. Chem.* **265,** 19208–19215.

Harris, M. E., and Hajduk, S. L. (1992). *Cell (Cambridge, Mass.)* **68,** 1–20.

Harris, M. E., Moore, D. R., and Hajduk, S. L. (1990). *J. Biol. Chem.* **265,** 11368–11376.

Hensgens, L. A., Brakenhoff, J., de Vries, B. F., Sloof, P., Tromp, M. C., Van Boom, J. H., and Benne, R. (1984). *Nucleic Acids Res.* **12,** 7327–7344.

Hertwig, O. (1893). "Zelle und die Gewebe," Jena. Verlag von Gustav Fischer, pp. 34–36.

Hoare, C. A. (1954). *J. Protozool.* **1,** 28–33.

Jasmer, D. P., Feagin, J. E., and Stuart, K. (1985). *Mol. Cell. Biol.* **5,** 3041–3047.

Jasmer, D. P., Feagin, J. E., Payne, M., and Stuart, K. (1987). *Mol. Biochem. Parasitol.* **22,** 259–272.

Jasmer, D. P., and Stuart, K. (1986a). *Mol. Biochem. Parasitol.* **18,** 257–269.

Jasmer, D. P., and Stuart, K. (1986b). *Mol. Biochem. Parasitol.* **18,** 321–331.

Kidane, G. Z., Hughes, D., and Simpson, L. (1984). *Gene* **27,** 265–277.

Koslowsky, D. J., Bhat, G. J., Read, L. K., and Stuart, K. (1991). *Cell (Cambridge, Mass.)* **67,** 537–546.

Koslowsky, D. J., Bhat, G. J., Perrollaz, A. L., Feagin, J. E., and Stuart, K. (1990). *Cell (Cambridge, Mass.)* **62,** 901–911.

Koslowsky, D. J., Göringer, H. U., Morales, T., and Stuart, K. (1992a). *Nature (London)* **356,** 807–809.

Koslowsky, D. J., Riley, G. R., Souza, A. E., Feagin, J. E., and Stuart, K. (1992b). *Mol. Cell. Biol.* **12,** 2043–2049.

Laskay, T., Kiessling, R., DeWit, T. F. R., and Wirth, D. F. (1991). *Mol. Biochem. Parasitol.* **44,** 279–286.

Maslov, D. A., Sturm, N. R., Miner, B. M., Groszynski, E. S., Peris, M., and Simpson, L. (1992). *Mol. Cell. Biol.* **12,** 56–67.

Michelotti, E. F., and Hajduk, S. L. (1987). *J. Biol. Chem.* **262,** 927–932.

Muhich, M. L., Neckelmann, N., and Simpson, L. (1985). *Nucleic Acids Res.* **13,** 3241–3260.

Muhich, M. L., Simpson, L., and Simpson, A. M. (1983). *Proc. Natl. Acad. Sci. U.S.A.* **80,** 4060–4064.

Ntambi, J. M., Marini, J. C., Bangs, J. D., Hajduk, S. L., Jimenez, H. E., Kitchin, P. A., Klein, V. A., Ryan, K. A., and Englund, P. T. (1984). *Mol. Biochem. Parasitol* **12,** 273–286.

Payne, M., Rothwell, V., Jasmer, D. P., Feagin, J. E., and Stuart, K. (1985). *Mol. Biochem. Parasitol.* **15,** 159–170.
Pollard, V. W., and Hajduk, S. L. (1991). *Mol. Cell. Biol.* **11,** 1668–1675.
Pollard, V. W., Rohrer, S. P., Michelotti, E. F., Hancock, K., and Hajduk, S. L. (1990). *Cell (Cambridge, Mass.)* **63,** 783–790.
Ray, D. S., Hines, J. C., Sugisaki, H., and Sheline, C. (1986). *Nucleic Acids Res.* **14,** 7953–7965.
Read, L. K., Corell, R. A., and Stuart, K. (1992a). *Nucleic Acids Res.* **20,** 2341–2347.
Read, L. K., Myler, P. J., and Stuart, K. (1992b). *J. Biol. Chem.* **267,** 1123–1128.
Riou, G., Gabillot, M., Douc-Rasy, S., and Kayser, A. (1982). *C. R. Acad. Sci. Paris* **294,** 439–442.
Riou, G., and Yot, P. (1977). *Biochemistry* **16,** 2390–2396.
Robinson, D. R., and Gull, K. (1991). *Nature (London)* **352,** 731–733.
Rohrer, S. P., Michelotti, E. F., Torri, A. F., and Hajduk, S. L. (1987). *Cell (Cambridge, Mass.)* **49,** 625–632.
Ryan, K. A., Shapiro, T. A., Rauch, C. A., and Englund, P. T. (1988). *Annu. Rev. Microbiol.* **42,** 339–358.
Shaw, J. M., Campbell, D., and Simpson, L. (1989). *Proc. Natl. Acad. Sci. U.S.A* **86,** 6220–6224.
Shaw, J. M., Feagin, J. E., Stuart, K., and Simpson, L. (1988). *Cell (Cambridge, Mass.)* **53,** 401–411.
Silver, L. E., Torri, A. F., and Hajduk, S. L. (1986). *Cell (Cambridge, Mass.)* **47,** 537–543.
Simpson, A. M., and Simpson, L. (1974). *J. Protozool.* **21,** 379–382.
Simpson, A. M., Suyama, Y., Dewes, H., Campbell, D. A., and Simpson, L. (1989). *Nucleic Acids Res.* **17,** 5427–5445.
Simpson, L. (1986). *Int. Rev. Cytol.* **99,** 119–179.
Simpson, L. (1987). *Annu. Rev. Microbiol.* **41,** 363–382.
Simpson, L. (1990). *Science* **250,** 512–513.
Simpson, L., Neckelmann, N., de la Cruz, V. F., and Muhich, M. (1985). *Nucleic Acids Res.* **130,** 5977–5993.
Simpson, L., Neckelmann, N., de la Cruz, V. F., Simpson, A. M., Feagin, J. E., Jasmer, D. P., and Stuart, K. (1987). *J. Biol. Chem.* **262,** 6182–6196.
Simpson, L., and Shaw, J. (1989). *Cell (Cambridge, Mass.)* **57,** 355–366.
Sloof, P., van den Burg, J., Voogd, A., and Benne, R. (1987). *Nucleic Acids Res.* **15,** 51–65.
Sloof, P., van den Burg, J., Voogd, A., Benne, R., Agostinelli, M., Borst, P., Gutell, R., and Noller, H. (1985). *Nucleic Acids Res.* **13,** 4171–4190.
Souza, A. E., Myler, P. J., and Stuart, K. (1992). *Mol. Cell. Biol.,* **12,** 2100–2107.
Stuart, K. (1971). *J. Cell Biol.* **49,** 189–195.
Stuart, K. (1983a). *Mol. Biochem. Parasitol.* **9,** 93–104.
Stuart, K. (1983b). *J. Cell Biochem.* **23,** 13–26.
Stuart, K. (1989a). *Parasitol. Today* **5,** 5–8.
Stuart, K. (1989b). *Exp. Parasitol.* **68,** 486–490.
Stuart, K. (1991). *Annu. Rev. Microbiol.* **45,** 327–344.
Stuart, K., Feagin, J. E., and Jasmer, D. P. (1985). *In* "Sequence Specificity in Transcription and Translation" (R. Calendar and L. Gold, eds.), pp. 621–631. A. R. Liss, New York.
Stuart, K., Feagin, J. E., and Jasmer, D. P. (1987). *In* "Molecular Strategies of Parasite Invasion" (N. Agabian, H. Goodman, and N. Nogueria, eds.), pp. 145–155. A. R. Liss, New York.
Stuart, K., and Gelvin, S. R. (1980). *Am. J. Trop. Med. Hyg.* **29,** 1075–1081.
Sturm, N. R., Degrave, W., Morel, C., and Simpson, L. (1989) *Mol. Biochem. Parasitol.* **33,** 205–214.

Sturm, N. R., and Simpson, L. (1990). *Cell (Cambridge, Mass.)* **61,** 879–884.
Sturm, N. R., and Simpson, L. (1991). *Nucleic Acids Res.* **19,** 6277–6281.
Suyama, Y. (1982). *In* "Mitochondrial Genes" (P. Slonimski, P. Borst, and G. Attardi, eds.), pp. 449–455. Cold Spring Harbor Laboratory, Cold Spring Harbor, N.Y.
van der Spek, H., Arts, G.-J., van den Burg, J., Sloof, P., and Benne, R. (1989). *Nucleic Acids Res.* **17,** 4876.
van der Spek, H., Arts, G.-J., Zwaal, R. R., van den Burg, J., Sloof, P., and Benne, R. (1991). *EMBO J.* **10,** 1217–1224.
van der Spek, H., Speijer, D., Arts, G.-J., van den Burg, J., Steeg, H., Sloof, P., and Benne, R. (1990). *EMBO J.* **9,** 257–262.
van der Spek, H., van den Burg, J., Croiset, A., van den Broek, M., Sloof, P., and Benne, R. (1988). *EMBO J.* **7,** 2509–2514.
Vickerman, K. (1971). *In* "Ecology and Physiology of Parasites" (A. Fallis, ed.), pp. 50–91. University of Toronto, Toronto.
Vickerman, K. (1985). *Br. Med. Bull.* **41,** 105–114.
Wallace, D. C. (1982). *Microbiol. Rev.* **46,** 208–240.
Wesley, R. D., and Simpson, L. (1973). *Biochim. Biophys. Acta* **319,** 267–280.
Wolstenholme, D. R., and Clary, D. O. (1985). *Genetics* **109,** 725–744.
Woo, P. T. K. (1977). *In* "Parasitic Protozoa" (J. P. Kreier, ed.), pp. 269–296. Academic Press, New York.

Evolution of Mitochondrial Genomes in Fungi

G. D. Clark-Walker

Molecular and Population Genetics Group, Research School of Biological Sciences, Australian National University, Canberra City 2601, Australia

I. Introduction

Sizes and structural complexity of fungal mitochondrial genomes occupy a middle ground between those of metazoa and higher plants. The organization of genes is not as frugal as that in animal mtDNA nor as loose as that in plant mitochondrial genomes. Although mtDNAs from some filamentous fungi match those from vertebrates in the number of genes they encode, this is not so for yeasts. However, in yeasts the relative poverty of mitochondrial genes is redressed in some species by a richness in structural complexity of intergenic regions and also by the presence of optional introns. In addition, some fungi have DNA or RNA "plasmids" in their mitochondria that are of two types depending on whether they are derived from the mitochondrial genome or lack sequence relatedness to the resident mtDNA (Nargang, 1985; Meinhardt et al., 1990). These elements, which apparently play no part in the biogenesis of mitochondria, will not be dealt with in this review.

Any discussion of fungal mitochondrial genome evolution must be considered in the context of the likely polyphyletic origins of fungi. Based on both morphological and biochemical characteristics, bicilliate water molds of the Oomycota are considered to have a separate origin from "true" fungi (Cavalier-Smith, 1986; Kwok et al., 1986; Dick, 1989). This view is supported by sequence comparisons which suggest that oomycetes have a closer affinity to chrysophytes than to other fungi (Gunderson et al., 1987; Forster et al., 1990). In this respect it is significant that both groups of organisms have tubular christae in their mitochondria whereas true fungi, including the unicilliate chytridiomycetes, have lamellar christae (Cavalier-Smith, 1986; Dick, 1989). Chytridiomycetes are also thought to be deeply divided from "higher" fungi of the Zygomycotina, Ascomycotina, and Basidiomycotina (Forster *et al.*, 1990). Although these higher fungi are joined by their common lack of cilia, they in turn are not closely related, as recent sequence comparisons indicate. (Forster *et al.*, 1990) In view of these considerations, fungal mtDNAs have two unusual properties. Sizes of mtDNAs from the disparate groups have similar ranges, generally 30–80 kb and between members of a single genus sizes can have a three- to fourfold variation (Hoeben and Clark-Walker, 1986; Bruns and Palmer, 1989). These results suggest that in some related species processes are occurring to increase mitochondrial genome size dispersion, whereas in some genera, mtDNA appears to be relatively sheltered from such changes.

The first part of this review will be concerned with describing sizes, gene topology, and structures contributing to size variation of mtDNAs. In the following section, mechanisms for generating length mutations and rearrangements will be presented together with a description of mtDNA codon variations. Finally, attention will be drawn to a mitochondrial genome structure that could be similar to the architecture of an ancestral mtDNA.

Although comparative studies are our main source for deducing likely pathways of change, this *modus operandi* can in a few instances be supplemented by experiments. Thus insight gained by experimentally demonstrating the feasibility of a particular mechanism, as outlined in a later section, can help in describing steps in fungal mtDNA evolution.

Other articles dealing with the evolution of fungal mitochondrial genomes can be found (Sederoff, 1984; Clark-Walker, 1985; Grossman and Hudspeth, 1985; Scazzocchio, 1986; Taylor, 1986; Wolf, 1987; Wolf and Del Guidice, 1988; Dujon and Belcour, 1989; Gray, 1989a).

II. Diversity in mtDNA Size and Structure

Before the advent of restriction enzymes and gel electrophoresis, fungal mtDNA size determination was done by electron microscopy (Hollenberg *et al.*, 1970; Clayton and Brambl, 1972; Clark-Walker and Gleason, 1973). Technical advances, such as those mentioned above and introduction of the dye bisbenzimid H33258, for mtDNA isolation, have facilitated studies. Consequently, mtDNA size estimates have been obtained from wide-ranging types of fungi. Nevertheless, gaps in our knowledge exist and the most obvious is that there is only one report of mtDNA from a zygomycete. *Phycomyces blakesleanus* mtDNA is recorded to be 25.6 kb as noted in a reference to unpublished work (Grossman and Hudspeth, 1985). Other difficulties are associated with obtaining sufficient quantities of fungi that are refractory to culture and in disrupting cell walls without destroying organelles. It is hoped that problems in mtDNA isolation and characterization will be overcome by the application of innovative techniques so that gaps in our knowledge will soon be filled.

A. Oomycetes and Hypochytridiomycetes

Two unusual features of mtDNA from these allied groups of fungi have been discovered. Since the initial report that the circular mtDNA in *Achyla ambisexualis* has inverted repeats (Hudspeth *et al.*, 1983), many instances of this architecture have been described in water molds (Table I). Second, genome complexity, namely single copy length, falls in a narrow range between 36.2 kb *(Phytophthora infestans)* and 45.3 kb *(P. megasperma)* (McNabb and Klassen, 1988). The presence of inverted repeats is not universal, as mtDNAs from *Apodachyla* and *Phytophthora* species apparently lack this structure. However, a more detailed examination of the mitochondrial genome in *P. megasperma* has shown a small inverted repeat of 0.5–0.9 kb. Whether this is a remnant or novel structure has yet to be determined (Shumard-Hudspeth and Hudspeth, 1990). Inverted repeats in *Phythium* mtDNAs are around 20 kb, whereas those in other genera are 10–15 kb.

The topology of mitochondrial genomes with inverted repeats is similar to that of chloroplast DNA in angiosperms, as the inverted repeats separate small and large single copy regions. Another similarity is that somersault isomerism occurs presumably by recombination through the inverted repeat (Boyd *et al.*, 1984). By analogy with chloroplast genomes from some legumes that lack an inverted repeat, the absence of the inverted repeat in mtDNA from some oomycetes is thought to be a derived character. This view is strengthened by the presence of an inverted repeat in mtDNA from *Hyphochytrium catenoides,* a member of the hyphochytriomycetes (McNabb *et al.*, 1988). As this group of fungi is separate from oomycetes (Kwok *et al.*, 1986), the presence of an inverted repeat in mtDNAs from both groups of water molds is strong evidence that this character has an ancient origin (McNabb and Klassen, 1988).

Identification of genes in oomycete mtDNA has not received the same attention as in other fungi. However, using probes, mainly from baker's yeast mtDNA, it has been shown that the inverted repeat of oomycetes hybridizes to both the large (LSU) and small (SSU) subunit ribosomal RNA genes (Shumard *et al.*, 1986). Juxtaposition of these two regions is also found in mtDNA of *Phytophthora* species that lack the inverted repeat (Forster *et al.*, 1988; Shumard-Hudspeth and Hudspeth, 1990). Regions hybridizing to cytochrome oxidase subunits 1, 2, and 3 (COX1, -2, and

TABLE I

Size of Fungal Mitochondrial DNA

Organism	Size[a] (kb)	Inverted repeat (kb)	Reference
Oomycetes			
Achlya ambisexualis	49.8	9.6–12.1	1
Achlya heterosexualis	51.3	9.9–11.2	2
Achlya klebsiana	50.7	9.7–10.9	3
Aplanopsis terrestris	45.0	8.1–9.0	4
Apodachlya brachynema	36.4	—	4
Apodachlya pyrifera	40.0	—	4
Leptolegnia caudata	52.4	8.7–13.5	4
Saprolegnia sp.	44.5[b]		5
Sapromyces elongatus	53.0	15.1–17.0	4
Phythium diclinum	70.0	28.2–29.5	6
Phythium irregulare	60.0	21.7–23.8	4
Phythium paddicum	61.3	19.4–24.1	4
Phythium torulosum	73.0	27.6–29.1	6
Phytophthora cryptogea	40.7	—	4
Phytophthora infestans	36.2	—	7
Phytophthora megasperma	41–45.3	0.5–0.9	8,9
Phytophthora parasitica	39.5	—	9
Hypochytridiomycetes			
Hypochytrium catenoides	54.0	13.5–14.5	10
Rhizidiomyces apophysatus	50.0	—	10

Note: References: 1, Hudspeth *et al.* (1983); 2, Shumard *et al.* (1986); 3, Boyd *et al.* (1984); 4, McNabb and Klassen (1988); 5, Clark-Walker and Gleason (1983); 6, McNabb *et al.* (1987); 7, Klimczak and Prell (1984); 8, Shumard-Hudspeth and Hudspeth (1990); 9, Forster *et al.* (1987); 10, McNabb *et al.* (1988).
[a] All mtDNA maps are circular.
[b] Calculated from contour length. (See Table III.)

-3), apo-cytochrome *b* (CYB), and ATPase subunit 9 (A9) have also been mapped in *Phytophthora* and *Achyla* mtDNAs (Forster *et al.*, 1988; Shumard Hudspeth and Hudspeth, 1990; Shumard *et al.*, 1986). Between strains of *P. megasperma* there is an inversion of COX1–COX2 regions, but this result, showing a rearrangement in mtDNA within a species, is tempered by knowledge that considerable divergence exists between organisms classified as *megasperma* thereby casting doubt on the coherence of this species (Forster *et al.*, 1988). In *Achyla* mtDNA two additional regions have been mapped that hybridize to the ATPase subunit 6 (A6) and the variant 1 protein (VAR1) of *Saccharomyces cerevisiae* mtDNA (Shumard *et al.*, 1986). The hybridization of the VAR1 probe in these species, as well as to mtDNA from the basidomycete *Coprinus stercorarius* (Weber *et al.*, 1986), is of interest, as these reports are the first indication that the VAR1 locus may occur in mtDNAs from organisms other than some budding yeasts.

As mentioned, one of the curious features of oomycete mitochondrial genomes is that there is only a small size range of single copy sequence. This suggests that the two most important contributors to size dispersion in other mitochondrial genomes, namely changes to intergenic regions and optional introns, do not occur with the same abundance in oomycetes. More detailed studies on genome structure, including the occurrence of introns and the frequency of mtDNA recombination processes, are needed to gain insight into likely causes of uniform mtDNA size in oomycetes.

B. Ascomycetes

The most detailed analysis of fungal mitochondrial genome structure has been done with ascomycetes. In a *tour de force,* the complete 94,192-bp sequence of mtDNA from a single strain of *Podospora anserina* has been obtained (Cummings *et al.*, 1990a). Sequences from several strains of baker's yeast *S. cerevisiae* have been assembled, representing over 90% of the genome (De Zamaroczy and Bernardi, 1986b). References to many reports of mtDNA sequences from the filamentous fungi *Neurospora crassa* and *Aspergillus nidulans* have been listed (Collins, 1990; Brown *et al.*, 1985; Brown, 1990). The precise location of the large genes and the total tRNA complement in mtDNA of *Candida (Torulopsis) glabrata* has been obtained by sequencing all intergenic regions of the 19-kb molecule (Clark-Walker *et al.*, 1985), and the total sequence of the 19-kb mitochondrial genome of *Schizosaccharomyces pombe* has been determined but the sequence has not been published (B. F. Lang and K. Wolf, personal communication). These comprehensive studies have provided a wealth of information for those interested in gene structure, intron structure, control signals for transcription, RNA processing, translation, and mtDNA evolution. Whether the processes of change deduced from these ascomycete mitochondrial genome data are general must await sequence analysis from fungi in other groups.

1. Filamentous Ascomycetes

Sizes of mitochondrial genomes in this subclass exhibit a fourfold variation from 26.7 kb of *Cephalosporium acremonium* to 115 kb for *Cochliobolus heterostrophus* though the majority of lengths lie between 30 and 80 kb (Table II, Fig. 1). Within species of *A. nidulans, N. crassa,* and *P. anserina,* intraspecific length mutations in mtDNA are mainly due to optional introns (Collins and Lambowitz, 1983; Kuck *et al.,* 1985; Dujon, 1989; Field *et al.,* 1989; Cummings *et al.,* 1990a). For example, strains of *N. crassa* can have four introns in COX1 or lack introns in this gene (Burger *et al.,* 1982; De Jonge and De Vries, 1983; Field *et al.,* 1989).

Intergeneric differences in mtDNA length also result primarily from introns. Thus, in *A. nidulans* mtDNA, single introns are present in genes for LSU and CYB, whereas three introns are found in COX1 (Waring *et al.,* 1984; Brown, 1990). By contrast, 60 kb of the 100.3 kb mtDNA from

TABLE II

Size of Fungal Mitochondrial DNA

Organism	Size[a] (kb)	Reference
Filamentous ascomycetes		
Aspergillus amstelodami	40.5	1
Aspergillus flavus	32–33	2
Aspergillus nidulans	33–40	3–6
Cephalosporium acremonium	26.7	7
Claviceps purpurea	44.9–51.3	8,9
Cochliobolus heterostrophus	115	10
Fusarium oxysporum	48.5–52.3	11,12
Neurospora crassa	60–73	13–17
Neurospora intermedia	65	15
Neurospora sitophila	62–71	15
Penicillium chrysogenum	48.4	18
	25.8	19
Penicillium urticae	27.2	19
Podospora anserina	87–100.3	20–22

[a] All mtDNA maps are circular.

Note: References: 1, Lazarus et al. (1980); 2, Moody and Tyler (1990); 3, Perez and Turner, (1975); 4, Earl *et al.* (1981); 5, Brown *et al.* (1985); 6, Brown (1990); 7, Minuth *et al.* (1982); 8, Tudzynski *et al.* (1983); 9, Tudzynski and Esser (1986); 10, Garber and Yoder (1984); 11, Marriott *et al.* (1984); 12, Kistler *et al.* (1987); 13, Agsteribbe *et al.* (1972); 14, Clayton and Brambl (1972); 15, Collins and Lambowitz, (1983); 16, Taylor *et al.* (1986); 17, Collins (1990); 18, Smith *et al.* (1984); 19, Sekiguchi *et al.* (1990); 20, Cummings *et al.* (1979a); 21, Kuck *et al.* (1985); 22, Cummings *et al.* (1990a).

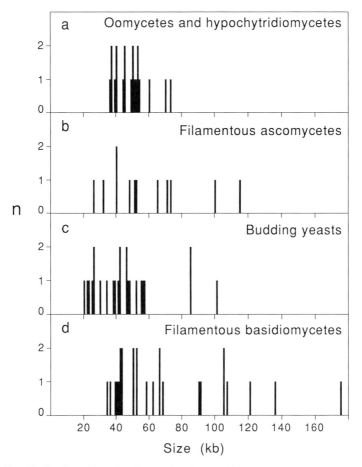

FIG. 1   Size distribution of fungal mtDNA taken from Tables I–IV. Where a size range exists for a species, only the largest value has been plotted.

race A of *P. anserina* is composed of introns (Cummings *et al.*, 1990a). This mtDNA contains 33 group I introns and 3 group II introns located in 10 genes. In addition to the three genes listed above for *A. nidulans* mtDNA, introns are also found in genes for NADH dehydrogenase subunits ND1, ND3, ND4, ND4L, ND5 and in genes for COX2 and ATPase subunit 6. Indeed, the only large genes that lack introns are COX3, ND2, ND6, and SSU.

With the exception of COX2, the same genes in some strains of *N. crassa* mtDNA contain introns (Field *et al.*, 1989; Collins, 1990). More than half the 38-kb size difference between the mtDNAs of *N. crassa* strain 74A (62 kb) and *P. anserina* race A can be accounted for by the remarkable mosaic nature of the COX1 gene in the latter organism that contains 16 introns totaling 22.9 kb (Cummings *et al.,* 1989, 1990a); however, as noted, the COX1 gene in *N. crassa* strain 74A lacks introns (Burger *et al.,* 1982; De Jonge and De Vries, 1983). Possible explanations for these observations will be considered later.

A further potential source of interspecific mtDNA length variation could come from the differential presence of genes. However, the mitochondrial genomes of *P. anserina, N. crassa,* and *A. nidulans* are not markedly different in this property. In addition to the major structural genes mentioned above that are common to all three genomes, there are 25 tRNA genes in *P. anserina* (Cummings *et al.,* 1989, 1990a), 27 in *N. crassa* (Collins, 1990), and 28 in *A. nidulans* (Brown *et al.,* 1985, 1990). A number of small unidentified reading frames have been found in the *P. anserina* molecule; whether these regions have a genetic function is still undecided. *Neurospora crassa* mtDNA has an intron in LSU that contains an open reading frame encoding ribosomal protein S5 (Burke and Raj Bhandary, 1982). This genome also contains five small unidentified reading frames (Collins, 1990) and an open reading frame of 633 codons (URFN) that could possibly code for a structural protein (Burger and Werner, 1986). In *A. nidulans,* URFA3, the only large unassigned reading frame, may code for a mitochondrial membrane protein (Brown *et al.,* 1989). Both *N. crassa* and *A. nidulans* mtDNAs contain an A9 sequence, but the active gene is apparently in the nucleus (Brown *et al.,* 1985; Van den Boogaart *et al.,* 1982). On the other hand, *P. anserina* mtDNA lacks an A9 sequence (Cummings *et al.,* 1990a). However, these small differences in the types of structural genes in the three mtDNAs do not contribute in a major way to length variation.

Another possible source of size dispersion of fungal mtDNAs is noncoding, intergenic sequence. Although changes in such regions play a large part in mtDNA size variation between yeasts, this is not the case in the three genomes under discussion. *Podospora anserina* mtDNA contains the largest intergenic region of 2210 bp between the CYB and the Cys tRNA genes. Smaller intergenic regions of 1491 and 801 bp are found between COX1 and Arg tRNA and between ORFC and A6, respectively. Small unidentified reading frames are also found in these latter regions. The remaining intergenic regions in *P. anserina* mtDNA are in general less than 500 bp. Likewise, intergenic regions in *N. crassa* mtDNA, with the possible exception of three regions between URFK and A8, between A6 and A9 (MAL), and between COX2 and URFL, are less than 1000 bp. In *A. nidulans* mtDNA the genes are also parsimoniously organized, as there is only about 6 kb of intergenic sequence (Brown *et al.,* 1985). Together these findings indicate that intergenic length variation does not contribute in a marked way to mtDNA size disparity between these filamentous ascomycetes.

Finally, it must be noted that small duplications, involving tRNA genes, have been found in all three mtDNAs. In *A. nidulans* mtDNA there is a 300-bp duplication containing a second copy of the asparagine tRNA gene and the 5′ end of the ATPase 8 sequence (Brown *et al.*, 1983), whereas a cysteine tRNA duplication is present on an 153-bp reiterated segment (Brown *et al.*, 1985). Duplicate copies of tRNA genes for cysteine and methionine are found in *N. crassa* mtDNA together with similar flanking sequences (Agsteribbe *et al.*, 1989). Two tRNA genes for methionine and valine are also duplicated in *P. anserina* mtDNA (Cummings *et al.*, 1990a), whereas in race A, the ND4L gene has within the intron a duplication of the 5′ exon (Cummings *et al.*, 1990b). However, these small duplications, involving only a few hundred base pairs, do not help to explain mtDNA size disparities. This leaves us with the conclusion that in these filamentous ascomycetes, optional introns are the chief factor in mitochondrial genome length variation.

Heterogeneity of mtDNA is not confined to size since gene order also varies between species. Comparison of gene topology in the three mtDNAs reveals extensive rearrangement, though some common gene linkages do occur. With *N. crassa* and *A. nidulans* mtDNAs the largest common cluster involves the ND1–ND4–A8–A6 region, whereas only the ND4 and A8 sequences are juxtaposed in *P. anserina*. On the other hand, *N. crassa* and *P. anserina* mtDNAs share a common gene order for the SSU, COX3, and ND6 sequences, whereas in *A. nidulans*, the latter two genes have been rearranged. Genes for ND4L and ND5 are clustered in all three genomes, whereas the grouping of LSU, ND2, and ND3 as well as CYB–COX1 is common to *P.anserina* and *N.crassa* mtDNAs. Downstream from the LSU there is a cluster of 11 or 12 tRNA genes that is almost identical in the *P. anserina* (Cummings *et al.*, 1990a), *A. nidulans* (Kochel *et al.*, 1981), and *N. crassa* mtDNAs (Yin *et al.*, 1982). A more interesting situation occurs upstream of LSU in *A. nidulans* and *P. anserina* mtDNAs. In the former genome there is a cluster of 9 tRNA genes, whereas in the latter mtDNA only the 3 most proximal tRNA genes are juxtaposed to the LSU and the other 6 genes occur as a block downstream of the COX3 gene (Cummings *et al.*, 1990a; Kochel *et al.*, 1981).

From these results it is apparent that mitochondrial genomes in fungi are susceptible to rearrangement. Indeed, trying to trace mtDNA evolution from topological analysis alone can be extremely misleading as exemplified by studies with yeasts.

2. Ascomycetous Yeasts

Apart from four notable large exceptions, sizes of mtDNAs in both budding yeasts and the sole fission yeast *Schizosaccharomyces pombe* lie between

18 and 60 kb (Table III; Fig. 1). The lower size limit of around 18 kb for both *S. pombe* and *C. glabrata* mitochondrial genomes points to a possible difference in gene content between these two genomes and those of the larger mtDNAs in filamentous ascomycetes. Indeed sequence analysis of mtDNAs from the two yeasts (Wolf, 1987; Wolf and Del Guidice, 1988; Clark-Walker *et al.*, 1985), and also the larger mtDNA in *S. cerevisiae* (De Zamaroczy and Bernardi, 1986b), shows that these genomes do not encode any of the seven NADH dehydrogenase subunits that are found in mtDNAs of filamentous ascomycetes. However, this may not be a general property of yeast mitochondrial genomes, as a fragment of *A. nidulans* mtDNA containing the ND1 sequence was found to hybridize to mtDNA from *Brettanomyces anomalus* (Hoeben and Clark-Walker, 1986). It has yet to be determined whether this result is indicative that functional NADH dehydrogenase subunit genes occur in some yeast mitochondrial genomes.

Genes that do occur in the *S. pombe, C. glabrata, Kluyveromyces lactis,* and *S. cerevisiae* mitochondrial genomes are those for LSU, SSU, COX1–3, CYB, and A6, -8, and -9. As well as these genes, *S. pombe* mtDNA encodes 25 tRNAs and an unidentified reading frame, urfa, of 227 codons that has no known counterpart in other mitochondrial genomes (Wolf, 1987). *Candida glabrata* mtDNA has 23 tRNAs and like the *S. cerevisiae genome,* it encodes a structural gene called VAR1 that is thought to specify a mitochondrial ribosomal protein (Ainley *et al.*, 1985; Clark-Walker *et al.*, 1985). Another shared characteristic of *C. glabrata* and *S. cerevisiae* mtDNAs is that a tRNA synthesis locus is found between genes for formyl methionine and proline tRNAs (Underbrink-Lyon *et al.*, 1983; Shu *et al.*, 1991).

Saccharomyces cerevisiae mtDNA, in addition to the large structural genes mentioned above, has 24 tRNAs and a number of unassigned reading frames (URFs, Grivell, 1990b) or open reading frames (ORFs, De Zamaroczy and Bernardi, 1986b) that have been given different numbers by different authors. These ORFs or URFs are of dubious genetic significance, as mutations have not been localized to these regions, they do not occur in *C.glabrata* or *S. pombe* mtDNAs, some are absent from various strains of *S. cerevisiae,* and one can be deleted without affecting mitochondrial biogenesis. Thus strains of *S. cerevisiae* exist that lack ORF4 (nomenclature of De Zamaroczy and Bernardi, 1986b) (Cobon *et al.*, 1982; Clark-Walker, 1989), ORF2 (Clark-Walker, 1989), and possibly ORF3 (Grivell, 1990b). Furthermore, ORF5 can be deleted without upsetting growth of *S. cerevisiae* on nonfermentable carbon sources (Skelly and Clark-Walker, 1990). The remaining ORF1, located downstream of COX2, is by default the only sequence that may still have a function. In this respect it has been reported recently that ORF1 is present in the 73-kb mtDNA of *S. douglasii* (Tian *et al.*, 1991).

TABLE III

Size of Fungal Mitochondrial DNA

Organism	Size (kb)	Reference
Ascomycetous budding yeasts		
Brettanomyces anomalus CBS 77	57.7	1
Brettanomyces custersianus CBS 4805	28.5	1
Brettanomyces custersii CBS 5512	101.1	1
Brettanomyces naardenensis CBS 6042	41.7	1
Candida (Torulopsis) glabrata	18.9–20.3	2,3,4
Candida maltosa	52	5
Candida parapsilosis[a]	20.5–30	6
Candida rhagii	22	6
Dekkera bruxellensis CBS 74	85.0	1
Dekkera intermedia CBS 4914	73.2	1
Eeniella nana CBS 1956	34.5	1
Hanseniaspora vineae CBS 2171	26.7	4
Hansenula mrakii	55	7
Hansenula petersonii	42	8
Hansenula saturnus	47	9
Hansenula wingei V 31B	26.1[b]*	10
Kloeckera africana CBS 277	27.1	11
Kluyveromyces lactis	34.9–39.5	10,12,13,14
Pachytichospora transvaalensis CBS 2186	41.4	4
Saccharomyces cerevisiae	68–85	15,16,17,18
Saccharomyces douglasii	73	19
Saccharomyces exiguus CBS 379	23.7	20
Saccharomyces telluris CBS 2685	38.4	4
Saccharomyces unisporus CBS 398	27.3	4
Torulaspora delbrueckii (T. colliculosa)	46.7*	21
Yarrowia (Saccharomycopsis) lipolytica	44–48.3	13,22
Ascomycetous fission yeast		
Schizosaccharomyces pombe	17.3–24.6	10,23,24

Note: References: 1, Hoeben and Clark-Walker (1986); 2, O'Connor *et al.* (1976); 3, Clark-Walker *et al.* (1980); 4, Clark-Walker *et al.* (1981a); 5, Kunze *et al.* (1986); 6, Camougrand *et al.* (1988); 7, Wesolowski and Fukuhara (1981); 8, Falcone (1984); 9, Lawson and Deters (1985); 10, O'Connor *et al.* (1975); 11, Clark-Walker *et al.* (1981b); 12, Sanders *et al.* (1974); 13, Wesolowski *et al.* (1981); 14, Hardy *et al.* (1989); 15, Hollenberg *et al.* (1970); 16, Christiansen and Christiansen (1976); 17, Sanders *et al.* (1977); 18, Evans and Clark-Walker (1985); 19, Tian *et al.* (1991); 20, Clark-Walker *et al.* (1983); 21, Kojo (1976); 22, Kuck *et al.* (1980); 23, Zimmer *et al.* (1984); 24, Zimmer *et al.* (1987).

[a] The identity of the yeast recorded as *C. parapsilosis* in an early study could not be confirmed (Ref. 10).

[b] Values marked with an asterisk have been obtained from contour lengths using a factor of 3.18 kb/μm (Ref. 4). All mtDNAs are circular except for *H. mrakii* and possibly *C. rhagii* (Kovac *et al.*, 1984, but see Camougrand *et al.*, 1988).

Other open reading frames exist in some optional introns of *S. cerevisiae* mtDNA. Some products of these ORFs play a role in the excision of introns and the formation of mature messenger RNAs for COX1 and CYB (Dujon and Belcour, 1989; Grivell, 1989; Wenzlau *et al.*, 1989). However, these open reading frames are "reflexive genes," at least in *S. cerevisiae*, because it has been shown by the construction of an intron-free mitochondrial genome that their only influence is on themselves (Seraphin *et al.*, 1987).

Support for the notion that mitochondrial introns are not necessary in nature comes from the observation that intervening sequences are absent from the CYB gene in *C. glabrata* (G. D. Clark-Walker, unpublished observation), *K. lactis* (Brunner and Coria, 1989), and the EF1 strain of *S. pombe* (Trinkl *et al.*, 1985). Introns are also absent from the LSU gene in *S. pombe* (Lang *et al.*, 1987) and the LSU in some strains of *S. cerevisiae* (Grivell, 1989; Dujon and Belcour, 1989). In view of these results it is curious that all yeast mitochondrial genomes examined to date contain a mosaic COX1 gene (Refs. Table III). However, this need not mean that the presence of introns in COX1 is necessary for proper gene expression. Both in some *N. crassa* wild-type strains and in the constructed intron-free *S. cerevisiae* strain mentioned above, COX1 expression occurs despite the absence of introns. Thus although introns seem to be a normal component of mtDNA structure, they do not appear to be necessary for normal gene expression.

Optional introns do contribute to both inter- and intraspecific size dispersion in yeast mitochondrial genomes. Intraspecific mtDNA variation is exemplified in strains of *S. cerevisiae* where size can vary from 68 to 85 kb due to the presence of 5 introns in the smallest molecule and 13 in the largest (Sanders *et al.*, 1977; Grivell, 1990b). Likewise in *S. pombe* (Wolf, 1987) and *C. glabrata* strains (Table III) (Clark-Walker *et al.*, 1981a) variation in mtDNA size is due to optional introns. Recently, it has been found that differences in mitochondrial genome lengths of 34.9–39.5 kb in *K. lactis* strains can be largely attributed to 3 extra introns in the COX1 gene (Skelly *et al.*, 1991). From these results it can be seen that the main source of intraspecific size variation in yeast mitochondrial genomes is optional introns.

Although optional introns play a part in both interspecific and intergeneric size variation in mtDNAs, a more significant contribution to disparity in length comes from intergenic regions as revealed by sequence analysis of *C. glabrata* and *S. cerevisiae* mitochondrial genomes. In the 19-kb mtDNA of *C. glabrata* there is approximately 2890 bp of intergenic sequence of 97% A + T that is distributed among 33 locations (Clark-Walker *et al.*, 1985) (9 large genes, 23 tRNA genes, and the tRNA synthesis locus that occupies 227 nucleotides, Shu *et al.*, 1991). By contrast, in *S.*

cerevisiae it is estimated that approximately 63% of the 85-kb long form mtDNA consists of intergenic regions (De Zamaroczy and Bernardi, 1986b, 1987). This value of 53.5 kb is provisional because not all the genome has been sequenced. The figure is also likely to be an underestimate, as it does not include the five ORFs that are questionable genes. Nevertheless, there is nearly 19 times more intergenic sequence in the *S. cerevisiae* mtDNA than in the smaller genome. Possible mechanisms involved in the creation of this disparity will be discussed later.

Another point to be made from the available data on mtDNA size is that within a number of genera, length can be quite variable. Thus within *Saccharomyces* lengths range from 23.7 kb in *S. exiguus* to 85 kb in baker's yeast (Table III). Likewise in *Hansenula* sizes range from 26.1 to 55 kb and in *Kluyveromyces* sizes vary from 34.9 kb in *K. lactis* to approximately 103 kb in *K. quercuum* (H. Phaff, A. Lachance, and G.D. Clark-Walker unpublished observations). However, the most striking variation occurs in the genus *Dekkera/Brettanomyces* (the latter genus is the anamorphic form) between *B. custersianus* (28.5 kb) and *B. custersii* (101 kb). These observations on size dispersion of mtDNAs in separate genera mean that mitochondrial genomes in yeasts are prone to length mutations. Indeed in three separate instances it would seem that some species in *Dekkera*, *Saccharomyces*, and *Kluyveromyces* have arisen with abnormally large mitochondrial genomes. In the case of *S. cerevisiae*, as discussed above, intergenic regions constitute around 63% of the 85-kb molecule. It remains to be determined whether the other large genomes mimic this architecture or whether, like the big genomes in filamentous ascomycetes, they contain remarkable quantities of introns.

In addition to length mutations, mitochondrial genomes of yeasts also show rearrangements (Clark-Walker, 1985). Mitochondrial gene topology is known for several species both from sequence analysis and by cross-hybridization mapping. Alignment of maps has shown that the most frequent linkage group involves COX1–A8–A6 that is present in mitochondrial genomes of *C. glabrata*, *S. cerevisiae*, *S. douglasii*, *S. exiguus*, *K. lactis*, the two smallest *Brettanomyces* mtDNAs, *E. nana*, and possibly *Kloeckera africana* (Refs. Table III). It must be mentioned that analysis for A8 in this cluster has not been undertaken in some instances (*S. exiguus*, *Brettanomyces* sp., *Eeniella nanna*, and *K. africana*). The absence of the COX1–A6 linkage group from *B. anomalus* and *B. custersii* mtDNAs indicates that these genomes have undergone rearrangement in this region since the *Dekkera/Brettanomyces* ancestor diverged from other yeasts (see below).

A more extensive linkage group involving CYB–COX1–A8–A6–A9–COX2 is shared between the two smallest genomes of *C. glabrata* and *S. exiguus* (Clark-Walker, 1985). In the two smallest *Brettanomyces* mtDNAs and in *E. nana* the CYB–COX1–A6 linkage group is also present. *Schizosaccharomyces pombe* mtDNA has a linkage of A9–COX2 (Wolf, 1987). From the combined information it can be argued that the linage group from CYB–COX2 could be ancient. The survival of this linkage group in the small mitochondrial genomes is thought to be significant in view of results obtained from analyzing mitochondrial genomes from the *Dekkera/Brettanomyces/Eeniella* group.

As noted, *Brettanomyces* is the anamorphic form of *Dekkera*. The genus *Eeniella* contains one species and has been newly created by reclassifying *Brettanomyces nana*, though the validity of this reassignment is problematic (Clark-Walker, *et al.*, 1987; Smith *et al.*, 1990). Species in the *Dekkera* genus are clearly related on physiological grounds; yet mtDNA size differs almost fourfold. The three smallest mtDNAs have the same gene order, whereas the three largest mitochondrial genomes from *B. anomalus, D. bruxellensis* (*D. intermedia* is now known to be a strain of *D. bruxellensis;* Clark-Walker *et al.*, 1987; Smith *et al.*, 1990), and *B. custersii* are rearranged in relation to the smaller genomes and among themselves. However, sequence analysis of COX2 from each of the six mtDNAs has shown that the three smallest genomes are the most diverged whereas the three largest are the most closely related (Clark-Walker *et al.*, 1987; Hoeben, (Table III) Weiller, and Clark-Walker, submitted). This apparently contradictory result is nevertheless reinforced by the observation that very few restriction sites are common to the three smallest genomes whereas many are shared among the larger molecules (Hoeben and Clark Walker, 1986).

Sequence rearrangement has also been reported between the large mtDNAs of the interfertile yeasts *S. cerevisiae* and *S. douglasii* whereby a segment of approximately 15 kb, containing the genes for COX3 and SSU, occurs at a different location in the two genomes (Tian *et al.*, 1991). The conclusion to be drawn from these data is that larger genomes are more prone to rearrangement than smaller ones. Viewed in this light, the conserved gene order between the small mtDNAs of *C. glabrata* and *S. exiguus* is likely to represent the topology in an ancestral genome. In other words, in trying to deduce the structure of likely ancestral molecules, more weight should be placed on shared gene order in smaller rather than larger mtDNAs.

Having reached what looks to be reasonably firm ground among the quicksand of mitochondrial genome topology, it is therefore somewhat disturbing to discover that rearrangement can occur even within a single species. Until now, a tenet has been that mitochondrial gene order is conserved within a species even though size can vary. However, while analyzing intron mobility in strains of *K. lactis* we discovered that 6 out of 14 strains in our possession had a rearranged mitochondrial genome (Skelly *et al.*, 1991) relative to mtDNA in our standard strain, K8 (Hardy *et al.,* 1989). Thus in strain W600B, a 10-kb translocation has shifted a segment containing A9–COX3–COX2 to between CYB and SSU, whereas in strain K8 this region is flanked by LSU and COX1 (Hardy *et al.,* 1989). This unusual result calls into question whether the strains harboring the two types of mtDNA have been misclassified. However, mating and sporulation are normal, no spore inviability is found, and the karyotype of the two strains is the same (R. Maleszka, personal communication). It must therefore be concluded that between interbreeding strains, normally considered to be a single species, constancy of gene order is not necessarily a reliable taxonomic yardstick. A further point to note from the above results with the 36- to 39 kb mtDNA is that size alone is not the critical determinant in predisposing a molecule to rearrangement; it is merely a pointer to underlying structural features that facilitate this process. Structural elements in DNA that facilitate recombination are short directly repeated sequences. The role played by these recombinogenic elements in gene shuffling will be discussed in a subsequent section.

In addition to preservation of linkages between large genes, some tRNA clusters have also been found to persist in the mtDNAs of *S. cerevisiae, C. glabrata,* and *K. lactis.* From sequence analysis of *K. lactis* mtDNA the positions of 22 tRNA genes have been located in one major region and three minor places (Wilson *et al.,* 1989; Hardy and Clark-Walker, 1990). Conserved tRNA gene order between all three yeast genomes is noted for Leu-Glu-Lys-Arg-Asp-Ser and f Met-Pro; while Tyr-Asn-Ala-Ile is common to *K. lactis* and *C. glabrata.* Curiously, in *S. cerevisiae* mtDNA, the tRNA for Thrl is on the opposite strand to all other genes. Although for *C. glabrata* mtDNA all genes are transcribed in one direction (Clark-Walker *et al.,* 1985), this is not the case for *B. anomalus* and *D. bruxellensis* mitochondrial genomes where rearrangements have produced genes in both directions (Hoeben and Clark-Walker, 1986). Inversions, however, are not frequently observed in mitochondrial genomes of yeasts and only one example of an inverted repeat has been reported (Clark-Walker *et al.,* 1981b). In the 27.1-kb mtDNA of *K. africana* a 4.3-kb segment has been inverted. This structure, which contains a truncated LSU gene, differs from the inverted repeats in oomycete mtDNAs by lacking single-copy sequence between the repeats. The isolated example of this structure in *K. africana* mtDNA suggests that this is a recently derived character. An inverted repeat has been found in *C. albicans* mtDNA, a yeast with basidiomycete affiliation, as discussed in the next section.

C. Basidiomycetes

Strains of *Agaricus bitorquis* contain the largest mitochondrial genome yet found in fungi (176.3 kb) (Hintz *et al.,* 1985). However, the majority of filamentous basidiomycetes have mtDNAs ranging in size from 35 to 70 kb (Table IV; Fig. 1), the smallest being found in *Gyrodon lividus* (34 kb) (Bruns and Palmer, 1989). As in ascomycetous yeasts, mtDNA size can vary from two- to threefold within a genus. For instance, mtDNA size ranges from 36 to 91 kb in *Coprinus* and from 36 to 121 kb in *Suillus* (Table IV). In a comprehensive study, it has been found that mtDNAs from 15 species of *Suillus* show small incremental changes in size between the limits noted above (Bruns *et al.*, 1988; Bruns and Palmer, 1989). Mitochondrial DNAs from one rust, *Ustilago cynodantis*, and the basidiomycetous yeast *C. albicans* have been found to have sizes of 76.5 and 41 kb, respectively (Mery-Drugeon *et al.*, 1981; Wills *et al.*, 1985). The latter genome has an inverted repeat of 5 kb separated by small and large unique copy regions. An inverted repeat of 4.6–9.2 kb is also seen in the 136-kb circular genome of *A. brunnescens*, but unlike its counterpart in oomycete mtDNAs, the LSU and SSU are not encoded in this region (Hintz et al., 1988).

Gene mapping studies using probes from *S. cerevisiae* mtDNA have only been undertaken in a few instances. In species of *Coprinus* (Weber, *et al.*, 1986; Economou and Casselton, 1989) and *Suillus* (Bruns *et al.*, 1988; Bruns and Palmer, 1989) and in *Agaricus brunnescens* mtDNAs (Hintz *et al.*, 1988), regions hybridizing to LSU, SSU, A6, A8, CYB, and COX1 have been detected. Additionally *Coprinus* and *Suillus* mtDNAs hybridize to COX2, COX3, and A9 (Economou and Casselton, 1989; Bruns and Palmer, 1989), whereas hybridization to the VAR1 probe from *S. cerevisiae* has been found with *C. stercorarius* mtDNA (Weber *et al.*, 1986). Furthermore, introns have been seen in the COX1 and LSU genes of *C. cinereus* (Economou *et al.*, 1987; Economou and Casselton, 1989) and *S. luteus* (Bruns and Palmer, 1989). Reports have not mentioned whether sequences for NADH dehydrogenase subunits have been observed but arguing from the similarity in the lower size limit of mtDNAs from filamentous ascomycetes and basidiomycetes (26–34 kb) the presence of such sequences can be anticipated.

Rearrangements of gene order are seen between the 43 and 91 kb mtDNAs of *Coprinus* species (Weber *et al.*, 1986) but only one topological change has been detected in the 15 mitochondrial genomes from *Suillus* species (Bruns *et al.*, 1988; Bruns and Palmer, 1989). An explanation for this latter result could be that *Suilius* mtDNAs lack recombinogenic sequences or, alternatively, enzymes catalyzing recombination are low or absent. Pertinent to the latter suggestion is that somersault isomerization about the inverted repeat of *Agaricus brunnescens* mtDNA has not been detected (Hintz *et al.*, 1988). This observation hints that basidiomycetes may be different from other fungi in the level of mitochondrial genome recombination.

TABLE IV

Size of Fungal Mitochrondrial DNA

Organism	Size[a] (kb)	Inverted repeat (kb)	Reference
Filamentous basidiomycetes			
Agaricus brunnescens	136	4.6–9.2	1,2
Agaricus bitorquis	148.5–176.3		1
Coprinus cinereus	36–43		3,4
Coprinus stercorarius	91.1		3
Gyrodon lividus	34		5
Gyrodon meruliodes	52		5
Paragyrodon sphaerosporus	39		5
Rhizopogon subcaerulescens	66		6
Schizophyllum commune	50.3–52.2		7
Suillus americanus	66		5
Suillus cavipes	36		5,8
Suillus grevillei	85		5
Suillus grisellus	121		5,8
Suillus luteus	106		5,8
Suillus ochraceoroseus	41		5
Suillus placidus	62		5
Suillus sinuspaulianus	50		5
Suillus spectabilis	50		5
Suillus spraguei (pictus)	104		5
Suillus subalutaceus	58		5
Suillus tridentinus	68		5,8
Suillus tomentosus	43		5
Suillus variegatus	40		5
Suillus viscidus	104		5
Rusts			
Ustilago cynodontis	76.5		9
Basidiomycetous yeasts			
Candida albicans	41	5	10
Histoplasma capsulatum	33–47		11

Note: References: 1, Hintz *et al.* (1985); 2, Hintz *et al.* (1988); 3, Weber *et al.* (1986); 4, Economou and Casselton (1989); 5, Bruns and Palmer, (1989); 6, Bruns *et al.* (1989); 7, Specht *et al.* (1983); 8, Bruns *et al.* (1988); 9, Mery-Drugeon *et al.* (1981); 10, Wills *et al.* (1985); 11, Vincent *et al.* (1986).

[a] All mtDNA maps are circular.

In view of the suggestion that change in genome size among *Suillus* species has been by incremental acquisition of short segments (Bruns and Palmer, 1989), it would be important to learn from sequence analysis whether such accretions have given rise to repetitive structures in intergenic regions of whether expansion has been by addition of introns. In the former case, rearrangements might be antiticipated in the larger genomes, whereas this need not be so if introns are responsible for length changes. On the other hand, the *Suillus* species could be much more closely related than their classification suggests thereby limiting the time for rearrangements to occur. Countermining this argument is that sequence divergence in the LSU, estimated from restriction sites, ranges up to 2.9% (Bruns and Palmer, 1989). Clearly more data are needed to resolve some of these questions.

The issues touched on above are also pertinent to a consideration of the structures responsible for the long genome sizes of *Agaricus* species that are the largest known among fungi. Could it be that these genomes combine both excessively large intergenic regions and abundant introns, features that underly large mitochondrial genomes in yeasts and filamentous ascomycetes, respectively?

III. Generation of Diversity in mtDNA

Macro structural changes to fungal mtDNAs can involve length mutations and rearrangements as reviewed above. In addition, a micro structural change has led to variation to the genetic code both between different groups of fungi and in relation to other organisms. Mechanisms for changing the genetic code have been suggested recently and will be discussed later. In the following sections, possible mechanisms for generating length mutations and rearrangements will be presented.

A. Length Mutations

1. Introns

Acquisition or loss of fungal mitochondrial introns has been reviewed extensively (Dujon, 1989; Dujon *et al.*, 1989; Lambowitz, 1989; Perlman and Butow, 1989; Grivell, 1990a) and will also be examined by others in accompanying chapters. Hence details of these processes will not be repeated here. However, in endeavoring to present a balanced view of mtDNA evolution, a few points need to be discussed in relation to intron gain or loss.

In considering pathways of intron gain (introns can be divided into groups I or II on structural grounds; Michel and Dujon, 1983) it has been found in crosses between organisms with mitochondrial genomes differing in the presence or absence of some key introns that progeny frequently have mtDNA with these introns. The best characterized example of this biased transmission concerns the omega locus in *S. cerevisiae* mtDNA. In crosses between strains differing by the presence of a 1132-bp group I intron in the LSU gene, genomes containing this intron are preferentially recovered (Dujon, 1989). Mobility of this intron has been shown to involve a double-stranded endonuclease, encoded in the open reading frame, that cleaves a specific sequence in the intron-free allele (Colleaux *et al.*, 1985; Macreadie *et al.*, 1985). Details of subsequent steps in the transfer have yet to be finalized. The biased propagation of the *S. cerevisiae* COXI.4 group I intron may likewise involve an endonuclease encoded by this element (Wenzlau *et al.*, 1989).

Biased transmission of introns can also be observed in mtDNA of other fungi. Unidirectional transmission of three inserts in a long form mtDNA has been observed in crosses between *A. nidulans* strains. At least two and possibly all three inserts are in the COX1 gene and are group I introns (Earl *et al.*, 1981; Waring *et al.*, 1984). A similar phenomenon has been observed in the basidiomycete *Coprinus cinereus* where crosses between strains differing in the presence of two inserts in their mtDNAs predominantly yield recombinants with both inserts (Economou *et al.*, 1987). Mapping has shown that at least one of the inserts is an optional COX1 intron but its classification has not been determined.

Group II introns can also be mobile in crosses. In *S. cerevisiae*, biased transmission has been found for the COX1.1 and COX1.2 introns (Meunier *et al.*, 1990) and similarly we have observed that the group II *K. lactis* COX1.1 intron is transmitted with a high frequency (Skelly *et al.*, 1991).

As noted above, some group I introns may be preferentially transmitted because they encode a site-specific endonuclease that cleaves at or near the insertion site. The transmission of other group I introns is viewed as a conversion event involving DNA recombination (Dujon, 1989). However, it has been suggested that biased transmission of group II introns may proceed by reversal of the intron splicing reaction that also must ultimately involve a DNA recombination event. The first step in the pathway has recently been demonstrated by *in vitro* experiments showing that the circular RNA excision product from the first intron of the *S. cerevisiae* CYB gene can integrate into RNA having fused flanking exonic sequences (Augustin *et al.*, 1990; Morl and Schmelzer, 1990). If this reaction also takes place *in vivo* then a DNA copy of the recombinant RNA may be made by reverse transcription, as open reading frames of group II introns encode a protein resembling retroviral reverse transcriptase (Michel and Lang, 1985; Xiong and Eickbush, 1988). Finally recombination of the DNA with the intronless genome would complete the process.

Although intron mobility has been demonstrated in crosses, thereby providing a mechanism for their spread, these results do not give insight into how introns are acquired by mitochondrial genomes of species that lack such elements. However, it has been suggested that mitochondrial genomes can gain both group I and group II introns by horizontal transfer from one species to another (Lang, 1984; Waring et al., 1984; Michel and Dujon, 1986). Supporting these suggestions are results from sequence analyses showing greater similarity between the S. pombe COX1.2 and the A. nidulans COX1.3 introns than base-matching in the overall exonic sequence (Lang, 1984; Waring et al., 1984). A more striking example of this phenomenon has been discovered by sequence determination of the COX1 gene in K. lactis mtDNA. The first intron has 96% base-matching to S. cerevisiae COX1.2, whereas flanking exonic regions show only 88% similarity (Hardy and Clark-Walker, 1991). This observation is the first case of possible horizontal transfer of a group II intron.

These observations raise questions about the origin of these mobile elements and the mechanism of their propagation. In the case of class II introns, sequence comparisons suggest that they are most closely related to retrotransposons that require a host promoter for their transcript (Xiong and Eickbush, 1988). Horizontal transmission of these elements could involve either a DNA or an RNA molecule. For group II introns the presence of an open reading frame encoding a reverse transcriptase would allow RNA to be the vehicle. Propagation of group I introns by an RNA molecule, on the other hand, may require the presence of a resident group II intron to supply the reverse transcriptase.

Transposition, meaning insertional translocation of a duplication, has also been invoked to explain the base-matching between the S. cerevisiae COX1.1 and COX1.2 introns (Bonitz et al., 1980). Likewise intron transposition is seen as a major contributing factor to the large number of these elements in P. anserina mtDNA (Cummings et al., 1990a). If transposition is occurring on such a scale then a question is raised about the nature of target sites for intron insertion. Hitherto it has been argued that introns insert as specific sites and this is clearly so for some introns of S. cerevisiae such as omega in LSU and COX1.4. However, more studies appear to be necessary to resolve whether intron homing is restricted to special cases (Dujon, 1989).

Intron loss from mitochondrial genomes can be achieved experimentally and the mechanism is thought to proceed by an RNA intermediate and reverse transcription (Gargouri et al., 1983). It has been observed in S.

cerevisiae that some intron splicing deficient mutants, resulting from lesions within the intron, can revert by excision of the intron and in some cases neighboring introns are excised (Jacq *et al.*, 1982; Gargouri *et al.*, 1983; Hill *et al.*, 1985; Seraphin *et al.*, 1987). Clean excision of introns has also been found in *S. pombe* mtDNA (Merlos-Lange *et al.*, 1987; Schafer *et al.*, 1990). Deletion of introns from the mitochondrial genome of *S. cerevisiae* has been shown to depend on the presence of COX1.1 and COX1.2 and by inference, on a reverse transcriptase encoded by these sequences (Levra-Juillet *et al.*, 1989). Thus mtDNAs polymorphic for introns, where some alleles lack introns such as COX1 of *N. crassa* or CYB of *S. pombe*, could have arisen by intron excision as outlined above. Indeed intron polymorphisms imply that an equilibrium exists between the gain and the loss of introns (Dujon, 1989). In some instances, for unknown reasons, the equilibrium may be tilted in favor of intron gain as exemplified by the extremely mosaic character of *P. anserina* mtDNA. In this context, intron gain or loss could be rationalized if changed phenotypes were detectable. Despite intron polymorphisms in many fungal mitochondrial genomes and the construction of an intron-free mtDNA in *S. cerevisiae*, phenotypic alterations have not been reported. Consequently, we are left with unsatisfactory alternative explanations. Intron polymorphisms may produce subtle changes allowing organisms to exploit different niches or they could be inconsequential. The same points apply to explain length mutations in intergenic regions.

2. Intergenic Regions

No less perplexing from a functional viewpoint is intergenic length variation. As described, the 53 kb of intergenic sequence in *S. cerevisiae* mtDNA is nearly 19 times that in *C. glabrata*. Although intergenic sequence in both genomes is composed largely of A+T, the *S. cerevisiae* molecule has other elements. In addition to the five open reading frames discussed previously, there are approximately 200 G+C rich sequences of 20–50 bp termed G+C clusters and seven or eight *ori* or *rep* elements of 280–300 bp (De Zamaroczy and Bernardi, 1986a,b). As well as their location in intergenic regions, G+C clusters have been found in genes for LSU (Dujon, 1980), SSU (Sor and Fukuhara, 1982), and VAR1 (Hudspeth *et al.*, 1984) and in introns (De Zamaroczy and Bernardi, 1986a,b; Weiller *et al.*, 1989). In an extensive analysis it has been found that 26 sites, scattered around the genome, are polymorphic for the presence of G+C clusters (Weiller *et al.*, 1989). This has enabled the identification of a target sequence, TAG, where G+C clusters of the M1 and M2 subclasses (a1 and a2 in the nomenclature of De Zamaroczy and Bernardi, (1986a) are inserted. As AG nucleotides flank G+C clusters, it appears that a 2-nucleotide staggered cut is made as part of the insertion mechanism. Two nucleotide duplications are also found at distinct insertion sites for the G and V subclasses of G+C clusters (Weiller *et al.*, 1989). Taken together these observations strongly suggest that G+C elements are mobile as originally proposed by Sor and Fukuhara (1982) and that numerous transpositions have resulted in the distribution of these elements to target sites around the genome. The transposition mechanism has yet to be described.

Analogous G+C-rich, repetitive sequence elements have been found in mtDNAs from *N. crassa*, characterized by the presence of palindromic Pst1 sites (Yin *et al.*, 1981, 1982) and in *K. lactis,* where *Sac*II sites are prevalent (Ragnini and Fukuhara, 1988). In *N. crassa*, the Pst1-palindrome has been found in numerous places flanking large genes and separating tRNA sequences (Yin *et al.*, 1981, 1982; Macino and Morelli, 1983). The suggestion that this element is mobile (Yin *et al.*, 1981) is strengthened by the observation that *Pst*1-palindromes have been found in the 3.6 kb Mauriceville plasmid that has little sequence similarity to the mitochondrial genome, although it is located in mitochondria (Nargang *et al.*, 1983).

Sequence comparison between similar introns in the COX1 gene of *A. nidulans* and *S. pombe*, described earlier, show that the former has a 37 bp G+C-rich insert that is absent in the latter (Waring *et al.*, 1984). As this insert is flanked by a 5-bp repeat, it seems likely that it is or was a type of mobile element.

The occurrence of G+C-rich sequences in *K. lactis* mtDNA is often, but not always, correlated with the presence of *Sac*II sites (Ragnini and Fukuhara, 1988). However, there is little sequence relatedness among G+C-rich elements apart from the hexanucleotide *Sac*II site (Ragnini and Fukuhara, 1988; Hardy and Clark-Walker, 1990; and unpublished observations). Nor are these elements flanked by short repetitive sequences that could suggest a target site. However, their occurrence within COX1 introns of *K. lactis* mtDNA and their absence from similar introns in *S. cerevisiae* mtDNA (Hardy and Clark-Walker, 1991) again suggests that some of these sequences are mobile. Curiously, very few *Sac*II sites have been detected in mtDNA from the closely related yeast *K. marxianus* (Ragnini and Fukuhara, 1988). This result could mean either that G+C-rich elements are few or absent or that the *Sac*II sites have changed, though the frequency and distribution of G+C clusters could be the same.

In view of the possibility that some *Kluyveromyces* species lack G+C-rich elements, it is relevant to note that such sequences are absent from *C. glabrata* mtDNA (Clark-Walker *et al.*, 1985). As this yeast is related to *S. cerevisiae* (Wong and Clark-Walker, 1990), the most reasonable interpretation of this observation is that a G+C-rich sequence element has never been acquired. Viewed in this context it would seem that *N. crassa* and *S. cerevisiae* mitochondrial genomes at least have been separately infected with G+C-rich short mobile elements that have proliferated to specific target sites around the genome (Weiller *et al.*, 1989). It cannot be excluded that the vehicle of infection is a plasmid, such as the Mauriceville species of *N. crassa*, or an intron that subsequently has been lost from some species or strains.

The *ori/rep* element in *S. cerevisiae* mtDNA can also be viewed as a mobile element that has recently infected the mitochondrial genome followed by its transposition to different sites. An alternative view is that *ori* elements have evolved from mitochondrial promoters in association with G+C-rich DNA (De Zamaroczy and Bernardi, 1987).

Another type of mobile DNA in the *S. cerevisiae* genome could be A+T-rich oligonucleotides, as it has been found that the SSU gene can be polymorphic for the presence of two 16-bp A+T inserts (Huttenhofer *et al.*, 1988). A similar conclusion regarding the mobility of A+T-rich oligonucleotides has also been reached by De Zamaroczy and Bernardi (1987) in their comprehensive analysis of the A+T spacer regions of *S. cerevisiae* and *C. glabrata* mitochondrial genomes. These authors have found that some A+T oligonucleotides are present at frequencies much higher than expected in random A+T sequence. They conclude that A+T spacer regions in baker's yeast mtDNA have been built by either recombination or replication slippage (slipped-strand mispairing). In a review on slipped-strand mispairing it has been emphasized that the generation of initial tandem repeats is more likely to occur by this mechanism than by unequal crossing-over (Levinson and Gutman, 1987). However, once a tandem repeat has been formed then unequal crossing-over could contribute to further expansion.

The formation of tandem repeats in *S. cerevisiae* mtDNA is thought to be a frequent event, as it has been found that a spontaneously arising polymorphism is due to the addition of two extra copies of a 14-bp sequence to a tandem repeat comprising six copies (Skelly and Clark-Walker, 1991). It remains unresolved whether this new structure has been produced by slipped-strand mispairing or unequal crossing-over. Further support for the idea that tandem repeat formation is a frequent event comes from a comparison of almost 6 kb of intergenic sequence between our strain of *S. cerevisiae* and others. We find that polymorphisms in tandem repeats of three or more base pairs occur every 400–500 bp (Skelly and Clark-Walker, 1991).

From the preceding information it would appear that intergenic regions in *S. cerevisiae* mtDNA have been built by a variety of different processes. If the mitochondrial genome of *C. glabrata* is taken to resemble the ancestral molecule (see below) then expansion of small A+T-rich segments may first occur by slipped-strand mispairing. The resulting tandem repeats may be enlarged further by unequal crossing-over, and base substitutions, as well as small additions or deletions, would alter their repetitive nature. Transposition and inversion of oligonucleotide segments probably occur as well as the acquisition of G+C clusters. These sequences may have been built by similar processes to A+T segments or they could have an external origin. However, it appears that once acquired, G+C elements are mobile and have been transposed to target sites within intergenic regions and, where tolerated, into some genes as well.

Although formation of large intergenic regions, by the pathway outlined above, would imply that large segments of sequence do not influence mitochondrial gene expression, until recently it has not been possible to approach this question. However, by creating two deletions of 3.7 and 5.0 kb in intergenic regions between A6 and Glu tRNA and between Pro tRNA and SSU genes respectively, it has been found that these alterations do not change the growth rate of *S. cerevisiae* on nonfermentable substrates (Skelly and Clark-Walker, 1990). Likewise large deletions from two other intergenic regions do not appear to affect mitochondrial gene expression (Clark-Walker, 1989). The conclusion to be drawn from these results is that large portions of intergenic regions in baker's yeast mtDNA are dispensable.

There is nevertheless an unfortunate consequence arising from intergenic sequence that has been made from repetitive elements. The mitochondrial genome in *S. cerevisiae* is destabilized because frequent excisions occur at short directly repeated sequences, mainly G+C clusters, resulting in the formation of defective mtDNA that is manifested in petite mutants. Strains of *S. cerevisiae* produce petite mutants spontaneously at approximately 1% per generation (Clark-Walker *et al.,* 1981a, and references therein; Dujon and Belcour, 1989). On the other hand the rate of spontaneous petite mutant formation in *C. glabrata* is four orders of magnitude less (Clark-Walker *et al.,* 1981a). This result suggests that the more frequent excision events in the larger genome are a consequence of the G+C repetitive elements that are absent from the smaller molecule. As any increase in the abundance of G+C elements might be expected to further destabilize the mtDNA, the mitochondrial genome in *S. cerevisiae* can be viewed as a product of processes leading to expansion of intergenic regions and to forces selecting against molecules that have become too unstable.

In the above discussion, emphasis has been placed on expansion of intergenic regions. The part played by loss of intergenic regions in the evolution of fungal mtDNA is more problematic. For instance, it can be argued that mitochondrial genomes below the median size have arisen by deletion. This notion would be supported if lengths are distributed on a normal curve, but the present data are insufficient to allow this conclusion to be made. The alternative view is that all genomes have arisen from a small ancestral form by acquisition of introns and expansion of intergenic regions.

Loss of intergenic regions may occur through recombination–excision processes or, for small regions, by slipped-strand mispairing (Levinson and Gutman, 1987). Deletions ranging from 500 to 1500 bp in mtDNA of a mutator strain of *S. pombe* have been suggested to occur by this process, although in these examples the regions involved are not intergenic (Ahne *et al.*, 1988). Large deletions in spacer regions of *S. cerevisiae* mtDNA can be obtained experimentally by intramolecular recombination as described below. This demonstration opens the possibility that a similar process may operate in nature.

B. Rearrangements

Rearrangements in circular molecules could take place by either of two pathways. Both avenues involve the formation of subgenomic species and excision from a recombinant intermediate containing a direct duplication (or as a further possibility, an inverted repeat). These two pathways can be classified either as transposition or subgenomic recombination.

In the transposition pathway a subgenomic molecule, arising from the wild-type by excision, recombines with another wild-type molecule at a distal or foreign site to form an intermediate with a direct duplication. Creation of an intermediate molecule, by the second pathway, involves the formation of two subgenomic molecules by separate excisions from the wild-type, as illustrated in Fig. 2. Recombination between the two subgenomic molecules at sequences within their shared region (segments 1 and 9) creates an intermediate with direct duplications that lacks a wild-type segment (segment 5). This latter feature ensures *a priori* that the intermediate cannot undergo excision of the foreign placed or alloposed duplication to restore the wild-type molecule. For the intermediate formed by transposition, precise excision of the transposed segment would restore the wild-type. Excision of the resident or symposed duplication in either case leads to a molecule with three novel junctions and a rearranged gene order. Intramolecular recombination need not remove the duplication precisely. If recombination occurs in single copy regions flanking the duplication then a deletion of wild-type sequence occurs or, alternatively, smaller duplications could persist if excision takes place internal to the ends of the repeat. Combinations of these two possibilities can also be imagined. Thus, as already described, small duplications in *P. anserina*, *N. crassa*, and *A. nidulans* mtDNAs could result from imprecise excision of a larger repeat.

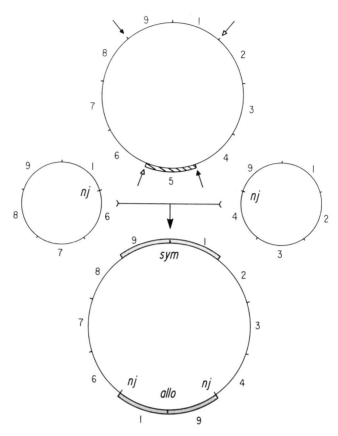

FIG. 2 Formation of a molecule containing direct repeats by the subgenomic recombination pathway. Excisions from the wild-type molecule take place at points indicated by the arrows to produce two subgenomic species that each contain a common region (1,9) and lack a sector (5). Sector 5 represents a dispensable intergenic region. Reciprocal products of the two excision events are not illustrated. Recombination between the two subgenomic species takes place at sites in the common sector to produce a molecule with a resident or symposed repeat and a foreign placed or alloposed repeat between two novel junctions (nj). Note that the alloposed repeat is not the result of a transposition, since it is bounded by sectors 4 and 6, whereas transposition would have divided sector 5. Excision of the alloposed repeat would lead to a molecule with wild-type topology but having a deletion of section 5. Loss of the symposed repeat would form a molecule with three novel junctions and a rearrangement. Depending on the location of the excision sites, a further deletion of wild-type sequence or a fused duplication could be produced. Other possibilities could occur if excision sites are located in unique and duplicated regions.

Evidence supporting the subgenomic pathway has been obtained experimentally in *S. cerevisiae* by a series of *in vivo* steps (Clark-Walker, 1989). In the first step, nascent spontaneous petites, containing newly formed subgenomic mtDNA molecules, are mated and zygotes that have a high frequency of spontaneous petite mutant formation are selected (>60% petites, hfp strains) (Oakley and Clark-Walker, 1978). Such hfp strains have mitochondrial genomes containing mixtures of complementary subgenomic species together with a recombinant molecule (Evans *et al.*, 1985; Evans and Clark-Walker, 1985). Revertants from hfp strains, which have a near normal level of petite production, contain mitochondrial genomes with either a rearranged (Clark-Walker, 1989) or a wild-type gene order (Skelly and Clark-Walker, 1990). In both cases segments of the wild-type mtDNA have been deleted from intergenic regions.

Sequencing novel junctions has shown that both the initial intramolecular recombinations and subsequent excisions take place at direct repeats in G+C clusters (Skelly and Clark-Walker, 1990; Clark-Walker, 1989). Likewise, direct G+C repeats are sites for an 11-kb deletion in the COX1 gene. This deletion occurs in about 10% of all manganese-induced mit⁻ mutants at this locus (Weiller *et al.*, 1991). These results confirm previous observations that G+C repeats act as sites for recombination in *S. cerevisiae* mtDNA (De Zamaroczy *et al.*, 1983; Dieckman and Gandy, 1987; Zinn *et al.*, 1988).

As no change in growth rate of strains containing rearranged genomes has been observed on nonfermentable substrate, it is concluded that mitochondrial gene expression has not been altered. This observation is not unexpected on two grounds. First, transcription occurs from several promoters around the genome and appears to be precessive, meaning that there are no termination sites. In other words, changing the gene order is unlikely to affect overall levels of transcripts because of multiple promotion and ongoing transcription. Second, mtDNA gene order can differ between naturally occurring, currently accepted strains of *K. lactis* without apparent change to growth rates on nonfermentable substrates (Skelly *et al.*, 1991). These observations suggest that rearrangements in mitochondrial genomes of yeasts, and perhaps other fungi, do not have noticeable consequences for the phenotype.

Investigations of mtDNA rearrangements in filamentous fungi have shown that formation of subgenomic molecules proceeds in a manner similar to that of deletions in *S. cerevisiae*. These observations have been obtained from examination of senescence in *P. anserina* (Kuck, 1989), stopper in *N. crassa,* and ragged growth in *A. amstelodami* (Wolf and Del Guidice, 1988; Dujon and Belcour, 1989). Senescent mycelia of *P. anserina* have been found to have defective mitochondria that contain subgenomic species of mtDNA (Stahl *et al.*, 1978; Cummings *et al.*, 1979b). Apart

from the species termed α, which is derived from intron COX1.1, most
subgenomic molecules that have been studied appear to be excised at an
11-bp repeat rich in G+C (Turker *et al.*, 1987).

Subgenomic molecules have also been found in various mutants of *N.
crassa* that included stopped growth or stopper (Mannella *et al.*, 1979;
Bertrand *et al.*, 1980) and in small amounts in a wild-type strain (Gross *et
al.*, 1984). In one case, production of subgenomic species has occurred by
recombination at duplicated Met tRNA genes (Gross *et al.*, 1984, 1989).
Other deletions have been found at G+C-rich repeats that can be as short
as 9 bp and are sometimes near Pst1-palindromes (De Vries *et al.*, 1986;
Almasan and Mishra, 1988: Gross *et al.*, 1989). It has been suggested that
these recombinogenic sites are found in regions where single-strand breaks
may occur as a consequence of hairpin loop formation (Almasan and
Mishra, 1988; Gross *et al.*, 1989). This may be important for initiating
recombination at these regions but details have not been unraveled. In a
ragged growth mutant of *A. amstelodami* two regions of mtDNA are prone
to intramolecular deletion (Lazarus *et al.*, 1980; Lazarus and Kuntzel,
1981). It remains an open question whether the resulting subgenomic
molecules have arisen by deletion at short direct repeats.

Intramolecular recombination must also be responsible for remnants of
mtDNA that are found in naturally occurring respiratory-deficient yeasts
such as *Candida sloofii* (Arthur *et al.*, 1978). This yeast may represent an
intermediate stage on the pathway to complete elimination of mtDNA.
For instance, some anaerobic fungi, such as *Neocallimastix* sp., found in
herbivore intestinal tracts, lack recognizable mitochondria (Heath *et al.*,
1983) or any trace of mtDNA (A. Brownlee, personal communication).

From the foregoing examples, recombinogenic sequences in fungal
mtDNAs are seen to be rich in G+C. This is not an exclusive trait,
however, as some defective mtDNA molecules in petite mutants of *S.
cerevisiae* can arise at A+T-rich repeats (De Zamaroczy *et al.*, 1983) and
likewise, a deletion in *K. lactis* mtDNA occurs at an 11-bp repeat that has
only one G:C bp (Hardy *et al.*, 1989). In this respect it needs to be
mentioned that recombinations have not been found at the A+T-rich
nonomer promoter and dodecamer RNA processing sequences that are at
scattered locations in mtDNAs of various yeasts (Tabak *et al.*, 1983; Clark-
Walker *et al.*, 1985). However, this result may reflect the rarity of such
events in the face of more frequent excisions at other structures. Clearly,
more studies are necessary to identify structural features in DNA that are
important in recombination. What is apparent is that subgenomic mtDNA
molecules are produced in a number of yeasts (Clark-Walker *et al.*, 1981a)
and filamentous ascomycetes. Because these molecules arise in most, if
not all, cases by excision at short direct repeats, there is a possibility that
they may reintegrate into wild-type mtDNA at further copies of repeated

sequences in different locations by a reversal of the excision process. In addition, because some of the short recombinogenic sequences may be oppositely oriented in the wild-type, as has been found for G+C clusters in *S. cerevisiae* mtDNA (De Zamaroczy and Bernardi, 1986b), integration of a subgenomic molecule could occur to produce an inverted segment. Such molecules have been produced experimentally in *S. cerevisiae* mtDNA (Evans and Clark-Walker, 1985) and examples of mtDNA molecules with inversions or inverted repeats have been described previously.

C. Codon Changes

As foreshadowed, changes from the universal code have been found in fungal mtDNAs (Fox, 1987). Most widespread is the change of the stop codon UGA to specify Trp. This alteration has been found in mtDNA of all ascomycetous budding yeasts (Table V) and in all four filamentous ascomycete mtDNAs so far examined, namely *A. nidulans, N. crassa,* and *P. anserina* (Fox, 1987) and for the ND1 gene in *Cephalosporium acremonium* (Penalva and Garcia, 1986). Indeed, in these latter organisms, alteration of UGA to specify Trp is the only change to the code. However, this change is not present in all fungal mtDNAs, as no alteration of the universal code has been found in COX3 of the basidiomycete *Schizophyl-*

TABLE V

Deviations from the Universal Code in mtDNA from Ascomycetous Budding Yeasts

Organism	Codon change		
	UGA (Term)	AUA (Ile)	CUN (Leu)
Brettanomyces custersianus	Trp	+	+[a]
Candida (Torulopsis) glabrata	Trp	Met	Thr
Hansenula saturnus	Trp	(+)	(+)
Kluyveromyces lactis	Trp	Met	—[b]
Saccharomyces cerevisiae	Trp	Met	Thr

Note: +, no change from universal code; (+), AUA use is uncertain as it only appears once; CUU codons appear three times in COX2 of *H. saturnus* at positions occupied by Leu in *S. cerevisiae* (Lawson and Deters, 1985).

[a] CUA occurs once in *B. custersianus* COX2 at a position occupied by Leu in *K. lactis* and *S. cerevisiae* COX2 (Hoeben and Clark-Walker, manuscript in preparation).

[b] CUN is not found in CYB, COX1, COX2, or A8 of *K. lactis* (Brunner and Coria, 1989; Hardy and Clark-Walker, 1990, 1991). ACN (Thr) occurs five times in *K. lactis* at sites occupied by CUN in *S. cerevisiae* (Brunner and Coria, 1989; Hardy and Clark-Walker, 1990, 1991).

lum commune (Phelps *et al.*, 1988). Even within ascomycetes it has been observed that UGA does not code for Trp in mtDNA of *S. pombe* except in two intron open reading frames and in urfa (Lang, 1984; Wolf and Del Guidice, 1988). As the mitochondrial-encoded tRNA specific for Trp contains the anticodon 3′ CCA 5′, it remains unclear whether urfa and the intron open reading frames are translated.

In ascomycetous budding yeasts the pattern of codon change is more complex (Table V). In addition to use of UGA to denote Trp, which is common to all five mitochondrial genomes, AUA has been found to encode Met and CUN (where N is any of the four bases) codes for Thr in *S. cerevisiae* (Fox, 1987). Similar changes have been found in *C. glabrata* (Ainley *et al.*, 1985; G. D. Clark-Walker, unpublished observations) and in *K. lactis* AUA codes for Met, but CUN codons have not been found (Table V) (Brunner and Coria, 1989; Hardy and Clark-Walker, 1990, 1991). It has also been observed in *C. glabrata* that Arg is specified by AGA only and that CGN is not used (Ainley *et al.*, 1985; Clark-Walker, unpublished observations). The absence of CGN codons for this yeast is not surprising, as only one tRNA gene for Arg has been found that contains the anticodon 3′ UCU 5′ (Clark-Walker, *et al.*, 1985). It remains to be determined whether *K. lactis* mtDNA encodes two Arg tRNAs, as appears likely, because five CGC codons have been found in COX1 intron ORFs (Hardy and Clark-Walker, 1991). Sequence determination of mtDNA from another species of *Kluyveromyces* has been undertaken on the LSU intron of *K. thermotolerans*. Although this analysis indicates that UGA encodes Trp, firm conclusions cannot be made about other possible changes from the universal code, as only one AUA and two CUU codons are present, the latter being at nonconserved sites in comparison to the cognate gene in *S. cerevisiae* (Jacquier and Dujon, 1983).

In COX2 of *B. custersianus*, CGU is used twice to encode Arg at positions occupied by this amino acid in *K. lactis* COX2 where it is encoded by AGA (Hardy and Clark-Walker, 1990; Hoeben Weiller and Clark-Walker, submitted). Likewise, CGU is used four times in COX2 of *H. saturnus* and three of these locations are at positions occupied by AGA in COX2 of *S. cerevisiae* (Lawson and Deters, 1985). Hence the mtDNA of *C. glabrata* differs from at least three other yeast mitochondrial genomes in lacking a second gene for Arg tRNA. For *C. glabrata* and *S. cerevisiae* the difference in tRNA content of their mtDNAs is confined to this gene, as there are 23 tRNA genes in the former yeast and 24 in the latter. Both genomes have two genes each for Ser and Thr as well as isoacceptor species for formyl Met and Met.

In general it seems that mtDNAs of yeasts and filamentous ascomycetes have sufficient tRNA genes for translation and that there is no need to import any tRNAs into miotchondria. The reduced number of tRNA genes, especially in yeasts, is made possible by employing an unmodified U in anticodons to enable two-of-three reading for codons employing all four bases at the third position (Fox, 1987; Wolf and Del Guidice, 1988).

The occurrence of different codons in mtDNAs from yeasts indicates that changes have taken place since divergence of these organisms from filamentous ascomycetes and fission yeasts. In view of the "codon capture" proposal for alteration of the genetic code without disrupting amino acids (Osawa and Jukes, 1989), it is easier to accept that such changes are not rare. Detailed proposals have been made for reassignment of AUA from isoleucine to methionine (Osawa, *et al.*, 1989) and for CUN from leucine to threonine (Osawa *et al.*, 1990). These suggestions incorporate the idea that directional mutation pressure (Sueoka, 1988) leads to the disappearance of a codon. In the case of fungal mtDNA, mutational pressure is toward A+T, although the mechanism causing this change is unknown.

Different codon usage in yeasts could also be important for yeast taxonomy. Further studies are necessary to establish whether codon changes are confined to related organisms. For example, it may be found that budding yeasts can be divided into two groups that resemble either *S. cerevisiae* or *B. custersianus* in their codon usage (Table V). Further subdivision may also exist with *K. lactis* being separated from *S. cerevisiae* and *C. glabrata* on the basis that CUN codons are not utilized in the former yeast.

IV. Conclusions

Attempts to determine a time frame for fungal mtDNA evolution are hampered by the lack of a fossil record. Hence studies will have to focus on relating rates of mtDNA base substitutions to changes in nuclear genes. In turn, an assumption will still have to be made that mutation rates of nuclear genes are similar to their counterparts in organisms with datable fossils. For instance, base substitutions in the nuclear-encoded cytochrome *c* gene of fungi can be obtained and mtDNA changes can also be examined. When this has been done for three yeasts, *C. glabrata, K. lactis,* and *S. cerevisiae* it has been found that base mismatching in mitochondrial genes is less than that in the nuclear-encoded cytochrome *c* (Clark-Walker, 1991). By contrast, examination of the same genes from mammals shows that mitochondrial genes have higher rates of base substitution than the cytochrome *c* gene and other nuclear genes (Brown *et al.*, 1979; Miyata *et al.*, 1982; Wallace *et al.*, 1987).

Anecdotal evidence supporting the notion that slow base substitution rates apply to other fungi comes from the observation that in the mapping studies described earlier, strong cross-hybridization has been obtained between mtDNA probes from *S. cerevisiae* and mitochondrial genomes from widely dispersed organisms. On the other hand, some nuclear genes of *S. cerevisiae* do not cross-hybridize with *K. lactis* nuclear DNA (G. D. Clark-Walker, unpublished observations) and sequence comparisons show that only 60–80% base-matching is found in a number of nuclear genes (Shuster *et al.*, 1987; Stark and Milner, 1989; Fournier *et al.*, 1990; Saliola *et al.*, 1990). Taken together these observations support the results that show yeast mtDNA genes have a slow rate of base substitution, and suggest that other fungal mtDNAs behave likewise. Studies similar to the ones described above will form the basis for answering whether phylogenetic trees, constructed from either nuclear or mitochondrial gene sequence comparisons, are congruent. If this proves so, then studies into the evolutionary relationships of fungi will be aided since mitochondrial genes are more accessible than those in nuclear DNA.

One aim of studies on fungal mtDNA evolution is to reconstruct a hypothetical ancestral molecule. Two approaches can be used. First, we can extrapolate from conserved structure in mtDNAs as described above, and, second, we can use as a model the architecture of genomes in prokaryotes. This second method is plausible because mitochondria are thought to have arisen from endosymbionts sharing a common ancestor with purple nonsulfur bacteria (Gray, 1989b). However, a potential difficulty with the latter approach is that the genome of organisms such as *Paracoccus denitrificans*, a species thought to be descended from an ancestor of mitochondria (Whatley *et al.*, 1979), may be now more streamlined. For example, genes in a present-day bacterium that are linked into operons and lack intervening sequences may once have been scattered and interrupted by introns. On the other hand, prokaryotes may have already refined their genomes by the time they started forming intracellular associations with early eukaryotic cells. This being the case, then prokaryotic operons may be reflected by linkage groups in mtDNAs. Quite remarkably, a strong case can be made that some bacterial linkage groups persist in the 19-kb *C. glabrata* mtDNA. In this genome COX2–COX3 are juxtaposed, whereas in *P. denitrificans* these genes have the same order and orientation in an operon although they are separated by three open reading frames (Raitio *et al.*, 1987). The gene for COX1 is at a separate locus in the bacterium (Raitio *et al.*, 1987), and in *C. glabrata* this gene is separated from COX2–COX3. Second, genes for ATPase subunits 6 and 9 are juxtaposed in the yeast's mtDNA, and in the *unc* or *atp* operon of *Escherichia coli* analogous genes have the same order and orientation (Downie *et al.*, 1981; Gay and Walker, 1981). Third, SSU and LSU ribosomal RNA genes have the same order and orientation in *C. glabrata* mtDNA as in *E. coli* rRNA operons (King *et al.,* 1986), although in the yeast these genes are separated by VAR1. On the other hand, examination of linkages and positions of tRNA genes in the *E. coli* chromosome (Komine *et al.,* 1990) has not shown any correlation with the location of equivalent genes in *C. glabrata* mtDNA.

Lending support to the idea that the mitochondrial genome in *C. glabrata* may resemble an ancestral molecule is the observation, discussed earlier, that smaller mtDNAs are less prone to rearrangement than larger forms. Following from this, it is concluded that larger mtDNAs do not represent intermediates on the pathway of genome economization, but are derived forms having undergone size increases and rearrangement. Even earlier, it can be argued, the endosymbiont's genome would have contracted due to transfer of genes to the nucleus and elimination of superfluous loci. During this phase, if the mitochondrial genome of *C. glabrata* can be taken as a guide, rearrangements must have been infrequent.

From the foregoing descriptions on the diverse sizes and structures of fungal mtDNAs it is apparent that the most detailed knowledge has been obtained from ascomycetes. However, mitochondrial genomes from different groups of fungi need to be studied to determine whether length mutations and rearrngements take place to the same extent as in yeasts and filamentous ascomycetes. Information from such studies would also help in trying to deduce the structure of an ancestral mitochondrial genome as outlined above. In other words, our present knowledge, gained from only a few organisms, could be unrepresentative of the majority of fungal mtDNAs.

Although mention has been made of enzymes catalyzing recombination events, our understanding of the role played by nuclear genes in mtDNA evolution is quite limited. It is conceivable that alterations to enzymes involved in DNA metabolism could have marked consequences for the stability or mutability of mtDNA. Thus rates of recombination, slipped-strand mispairing, and base substitution could be anticipated to change. These intrinsic properties, coupled with chance infections from introns and short mobile elements, could play a significant part in driving mitochondrial genomes down different evolutionary paths.

Acknowledgments

I thank Chris Hardy, Ryszard Maleszka, and Georg Weiller for critical comments; Alan Brownlee, Adrienne Hardham, and Patrick Skelly for helpful discussions; and many colleagues for manuscripts.

References

Agsteribbe, E., Hartog, M., and De Vries, H. (1989). *Curr. Genet.* **15**, 57–62.

Agsteribbe, E., Kroon, A. M., and Van Bruggen, E. F. J. (1972). *Biochim. Biophys. Acta* **269**, 299–303.

Ahne, M., Muller-Derlich, J., Merlos-Lange, A. M., Kanbay, F., Wolf, K., and Lang, B. F. (1988). *J. Mol. Biol.* **202**, 725–734.

Ainley, W. M., Macreadie, I. G., and Butow, R. A. (1985). *J. Mol. Biol.* **184**, 565–576.

Almasan, A., and Mishra, N. C. (1988). *Genetics* **120**, 935–945.

Arthur, H., Watson, K., McArthur, C. R., and Clark-Walker, G. D. (1978). *Nature (London)* **271**, 750–752.

Augustin, S., Muller, M. W., and Schweyen, R. J. (1990). *Nature (London)* **343**, 383–386.

Bertrand, H., Collins, R. A., Stohl, L. L., Goewert, R. R., and Lambowitz, A. M. (1980). *Proc. Natl. Acad. Sci. U.S.A.* **77**, 6032–6036.

Bonitz, S. G., Coruzzi, G., Thalenfeld, B. E., Tzagoloff, A., and Macino, G. (1980). *J. Biol. Chem.* **225**, 11927–11941.

Boyd, D. A., Hobman, T. C., Gruenke, S. A., and Klassen, G. R. (1984). *Can. J. Biochem. Cell Biol.* **62**, 571–576.

Brown, T. A. (1990). *In* "Genetic Maps, Locus Maps of Complex Genomes: Book 3" (S. J. O'Brien, ed.), pp. 109–110. Cold Spring Harbor Laboratory, Cold Spring Harbor, NY.

Brown, T. A., Constable, A., Waring, R. B., Scazzocchio, C., and Davies, R. W. (1989). *Nucleic Acids Res.* **14**, 5838.

Brown, T. A., Davies, R. W., Waring, R. B., Ray, J. A., and Scazzocchio, C. (1983). *Nature (London)* **302**, 721–723.

Brown, T. A., Waring, R. B., Scazzocchio, C., and Davies, R. W. (1985). *Curr. Genet.* **9**, 113–117.

Brown, W. M., George, M., Jr., and Wilson, A. C. (1979). *Proc. Natl. Acad. Sci. U.S.A.* **76**, 1967–1971.

Brunner, A., and Coria, R. (1989). *Yeast* **5**, 209–218.

Bruns, T. D., Fogel, R., White, T. J., and Palmer, J. D. (1989). *Nature (London)* **339**, 140–142.

Bruns, T. D., and Palmer, J. D. (1989). *J. Mol. Evol.* **28**, 349–362.

Bruns, T. D., Palmer, J. D., Shumard, D. S., Grossman, L. I., and Hudspeth, M. E. S. (1988). *Curr. Genet.* **13**, 49–56.

Burger, G., Scriven, C., Machleidt, W., and Werner, S. (1982). *EMBO J.* **1**, 1385–1391.

Burger, G., and Werner, S. (1986). *J. Mol. Biol.* **191**, 589–599.

Burke, J. M., and Raj Bhandary, V. L. (1982). *Cell (Cambridge, Mass.)* **31**, 509–520.

Camougrand, N., Mila, B., Velours, G., Lazowska, J., and Guerin, M. (1988). *Curr. Genet.* **13**, 445–449.

Cavalier-Smith, T. (1986). *In* "Evolutionary Biology of the Fungi". (A. D. M. Rayner, C. M. Brasier, and D. Moore, eds.), pp 339–353. Cambridge Univ. Press, Cambridge.

Christiansen, G., and Christiansen, C. (1976). *Nucleic Acids Res.* **3**, 465–476.

Clark-Walker, G. D. (1985). *In* "The Evolution of Genome Size" (T. Cavalier-Smith, ed.), pp. 277–297. Wiley, Chichester, UK.

Clark-Walker, G. D. (1989). *Proc. Natl. Acad. Sci. U.S.A.* **86**, 8847–8851.

Clark-Walker, G. D. (1991). *Curr. Genet.* **20**, 195–198.

Clark-Walker, G. D., and Gleason, F. H. (1983). *Arch Mikrobiol.* **92**, 209–216.

Clark-Walker, G. D., Hoeben, P., Plazinska, A., Smith, P. K., and Wimmer, E. (1987). *In* "The Expanding Realm of Yeast-like Fungi". (G. S. De Hoog, M. Th. Smith, and A. C. M. Weijman, eds.), pp. 259–266. Elsevier Science, Amsterdam.

Clark-Walker, G. D., McArthur, C. R., and Daley, D. J. (1981a). *Curr. Genet.* **4**, 7–12.

Clark-Walker, G. D., McArthur, C. R., and Sriprakash, K. S. (1981b). *J. Mol. Biol.* **147**, 399–415.

Clark-Walker, G. D., McArthur, C. R., and Sriprakash, K. S. (1983). *J. Mol. Evol.* **19**, 333–341.

Clark-Walker, G. D., McArthur, C. R., and Sriprakash, K. S. (1985). *EMBO J.* **4**, 465–473.

Clark-Walker, G. D., Sriprakash, K. S., McArthur, C. R., and Azad, A. A. (1980). *Curr. Genet.* **1**, 209–217.

Clayton, D. A., and Brambl, R. M., (1982). *Biochem. Biophys. Res. Commun.* **46**, 1477–1482.

Cobon, G. S., Beilharz, M. W., Linnane, A. W., and Nagley, P. (1982). *Curr. Genet.* **5**, 97–107.

Colleaux, L., d'Auriol, L., Galibert, F., and Dujon, B. (1985). *Proc. Natl. Acad. Sci. U.S.A.* **85**, 6022–6026.

Collins, R. A. (1990). *In* "Genetic Maps, Locus Maps of Complex Genomes: Book 3" (S. J. O'Brien, ed.), pp. 19–21. Cold Spring Harbor Laboratory, Cold Spring Harbor, NY.

Collins, R. A., and Lambowitz, A. M. (1983). *Plasmid* **9**, 53–70.

Cummings, D. J., Belcour, L., and Grandchamp, C. (1979a). *Mol. Gen. Genet.* **171**, 229–238.

Cummings, D. J., Belcour, L., and Grandchamp, C. (1979b). *Mol. Gen. Genet.* **171**, 239–250.

Cummings, D. J., McNally, K. L, Domenico, J. M., and Matsuura, E. T. (1990a). *Curr. Genet.* **17**, 375–402.

Cummings, D. J., Michel, F., Domenico, J. M., and McNally, K. L., (1990b). *J. Mol. Biol.* **212**, 269–286.

Cummings, D. J., Michel, F., and McNally, K. L. (1989). *Curr. Genet.* **16**, 381–406.

De Jonge, J., and De Vries, H. (1983). *Curr. Genet.* **7**, 21–28.

De Vries, H., Alzner-De Weerd, B., Breitenberger, C. A., Chang, D. D., De Jonge, J. C., and RajBhandary, L. (1986). *EMBO J.* **5**, 779–785.

De Zamaroczy, M., and Bernardi, G. (1986a). *Gene* **41**, 1–22.

De Zamaroczy, M., and Bernardi, G. (1986b). *Gene* **47**, 155–177.

De Zamaroczy, M., and Bernardi, G. (1987). *Gene* **54**, 1–22.

De Zamaroczy, M., Faugeron-Fonty, G., and Bernardi, G. (1983). *Gene* **21**, 193–202.

Dick, M. W. (1989). *In* "Handbook of Protoctista" (L. Margulis, J. O. Corliss, M. Melkonian, and D. J. Chapman, eds.), pp. 661–685. Jones and Bartlett, Boston.

Dieckman, C. L., and Gandy, B. (1987). *EMBO J.* **6**, 4197–4203.

Downie, J. A., Cox, G. B., Langman, L., Ash, G., Becker, M., and Gibson, G. (1981). *J. Bacteriol.* **145**, 200–210.

Dujon, B. (1980). *Cell* (*Cambridge, Mass.*) **20**, 185–197.

Dujon, B. (1989). *Gene* **82**, 91–114.

Dujon, B., and Belcour, L. (1989). *In* "Mobile DNA" (D. E. Berg and M. M. Howe, eds.), pp. 861–878. Academic Press, New York.

Dujon, B., Belfort, M., Butow, R. A., Jacq, C., Lemieux, C., Perlman, P. S., and Vogt, V. M. (1989). *Gene* **82**, 115–118.

Earl, A. J., Turner, G., Croft, J. H., Dales, B. G., Lazarus, C. M., Lunsdorf, H., and Kuntzel, H. (1981). *Curr. Genet.* **3**, 221–228.

Economou, A., and Casselton, L.-A. (1989). *Curr. Genet.* **16**, 41–46.

Economou, A., Lees, V., Pukkila, P. J., Zolan, M. E., and Casselton, L. A. (1987). *Curr. Genet.* **11**, 513–519.

Evans, R. J., and Clark-Walker, G. D. (1985). *Genetics*, **111**, 403–432.

Evans, R. J., Oakley, K. M., and Clark-Walker, G. D. (1985). *Genetics* **111**, 389–402.

Falcone, C. (1984). *Curr. Genet.* **8**, 449–455.

Field, D. J., Sommerfield, A., Saville, B. J., and Collins, R. A. (1989). *Nucleic Acids Res.* **17**, 9087–9099.

Forster, H., Coffey, M. D., Elwood, H., and Sogin, M. L. (1990). *Mycologia* **82**, 306–312.

Forster, H., Kinscherf, T. G., Leong, S. A., and Maxwell, D. P. (1987). *Curr. Genet.* **12**, 215–218.

Forster, H., Kinscherf, T. G., Leong, S. A., and Maxwell, D. P. (1988). *Mycologia,* **80**, 466–478.

Fournier, A., Fleer, R., Yeh, P., and Mayaux, J.-F. (1990). *Nucleic Acids Res.* **18**, 365.

Fox, T. D. (1987). *Annu. Rev. Genet.* **21**, 67–91.

Garber, R. C., and Yoder, D. C. (1984). *Curr. Genet.* **8**, 621–628.

Gargouri, A., Lazowska, J., and Slonimski, P. P. (1983). *In P* "Mitochondria 1983" (R. J. Schweyen, K. Wolf, and F. Kaudewitz, eds.), pp. 259–268. de Gruyter, Berlin.

Gay, N. J., and Walker, J. E. (1981). *Nucleic Acids Res.* **9**, 3919–3926.

Gray, M. W. (1989a). *Annu. Rev. Cell Biol.* **5**, 25–50.

Gray, M. W. (1989b). *Trends Genet.* **5**, 294–299.

Grivell, L. A. (1989). *Eur. J. Biochem.* **182**, 477–493.

Grivell, L. A. (1990a). *Nature (London),* **344**, 110–111.

Grivell, L. A. (1990b). *In* "Genetic Maps, Locus Maps of Complex Genomes: Book 3" (S. J. O'Brien, ed.), pp. 50–57. Cold Spring Harbor Laboratory, Cold Spring Harbor, NY.

Gross, S. R., Hsieh, T.-S., and Levine, P. H. (1984). *Cell (Cambridge, Mass.)* **38**, 233–239.

Gross, S. R., Levine, P. H., Metzger, S., and Glaser, G. (1989). *Genetics* **121**, 693–701.

Grossman, L. I., and Hudspeth, M. E. S. (1985). *In* "Gene Manipulation in Fungi" (J. W. Bennett and L. L. Lasure, eds.), pp 65–103. Academic Press, New York.

Gunderson, J. H., Elwood, H., Ingold, A., Kindle, K., and Sogin, M. L. (1987). *Proc. Natl. Acad. Sci. U.S.A.* **84**, 5823–5827.

Hardy, C. M, and Clark-Walker, G. D. (1990). *Yeast* **6**, 403–410.

Hardy, C. M., and Clark-Walker, G. D. (1991). *Curr. Genet.* **20**, 99–114.

Hardy, C. M., Galeotti, C. L., and Clark-Walker, G. D. (1989). *Curr. Genet.* **16**, 419–427.

Heath, I. B., Bauchop, T., and Skipp, R. A. (1983). *Can. J. Bot.* **61**, 295–307.

Hill, J., McGraw, P., and Tzagoloff, A. (1985). *J. Biol. Chem.* **260**, 3235–3238.

Hintz, W. E., Anderson, J. B., and Horgen, P. A. (1988). *Curr. Genet.* **14**, 43–49.

Hintz, W. E., Mohan, M., Anderson, J. B., and Horgen, P. A. (1985). *Curr. Genet.* **9**, 127–132.

Hoeben, P., and Clark-Walker, G. D. (1986). *Curr. Genet.* **10** 371–379.

Hollenberg, C. P., Borst, P., and Van Bruggen, E. F. J., (1970). *Biochim. Biophys. Acta* **209**, 1–15.

Hudspeth, M. E. S., Shumard, D. S., Bradford, C. J. R., and Grossman, L. I. (1983). *Proc. Natl. Acad. Sci. U.S.A.* **80**, 142–146.

Hudspeth, M. E. S., Vincent, R. D., Perlman, P. S., Shumard, D. S., Treisman, L. O., and Grossman, L. I. (1984). *Proc. Natl. Acad. Sci. U.S.A.* **81**, 3148–3152.

Huttenhofer, A., Sakai, H., and Weiss-Brummer, B. (1988). *Nucleic Acids Res.* **17**, 8665–8674.

Jacq, C., Pajot, P., Lazowska, J., Dujardin, G., Claisse, M., Groudinsky, O., de la Salle, H., Grandchamp, C., Labouesse, M., Gargouri, A., Guiard, B., Spyridakis, A., Dreyfus, M. and Slonimski, P. P. (1982). *In* "Mitochondrial Genes" (P. P. Slonimski, P. Borst, and G. Attardi, eds.), pp. 155–183. Cold Spring Harbor Laboratory, Cold Spring Harbor, NY.

Jacquier, A., and Dujon, B. (1983). *Mol. Gen. Genet.* **192**, 487–499.

King, T. C., Sirdeskmukh, R., and Schlessinger, D. (1986). *Microbiol. Rev.* **50**, 428–451.

Kistler, H. C., Bosland, P. W., Benny, U., Leong, S. and Williams, P. H. (1987). *Phytopathology* **77**, 1289–1293.

Klimczak, L. J., and Prell, H. H. (1984). *Curr. Genet.* **8**, 323–326.

Kochel, H. G., Lazarus, C. M., Basak, N., and Kuntzel, H. (1981). *Cell (Cambridge, Mass.)* **23**, 625–633.

Kojo, H. (1976). *FEBS lett.* **67**, 134–136.

Komine, Y., Adachi, T., Inokuchi, H., and Ozeki, H. (1990). *J. Mol. Biol.* **212,** 579–598.
Kovac, L., Lazowska, J., and Slonimski, P. P. (1984). *Mol. Gen. Genet.* **197,** 420–424.
Kuck, U. (1989). *Exp. Mycol.* **13,** 111–120.
Kuck, U., Kappelhoff, B., and Esser, K. (1985). *Curr. Genet.* **10,** 59–67.
Kuck, U., Stahl, U., Lhermitte, A., and Esser, K. (1980). *Curr. Genet.* **2,** 97–101.
Kunze, G., Bode, R., and Birnbaum, D. (1986). *Curr. Genet.* **10,** 527–530.
Kwok, S., White, T. J., and Taylor, J. W. (1986). *Exp. Mycol.* **10,** 196–204.
Lambowitz, A. M. (1989). *Cell (Cambridge, Mass.)* **56,** 323–326.
Lang, B. F. (1984). *EMBO J.* **3,** 2129–2136.
Lang, B. F., Cedergren, R., and Gray, M. W. (1987). *Eur. J. Biochem.* **169,** 527–537.
Lawson, J. E., and Deters, D. W. (1985). *Curr. Genet.* **9,** 345–350.
Lazarus, C. M., Earl, A. J., Turner, G., and Kuntzel, H. (1980). *Eur. J. Biochem.* **106,** 633–641.
Lazarus, C. M., and Kuntzel, (1981). *Curr. Genet.* **4,** 99–107.
Levinson, G., and Gutman, G. A. (1987). *Mol. Biol. Evol.* **4,** 203–221.
Levra-Juillet, E., Boulet, A., Seraphin, B., Simon, M., and Faye, G. (1989). *Mol. Gen. Genet.* **217,** 168–171.
McNabb, S. A., Boyd, D. A., Belkhiri, A., Dick, M. W., and Klassen, G. R. (1987). *Curr. Genet.* **12,** 205–208.
McNabb, S. A., Eros, R. W., and Klassen, G. R. (1988). *Can. J. Bot.* **66,** 2377–2379.
McNabb, S. A., and Klassen, G. R. (1988). *Exp. Mycol.* **12,** 233–242.
Macino, G., and Morelli, G. (1983). *J. Biol. Chem.* **258,** 13230–13235.
Macreadie, I. G., Scott, R. M., Zinn, A. R., and Butow, R. A. (1985). *Cell (Cambridge, Mass.)* **41,** 395–402.
Mannella, C. A., Goewert, R. R., and Lambowitz, A. M. (1979). *Cell (Cambridge, Mass.)* **18,** 1197–1207.
Marriott, A. C., Archer, S. A., and Buck, K. W. (1984). *J. Gen. Microbiol.* **130,** 3001–3008.
Meinhardt, F., Kempken, F., Kamper, J., and Esser, K. (1990). Curr. Genet. **17,** 89–95.
Merlos-Lange, A. M., Kanbay, F., Zimmer, M., and Wolf, K. (1987). *Mol. Gen. Genet.* **206,** 273–278.
Mery-Drugeon, E., Crouse, E. J., Schmitt, J. M., Bohnert, H.-J., and Bernardi, G. (1981). *Eur. J. Biochem.* **144,** 577–583.
Meunier, B., Tian, G. L., Macadre, C., Slonimski, P. P., and Lazowska, J. (1990). *In* "Structure, Function, and Biogenesis of Energy Transfer Systems" (E. Quagliariello, S. Papa, F. Palmieri, and C. Saccone, eds.), pp. 169–174. Elsevier Science, Amsterdam.
Michel, F., and Dujon, B. (1983). *EMBO J.* **2,** 33–38.
Michel, F., and Dujon, B. (1986). *Cell (Cambridge, Mass.)* **46,** 323.
Michel, F., and Lang, B. F. (1985). *Nature (London)* **316,** 641–643.
Minuth, W., Tudzynski, P., and Esser, K. (1982). *Curr. Genet.* **5,** 227–231.
Miyata, T., Hayashida, H., Kikuno, R., Hasegawa, M., Kobayashi, M., and Koike, K. (1982). *J. Mol. Evol.* **19,** 28–35.
Moody, S. F., and Tyler, B. M. (1990). *Appl. Environ. Microbiol.* **56,** 2441–2452.
Morl, M., and Schmelzer, C. (1990). *Cell (Cambridge, Mass.)* **60,** 629–636.
Nargang, F. E. (1985). *Exp. Mycol.* **9,** 285–293.
Nargang, F. E., Bell, J. B., Stohl, L. L., and Lambowitz, A. M. (1983). *J. Biol. Chem.* **258,** 4257–4260.
Oakley, K. M., and Clark-Walker, G. D. (1978). *Genetics* **90,** 517–530.
O'Connor, R. M., McArthur, C. R., and Clark-Walker, G. D. (1975). *Eur. J. Biochem.* **53,** 137–144.
O'Connor, R. M., McArthur, C. R., and Clark-Walker, G. D. (1976). *J. Bacteriol.* **126,** 959–968.

Osawa, S., Collins, R. A., Ohama, T., Jukes, T. H., and Watanabe, K. (1990). *J. Mol. Evol.* **30**, 322–328.

Osawa, S., and Jukes, T. H. (1989). *J. Mol. Evol.* **28**, 271–278.

Osawa, S., Ohama, T., Jukes, T. H., Watanabe, K., and Yokoyama, S. (1989). *J. Mol. Evol.* **29**, 373–380.

Penalva, M. A., and Garcia, J. L. (1986). *Curr. Genet.* **10**, 797–801.

Perez, M. J. L., and Turner, G. (1975). *FEBS Lett.* **58,**159–163.

Perlman, P. S., and Butow, R. A. (1989). *Science* **246**, 1106–1109.

Phelps, L. G., Burke, J. M., Ullrich, R. C., and Novotny, C. P. (1988). *Curr. Genet.* **14**, 401–403.

Ragnini, A., and Fukuhara, H. (1988). *Nucleic Acids Res.* **16**, 8433–8442.

Raito, M., Jalli, T., and Saraste, M. (1987). *EMBO J.* **6**, 2825–2833.

Saliola, M., Shuster, J. R., and Falcone, C. (1990). *Yeast* **6**, 193–204.

Sanders, J. P. M., Heyting, C., Verbeet, M. P., Meijlink, F. C. P. W., and Borst, P. (1977). *Mol. Gen. Genet.* **157**, 239–261.

Sanders, J. P. M., Weijers, P. J., Groot, G. S. P., and Borst, P. (1974). *Biochim. Biophys. Acta* **374**, 136–144.

Scazzocchio, C. (1986). *In* "Evolutionary Biology of the Fungi" (A. D. M. Rayner, C. M. Brasier, and D. Moore, eds.) , pp. 53–73. Cambridge Univ. Press, Cambridge.

Schafer, B., Merlos-Lange, A. M., Auderl, C., Welser, F., and Wolf, K. (1991). *Mol. Gen. Genet.* **225**, 158–167.

Sederoff, R. R. (1984). *Adv. Genet.* **22**, 1–108.

Sekiguchi, J., Ohsaki, T., Yamamoto, H., Koichi, K., and Shida, T. (1990). *J. Gen. Microbiol.* **136**, 535–543.

Seraphin, B., Boulet, A., Simon, F., and Faye, G. (1987). *Proc. Natl. Acad. Sci. U.S.A.* **84**, 6810–6814.

Shu, H.-H., Wise, C. A., Clark-Walker, G. D., and Martin, N. C. (1991). *Mol. Cell. Biol.* **11**, 1662–1667.

Shumard, D. S., Grossman, L. I., and Hudspeth, M. E. S., (1986). *Mol. Gen. Genet.* **202**, 16–23.

Shumard-Hudspeth, D. S., and Hudspeth, M. E. S. (1990). *Curr. Genet.* **17**, 413–415.

Shuster, J. R., Moyer, D., and Irvine, B. (1987). *Nucleic Acids Res.* **15**, 8573.

Skelly, P. J., and Clark-Walker, G. D. (1990). *Mol. Cell. Biol.* **10**, 1530–1537.

Skelly, P. J., and Clark-Walker, G. D. (1991). *J. Mol. Evol.* **32**, 396–404.

Skelly, P. J., Hardy, C. M., and Clark-Walker, G. D. (1991). *Curr. Genet.* **20**, 115–120.

Smith, M. Th., Yamazaki, M., and Poot, G. A. (1990). *Yeast* **6**, 299–310.

Smith, T. M., Saunders, G., Stacey, L. M., and Holt, G. (1984). *J. Biotechnol.* **1**, 37–46.

Sor, F., and Fukuhara, H. (1982). *Nucleic Acids Res.* **10**, 1625–1633.

Specht, C. A., Novotny, C. P., and Ullrich, R. C. (1983). *Exp. Mycol.* **7**, 336–343.

Stahl, U., Lemke, P., Tudzynski, P., Kuck, U., and Esser, K. (1978). *Mol. Gen. Genet.* **162**, 341–343.

Stark, M. J. R., and Milner, J. S. (1989). *Yeast* **5**, 35–50.

Seuoka, N. (1988). *Proc. Natl. Acad. Sci. U.S.A.* **85**, 2653–2657.

Tabak, H. F., Grivell, L. A. and Borst, F. (1983). *CRC Crit. Rev. Biochem.* **14**, 297–317.

Taylor, J. W. (1986). *Exp. Mycol.* **10**, 259–269.

Taylor, J. W., Smolich, B. D., and May, G. (1986). *Evolution* **40**, 716–739.

Tian, G.-L., Macadre, C., Kruszewska, A., Szczesniak, B., Ragnini, A., Grisanti, P., Rinaldi, T., Palleschi, C., Frontali, L., Slonimski, P. P., and Lazowska, J. (1991). *J. Mol. Biol.* **218**, 735–746.

Trinkl, H., Lang, B. F., and Wolf, K. (1985). *Mol. Gen. Genet.* **198**, 360–363.

Tudzynski, P., Duvell, A., and Esser, K. (1983). *Curr. Genet.* **7**, 145–150.

Tudzynski, P., and Esser, K. (1986). *Curr. Genet.* **10**, 463–467.

Turker, M. S., Domenico, J. M., and Cummings, D. J. (1987). *J. Mol. Biol.* **198**, 171–185.

Underbrink-Lyon, K., Miller, D. L., Ross, N. A., Fukuhara, H., and Martin, N. C. (1983). *Mol. Gen. Genet.* **191**, 512–518.

Van den Boogaart, P., Samallo, J., and Agsteribbe, E. (1982). *Nature (London)* **298**, 187–189.

Vincent, R. D., Goewert, R., Goldman, W. E., Kobayashi, G. S., Lambowitz, A. M., and Medoff, G. (1986). *J. Bacteriol.* **165**, 813–818.

Wallace, D. C., Ye, J., Neckelmann, S. N., Singh, G., Webster, K. A., and Greenberg, B. D. (1987). *Curr. Genet.* **12**, 81–90.

Waring, R. B., Brown, T. A., Ray, J. A., Scazzocchio, C., and Davies, R. W. (1984). *EMBO J.* **3**, 2121–2128.

Weber, C. A., Hudspeth, M. E. S., Moore, G. P., and Grossman, L. I. (1986). *Curr. Genet.* **10**, 515–525.

Weiller, G., Bruckner, H., Kim, S. O., Pratje, E., and Schweyen, R. J. (1991). *Mol. Gen. Genet.* **226**, 233–240.

Weiller, G., Schueller, C. M. E., and Schweyen, R. J. (1989). *Mol. Gen. Genet.* **218**, 272–283.

Wenzlau, J. M., Saldanha, R. J., Butow, R. A., and Perlman, P. S. (1989). *Cell* **56**, 421–430.

Wesolowski, M., Algeri, A., and Fukuhara, H. (1981). *Curr. Genet.* **3**, 157–162.

Wesolowski, M., and Fukuhara, H. (1981). *Curr. Genet.* **1**, 387–393.

Whatley, J. M., John, P., and Whatley, F. R. (1979). *Proc. R. Soc. London B* **204**, 165-187.

Wills, J. W., Troutman, W. B., and Riggsby, W. S. (1985). *J. Bacteriol.* **164**, 7–13.

Wilson, C., Ragnini, A., and Fukuhara, H. (1989). *Nucleic Acids Res.* **17**, 4485–4491.

Wolf, K., (1987). *In* "Gene Structure in Eukaryotic Microbes" (J. R. Kinghorn, ed.), pp. 69–91. IRL Press, Oxford.

Wolf, K., and Del Giudice, L. (1988). *Adv. Genet.* **25**, 185–308.

Wong, O. C., and Clark-Walker, G. D. (1990). *Nucleic Acids Res.* **18**, 1888.

Xiong, Y., and Eickbush, T. H. (1988). *Mol. Biol. Evol.* **5**, 675–690.

Yin, S., Burke, J., Chang, D. D., Browning, K. S., Heckman, J. E., Alzner-DeWeerd, B., Potter, M. J., and RajBhandary, U. L. (1982). *In* "Mitochondrial Genes" (P. Slonimski, P. Borst, and G. Attardi, eds.) pp. 361–373. Cold Spring Harbor Laboratory, Cold Spring Harbor, NY.

Yin, S., Heckman, J., and RajBhandary, U. L. (1981). *Cell (Cambridge, Mass.)* **26**, 325–332.

Zimmer, M., Luckeman, G., Lang, B. F., and Wolf, K. (1984). *Mol. Gen. Genet.* **196**, 473–481.

Zimmer, M., Welser, F., Oraler, G., and Wolf, K. (1987). *Curr. Genet.* **12**, 329–336.

Zinn, A. R., Pohlman, J. K., Perlman, P. S., and Butow, R. A. (1988). *Proc. Natl. Acad. Sci. U.S.A.* **85**, 2686–2690.

Structure and Function of the Higher Plant Mitochondrial Genome

Maureen R. Hanson and Otto Folkerts[1]
Section of Genetics and Development, Cornell University,
Ithaca, New York 14853

I. Introduction

The high frequency of recombination is perhaps the most unique feature of higher plant mitochondrial genomes in contrast to those of other species. Other characteristics of particular interest include a large coding capacity, the encoding of genes present in the nuclear genome of other eukaryotes, the existence of mutations which disrupt growth and pollen development, the phenomenon of RNA editing, *trans*-splicing, the import of tRNAs from the cytoplasm, and rapid alteration in genome organization. Although this review will touch on all of these features of plant mitochondrial genomes, emphasis will be placed on genome structure. A number of reviews have covered general aspects of plant mitochondrial genomes, cytoplasmic male sterility, RNA editing, tRNAs and encoded genes, etc, (Laughnan and Gabay-Laughnan, 1983; Hanson and Conde, 1985; Hanson *et al.*, 1985; Pring and Londsdale, 1985; Eckenrode and Levings, 1986; Maliga, 1986; Newton, 1988; Lonsdale, 1987; Levings and Brown, 1989; Levings and Dewey, 1988; Fauron *et al.*, 1991; Hanson *et al.*, 1989, 1991; Hanson, 1991; Schuster *et al.*, 1987, 1991). The evolution of higher plant mitochondrial genomes has been reviewed elsewhere (Gray, 1989a,b; Palmer *et al.*, 1990; Palmer, 1990, 1991).

II. Abnormal Phenotypes Specified by the Mitochondrial Genome

Mitochondrial lesions have revealed unexpected roles for the mitochondrial genome. There are three mutant phenotypes that have been ascribed

[1] Current address: Agricultural Biotechnology Laboratory, DowElanco, Midland, MI 48674.

129

to mitochondrial genomes: male sterility, sensitivity to a fungal toxin, and abnormal growth with striping of leaves (Newton, 1988). The existence of plants exhibiting male sterility or green/nongreen striped phenotypes that are inherited cytoplasmically indicates that mitochondria have a role in pollen and chloroplast development that is not yet understood. While cytoplasmic male sterility (CMS) is a trait described in many different species, at present nonchromosomal stripe mutants have been characterized only in maize. However, because the leaf striping at first glance appears due to a defect in the chloroplast genome, such mutants in other species may have been overlooked as candidates for mutations in mitochondrial genomes.

Mitochondrial mutations causing male sterility are relatively benign defects, resulting primarily in the prevention of normal reproductive development. Although floral morphology is affected in CMS genotypes of certain species, in most cases the CMS plant appears to be completely vigorous and normal but for the absence of pollen (reviewed by Laser and Lersten, 1972; Hanson and Conde, 1985). Most CMS genotypes are completely stable over many generations; all mitochondria in the plant appear to carry the mutated mtDNA. In contrast, the nonchromosomal stripe mutants that have been characterized have severe lesions in mitochondrial genes, and the plants are not homoplasmic with respect to the mutated mtDNA. Instead, these mutants are heteroplasmic, carrying both mutant and wild-type forms of the mitochondrial genome (Newton and Coe, 1986). The more severely affected tissue appears to carry a greater proportion of the mutant mtDNA. In this regard, the nonchromosomal stripe mutants are analogous to the human patients with mitochondrial myopathies. These patients also carry mutations in known mitochondrial genes, and more severely affected patients contain greater proportions of mutant mtDNA (Wallace et al., 1988; Wallace, 1989; Goto et al., 1990).

A. Cytoplasmic Male Sterility

Because this phenomenon has been the subject of many reviews (Laughnan and Gabay-Laughnan, 1983; Hanson and Conde, 1985; Pring and Lonsdale, 1985; Levings, 1990; Hanson et al., 1985, 1988, 1989, 1991), the mutations thought to cause CMS will be described only briefly here. In a later section, the creation of CMS-causing DNAs and reversion to fertility by mitochondrial genome recombination and reorganization will be considered.

The first two mitochondrial loci shown to encode cytoplasmic male sterility were in maize CMS-T and in Petunia (Fig. 1) (Dewey et al., 1986; Young and Hanson, 1987; Rasmussen and Hanson, 1989; Stamper et

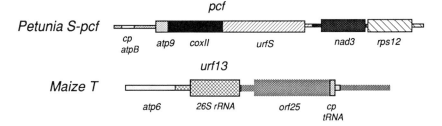

FIG. 1   CMS-associated loci in *Petunia* and maize. The *Petunia S-pcf* locus encompasses three genes (*pcf, nad3, rps12*) and contains a portion of the chloroplast *atpB* gene in the 5′ flanking region. In maize T, the locus is comprises *urf13* and *orf25*, which contains a portion of a chloroplast tRNA gene at its 3′ end. The *pcf* gene is derived from *atp9, coxII*, and an unidentified reading frame (urf) termed *urfS*, whereas *urf13* is derived from the 3′ flanking and coding region of the 26S rRNA gene.

al., 1987). Both these loci contain abnormal genes derived from multiple recombination events. In both cases, the abnormal genes are upstream of apparently normal mitochondrial genes. The possibility that disruption of the downstream genes plays a role in the sterile phenotype has not absolutely been ruled out, though most attention has been focused on the expression of the abnormal genes.

The gene products of maize *urf13* and *Petunia pcf*, which are 13- and 25-KD proteins, respectively, have been detected with antibodies (Dewey *et al.*, 1987; Wise *et al.*, 1987a; Nivison and Hanson, 1989). The maize *urf13* gene predicts a 13 kDa protein. Immunological data indicate that the 25-kDa *pcf* gene product derives from N-terminal processing. The maize 13-kDa protein fractionates as an integral membrane protein, whereas the 25-kDa *Petunia* protein appears in both the membrane and soluble fractions (Dewey *et al.*, 1987; Korth *et al.*, 1991; Nivison and Hanson, 1989).

Nuclear genes termed nuclear fertility restorer (Rf) alleles, which affect the phenotype of lines carrying the CMS mutations, are known in many species. In *Petunia*, a single dominant nuclear allele (Rf) can confer fertility on a line carrying the CMS-encoding mtDNA, whereas in maize CMS-T, two nuclear alleles *Rf1* and *Rf2* must both be present to restore fertility to a CMS line (reviewed by Hanson and Conde, 1985; Pring and Lonsdale, 1985). Transcripts of the maize CMS-T *urf13* locus and the *Petunia S-pcf* locus differ in amount between nonrestored and restored lines (Kennell and Pring, 1989; Kennel *et al.*, 1987; Pruitt and Hanson, 1991). Restored lines of both maize CMS-T and *Petunia* contain abundant transcripts that could encode the 13- and 25-kDa gene products, yet these proteins are greatly reduced in lines carrying restorer alleles (Forde and Leaver, 1980; Forde *et al.*, 1978; Dewey *et al.*, 1987; Wise *et al.*, 1987a, Nivison and

Hanson, 1989). A particular mystery concerning CMS-T is the function of the *Rf2* allele. Maize plants carrying the *urf13* gene and the *Rf1* alleles contain much reduced levels of the 13-kDa gene product but are male sterile (Dewey *et al.*, 1987); only when *Rf2* is also present are the plants fertile.

Despite the identification of loci encoding CMS, the molecular mechanism of the disruption in pollen development is still an unsolved question. In *Petunia*, CMS and fertile lines backcrossed to carry the same nuclear background ("isonuclear" lines) differ with respect to partioning of electron transport. CMS lines exhibit significantly lower transport through the cyanide-resistant alternative pathway than do fertile lines. Fertility-restored lines exhibit wild-type levels of alternative pathway activity (Connett and Hanson, 1990). Because the same nuclear restorer allele restores fertility and alternative pathway activity and reduces the 25-kDa protein abundance, one hypothesis is that the *pcf* gene product directly disrupts the regulation of alternative oxidase, which would then be proposed to be critical for pollen development. However, it is also possible that the reduction in alternative oxidase activity is a secondary effect of a CMS-encoded defect in some other critical activity.

In maize, the mechanism of pollen disruption is also unknown, though the membrane localization of the 13-kDa product and its role in toxin sensitivity has led to the suggestion that the protein causes mitochondrial membrane disruption in anthers. The 13-kDa protein has been shown to cause the characteristic sensitivity of maize CMS-T plants to a fungal toxin and the insecticide methomyl. When the 13-kDa protein is expressed in *Escherichia coli*, the bacterium becomes sensitive to these compounds (Dewey *et al.*, 1988, Braun *et al.*, 1989; Korth *et al.*, 1991). Evidently, in the presence of either one, the 13-kDa protein forms a membrane channel, effectively permeabilizing the membrane. One hypothesis is that the 13-kDa protein can be present harmlessly throughout the plant in the absence of fungal toxin or methomyl, but that in the anthers, some natural compound mimics toxin action, causing membrane defects (Flavell, 1974). Clearly, further analysis is necessary to discern whether the expression of 13-kDa protein truly causes pollen developmental disruption by permeabilizing mitochondrial membranes in reproductive tissue.

Since the description of the maize and *Petunia* CMS-encoding loci, efforts have also been made to identify CMS-encoding genes in other species. The discovery of abnormal open reading frames in both loci may have provided a clue for locating mutant genes in other species. However, because novel recombinant genes are common in mtDNAs (see Section III,E,1), a chimeric structure alone will not be sufficient to identify a CMS-encoding gene. It is essential to link the suspect gene to the phenotype in some way. This could be done in maize by finding that the *urf13* gene was deleted or altered in fertile revertants (Rottman *et al.*, 1987; Wise *et al.*, 1987b, Fauron *et al.*, 1990a,b) and that its gene products were affected by a nuclear restorer gene (Wise *et al.*, 1987a; Dewey *et al.*, 1987). In *Petunia*, the *S-pcf* locus segregated with the phenotype in somatic hybrid plants carrying recombinant mtDNAs (Boeshore *et al.*, 1985) and its gene products were also affected by the *Rf* allele (Nivison and Hanson, 1989; Pruitt and Hanson, 1991).

In several other species, abnormal loci that may encode CMS have been identified. Makaroff and Palmer (1988) found three rearrangements in the *Raphanus sativa* Ogura cytoplasm relative to the normal type that were in close proximity to mitochondrial genes (*coxI, atp6, and atpA*). Subsequent analysis ruled out a role of the *coxI* rearrangement in CMS, since it could be found in related fertile species (Makaroff *et al.*, 1991). An alteration in *atpA* transcription as a result of the rearrangement was also not found to be strictly correlated with CMS (Makaroff *et al.*, 1991). However, the Ogura CMS *atp6* gene was found to be cotranscribed with an upstream 105-codon unidentified reading frame (Makaroff *et al.*, 1989). Because the nuclear restorer gene has no obvious effect on transcription of this complex *atp6* locus (Markaroff and Palmer, 1988), further analysis at the protein level will be needed to determine whether it is correlated with CMS. *Atp6* is also implicated in CMS in rice. A normal *atp6* gene is present in rice CMS-Bo, as well as a chimeric gene termed *urf-rmc*, which contains the first 170 N-terminal codons of *atp6* fused to an unidentified reading frame. Transcription of *urf-rmc* is altered in fertility-restored lines (Kadowaki *et al.*, 1990).

The maize CMS-C mitochondrial genome contains several aberrant genes, none of which have been definitively correlated with the sterile phenotype. The CMS-C *atp6* gene has an N-terminal fusion with 13 codons of *atp9* and 141 codons derived from chloroplast DNA, whereas the *atp6* 5' flanking region and a portion of the *atp6* coding region are fused to *coxII* (Levings and Dewey, 1988; Fragoso *et al.*, 1989).

In sunflower (*Helianthus annus*) the difference between the normal and the CMS mitochondrial genomes are restricted to one region next to the *atpA* gene (Siculella and Palmer, 1988). A 12-kb inversion is found adjacent to *atpA*, and 5kb of DNA present at one endpoint of the inversion in the CMS genome is absent in the fertile genome. Further sequencing revealed that *atpA* in CMS lines is cotranscribed with another open reading frame, *urf522* (Kohler *et al.*, 1991). Whether the 533 nucleotides of *urf522* encode a 16-kDa protein found in CMS but not the fertile progenitor sunflower lines (Horn *et al.*, 1991) is an intriguing possibility, but has not yet been verified.

The most unique effect of a nuclear restorer locus on a putative CMS-encoding DNA was found by Mackenzie and Chase (1990) in *Phaeseolus vulgaris*. Fertility-restored lines permanently lose a mtDNA region present in nonrestored CMS lines, and all progeny are fertile. This region contains a mtDNA sequence not found in normal fertile lines. In fertility-restored lines of all other species that have been analyzed at the molecular level, the restoration is not accompanied by loss of DNA, and progeny of a restored plant are sterile if they do not carry the restorer allele.

In maize CMS-S two linear plasmids, S1 (6.4 kb) and S2 (5.4 kb), have been observed. These plasmids were not found in the N, C, and T cytoplasms, sparking interest in them as possible causal factors of sterility. However, at present the correlation of these plasmids and expression of their genes with the CMS phenotype remains tenuous. No effect of the single dominant CMS-S nuclear restorer allele *Rf3* has been detected on proteins synthesized by CMS-S mitochondria, though proteins are synthesized in CMS-S lines that are not found in other maize genotypes (Forde and Leaver, 1980). The S plasmids are retained in *Rf3*-restored lines (Laughnan and Gabay-Laughnan, 1983). There is evidence that a 187-nt sequence repeated at the terminus of these plasmids can recombine with an identical sequence present in the main mitochondrial genome. Such a recombination event results in integration and linearization of the mtDNA (Levings *et al.*, 1980; Schardl *et al.*, 1985). In some fertile revertants, free S1 and S2 are lost by integration into the main mtDNA (Schardl *et al.*, 1984, 1985). However, free S2 and/or free S1 is maintained in some revertants (Escote *et al.*, 1985; Ishige *et al.*, 1985). Thus whether S1 and/or S2 is responsible for CMS is still an open question. A CMS-associated locus present in the main mtDNA in CMS-S could remain undetected.

B. Nonchromosomal Stripe Mutants

A maize mutant exhibiting abnormal morphology, poor growth, and leaf striping was first reported by Shumway and Bauman (1967). Two such nonchromosomal (NCS) mutations were since described by Coe (1983) and were shown by Newton and Coe (1986) to carry mtDNA alterations. From the present data, it appears likely that NCS mutations are altered in essential mitochondrial genes, and mutant plants survive because they carry a mixture of wild-type and mutant mtDNAs. The NCS2 mutant, which has an alteration in the *nad4* gene (K. Newton, personal communication), exhibits reduction in a 24-kDa polypeptide synthesized by wild-type mitochondria (Feiler and Newton, 1987). The NCS3 mutant has an alteration in the cotranscribed ribosomal protein S3/L16 genes (Hunt and Newton, 1991), whereas NCS5 and NCS6 have different deletions in the *coxII* gene (Lauer *et al.*, 1990; Newton *et al.*, 1990). Several NCS mutations have been traced to recombination at short repeated sequences, as discussed in Section III,D,3.

The yellow and white striping of leaves, which is accompanied by plastid structural abnormalities and reduced carbon fixation, remains an unexplained but extremely intriguing facet of the nonchromosal stripe mutations. Evidently, defects in mitochondrial function can affect chloroplast morphology and function (Newton *et al.*, 1989). Nonchromosomal stripe mutations may thus be a means to examine interactions between chloroplasts and mitochondria.

III. Structure of the Plant Mitochondrial Genome

A. Physical Characterization

Early characterization of plant mtDNA by physical methods revealed that the buoyant density of mtDNA and therefore the G+C values were relatively constant even in widely divergent species (Wells and Ingle, 1970; Bailey-Serres *et al.*, 1987). The similarity in GC content is in contrast to the kinetic complexity of mitochondrial genomes. In 1981 Ward *et al.* reported unexpected results from renaturation kinetics: different species in the cucurbit family ranged in size from 300 kb for watermelon to 2400 kb for muskmelon. The cucurbit mtDNAs were shown to contain less than 5 to 10% repetitive sequences. The values determined for the smaller cucurbit genomes by reassociation experiments agreed reasonably well with genome size estimates based on summation of restriction fragment sizes (Ward *et al.*, 1981), and the complexity of restriction digests of cucurbits (Stern and Newton, 1985) is consistent with the renaturation data.

Plant mtDNA preparations have also been analyzed by electron microscopy, with variable results. Despite one early, currently discounted, report describing the extreme uniformity of the pea mtDNA (30-μm circles, Kolodner and Tewari, 1972), most plant mtDNA preparations have been shown to consist of a large range of circular and linear molecules (Synenki *et al.*, 1978; Sparks and Dale, 1980; Levings *et al.*, 1979; Levings and Sederoff, 1983; Kool *et al.*, 1985; Fontarneau and Hernandez-Yago, 1982; Bailey-Serres *et al.*, 1987). As will be seen below, the circular molecules that are predicted from restriction maps of many genomes are larger than 200 kb, yet the molecules seen in the electron microscope are smaller, generally ranging from a few kilobases to 100 kb. The lack of circular molecules over 200 kb is likely due to breakage during isolation and preparation for electron microscopy. However, circular forms less than 100 kb, which would be expected to be less fragile and are also predicted to be abundant according to mapping data of several species, form only a small percentage of the DNA seen by microscopy.

Circular molecules less than 20 kb in size have been seen by several investigators in the microscope and upon electrophoresis (Brennicke and Blanz, 1982; Bailey-Serres *et al.*, 1987; Kool *et al.*, 1985). These molecules are lower in abundance than the so-called mitochondrial plasmids (see Section III,B). Although linear fragments could easily be generated from breakage of mitochondrial chromosomes during the isolation, it is not likely that small circular molecules similarly arose as artifacts. In *Petunia*, small molecules are seen in the microscope, but are not predicted from mapping data of the main mtDNA (Kool *et al.*, 1985; Folkerts and Hanson, 1989). Some of these molecules may represent rare recombination events between sequences present nearby on larger molecules. Small circular molecules carrying *atpA* pseudogenes that may have resulted from homologous recombination events have been observed in *Oenothera* (Schuster and Brennicke, 1986).

The interpretation of electrophoretic patterns of plant mtDNAs on pulse-field gels is not straightforward (our unpublished data and several personal communications). Levy *et al.* (1991) examined mtDNA in suspension cultures of Black Mexican Sweet (BMS) maize. When DNA prepared from protoplast-derived mitochondria lysed in agarose was electrophoresed in CHEF (Chu *et al.*, 1986) and Eckhardt (1978) gels, most DNA remained in the wells, which the authors interpreted to indicate the presence of large relaxed circular molecules. Partial digestion with *Sfi*I permitted visualization of large restriction fragments, though none were in the 500- to 600-kb range expected for a maize mastercircle according to the restriction map (see Section III,C,3). A 120-kb circular DNA molecule, detectable on Southern transfer from Eckhardt gels, was found to be colinear with the maize B37N genome map, except at the points of circularization, and was estimated to represent 10–20% of the mtDNA (Levy *et al.*, 1991).

When watermelon mtDNA preparations were examined by fluorescence microscopy of pulsed-field gels, Bendich and Smith (1990) observed linear molecules longer than 1200 kb present near the top of the gel. No DNA of the predicted genomic size of 330 kb was seen; instead most of the DNA migrated as linear molecules of 50–100 kb, as calibrated by yeast chromosomes and λ phage multimers (Bendich and Smith, 1990). W. Hauswirth (personal communication) has also observed apparently linear molecules upon electrophoresis of *Brassica* mtDNA, rather than the expected circular molecules.

It is evident that a major inconsistency exists between present microscopic and electrophoretic characterizations of mtDNAs and restriction mapping analysis. When a genome map is assembled only from cosmid walking, a linkage group could possibly be constructed that does not correspond to the major configuration of the genome *in vivo*. Alternatively, it may be that the physical state of plant mtDNA *in vivo*—perhaps complexed with proteins or associated with membranes—prevents isolation of intact molecules by current methods, and the predictions of molecular configuration made by restriction mapping methods may turn out to represent the actual genome organization *in vivo*.

B. Mitochondrial Plasmids

Relatively abundant small nucleic acid molecules (linear and circular DNA as well as double-stranded RNAs) have been observed in mitochondria of several plant species. Maize, sorghum, broad bean, and *Brassica* mitochondria have been shown to contain small linear or circular molecules lacking sequence homology with the main mtDNA; they are currently thought to be autonomously replicating mtDNA molecules (Kemble and Bedbrook, 1980; Dale, 1981; Dale *et al.*, 1981; Palmer *et al.*, 1983; Schardl *et al.*, 1984, 1985; Chase and Pring, 1985; Bailey-Serres *et al.*, 1987; Turpen *et al.*, 1987; Wahleithner and Wolstenholme, 1988a).

Nucleotide sequences of many minicircles and minilinear molecules are available (Levings and Sederoff, 1983; Hansen and Marcker, 1984; Ludwig, *et al.*, 1985; Paillard *et al.*, 1985; Sederoff *et al.*, 1986; Wahleithner and Wolstenholme, 1987, 1988a; Smith and Pring, 1987; Zabala and Walbot, 1988). Several such linear molecules carry terminal proteins (Erickson *et al.*, 1985; Kemble and Thompson, 1982) or open reading frames of unknown function. Protein products of the maize S plasmids do exist, as shown by immunological studies using antisera raised to S1 ORF3 and S2 ORF1 (Manson *et al.*, 1986; Zabala *et al.*, 1987; Zabala and Walbot, 1988). Only for a 2.3-kb maize plasmid has sequencing revealed a possible reason for the presence of the plasmid in all maize lines examined. This plasmid contains the only known functional copy of a tRNA[Trp] gene (Leon *et al.*, 1989).

In mitochondria of several species, automously replicating single- and double-stranded RNAs have been reported (Finnegan and Brown, 1986; Kemble et al., 1986; Rogers *et al.*, 1987; Fairbanks *et al.*, 1988; Kim and Klassen, 1989). Recently Monroy *et al.* (1990) showed that *Brassica* double-stranded RNAs were not actually associated with mitochondria, contrary to a prior report (Kemble *et al.*, 1986).

C. Physical Mapping of Mitochondrial Genomes by Restriction Mapping

Restriction site maps have now been produced for a number of mitochondrial genomes (see Table I). The present convention for describing a

TABLE I

Multipartite Organization of Mitochondrial Genomes Predicted by Complete or Partial Restriction Maps

Species	Mastercircle	Recombination repeats	Subgenomic circles	Reference
Brassica campestris (turnip)	218	Pair direct, 2 kb	83, 135 kb	Palmer and Shields, 1984
B. napus	221	Pair direct, 2 kb	98, 123 kb	Palmer and Herbon, 1988
B. nigra (black mustard)	231	Pair direct, 7 kb	96, 135 kb	Palmer and Herbon, 1986
B. hirta (white mustard)	208	None		Palmer and Herbon, 1987
B. oleracea (cauliflower)	219	Pair direct, 2 kb	170, 49 kb	Palmer and Herbon, 1988 Chetrit *et al.*, 1984
Raphanus sativa (radish)	242	Pair direct, 17 kb	103, 139 kb	Palmer and Herbon, 1986
R. sativa Ogura cytoplasm	257	Pair direct, 10 kb	127, 130 kb	Makaroff and Palmer, 1988
Spinacea oleracea (spinach)	327	Pair direct, 6 kb	93, 234 kb	Stern and Palmer, 1986
Triticum vulgare (wheat)	430	2 trios, 8 pairs	Many	Quetier *et al.*, 1984, 1985; Lejeune *et al.*, 1987; Falconet *et al.*, 1985; Bonen and Bird, 1988

Rye	300	At least 1 trio		Coulthart et al., 1990
Helianthus annus (normal sunflower)	300	Pair direct, 12 kb	64, 236 kb	Siculella and Palmer, 1988
Sunflower CMS	305	Pair direct, 12 kb	64, 241 kb	Siculella and Palmer, 1988
Maize N	570	5 pairs direct: 0.5, 0.7, 5.2, 11, 14 kb; 1 pair inverted: 14 kb	Many	Lonsdale et al., 1984; Fauron et al., 1989; Havlik and Fauron, 1988
Maize CMS-T	540	4 direct: 0.7, 4, 6, 11 kb; 1 trio; 1.5 kb direct and indirect	Many	Fauron et al., 1989
Maize CMS-T Revertant	705	2 pairs direct: 4.6, 11 kb; 2 trios: 0.7, 6 kb direct and indirect; 1 quadruple with 1.5 kb direct and indirect	Many	Fauron et al., 1990a, 1991
Beta vulgaris (sugarbeet)	386	5 pairs	Many	Brears and Lonsdale, 1988
Petunia fertile 3704	443	1 trio: 5 kb direct and indirect	4, 443 and 2, 244 kb; isomers, 199 kb	Folkerts and Hanson, 1989
Petunia CMS 3688	807	1 quadruple: 1.2 kb direct and indirect	398, 407, 266	Folkerts and Hanson, 1991

mapped mitochondrial genome is a large "mastercircle" which is the smallest size that can contain all of the mapped DNA. Some of the smaller genomes have been mapped by double-digestion/hybridization methods, but the complexity of the larger genomes has required cosmid libraries and walking techniques. Individual cosmids could represent low-abundance configurations of mtDNA. Thus the circular maps obtained by cosmid walking in libraries of mtDNAs may not be the primary configuration of the genome, unless the investigator has carefully checked each cosmid clone by hybridization to total mtDNA. The maps of genomes produced directly by double-digestion and hybridization are more difficult to discount than the cosmid walking maps.

1. Recombination Repeats—Definition and Discovery

Analysis of plant mtDNAs has revealed the presence of recombination repeats, which have been defined by Stern and Palmer (1984b) as those sequences that are present in two copies relative to other sequences in the genome, but occur in four genomic environments as the result of apparent recombination occurring between the copies. The hallmark of a recombination repeat is its location in these four genomic environments, which can be detected by digesting mtDNA with an enzyme that lacks a site in the repeat, followed by hybridization with a probe internal to the repeat (Fig. 2). Each of the resultant four fragments represents one of the genomic

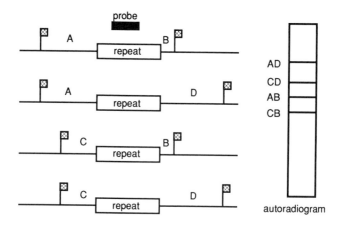

FIG. 2 Structure of four recombination repeat versions. Four hypothetical restriction maps surrounding a sequence repeated twice in a mitochondrial genome. The finding of four such mtDNA regions would identify the repeat as a "recombination repeat," because the four regions could be derived by recombination. Flags indicate a restriction site, and letters designate different DNA regions flanking the repeat. If a probe internal to the repeat is labeled and hybridized to mtDNA digested with the restriction enzyme shown, the resultant autoradiogram would exhibit four hybridizing bands corresponding to each repeat environment.

environments or repeat "versions." These repeat versions are four combinations of sequences flanking the repeat copies. In two mapped genomes, *Petunia* line 3704 and maize CMS-T, instead of two, there are three copies of the same sequence. A trio of repeats predicts nine genomic environments, and nine repeat versions have indeed been observed (Folkerts and Hanson, 1989; Fauron *et al.*, 1989).

The first recombination repeat was identified by Palmer and Shields (1984) in the mapping of the *Brassica campestris* mitochondrial genome. Subsequently, recombination repeats have been shown to be present in all genomes mapped to date, with the exception of *B. hirta* (Palmer and Herbon, 1987) and *Marchantia polymorpha* (Ohyama *et al.*, 1990). In order to detect a recombination repeat, a restriction enzyme that does not cut within the repeat must be used. Most known recombination repeats are in the range of 1–10 kb; the smallest known is 266 bp (Schuster and Brennicke, 1986). If a repeat is very large, the standard digestion/ hybridization method for investigating genomic environments cannot be used. Two genomes contain very large repeats—a maize CMS-T fertile revertant genome carries a 165-kb repeat (Fauron *et al.*, 1990a) and the *Petunia* CMS genome contains a 252-kb repeat (Folkerts and Hanson, 1991). Recombination in these large repeats could go undetected.

Not all repeated sequences in plant mtDNAs are recombination repeats. In addition to repeats which recombine, repeated sequences that are present in two but not four environments have been observed. For example, we have not been able to detect any evidence of recombination of the two *atp9* genes in *Petunia* line 3704 (Rothenberg and Hanson, 1987a), though the appropriate restriction digestions and hybridizations were performed (Rothenberg and Hanson, 1987b). A sequence repeated upstream of several *Petunia* mitochondrial genes does not undergo frequent recombination (Folkerts and Hanson, 1989; Pruitt and Hanson, 1989). Such a repeated sequence has also been observed in *Oenothera* (Schuster and Brennicke, 1987c). A number of minor repeat families, 100 to 450 nt in size, have also been observed in *Brassica* (Shirzadegan *et al.*, 1991).

An actual ongoing process of recombination to produce the multiple genomic environments has not yet been formally demonstrated. However, the existence of DNA molecules that contain a repeat flanked by different sequences is consistent with recombination, and is currently the accepted model. For convenience in discussing recombination repeats, each different genomic environment will be referred to as a repeat "version."

2. Structure of Recombination Repeats in Various Species

Brassica campestris and *B. oleracea* both have 2-kb repeated sequences, which are indistinguishable by hybridization analysis. In both species the same repeat versions (having the same combinations of flanking sequences) exist on the mastercircle. This contrasts with the mitochondrial genomes of *B. nigra, R. sativa* (normal and Ogura CMS), which contain the 2-kb sequence as a single copy sequence (Palmer and Shields, 1984; Stern and Palmer, 1984b; Palmer, 1988).

In both *Raphanus* mitochondrial genomes a 10-kb sequence is repeated, and a 7-kb repeat, which is contained entirely within the 10-kb repeat of *Raphanus*, was observed in *B. nigra* (Palmer and Herbon, 1986; Makaroff and Palmer, 1988). Interestingly, the repeat versions that occur on the mastercircle in *B. nigra* were found on the subgenomic circle in *R. sativa* (both the CMS and the normal cytoplasm) and vice versa. Such a different distribution of repeat versions could be the result from two sets of inversions postulated to have occurred during the evolutionary divergence of the two cytoplasms.

In the two different maize mtDNAs mapped to date [normal (N) and CMS-T] many recombination repeats were identified, but only one recombination repeat is in common between the two genomes (Fauron and Havlik, 1989; Fauron *et al.*, 1989). The N cytoplasm of maize contains five direct repeats and one inverted repeat of 14 kb (Lonsdale *et al.*, 1984). The T cytoplasm contains three direct repeats occurring in two copies and one repeat that occurs in three copies. Two of the trio repeat copies are in direct orientation, whereas the third repeat is in an inverted orientation relative to the other two. The recombination repeat in common between the two cytoplasms in the 11-kb repeat of the T cytoplasm, which corresponds to the 10-kb repeat of the N cytoplasm according to hybridization and restriction analysis (Fauron *et al.*, 1989).

In several genomes, known genes flank or extend into the recombination repeat. In wheat (*Triticum aestivum*), both the *rrn18* and the *rrn26* genes occur in nine different genomic environments, resulting from recombination occurring between any two or three copies of a repeat adjacent to *rrn18* (Lejeune *et al.*, 1987) or encompassing the gene *rrn26* (Falconet *et al.*, 1985). The *atp6* gene is part of a 1.4-kb repeat present in two copies in the wheat mitochondrial genome. As a result, the wheat *atp6* gene occurs in four genomic environments (Bonen and Bird, 1988). In pea, *Pisum sativum*, the *atpA* gene is contained in a 1.7-kb recombination repeat (Morikami and Nakamura, 1987). The *rrn26* gene is present in the 6-kb spinach recombination repeat (Stern and Palmer, 1986). A 6-kb repeat in sugarbeet partially overlaps with the *rrn26* gene (Brears and Lonsdale, 1988). The 5′ transcript termini of the *Petunia coxII* genes are located in a recombination repeat (Pruitt and Hanson, 1989), and the *Petunia rps19* gene is located within the repeat (Conklin and Hanson, 1991). The *coxII* genes are contained in the *B. campestris* and *B. oleraceae* 2-kb recombination repeats (Stern and Palmer, 1984b). In *Oenothera*, one sequence block of 657 bp was found upstream of the *coxI* and *coxII* genes. Both genes appear to be transcribed from an identical promoter situated in this repeated element. Recombination within this repeat produces four combinations of flanking sequences. Each gene, therefore, occurs in two genomic contexts (Hiesel *et al.*, 1987).

3. Genomic Configurations Predicted by the Recombination Repeat Versions

The multiple genomic environments engendered by recombination at repeats have been detected in two ways: hybridization analysis and cosmid mapping. When four different cosmids are found to contain a repeated sequence flanked by different combinations of sequences, further verification by hybridization is necessary to rule out cloning artifacts or cloning of rare recombined regions. For the larger mapped genomes, hybridization of a repeat probe has been essential to confirm that the repeat version actually exists in abundance *in vivo*. In the simple, single-repeat *Brassica* genomes, repeat-containing fragments were present in substoichiometric amounts relative to each entirely unique fragment on ethidium bromide gels (Palmer and Shields, 1984; Stern and Palmer, 1984b), as would be predicted from a multipartite model of the plant mitochondrial genome.

Depending on their orientation (direct or indirect), recombination repeats are predicted to give rise to different isomers of the main mitochondrial genome or produce subgenomic cirlces. The *Petunia* line 3704 mitochondrial genome, because it contains three copies of one repeat, two in direct orientation and a third in indirect orientation, can serve as an example of the different configurations predicted by the existence of the nine genomic environments of the recombination repeat (Fig. 3). In the more complex genomes with multiple recombination repeats, such as maize, a pair of repeats cannot always be described as direct or indirect, since their orientation may shift from one subgenomic circle to another.

In the case of a genome such as *Petunia* line 3704, where full-size mastercircle isomers are predicted, a single isomer must be selected arbitrarily as the map "standard." In those species where more than two copies of a single recombination repeat exist, or where more than one different sequence is active in recombination, the number of predicted subgenomic molecules and isomers can quickly grow very large. A summary of the known recombination repeats, the mastercircle size, and the subgenomic circles they predict are provided in Table I.

If the configuration of mtDNA is indeed multipartite, then it is reasonable to consider the question of the number of molecules carried by a single mitochondrion. The amount of mtDNA per plant mitochondrion has not been studied in many species. One report by Bendich and Gauriloff (1984) indicates that individual mitochondria carry one genome equivalent

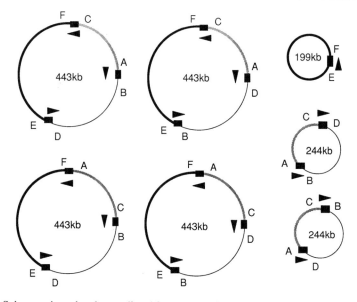

FIG. 3   Subgenomic molecules predicted for a mastercircle (*P. hybrida* line 3704) that carries a trio of repeats (Folkerts and Hanson, 1989). The physical map produced by the cosmid walking method indicates three copies of a sequence (shown as black box) that are in direct and inverted orientation with respect to each other (arrows). Four full-size isomers result from recombination of the inverted repeats, whereas smaller subgenomic molecules result from recombination of direct repeats. Letters indicate the different sequences flanking the repeats, and arcs of different line width represent regions between repeats.

or less on the average. The concept of an individual mitochondrion is also problematical, since mitochondria in yeast (Stevens, 1981), and probably in plants, appear to be a syncytium, in which individual bodies are continuously fusing and splitting apart. The number and size of discrete mitochondrial bodies seen in the microscope vary in different tissues (Juniper and Clowes, 1965; Warmke and Lee, 1978; Bendich and Gauriloff, 1984). It is possible that a discrete mitochondrial body at one particular moment may contain less than one genomic equivalent of mtDNA; however, because such a body could fuse with many others over a short time span, a complete complement of genes would, in effect, be present.

4. Mechanism of Recombination of Recombination Repeats

For the ensuing discussion, it will be assumed that the recombination products detected in cosmid libraries are the result of an active recombination process. As described earlier, the presence of these recombination products is readily explained by ongoing recombination at the so-called recombination repeats, but there is so far no proof that these molecules are maintained by recombination, rather than by replication of molecules formed by recombination events that occurred in the historical past.

Two contrasting hypotheses have been put forward for the mechanism of recombination across the repeats: simple homologous recombination or recombination requiring specific sequences. Because no *in vitro* recombination system yet exists, these theories have not yet been experimentally tested. However, there is evidence in favor of some specificity of recombination.

As mentioned above, some but not all known repeated sequences in plant mtDNAs are present in the multiple environments that would result from recombination. Both the *P. hybrida* and the *R. sativa* mitochondrial genomes contain repeated *atp9* genes that do not serve as substrates for recombination (Rothenberg and Hanson, 1987b, 1988; Makaroff and Palmer, 1988). In the case of the *P. hybrida atp9* genes, 414 bp of 5′ flanking sequence, coding region, and 3′ flanking sequence were duplicated (Rothenberg and Hanson, 1988). The broad bean (*Vicia faba*) mitochondrial genome contains a 469-nt sequence duplication containing the *rps14* coding sequence and 5′ flank that also does not serve as a recombination repeat (Wahleithner and Wolstenholme, 1988b). In both cases, the length of the sequence duplication is considerably more than the length of the smallest recombination repeat characterized to date, the 266-bp *atpA* duplication in *Oenothera* (Schuster and Brennicke, 1986). This indicates that absence of recombination at certain repeated sequences is not due to limited length of the sequence duplication.

Clearly, recombination repeats and repeated sequences that do not recombine are being distinguished from one another; this would suggest that the mechanism of recombination across repeats has some site-specificity. However, the poor conservation of recombination repeat sequences between closely related mitochondrial genomes has not permitted detection of any consensus sequence for recombination. From analysis of the complete sequences of several repeats (5.27 kb repeat of maize, Houchins *et al.*, 1986; *atpA* repeat in pea, Morikami and Nakamura, 1987), there is also no evidence that recombination repeats encode their own recombinases. Cryptic, degenerate target sequences could exist and yet still be undetected. An alternative model combining elements of site-specific and general mechanisms can also be envisioned. Perhaps recombination occurs within very short specific target sequences only when those are present in larger regions of similarity.

5. Function of Recombination Repeats

Recombination repeats in plant mtDNAs are not in general an essential feature, as one plant mitochondrial genome (*Brassica hirta*) does not carry any. Whether repeats that recombine are essential in the genomes where they reside is unknown. No known function has been ascribed to any of the characterized repeats, other than the functions of the genes contained within them. Recombination repeats could serve to alter the regulation of genes in the flanking sequences, or the genes contained within the recombination sequences. Transcription start sites are known to occur within recombination repeats (Pruitt and Hanson, 1989). One way that repeat recombination could regulate gene expression is by altering the adundance of the different repeat environments in different tissues or developmental times. Although there is no evidence that the abundance of the particular repeat versions vary from one tissue to another in plants, this question has not been thoroughly investigated. Mitochondrial gene expression can differ in different tissues; proteins synthesized by mito-chondria isolated from various maize tissues are not identical (Newton and Walbot, 1985).

Another possible function of the recombination repeats is the creation of subgenomic molecules that may give the mitochondrial genome more viable options for rearrangement. The function of recombination repeats in massive rearrangements of mitochondrial genomes will be discussed further below.

D. Origins of Replication

Because the actual configuration of the mitochondrial genome *in vivo* has not been demonstrated directly, the relative proportions of subgenomic circles have not been directly measured. Nevertheless, the stoichiometry of the recombination repeats determined by hybridization or direct exami-nation of DNA in some of the simpler genomes (Palmer and Shields, 1984; Folkerts and Hanson, 1989) suggests that the different versions of the repeats are relatively abundant. Multiple subgenomic circles could be maintained by an ongoing process of recombination, or by independent replication of different circles, or by both.

Whether the predicted subgenomic circles can replicate independently depends partly on the location and number of origins of replication. At present only two putative origins of replication have been mapped on a plant mitochondrial mastercircle, on *Petunia* fertile line 3704 (DeHaas *et al.*, 1992). Four *Petunia* mtDNA sequences were first chosen for analysis by their functioning as ARS sequences in yeast. Two of the sequences could be shown to function in a *Petunia* mitochondrial *in vitro* replication system (DeHaas *et al.*, 1992). If these are the only replication origins in *Petunia* mtDNA, then not all subgenomic molecules could replicate independently, since the predicted 199-kb circle (Fig. 3) would not carry an origin. Future analyses of origins of replication in other species where mtDNA maps are available may reveal whether only the master circle replicates.

Another source of mitochondrial replication origins is the small plasmid-like molecules found in many genomes. Because these DNAs are usually autonomous of the mastercircular DNA, they must carry their own replication origins. Replication origins of two broad bean plasmids have been mapped by examining replicative intermediated by electron microscopy (Wahleithner and Wolstenholme, 1988a). Further characterization of replication origins on the primary mtDNA molecules as well as plasmids will be valuable for understanding how plant mitochondrial genome structure is maintained.

E. Rare Recombination Events

Recombination across recombination repeats is thought to occur frequently, which, along with replication, results in an equilibrium of the genomic environments. Sequence analysis of mitochondrial genes and genomic regions provides evidence that infrequent recombination events also occur to give rise to chimeric mitochondrial genes and to the incorporation of DNA from the nuclear and chloroplast genomes. Because they often cannot be detected by standard hybridization analysis or have occurred in evolutionary time, these recombination events are considered rare in contrast to those occurring at the recombination repeats.

1. Chimeric Genes

Chimeric genes are present in many mitochondrial genomes (Table II). In some cases, by analyzing homologous genes in the same or closely related genomes, short regions of homology can be detected where recombination may have occurred to produce the chimeric gene. For example, there is a short stretch of similarity between the normal *Petunia atp9* and *coxII* genes, which may have been involved in the recombination event which produced the chimeric *pcf* gene, where the *atp9* and *coxII* coding regions are fused (Pruitt and Hanson, 1989).

If reciprocal recombination has occurred, two chimeric regions should be detected, but usually only one of the predicted regions can be located. Either the original event was nonreciprocal or the other product was lost. Whatever original event may have occurred, the detected chimeric gene or locus is now stable in its resident genome.

TABLE II

Chimeric Genes and Pseudogenes in Plant Mitochondrial Genomes

Gene	Description	Reference
Wheat *atp6*	N-terminal extension of reading frame	Bonen and Bird, 1988
Tobacco *atp6*	N-terminal extension of reading frame	Bland *et al.*, 1987
Maize *coxII*	N-terminal fusion with 122 bp of *atp6*	Dewey *et al.*, 1986
Sorghum 9E *coxI*	C-terminal extension of 101 codons	Bailey-Serres *et al.*, 1986a,b
Petunia coxII-2	C-terminal extension of 56 codons	Pruitt and Hanson, 1989
Petunia pcf	*atp9/coxII/urf*[a]	Young and Hanson, 1987
Petunia atp9 (two genes)	One pseudogene One fusion with a 5' *urf* and a 3' *urf*	P. L. Conklin and M. R. Hanson, unpublished
Soybean *atp6* (two genes)	Replacement of sequence with 27 bp of *atp9* Amino-terminal fusion with 113 bp of *coxII*	Grabau *et al.*, 1988
Rice *urf-rmc*	*atp6* with C-terminal *urf*	Kadowaki *et al.*, 1990
Oenothera atpA	Three pseudogenes	Schuster and Brennicke, 1986
Maize CMS-C *atp6* *coxII*	N-terminal fusion with *atp9* and cpDNA N-terminal fusion with *atp6*	Levings and Dewey, 1989 Fragoso *et al.*, 1989
Maize CMS-T *urf13*	Coding and flanking region 26S rRNA gene	Dewey *et al.*, 1986
Pea *atpA*	*atpA* with C-terminal open reading frame	Morikami and Nakamura, 1987
Oenothera rps13	889 bp 23S rRNA plus 3' portion of *rps13*	Schuster and Brennicke, 1987a
Arabidopsis tRNA[tyr]	*tRNA*[phe] gene or pseudogene with sequence upstream of *tRNA*[ala] gene	Chen *et al.*, 1989

[a] Unidentified reading frame.

2. Chloroplast and Nuclear DNA Sequences in Mitochondrial Genomes

Chloroplast DNA (cpDNA) sequences are commonly found in mitochondrial genomes (Stern and Palmer, 1984a). The first well-documented report of chloroplast DNA in plant mtDNA showed that the gene for the large subunit of ribulose bisphosphate carboxylase/oxygenase and a 12-kb cpDNA fragment containing several chloroplast genes were present in the maize N mitochondrial genome (Stern and Lonsdale, 1982; Lonsdale *et al.*, 1983). Since then cpDNA has been found in transcribed flanking regions (*pcf*, Fig. 1) or in coding regions (maize *orf25*, Fig. 1). Both transcribed genes and nontranscribed cpDNA sequences have been found in mtDNA (Stern and Lonsdale, 1982; Jubier *et al.*, 1990; Marechal *et al.*, 1987; Wintz *et al.*, 1988a; Moon *et al.*, 1988; Fejes *et al.*, 1988). Presumably these bits of chloroplast DNA recombined with mtDNA and became fixed in the mitochondrial genome. Although chloroplast genes incorporated into mtDNAs are usually incomplete and thought to be nonfunctional, it appears that chloroplast tRNA genes may have entered the mitochondrial genome and become functional components (see Section IV,B).

The extent of chloroplast DNA incorporation in the mitochondrial genomes of several crucifer species has been estimated by Nugent and Palmer (1988). *Brassica campestris, B. hirta,* and *R. sativa* contain approximately the same 12 to 14 kb of sequence homologous to chloroplast DNA divided over 11 different locations. *Crambe abyssinica* was shown to contain a similar amount of chloroplast-homologous sequence, but two other crucifer species, *Arabidopsis thaliana* and *Capsella brusa-pastoris,* contain significantly less (5 to 7 kb) sequence homologous to chloroplast DNA (Nugent and Palmer, 1988).

Several DNA regions containing sequences of nuclear and chloroplast origin were identified in *Oenothera* using randomly primed cDNA (r-cDNA) probes made against cytoplasmic or chloroplast RNA. One region hybridizing with the chloroplast r-cDNA probe was shown to contain part of the plastid 23S rRNA gene, the intergenic region, and the plastid 4.5S rRNA gene. A second region hybridized with both nuclear and chloroplast r-cDNA (Schuster and Brennicke, 1987a). Sequence analysis of this region showed the presence of part of the plastid *rps4* gene (ribosomal small subunit protein 4), a plastid tRNA[Ser] gene, part of the nuclear 18S rRNA gene, and an open reading frame that showed amino acid sequence similarity to retroviral reverse transcriptases (Schuster and Brennicke, 1987b). Further analyses are likely to reveal additional examples of nuclear DNA in mtDNA, as well as mtDNA in nuclear DNA.

3. Genomes of Fertile Revertant Maize Lines and Nonchromosomal Stripe Mutants

Massive genomic rearrangements can usually only be inferred by comparison of closely related genotypes of a particular species. In contrast, plants that have reverted to fertility from a CMS background, or have undergone recombination to produce NCS mutations, provide a unique opportunity to examine the immediate progenitors and products of recombination events that reorganize mtDNA.

Maize CMS-T plants regenerated from cell culture have provided a source of fertile revertant plants (Brettell *et al.*, 1979, 1980; Gengenbach *et al.*, 1977, 1981). A number of individual revertant lines have now been analyzed; they are characterized by a deletion of the *urf13* gene as the result of recombination between either a 127-bp repeat or a 58-bp repeat (Rottman *et al.*, 1987; Fauron *et al.*, 1987; Fauron *et al.*, 1990a,b). There is no evidence that these sequences act as substrates for recombination in normal plant tissue; it is possible that the cell culture environment either increased the abundance of rare recombinant molecules and/or increased the likelihood that the proposed novel recombinants would become fixed.

The mitochondrial genomes of spontaneous NCS mutants have not been mapped, but molecular analysis of several of the mutations indicates that recombination has occurred between short repeated regions present in distant locations in the genome. Only a 6-nt repeat was observed in the fragments that are proposed to have recombined to produce the *coxII* deletion in NCS5 (Newton *et al.*, 1990), whereas a 12-nt repeat and a 36-nt repeat were involved in the recombination events leading to the NCS3 and NCS6 mutations, respectively (Hunt and Newton, 1991; Lauer *et al.*, 1990).

4. Rearrangements in Normal Genomes

Recombination at shorts repeats may also be responsible for re-arrangements between closely related genomes, and for the production of gene duplications and pseudogenes in mitochondrial genomes of normal plants. Extensive sequence analysis indicated that several short repeated sequences have been involved in the duplication of the wheat tRNA[Pro] gene (Joyce *et al.*, 1988a).

A number of short repeats have been implicated in rearrangement in *Oenothera*. Within the coding sequence of *rrn26* and approximately 7.5 kb downstream, two copies of the decanucleotide GGAAGCAGCC were detected. Recombination between these two short repeats was postulated to lead to the formation of a 7.5-kb circular molecule that contains the 3′ half of the *rrn26* gene and its downstream sequences. The reciprocal product of this recombination event could not be detected (Manna and Brennicke, 1986). An imperfect indirect decanucleotide repeat was found at the borders of rearrangement sites near the *atpA* gene and three pseudogenes in *Oenothera* (Schuster and Brennicke, 1986). The sequence at a recombination site in the *Oenothera rrn18–rrn5* spacer showed a 7 out of 10 match with the decanucleotide sequence involved in the *rrn26* rearrangement (see above, Schuster *et al.*, 1987). A truncated copy of *nad5* was derived by recombination within the first intron; sequences around this recombination site did not at all resemble the *rrn26* decanucleotide.

The border sequences of a truncated copy of the *Oenothera atpA* gene also show sequence similarity to the *rrn26* decanucleotide repeat (6 out of 10 match, Schuster *et al.*, 1987). The similarity between sequences at some of the *Oenothera* rearrangement junctions may be indicative of some sequence specificity in recombination events in the *Oenothera* mitochondrial genome.

5. Recombination during Tissue Culture and Generation of Somatic Hybrids

Recombination of mitochondrial genomes usually occurs during the production of somatic hybrid lines (for reviews, see Hanson, 1984; Hanson *et al.*, 1985; Maliga, 1986). Mitochondria are thought to fuse following protoplast fusion, permitting genetic recombination that is prevented by maternal inheritance in normal sexual crosses. Mitochondrial genome recombination is followed by a process of sorting out during cell multiplication. By the time plants are regenerated, in most cases the mitochondrial genome is genetically stable (homoplasmic), though sometimes another generation or more is necessary to achieve mtDNA stability in progeny of plants carrying multiple mitochondrial genomes. Progeny of an individual somatic hybrid plant usually have the same recombinant mitochondrial genome, which differs from those of either of the parental protoplasts' genomes.

Recombination following protoplast fusion was first detected by comparing restriction patterns of mitochondrial genomes of somatic hybrid and parental tobacco lines. Restriction fragments unique to the somatic hybrid genomes could be detected (Belliard *et al.*, 1979; Nagy *et al.*, 1983). *Petunia* somatic hybrids were shown to contain recombinant mitochondrial genomes both by inspection of restriction digests and by hybridization with randomly selected probes (Boeshore *et al.*, 1983). Somatic hybrids' mtDNAs, like fertile revertants' mtDNAs, provide an opportunity to examine reorganizational recombination events in both progenitor and product.

Only a few recombined regions in somatic hybrids have been characterized in detail (Rothenberg *et al.*, 1985). In one example, the origin of a novel *atp9* gene in a *Petunia* somatic hybrid was analyzed. Gene 133*atp9*-R (in the somatic hybrid line 13-133) is flanked by the upstream region of 04*atp9*-2 (the second *atp9* copy from the 3704 parent) and the downstream region of 88*atp9*-1 (the first *atp9* copy from the 3688 parent). Although the *atp9* gene coding region evidently served as a substrate in recombination that occurred following protoplast fusion, recombination could not be detected between the two *atp9* genes in the 3704 mitochondrial genome in normal leaf tissue (Rothenberg and Hanson, 1987b, 1988). Recombinant mtDNA regions have also been characterized in *Brassica* somatic hybrids (Morgan and Maliga, 1987) and have been observed by hybridization in pearl millet (Ozias-Akins *et al.*, 1987).

At least some of the changes found in somatic hybrids may not be directly due to placing two different mitochondrial genomes in contact, but may instead result indirectly from the cell culture steps involved in producing somatic hybrids. As described previously, cell culture of maize CMS-T lines resulted in the production of fertile revertants with altered mtDNAs. There are several other reports of altered mitochondrial genomes in cell cultures in comparison to the original plants (Dale *et al.*, 1981: Hartmann *et al.*, 1987; Rode *et al.*, 1987; Negruk *et al.*, 1986; Shirzadegan *et al.*, 1989). Amplification of small mtDNA molecules has been observed in tobacco suspension culture mitochondria (Sparks and Dale, 1980; Grayburn and Bendich, 1987). In a thoroughly analyzed example, a 2-year-old *B. campestris* cell culture was found to contain extensively rearranged mtDNA and had lost a 11.3-kb linear mtDNA plasmid (Shirzadegan *et al.*, 1989).

Mitochondrial genome rearrangement is not an inevitable result of cell culture, however. Cell cultures of a number of species have been reported to exhibit stable mtDNAs (Galun *et al.*, 1982; Nagy *et al.*, 1983; Hanson, 1984; Kool *et al.*, 1985). For example, in *Petunia*, cell cultures have been a reliable source of mtDNA preparations over many years (our unpublished data; Kool *et al.*, 1985).

6. Sublimons

The products predicted from recombination at recombination repeats exist a high enough levels to allow their detection with standard methods. The presence of chimeric genes and rearranged genomes suggests that other recombination events sometimes occur, and can sometimes become fixed in a genome.

Recombination at sequences other than recombination repeats is usually not detectable. However, a few reports have provided evidence for rare recombinant molecules present in low abundance relative to the main mtDNA. Small *et al.* (1987) detected rare molecules exhibiting recombination involving the maize *atpA* gene. This gene occurs twice in both the normal and the CMS-S genomes, where both copies are of approximately equal abundance. In the N genome the two copies diverge in their 3' flanking sequence and are identified by different size restriction fragments in Southern blots of mtDNA (termed type 1 and type 2 according to the fragment size). Similarly, the CMS-S genome carries type 2 and type 3 forms of the *atpA* gene. In an extensive survey of different maize genotypes with different cytoplasms, Small *et al.* (1987) observed that several

N cytoplasms contained, at low levels, *atpA* types characterisitc of either the CMS-S cytoplasm (type 3), or the CMS-T cytoplasm (type 4). In addition, not all N cytoplasms contained equal levels of types 1 and 2, and not all S cytoplasms contained equal levels of type 2 and 3. Since, in all cases the rare forms of *atpA* could only be detected after prolonged exposure of Southern blots, they presumably resided on DNA molecules present substoichiometric relative to the main mitochondrial genome. The term "sublimons" was introduced to describe such low-abundance mtDNA molecules.

Sublimons could also be detected in *Brassica* mtDNA by Shirdazegan *et al.* (1991). These investigators found some of the unusual arrrangements detected in the cell culture to be present at low levels in plant DNA, implying that infrequently occurring mtDNA arrangements may have become stabilized and amplified in culture. Conceivably, tissue culture and subsequent regeneration of plants could allow certin sublimons to become established in the mitochondrial genome at high levels. Alternatively, recombination at short repeated sequences may have been promoted in tissue culture by an unknown mechanism. Following protoplast fusion, mitochondrial genomes recombine at sequences that normally do not act as recombination substrates (Rothenberg and Hanson, 1987b, 1988). Perhaps as a result of environmental stress during tissue culture, certain recombination or DNA repair systems are induced that normally are not active in plant mitochondria.

Sublimons could theoretically carry recombinant or chimeric genes, which could possibly undergo further recombination without impairing normal mitochondrial function. The two fused genes involved in CMS in maize CMS-T and *Petunia* could have originated on sublimons before they became established in the main mtDNA.

Sublimons may be the substrates from which new mitochondrial genomic configurations are generated. The levels of substoichiometric DNA molecules are dependent on the rate by which they are generated, and the efficiency by which they are maintained (i.e., replicated) in the whole population of mtDNA molecules. Nuclear genes may affect both generation and maintenance of sublimons. Different types of novel DNAs in CMS-S revertants were obtained depending on the nuclear background (Small *et al.*, 1988).

7. Models of Recombination Events That Result in Rearrangement of Mitochondrial Genomes

In the previous sections were documented numerous examples of infrequent recombination events at short repeats. These rare recombination events were found through analyses of chimeric genes, pseudogenes, gene duplications, and revertant and mutant genomes. Here we will consider how rare recombination at short repeats as well as frequent recombination at recombination repeats can facilitate the rearrangements of mitochondrial genomes that are seen between closely related species, or between mutants and progenitors.

Small *et al.* (1989) proposed that rare recombination of two short repeated sequences that do not normally serve as recombination repeats could give rise to new mtDNA molecules, which then could become the primary genome in a particular line or species. A diagram of recombination between two short repeats, leading to reorganization of a genome, is shown in Fig. 4A. This model requires three infrequent recombination events. Note that this mechanism results in the deletion of mtDNA that is present in the progenitor region. The size of the deletion is determined by the distance between the two different repeats that undergo the rare recombination events. Thus, short repeats that are in closer proximity are more likely to result in a deletion that could be tolerated.

A modification of this model, in which only a single rare recombination event is required, is proposed in Fig. 4B. In this model, a subgenomic molecule derived from recombination at a recombination repeat undergoes recombination with a rare subgenomic molecule that results from a rare recombination event at a short repeat. Note that this mechanism effectively increases the size of the recombination repeat. Also, deletion of DNA present in the progenitor genome results when one of the subgenomic molecules is lost (Fig. 4B). Such deletions may explain differences in DNA content between genomes of closely related species, as has been found in *Brassica,* maize, and *Petunia* (Palmer and Herbon, 1988; Fauron and Havlik, 1989; Folkerts and Hanson, 1991).

The necessity for only one rare recombination event, instead of two, in order to produce a novel genome, may enhance the probability of such a rearrangement through this second mechanism. In fact, the CMS-T fertile revertant genome mapped by Fauron *et al.* (1990a,b) fits this modified model rather than the triple rare recombination model of Small *et al.* (1989). One of the two repeats evidently involved in the generation of the CMS-T fertile revertant genome is a recombination repeat, whereas the other repeat is a short repeat that undergoes only rare recombination. The two short repeats in CMS-T are in close proximity, permitting the deletion of the DNA between the two repeats, and therefore the loss of the CMS-associated *urf13* gene.

A third model can be produced (Fig. 5), which requires yet another recombination event, but rescues all of the DNA present in the progenitor genome. In this model, recombination between abundant subgenomic circles and a subgenomic molecule derived from a rare recombination event occurs as described in the single rare recombination model. However,

A. Triple rare recombination model

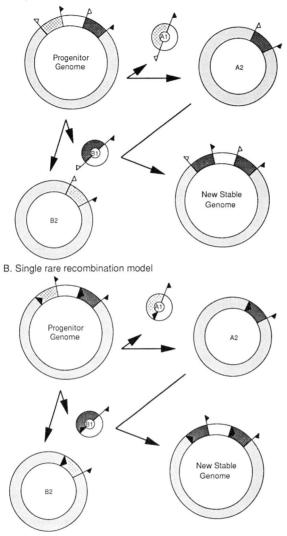

B. Single rare recombination model

FIG. 4   Rearrangements leading to new mitochondrial genome organizations. (A) Triple rare recombination model as originally proposed by Small *et al.* (1989). Two pairs of short direct repeats (solid black flags and open black flags) recombine to produce four different subgenomic molecules. Molecules A2 and B1 recombine to produce a new stable genome that lacks the hatched region found in the progenitor genome. (B) Single rare recombination model. Recombination at a recombination repeat (black triangles) produces subgenomic molecules A1 and A2. Recombination at a pair of short repeats (solid black flags) produces another two subgenomic molecules, B1 and B2. Recombination of A2 and B1 at the recombination repeat produces a new stable genome that lacks the hatched region of the progenitor genome.

Quadruple recombination model

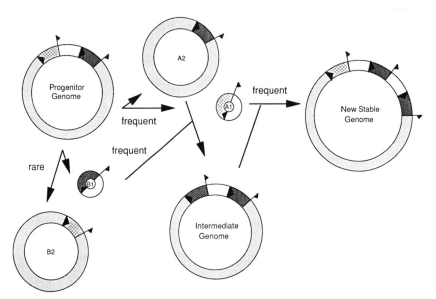

FIG. 5   Quadruple recombination model for rearrangement of mitochondrial genomes. Recombination at direct recombination repeats (black triangles) produces subgenomic molecules A1 and A2. Recombination at two short repeats (solid black flags) produces subgenomic molecules B1 and B2. Molecules A2 and B1 recombine at the recombination repeat to produce an intermediate genome that recombines again at the recombination repeat with molecule A1. A new stable genome is produced that carries all of the regions of the progenitor genome, along with two duplicated regions (dark filled region and white region) and an extra copy of the recombination repeat and the short repeat.

instead of the loss of a subgenomic molecule, another recombination event occurs at the recombination repeat, generating a master circle carrying three such repeats. Such a mechanism allows genomic rearrangement when the short repeat is at a distance from the recombination repeat, since deletion of a large mtDNA region is avoided. The presence of trios of repeats in several mitochondrial genomes, such as *Petunia* line 3704 (Fig. 2, Folkerts and Hanson, 1989), wheat (Lejeune *et al.*, 1987), rye (Coulthart *et al.*, 1990), and maize CMS-T (Fauron *et al.*, 1989), could be explained by this model.

These models illustrate how the duplications and deletions found in plant mitochondrial genomes may have been created. A further implication of these models is that accumulation of noncoding DNA—derived from the chloroplast or nuclear genome, or by duplication followed by divergence—could enhance the flexibility of a species' mitochondrial genome.

In other words, a plant carrying mtDNA that has stretches of noncoding DNA could tolerate the amplification of a deletion-bearing novel genome resulting from a rare recombination event, when nonessential DNA can be deleted without detriment.

IV. Genes of the Mitochondrial Genome and Their Genomic Locations

Transfer of chloroplast and nuclear DNA sequences to the mitochondrion, and duplication of sequences already present, can account for some of the heterogeneity and size variation found among mitochondrial genomes of different species. In addition, plant mitochondrial genomes encode genes not found in animal and fungal mtDNA. This review will not attempt a complete listing of all known encoded genes, since it would be rendered quickly obsolete by the complete sequence of the *Marchantia polymorpha* mitochondrial genome which will soon be available (K. Ohyama, personal communication).

At a gross genomic level there is no conservation of gene order between any mitochondrial genome of distantly related species. However, some regional similarities exist. The 5S rRNA gene (*rrn5*) is found directly downstream of the gene for the 18S rRNA (*rrn18*) in all organisms. In *Oenothera* the 18S and 5S rRNA genes are found upstream of and cotranscribed with the gene for NADH dehydrogenase subunit 5 (*nad5*, Wissinger *et al.*, 1988). The genes for subunit 3 of NADH dehydrogenase (*nad3*) and ribosomal protein S12 (*rps12*) were found closely linked and cotranscribed in *Petunia*, wheat, and maize. In *Petunia*, the genes are separated by 47 bp, in wheat and maize by 46 bp. The intergenic region was found to be 50% identical among the three species (Hanson *et al.*, 1989; Gualberto *et al.*, 1988). In several species exons of subunit B of NADH dehydrogenase (*nad1*) were found relatively close to the gene for yet another ribosomal protein, *rps13*. In *Oenothera*, maize, wheat, and tobacco, the *nad1* exon B is found downstream, but at varying distances, from the *rps13* gene (Bland *et al.*, 1986; Schuster *et al.*, 1987; Bonen, 1987). Linkage between *rps13* and *nad1* was also demonstrated in several *Brassica* species (Makaroff and Palmer, 1988).

A. Protein-coding genes

Through cloning and DNA sequence analysis, many higher plant mitochondrial genes have been identified (for a recent listing, see Lonsdale,

1988). Analogous to mammalian mitochondrial genomes (Attardi, 1985), plant mtDNAs were shown to encode components of the four major complexes of the respiratory chain. However, several subunits not encoded in mammalian mitochondrial genomes were identified. Maize mitochondria synthesize subunit alpha of the ATP synthase F_1 complex (Hack and Leaver, 1983), which is a nuclear gene product in other organisms. The maize gene for this subunit was cloned and sequenced (Braun and Levings, 1985; Isaac et al., 1985) and has been used to demonstrate the presence of a homologous gene in many other plant mitochondrial genomes. The gene for ATP synthase subunit 9 (F_0 complex subunit 9, or proteolipid subunit) was cloned and sequenced from the P. hybrida and maize mitochondrial genomes and subsequently shown to be present in all other higher plant mitochondrial genomes analyzed (Young et al., 1986; Dewey et al., 1986). This gene is not encoded by mammalian mitochondrial genomes. In Neurospora crassa, atp9 sequences are present in both nuclear and mitochondrial genomes, but the gene product is synthesized in the cytosol, and subsequently imported into the mitochondrion (Attardi, 1985; Van den Bogaart et al., 1982).

Plant mitochondria have been shown to contain several genes not previously detected in any mitochondrial genomes. Bland et al. (1986) first reported the identification of a gene with sequence similarity to E. coli and chloroplast ribosomal protein S13 (rps13). This gene is located downstream from atp9 in tobacco, and in turn upstream from the first exon of the plant nad1 gene in maize, tobacco, Oenothera, and wheat. Interestingly, rps13 was not found in the broad bean mitochondrial genome (Wahleithner and Wolstenholme, 1988a) and is present in wheat, but not transcribed (Bonen, 1987). An open reading frame with sequence similarity to ribosomal protein S14 was identified in broad bean; however, this gene was not detected in maize (Wahleithner and Wolstenholme, 1988b). In addition to rps13 and rps14, a gene with sequence identity to ribosomal protein S12 (rps12) was detected in P. hybrida, maize, and wheat mitochondrial genomes (Hanson et al., 1989; Gaulberto et al., 1989), and an rps19 gene has been found in P. hybrida (Conklin and Hanson, 1991). Further immunological studies are needed to determine whether the ribosomal protein genes are expressed. However, the fact that a mutation in the maize S3/L16 locus is present in an NCS mutant suggests that one or more of these ribosomal protein genes are needed for normal mitochondrial function (Hunt and Newton, 1991).

Recently, open reading frames with sequence similarity to reverse transcriptase have been identified in several species. The Oenothera reverse transcriptase (see Section III,E,2) has high sequence similarity to several retroviral reverse transcriptases, but is probably not an indigenous mitochondrial gene. Its origin is believed to be the nuclear genome (Schuster and Brennicke, 1987a). Wahleithner *et al.* (1990) have identified a reverse transcriptase-like open reading frame in the mitochondrial genomes of broad bean, maize, and soybean situated downstream from *nad1* exon D.

B. Transfer-RNA Genes

Genes for plant mitochondrial tRNAs can be grouped into three types: (1) so-called "native" tRNAs that share 60–80% sequence similarity with eubacterial tRNAs (2) chloroplast-like tRNAs with 90–100% sequence similarity to chloroplast tRNAs, and (3) nuclear-encoded tRNAs which are imported from the cytoplasm. Unlike most chloroplast sequences found in mtDNA, a number of the chloroplast-like tRNAs have been shown to be transcribed and are likely to function as genuine mitochondrial genes (Binder *et al.*, 1990; Marechal *et al.*, 1985, 1986; Marechal-Drouard *et al.*, 1988; Wintz *et al.*, 1988a; Joyce *et al.*, 1988a; Izuchi *et al.*, 1990; Joyce and Gray, 1989). Four bean tRNAs have been shown to be nuclear-encoded. Two such tRNAs were found to be identical to the corresponding cytoplasmic tRNAs except for a post-transcriptional methylation (Marechal-Drouard *et al.*, 1988). As only 16 tRNA genes have been identified in maize, and at least 23 are needed, it is likely that maize mitochondria also import nuclear-encoded tRNAs (Sangare *et al.*, 1990).

Although some tRNA genes are found in close proximity to protein-coding genes or to other tRNAs (Grabau, 1987; Makaroff *et al.*, 1989; Chen *et al.*, 1989), most are found by themselves. The tRNA genes are dispersed over the mtDNA map of maize (Sangare *et al.*, 1990) and *Petunia* (J.-M. Grienenberger, personal communication). The location of one mitochondrial tRNA gene, maize tRNA^{trp}, is of particular interest; it is found only on a 2.3-kb linear plasmid (Leon *et al.*, 1989).

The plant mitochondrial tRNAs that have been characterized exhibit sequences that can be folded into the standard cloverleaf secondary structure. Both initiator (Met$_F$) and elongator methionine tRNAs have been identified (Gray and Spencer, 1983; Parks *et al.*, 1984; Gottschalk and Brennicke, 1985; Marechal *et al.*, 1986; Wintz *et al.*, 1988b). The anticodons of the tRNAs sequenced are those predicted from the standard genetic code, with the exception of a CAT anticodon found in the potato isoleucine tRNA gene. The C residue in the anticodon is modified to the unusual base lysidine, which pairs with A, changing the amino acid specificity of the tRNA from the predicted methionine to isoleucine (Weber *et al.*, 1990).

V. Transcription, RNA Processing, and Translation

A. Transcription Initiation, Processing,
   and Termination Sites

In general, plant mitochondrial genes exhibit complex transcription patterns; multiple 5' termini and sometimes multiple 3' termini are present. RNA processing at the 5 ends of transcripts is responsible for some of the multiple termini, according to studies of the maize 26S and 18S genes (Mulligan et al., 1988a) and the T-*urf13* locus of maize (Kennell et al., 1987). Sequences carrying inverted repeats that can form stem–loop structures have been found upstream and downstream of several mitochondrial genes, including *Petunia atp9* and broad bean *atp6* (Rothenberg and Hanson, 1987a; Macfarlane et al., 1990). The functions of these sequences are unknown. Transcript termini map to some, but not all, of the sequences that could potentially form secondary structures (Schuster et al., 1986; Bland et al., 1986; Wissinger et al., 1988; Macfarlane et al., 1990; Rothenberg and Hanson, 1987a). Whether inverted repeats are involved in transcript processing and stability, as they are in chloroplast RNAs (Stern and Gruissem, 1987; Stern et al., 1989), merits investigation.

Most studies of 5' termini of mitochondrial gene transcripts have been performed with techniques that cannot distinguish primary transcripts from processed transcripts. Thus, when consensus sequences for 5' transcript termini were originally proposed by Young et al., (1986) and Schuster et al., 1986), it was not known whether the termini represented initiation or processing points. Subsequently, Mulligan et al. (1988a,b, 1991) have mapped initiation sites of maize gene transcripts, and measured transcription rates. An AT-rich consensus sequence (A/TCRTAG/TAA/TAAA), similar to a conserved sequence at transcript termini observed by Young et al. (1986) and Schuster et al. (1986), is present at transcript termini maize genes that have high rates of transcription. It appears likely that termini which map to sequences similar to the AT-rich consensus will represent initiation sites. Mulligan et al. (1990) pointed out the existence of several G(A/T)3-4 motifs present nearby upstream of most initiation sites. Termini of maize genes that are transcribed at lower rates map to sequences that do not fit the consensus readily. Many of these latter sequences carry a CCT/CTT motif also found at transcript termini of *Petunia atp9*, maize *coxI*, and wheat *coxII*,as described by Young et al. (1986). There may be some, as yet known, functional difference in the initiation sites that adhere to the AT-rich consensus versus those carrying the CTT/CTT motif. For example, perhaps different proteins interact with these two types of sequences.

Sequences that regulate transcription of plant mitochondrial tRNAs have been postulated by comparative analysis. For a monocotyledon, a purine-rich tRNA promoter motif containing a consensus sequence AA-GANRR has been proposed, whereas the consensus AAGAARAG has been proposed for tRNA genes of a dicotyledon (Joyce et al., 1988a; Binder et al., 1990). Sequence boxes A and B which form the internal promoter for RNA polymerase III in eukaryotic nuclear tRNAs (Galli et al., 1981) are present in most plant mitochondrial sequences. The typical −10 and −35 chloroplast tRNA promoter region sequences are found upstream of some chloroplast-like tRNAs, but not upstream of the native mitochondrial tRNA genes (Joyce et al., 1988b). Sequences 3′ to tRNAs that could serve as termination signals have not been identified.

Two recently developed systems that can process dicotyledon (Oenothera) or monocotyledon (wheat) mitochondrial tRNAs precursors in vitro should permit dissection of the sequences and enzyme activities needed for processing. The plant mitochondrial RNAse P-like activities can process both native and chloroplast-like precursor tRNAs, further establishing the chloroplast-like tRNA genes as functional mitochondrial genes (Marchfelder, et al., 1990; Hanic-Joyce and Gray, 1990).

B. Post-transcriptional Regulation of Transcript Abundance

About 30% of one of the smaller known mitochondrial genomes, the 218-kb B. campestris mtDNA, hybridizes with abundant transcripts, and additional regions hybridize with low-abundance transcripts (Makaroff and Palmer, 1987). Until recently, most analyses of plant mitochondrial transcripts examined only steady-state abundance.

Two reports concerning maize mtDNA transcription rates provide the new information that mtDNA regions which do not exhibit detectable steady-state amounts of transcripts are transcribed, and that transcription rate and transcript abundance are not correlated. Finnegan and Brown (1990) pulse-labeled maize mtDNA in vivo or in isolated mitochondria and detected RNAs hybridizing to mtDNA regions which did not hybridize to mtRNA that was extracted and then labeled in vitro. Mulligan et al. (1991) performed careful comparisons of steady-state abundance of transcripts of specific genes vs. their transcriptional activity in lysed mitochondria and demonstrated a lack of correlation that must result from post-transcriptional processes.

C. Intron Splicing

Another form of RNA processing in plant mitochondria is intron splicing. The maize coxII gene was the first one reported with an intron (Fox and

Lever, 1981); since then several other plant mitochondrial genes have been shown to contain one or several introns (Stern *et al.*, 1986; Wintz *et al.*, 1989; Wissinger *et al.*, 1988; Ecke *et al.*, 1990). In the case of *coxII*, the intron is optional, present in all monocotyledons investigated to date, but not in all dicotyledons (Moon *et al.*, 1985; Hiesel *et al.*, 1990). The intron location is the same in all cases studied. The size of the intron varies in different species because of insertions in domain IV of the predicted group II intron structure (Fox and Leaver, 1981; Bonen *et al.*, 1984; Kao *et al.*, 1984; Gallerani *et al.*, 1987; Turano *et al.*, 1987; Pruitt and Hanson, 1989). In addition to *cis*-splicing in *coxII* and several other mitochondrial genes, the *nad1* gene of wheat (Chapdelaine and Bonen, 1991), *Oenothera* (Wissinger *et al.*, 1991), and *Petunia* (Conklin *et al.*, 1991; Sutton *et al.*, 1991) is *trans*-spliced. Five *nad1* exons are present, requiring multiple *cis*- and *trans*-splicing events.

D. RNA Editing

The discovery of RNA editing in plant mitochondria in 1989 (Gualberto *et al.*, 1989; Covello and Gray, 1989; Hiesel *et al.*, 1989) makes it necessary to reevaluate all previous reports in which the DNA sequence was presumed to specify the protein sequence. It now appears that the RNAs of most protein-coding mitochondrial genes are edited. When cDNAs of mitochondrial genes are examined, Ts are frequently found at a small number of sites where Cs are present in the genomic sequence.

Previously, the plant mitochondrial genetic code had been proposed to differ from the standard code by the specification of tryptophan by CGG, normally an arginine codon. When the maize *coxII* protein predicted by the genomic sequence was compared to those of other species, CGG codons were present at positions where tryptophan was expected (Fox and Leaver, 1981). However, in other plant mitochondrial genes, sometimes CGG codons were present where arginine would be predicted. This puzzling observation is now explained by the editing of some, but not all, CGG codons into UGG, a tryptophan codon.

Editing of transcripts appears to be confined primarily to coding regions, though editing has been observed in at a few sites in flanking untranslated regions (Covello and Gray, 1989; Schuster *et al.*, 1990a, 1991). Editing has the potential to create an AUG start codon; one such occurrence has been reported in wheat *nad1* (Chapdelaine and Bonen, 1991). Editing can also create a stop codon; in *Petunia atp9* (Wintz and Hanson, 1990), *Oenothera atp9* (Schuster and Brennicke, 1990), and wheat *atp9* (Begu *et al.*, 1990; Nowak and Kuck, 1990), a stop codon is introduced in transcripts, shortening the reading frame slightly. A C to U conversion could not remove a stop codon. However, a few U to C conversions have been reported (Gualberto *et al.*, 1990; Hiesel *et al.*, 1990; Schuster *et al.*, 1990a), and such an event could change the length of a reading frame by converting a stop codon into one specifying an amino acid.

A site where a C in the genomic sequence has once been found to carry a T in cDNA is usually referred to as an editing site, or edit site. Transcripts represented by cDNAs that carry Ts at some, but not all, potential edit sites are usually referred to as partially edited, whereas cDNAs that carry Ts at all potential edit sites are considered to represent "fully edited" transcripts. The current interpretation of partially edited transcripts is that they represent intermediates in an editing process that is not complete by the end of transcription (Schuster *et al.*, 1990a; Wissinger *et al.*, 1990; Covello and Gray, 1990; Gualberto *et al.*, 1989).

Different genes vary in the proportion of transcripts that lack Ts at potential edit sites. Transcripts of some genes all appear to be fully edited (for example, *Petunia atp9*; Wintz and Hanson, 1990), whereas partially edited transcripts are found for other genes, including *nad3, rps13,* and *coxII* (Schuster *et al.*, 1990c; Wissinger *et al.*, 1990; Sutton *et al.*, 1991; Yang and Mulligan, 1991). RNA editing can precede either *cis-* or *trans-* splicing (Sutton *et al.*, 1991; Yang and Mulligan, 1991). Whether transcripts that do not have Ts at all potential edit sites are translated is not yet known (see below).

At this writing, little is known of the mechanism of RNA editing in plant mitochondria. Only rather degenerate hypothetical sequence similarities around edit sites can be proposed (Gualberto *et al.*, 1990). From analysis of partially edited transcripts, there does not appear to be an obvious directionality to the process. Although certain edit sites are less likely than others to carry a T in a cDNA clone of a particular gene, unedited sites in cDNA clones are found in both 5' and 3' regions of coding regions (Schuster *et al.*, 1990c; Wissinger *et al.*, 1990; Sutton *et al.*, 1991; Yang and Mulligan, 1991).

E. Initiation of Protein Synthesis

How plant mitochondrial ribosomes recognize the appropriate AUG codon for initiation is not known. A number of cotranscribed genes are known (e.g., *nad3/rps12* in wheat and *Petunia*, Gualberto *et al.*, 1988; Hanson *et al.*, 1989), so presumably there is some way that a ribosome can recognize a second initiating AUG in a transcript. A putative ribosome binding site, similar to the Shine-Dalgarno sequence found in bacterial genes, has been proposed (Dawson *et al.*, 1984). Upstream of a number of plant mitochondrial genes can be found a sequence at −10 to −30 that has some complementarity to an octanucleotide at the 3' and of 18S rRNA, but the degree of complementarity is often less than that observed in *E. coli* genes, and the position is more variable (Schuster and Brennicke, 1986; Hanson *et al.*, 1988). In the absence of an *in vitro* translation system, the hypothetical ribosome binding site has not been directly tested.

Another mystery concerns how mature transcripts are distinguished from unspliced or incompletely edited transcripts, if indeed such discrimination does occur. If unedited or partially edited transcripts are translated, than proteins with aberrant amino acid composition would be produced. At this writing it is not known whether proteins that result from translation of partially edited transcripts exist in detectable amounts. If ribosomes do not discriminate against partially edited transcripts, then multiple forms of many mitochondrially encoded proteins must be present.

VI. Summary and Conclusions

The propensity of plant mDNA to recombine is, on balance, a convenience for the investigator, despite the fact that the presence of recombination repeats makes restriction mapping of ordered cosmids of mtDNAs of certain species much more difficult. However, were it not for the tendency of mutants (CMS and NCS) and CMS revertants to arise through recombination of short repeats, genes involved in CMS and NCS mutations might yet have gone undetected. Single base changes could theoretically produce a mutant phenotype and would be very difficult to detect. The gross alterations in mtDNAs have occurred in CMS lines of several species and independent NCS mutants as a result of short repeat recombination has made these mutations accessible to the investigator. Hunting for CMS- and NCS-associated genes in additional species is likely not to be as difficult as might have been expected.

Plant mitochondria retain many mysteries. Technical advances may be necessary to resolve the discrepancy between restriction maps and physical characterization of isolated mtDNA. *In vitro* systems for dissecting transcription initiation, control of transcription rate, replication, and tRNA processing have already been achieved. Other processes, such as editing, intron splicing, and protein synthesis, should yield to analysis by *in vitro* systems, when they are established. Soon most of the protein-coding sequences of plant mitochondria will be known, with the completion of the sequencing of the *Marchantia* mtDNA. A large number of unidentified reading frames can be expected, which will require much effort in order to assign function. A method for introduction of DNA into mitochondria, currently hampered by the lack of selectable markers, will be essential to carry analyses further and to exploit our knowledge of mtDNA for genetic engineering.

Acknowledgments

Research concerning plant mitochondria in M.R.H.'s laboratory has been sponsored by the McKnight Foundation, the U.S.–Israel BARD fund, the DOE Biosciences program, the USDA Competitive Grants Program, the Cornell Biotechnology Program, and Hatch funds. O.F. received a McKnight Foundation traineeship in Plant Reproductive Biology. We thank many colleagues for providing reprints and unpublished information; Linda Narde for assistance in manuscript preparation; and Cassie Conley, Patricia Conklin, David Stern, Claudia Sutton, and Henri Wintz for helpful suggestions.

References

Attardi, G. (1985). *Int. Rev. Cytol.* **93,** 93–145.
Bailey-Serres, J., Dixon, L. K., Liddell, A. D., and Leaver, C. J. (1986a). *Theor. Appl. Genet.* **73,** 252–260.
Bailey-Serres, J., Hanson, D. K., Fox, T. D., and Leaver, C. J. (1986b). *Cell (Cambridge, Mass.)* **47,** 567–576.
Bailey-Serres, J., Leroy, P., Jones, S. S., Wahleithner, J. A., and Wolstenholme, D. R. (1987). *Curr. Genet.* **12,** 49–53.
Begu, D., Graves, P.-V., Domec, D., Arselin, G., Litvak, S., and Araya, A. (1990). *Plant Cell* **2,** 1283–1290.
Belliard, G., Vedel, F., and Pelletier, G. (1979). *Nature (London)* **281,** 401–403.
Bendich, A. J., and Gauriloff, L. P. (1984). *Protoplasma* **119,** 1–7.
Bendich, A. J., and Smith, S. B. (1990). *Curr. Genet.* **17,** 421–425.
Binder, S., Schuster, W., Grienenberger, J.-M., Weil, J.-H., and A. Brennicke (1990). *Curr. Genet.* **17,** 353–358.
Bland, M. M., Levings, C. S., III, and Matzinger, D. F. (1986). *Mol. Gen. Genet.* **204,** 8–16.
Bland, M. M., Levings, C. S., III, and Matzinger, D. F. (1987). *Curr. Genet.* **12,** 475–481.
Boeshore, M. L., Hanson, M. R., and Izhar, S. (1985). *Plant Mol. Biol.* **4,** 125–132.
Boeshore, M. L., Lifshitz, I., Hanson, M. R., and Izhar, S. (1983). *Mol. Gen. Genet.* **190,** 459–467.
Bonen, L. (1987). *Nucleic Acids Res.* **15,** 10393–10404.
Bonen, L., and Bird, S. (1988). *Gene* **73,** 47–56.
Bonen, L., Boer, P. H., and Gray, M. W. (1984). *EMBO J.* **3,** 2531–2536.
Braun, C. J., and Levings, C. S., III (1985). *Plant Physiol.* **79,** 571–577.
Braun, C. J., Siedow, J. N., Williams, M. E., and Levings, C. S., III (1989). *Proc. Natl. Acad. Sci U.S.A.* **86,** 4435–4439.
Brears, T., and Lonsdale, D. M. (1988). *Mol. Gen. Genet.* **214,** 514–522.
Brennicke, A., and Blanze, P. (1982). *Mol. Gen. Genet.* **187,** 461–466.
Brettell, R. I. S., Goddard, B. V. D., and Ingram, D. S. (1979). *Maydica* **24,** 203–213.
Brettell, R. I. S., Thomas, E., and Ingram, D. S. (1980). *Theor. Appl. Genet.* **58,** 55–58.
Chapedelaine, Y., and Bonen, L. (1991). *Cell (Cambridge, Mass.)* **65,** 465–472.
Chase, C. D., and Pring, D. R. (1985). *Plant Mol. Biol.* **5,** 303–311.

Chen, H.-C., Wintz, H., Weil, J.-H., and Pillay, D. T. N. (1989). *Nucleic Acids Res.* **17,** 2613–2621.

Chetrit, P., Mathieu, C., Muller, J. P., and Vedel, F. (1984). *Curr. Genet.* **8,** 413–421.

Chu, G., Vollrath, D., and Davis, R. W. (1986). *Science* **234,** 1582–1585.

Coe, E. H., Jr. (1983). *Maydica* **28,** 151–167.

Conklin, P. L., and Hanson, M. R. (1991). *Nucleic Acids Res.* **19,** 2701–2705.

Conklin, P. L., Wilson, R. K., and Hanson, M. R. (1991). *Genes Dev.* **5,** 1407–1415.

Connett, M., and Hanson, M. R. (1990). *Plant Physicol.* **93,**1634–1640.

Coulthart, M. B., Huh, G. S., and Gray, M. W. (1990). *Curr. Genet.* **17,** 339–346.

Covello, P. S., and Gray, M. W. (1989). *Nature* (London) **341,** 662–666.

Covello, P. S., and Gray, M. W. (1990). *Nucleic Acids Res.* **18,** 5189–5196.

Dale, R. M. K., (1981). *Proc. Natl. Acad. Sci. U.S.A.* **78,** 4453–4457.

Dale, R. M. K., Duesing, J. H., and Keene, D. (1981). *Nucleic Acids Res.* **9,** 4583–4593.

Dawson, A. J., Jones, V. P., and Leaver, C. J. (1984). *EMBO J.* **3,** 2107–2113.

DeHaas, J. M., Hille, J., Kors, F., Van der Meer, B., Kool, A. J., Folkerts, O., and Nijkamp, H. J. J. (1991). *Curr. Genet.* **20,** 503–514.

Dewey, R. E., Levings, C. S., III, and Timothy, D. H. (1986). *Cell (Cambridge, Mass.)* **44,** 439–449.

Dewey, R. E., Siedow, J. N., Timothy, D. H., and Levings, C. S., III (1988). *Science* **239,** 293–295.

Dewey, R. E., Timothy, D. H., and Levings, C. S., III (1987). *Proc. Natl. Acad. Sci. U.S.A.* **84,** 5374–5378.

Ecke, W., Schmitz, U., and Michaelis, G. (1990). *Curr. Genet.* **18,** 133–139.

Eckenrode, V. K., and Levings, C. S., III (1986). In *Vitro Cell. Dev. Biol.* **22,** 169–176.

Eckhardt, T. (1978). *Plasmid* **1,** 584–588.

Erickson, L., Beversdorf, W. D., and Paul, K. P. (1985). *Curr. Genet.* **9,** 676–682.

Escote, L. J., Gabay-Laughnan, S. J., and Laughnan, J. R. (1985). *Plasmid* **14,** 264–267.

Fairbanks, D. J., Smith, S. E., and Brown, J. K. (1988). *Theor. Appl. Genet.* **76,** 619–622.

Falconet, D., Delorme, S., Lejeune, B., Sevignac, M., Delcher, E., Bazetoux, S., and Quetier, F. (1985). *Curr. Genet.* **9,** 169–174.

Fauron, C. M.-R., and Havlik, M. (1988). *Nucleic Acids Res* **16,** 10395–10396.

Fauron, C. M.-R., and Havlik, M. (1989). *Curr. Genet.* **15,** 149–154.

Fauron, C. M.-R., Abbott, A. G., Brettell, R. I., and Gesteland, R. F. (1987). *Curr. Genet.* **11,** 339–346.

Fauron, C. M.-R., Havlik, M., Lonsdale, D., and Nichols, L. (1989). *Mol. Gen. Genet.* **216,** 395–401.

Fauron, C. M.-R., Havlik, M., and Brettel, R. I. S. (1990a). *Genetics* **124,** 423–428.

Fauron, C. M.-R., Havlik, M., Hafezi, S., Brettell, R. I. S., and Albertson, M. (1990b). *Theor. Appl. Genet.* **79,** 593–599.

Fauron, C., Havlik, M., and Casper, M. (1991). *In* "Plant Molecular Biology" (R. G. Herrmann and B. Larkins, eds.), in press.

Feiler, H. S., and Newton, K. J. (1987). *EMBO J.* **6,** 1535–1539.

Fejes, E., Masters, B. S., McCarty, D. M., and Hauswirth, W. W. (1988). *Curr. Genet.* **13,** 509–515.

Finnegan, P. M., and Brown, G. G. (1986). *Proc. Natl. Acad. Sci. U.S.A.* **83,** 5175–5179.

Finnegan, P. M., and Brown, G. G. (1990). *Plant Cell* **2,** 71–83.

Flavell, R. (1974). *Plant Sci. Lett.* **3,** 259–263.

Folkerts, O., and Hanson, M. R. (1989). *Nucleic Acid. Res.* **17,** 7345–7357.

Folkerts, O., and Hanson, M. R. (1991). *Genetics* **129,** 885–895.

Fontarnau, A., and Hernandez-Yago, Jr. (1982). *Plant Physiol.* **70,** 1678–1682.

Forde, B. G., and Leaver, C. J. (1980). *Proc. Natl. Acad. Sci. U.S.A.* **77,** 418–422.

Forde, B. G., Oliver, R. J. C., and Leaver, C. J. (1978). *Proc. Natl. Acad. Sci. U.S.A.* **75**, 3841–3845.

Fox, T. D., and Leaver, C. J. (1981). *Cell (Cambridge, Mass.)* **26**, 315–323.

Fragoso, L. L., Nichols, S. E., and Levings, C. S., III (1989). *Genome* **31**, 160–168.

Gallerani, R., DeBenedetto, C., Siclella, L., Perrotta, C. M., and Ceci, L. R. (1987). *In* "From Enzyme Adaptation to Natural Philosophy: Heritage from Jacques Monod" (E. Quagliariello, G. Bernardi, and A. Ullmann, eds.), pp. 149–158, Elsevier Science, Amsterdam.

Galli, G., Hofstetter, H., and Brinstiel, M. L. (1981). *Nature (London)* **294**, 626–631.

Galun, E., Arzee-Gonen, P., Fluhr, R., Edelman, M., and Aviv, D. (1982). *Mol. Gen. Genet.* **186**, 50–56.

Gengenbach, B. G., Connelly, J. A., Pring, D. R., and Conde, M. F. (1981). *Theor. Appl. Genet.* **59**, 161–167.

Gengenbach, B. G., Green, C. E., and Donovan, C. M. (1977). *Proc. Natl. Acad. Sci. U.S.A.* **74**, 5113–5117.

Goto, Y-i., Nonaka, I., and Horai, S. (1990). *Nature (London)* **348**, 651–653.

Gottschalk, M., and Brennicke, A. (1985). *Curr. Genet.* **9**, 165–168.

Grabau, E. A. (1987). *Curr. Genet.* **11**, 287–293.

Grabau, E., Havlik, M., and Gesteland, R. (1988). *Curr. Genet.* **13**, 83–89.

Gray, M. W. (1989a). *Annu. Rev. Cell Biol.* **5**, 25–50.

Gray, M. W. (1989b). *Trends Genet.* **5**, 294–299.

Gray, M. W., and Spencer, D. F. (1983). *FEBS Lett.* **161**, 323–327.

Grayburn, W. S., and Bendich, A. J. (1987). *Curr. Genet.* **12**, 257–261.

Gualberto, J. M., Lamattina, L., Bonnard, G., Weil, J.-H., and Grienenberger, J.-M. (1989). *Nature (London)* **341**, 660–662.

Gualberto, J. M., Weil, J.-H., and Grienenberg, J.-M. (1990). *Nucleic Acids Res.* **18**, 3771–3776.

Gualberto, J. M., Wintz, H., Weil, J.-H., and Grienenberger, J.-M. (1988). *Mol. Gen. Genet.* **215**, 118–127.

Hack, E., and Leaver, C. J. (1983). *EMBO J.* **2**, 1783–1789.

Hanic-Joyce, P. J., and Gray, M. W. (1990). *J. Biol. Chem.* **265**, 13782–13791.

Hansen, B. M., and Marcker, K. A. (1984). *Nucleic Acids Res.* **12**, 4747–4757.

Hanson, M. R. (1984). *Oxford Surv. Plant Mol. Cell Biol.* **1**, 33–52.

Hanson, M. R. (1991). *Annu. Rev. Genet.* **25**, 461–486.

Hanson, M. R., and Conde, M. F. (1985). *Int. Rev. Cytol.* **94**, 213–267.

Hanson, M. R., Connett, M. B., Folkerts, O., Izhar, S., McEvoy, S. M., Nivison, H. T., and Pruitt, K. D. (1991). "Plant Molecular Biology" (R. G. Herrmann and B. Larkins, eds.). Plenum, New York, in press.

Hanson, M. R., Pruitt, K. D., and Nivison, H. T. (1989). *Oxford Surv. Plant Mol. Cell Biol.* **6**, 61–85.

Hanson, M. R., Rothenberg, M., Boeshore, M. L., and Nivison, H. T. (1985). *In* "Biotechnology in Plant Science: Relevance to Agriculture in the Eighties" (P. Day, M. Zaitlin, and A. Hollaender, eds.), pp. 129–144. Academic Press, New York.

Hanson, M. R., Young, E. G., and Rothenberg, M. (1988). *Phil. Trans. R. Soc. Lond. B.* **319**, 199–208.

Hartmann, C. deBuyser, J., Jenry, Y., Falconet, D., Lejeune, B., Benslimane, A., Quetier, F., and Rode, A. (1987). *Plant Sci.* **53**, 191–198.

Hiesel, R., Schobel, W., Schuster, W., and Brennicke, A. (1987). *EMBO J.* **6**, 29–34.

Hiesel, R., Wissinger, B., and Brennicke, A. (1990). *Curr. Genet.* **18**, 371–375.

Hiesel, R., Wissinger, B., Schuster, W., and Brennicke, A. (1989). *Science* **246**, 1632–1634.

Horn, R., Kohler, R. H., and Zetsche, K. (1991). *Plant Mol. Biol.* **17**, 29–36.

Houchins, J. P., Ginsburg, H., Rohrbaugh, M., Dale, R. M. K., Schardl, C. L., Hodge, T. P., and Lonsdale, D. M. (1986). *EMBO J.* **5**, 2781–2788.

Hunt, M. D., and Newton, K. J. (1991) *EMBO J.* **10**, 1045–1052.

Isaac, P. G., Brennicke, A., Dunbar, S. M., and Leaver, C. J. (1985). *Curr. Genet.* **10**, 321–328.

Ishige, T., Storey, K. K., and Gengenbach, B. G. (1985). *Jpn J. Breed.* **35**, 285–291.

Izuchi, S., Terachi, T., Sakamoto, M., Mikami, T., and Sugita, M. (1990). *Curr. Genet.* **18**, 239–243.

Joyce, P. B. M., and Gray, M. W. (1989). *Nucleic Acids Res.* **17**, 5461–5476.

Joyce, P. B. M., Spencer, D. F., Bonen, L., and Gray, M. W. (1988a). *Plant Mol. Biol.* **10**, 251–262.

Joyce, P. B. M., Spencer, D. F., and Gray, M. W. (1988b). *Plant Mol. Biol.* **11**, 833–843.

Jubier, M.-F., Lucas, H., Delcher, E., Hartmann, C., Quetier, F., and Lejeune, B. (1990). *Curr. Genet.* **71**, 523–528.

Juniper, B. E., and Clowes, F. A. L. (1965). *Nature (London)* **208**, 864–865.

Kadowaki, K., Takeshi, S., and Kazama, S. (1990). *Mol. Gen. Genet.* **224**, 10–16.

Kao, T. H., Moon, E., and Wu, R. (1984). *Nucleic Acids. Res.* **12**, 7305–7315.

Kemble, R. J., and Bedbrook, J. R. (1980). *Nature (London)* **284**, 565–566.

Kemble, R. J., Carlson, J. E., Erickson, L. R., Sernyk, J. L., and Thompson, D. J. (1986). *Mol. Gen. Genet.* **205**, 183–185.

Kemble, R. J., and Thompson, R. D. (1982). *Nucleic Acids. Res.* **10**, 8181–8190.

Kennell, J. C., and Pring, D. R. (1989). *Mol. Gen. Genet.* **216**, 16–24.

Kennell, J. C., Wise, R., and Pring, D. R. (1987). *Mol. Gen. Genet.* **210**, 399–406.

Kim, W. K., and Klassen, G. R. (1989). *Curr. Genet.* **15**, 161–166.

Kohler, R. H., Horn R., Lossl, A., and Zetsche, K. (1991). *Mol. Gen. Genet.* **227**, 369–371.

Kolodner, R., and Tweari, K. K. (1972). *Proc. Natl. Acad. Sci. U.S.A.* **69**, 1830–1834.

Kool, A. J. de Haas, J. M., Mol, J. N. M., and Van Marrewijk, G. A. M. (1985). *Theor. Appl. Genet.* **69**, 223–233.

Korth, K. L., Struck, F., Kaspi, C. I., Siedow, J. N., and Levings, C. S., III (1991). "Plant Molecular Biology" (R. G. Herrmann and B. Larkins, eds.), pp. 375–381. Plenum, New York.

Laser, K. D., and Lersten, N. R. (1972). *Bot. Rev.* **38**, 425–454.

Lauer, M. Knudsen, C., Newton, K. J., Gabay-Laughnan, S., and Laughnan, J. R. (1990). *New Biol.* **2**, 179–186.

Laughnan, J. R., and Gabay-Laughnan, S. (1983). *Annu. Rev. Genet.* **17**, 27–48.

Lejeune, B., Delorme, S., Delcher, E., and Quetier, F. (1987). *Plant Physiol. Biochem.* **25**, 227–233.

Leon, P. Walbot, V., and Bedinger, P. (1989). *Nucleic Acids Res.* **17**, 4089–4099.

Levings, C. S., III (1990). *Science* **250**, 942–947.

Levings, C. S., III, and Brown, G. G. (1989). *Cell (Cambridge, Mass.)* **56**, 171–179.

Levings, C. S., III, and Dewey, R. E. (1988). *Phil. Trans. R. Soc., Lond. B* **319**, 93–102.

Levings, C. S., III, Kim, B. C., Pring, D. R., Conde, M. F., Mans, R. J., Laughnan, J. R., and Gabay-Laughnan, S. J. (1980). *Science* **209**, 1021–1023.

Levings, C. S., III, and Sederoff, R. R. (1983). *Proc. Nat. Acad. Sci. U.S.A.* **80**, 4055–4059.

Levings, C. S., III, Shah, D. m., Hu, W. W. L., Pring, D. R., and Timothy, D. H. (1979). *In* "Extrachromosomal DNA, ICN–UCLA Symposia on Molecular and Cellular Biology" D. Cummings, P. Borst, I. B. David, S. M. Weissman and C. F. Fox, eds.) Vol. 15, pp. 63–73. Academic Press, New York.

Levy, A. A., Andre, C., and Walbot, V. (1991). *Genetics* **128**, 417–424.

Lonsdale, D. M. (1987). *Plant Physiol. Biochem.* **25**, 265–271.

Lonsdale, D. M. (1988). *Plant Mol. Biol. Rep.* **6**, 266–273.

Lonsdale, D. M., Hodge, T. O., and Fauron, C. M.-R. (1984). *Nucleic Acids Res.* **12**, 9249–9261.

Lonsdale, D. M., Hodge, T. P., Howe, C., and Stern, D. B. (1983). *Cell* (*Cambridge, Mass.*) **34**, 1007–1014.

Ludwig, S. R., Pohlman, R. F., Vieira, J., Smith, A. G., and Messing, J. (1985). *Gene* **38**, 131–138.

Macfarlane, J. L., Wahleithner, J. A., and Wolstenholme, D. R. (1990). *Curr. Genet.* **18**, 87–91.

Mackenzie, S. A., and Chase, C. D. (1990). *Plant Cell* **2**, 905–912.

Makaroff, C. A., Apel, I. J., and Palmer, J. D. (1989). *J. Biol. Chem.* **264**, 11,706–11,713.

Makaroff, C. A., Apel, I. J, and Palmer, J. D. (1991). *Plant Mol. Biol.* **15**, 735–746.

Makaroff, C. A., and Palmer, J. D. (1987). *Nucleic Acids Res.* **15**, 5141–5156.

Makaroff, C. A., and Palmer, J. D. (1988). *Mol. Cell. Biol.* **8**, 1474–1480.

Maliga, P. (1986). *In* "Molecular Developmental Biology" (L. Bogard, ed.), pp. 45–53. A. R. Liss, New York.

Manna, E., and Brennicke, A. (1986). *Mol. Gen. Genet.* **203**, 377–381.

Manson, J. C., Liddell, A. D., Leaver, C. J., and Murray, K. (1986). *Mol. Gen. Genet.* **203**, 377–381.

Marchfelder, A., Schuster, W., and Brennicke, A. (1990). *Nucleic Acids Res.* **18**, 1401–1406.

Marechal, L., Guillemaut, P., Grienenberger, J-M., Jeannin, G., and Weil, J.-H. (1985). *Nucleic Acids Res.* **13**, 4411–4416.

Marechal, L., Guillemaut, P., Grienenberger, J. M., Jeannin, G., and Weil, J.-H. (1986). *Plant Mol. Biol.* **7**, 245,253.

Marechal, L., Runeberg-Roos, P., Grienenberger, J. M., Colin, J., Weil, J. H., Lejeune, B., Quetier, F., and Lonsdale, D. M. (1987). *Curr. Genet.* **12**, 91–98.

Marechal-Drouard, L., Weil, J.-H., and Guillemaut, P. (1988). *Nucleic Acids Res.* **16**, 4777–4788.

Monroy, A. F., Gao, C., Zhang, M., and Brown, G. G. (1990). *Curr. Genet.* **17**, 427–431.

Moon, E., Kao, T.-H., and Wu, R. (1985). *Nucleic Acids Res.* **13**, 3195–3211.

Moon, E., Kao, T.-H., and Wu, R. (1988). *Mol. Gen. Genet.* **213**, 247–253.

Morgan, A., and Maliga, P. (1987). *Mol. Gen. Genet.* **209**, 240–246.

Morikami, A., and Nakamura, K. (1987). *J. Biochem.* **101**, 967–976.

Mulligan, R. M., Lau, G. T., and Walbot, V. (1988a). *Proc. Natl. Acad. Sci. U.S.A.* **85**, 7998–8002.

Mulligan, M. R., Maloney, A. P., and Walbot, V. (1988b). *Mol. Gen. Genet.* **211**, 373–380.

Mulligan, R. M., Leon, P., and Walbot, V. (1991). *Mol. Cell. Biol.* **11**, 533–543.

Nagy, F., Lazar, G., Menczel, L., and Maliga, P. (1983). *Theor. Appl. Genet.* **66**, 203–207.

Negruk, V. I., Eisner, G. I., Redichkina, T. D., Dumanskaya, N. N., Cherny, D. I., Alexandrov, A. A., Shemyakin, M. F., and Butenko, R. G. (1986). *Theor. Appl. Genet.* **72**, 541–547.

Newton, K. J. (1988). *Annu. Rev. Plant Physiol. Plant Mol. Biol.* **39**, 503–532.

Newton, K. J., and Coe, E. H., Jr. (1986). *Proc. Natl. Acad. Sci. U.S.A.* **83**, 7363–7366.

Newton, K. J., Coe, E. H., Jr., Gabay-Laughnan, S., and Laughnan, J. (1989). *Maydica* **34**, 291–296.

Newton, K. J., Knudsen, C., Gabay-Laughnan, S., and Laughnan, J. (1990). *Plant Cell* **2**, 107–113.

Newton, K. J., and Walbot, V. (1985). *Proc. Natl. Acad. Sci. U.S.A.* **82**, 6879–6883.

Nivison, H. T., and Hanson, M. R. (1989). *Plant Cell* **1**, 1121–1130.

Nowak, C., and Kuck, U. (1990). *Nucleic Acids Res.* **18**, 7164.

Nugent, J. M., and Palmer, J. D. (1988). *Curr. Genet.* **14**, 501–509.

Ohyama, K., Ogura, Y., Oda, K., Yamato, K., Ohta, E., Nakamura, Y., Takemura, M., Nosato, N., Akashi, K., Kanegae, T., and Yamada, Y. (1990). *In* "Proceedings of the International Symposium on Evolution of Life." Springer-Verlag, Tokyo.

Ozias-Akins, P., Pring, D. R., and Vasil, I. K. (1987). *Theor. Appl. Genet.* **74**, 15–20.

170   MAUREEN R. HANSON AND OTTO FOLKERTS

uffttt

Paillard, M., Sederoff, R. R., and Levings, C. S., III (1985). *EMBO J.* **4,** 1125–1128.
Palmer, J. D. (1988). *Genetics* **118,** 341–351.
Palmer, J. D. (1990). *Trends Genet.* **6,** 155–120.
Palmer, J. D. (1992). In "Plant Gene Research," Vol. 6, "Organelles" (R. G. Herrmann, ed.), pp. 99–133. Springer-Verlag, Berlin.
Palmer, J. D., and Herbon, L. A. (1986). *Nucleic Acids Res.* **14,** 9755–9764.
Palmer, J. D., and Herbon, L. A. (1987). *Curr. Genet.* **11,** 565–570.
Palmer, J. D., and Herbon, L. A. (1988). *J. Mol. Evol.* **28,** 87–97.
Palmer, J. D., Markaroff, C. A., Apel, I. J., and Shirzadegan, M. (1990). In "Molecular Evolution (M. T. Clegg and S. J. O'Brien, eds.)pp. 85–96. A. R. Liss, New York.
Palmer, J. D., and Shields, C. R. (1984). *Nature (London)* **307,** 437–440.
Palmer, J. D., Shields, C. R., Cohen, D. G., and Orton, T. J. (1983). *Nature (London)* **301,** 725–728.
Parks, T. D., Dougherty, G., Levings, C. S., III, and Timothy, D. H. (1984). *Curr. Genet.* **9,** 517–519.
Pring, D. R., and Lonsdale, D. M. (1985). *Int. Rev. Cytol.* **97,** 1–46.
Pruitt, K. D., and Hanson, M. R. (1989). *Curr. Genet.* **16,** 281–291.
Pruitt, K. D., and Hanson, M. R. (1991). *Mol. Gen. Genet.* **19,** 191–197.
Quetier, F., Lejeune, B., Delorme, S., and Falconet, D. (1984). In "Higher Plant Cell Respiration, Ency. Plant Phys." Vol. 18, pp. 25–34. Springer-Verlag, Berlin.
Quetier, F., Lejeune, B., Delorme, S., Falconet, D., and Jubier, M. F. (1985). In "Molecular Form and Function of the Plant Genome," Vol. 83, pp. 413–420, Plenum, New York.
Rasmussen, J., and Hanson, M. R. (1989). *Mol. Gen. Genet.* **215,** 332–336.
Rode, A., Hartmann, C., Falconet, D., Lejeune, B., Quetier, F., Benslimane, A., Henry, Y., and de Buyser, J. (1987). *Curr. Genet.* **12,** 369–376.
Rogers, H. J., Buck, K. W., and Brasier, C. M. (1987). *Nature (London)* **329,** 558–561.
Rothenberg, M., Boeshore, M. L., Hanson, M. R., and Izhar, S. (1985). *Curr. Genet.* **9,** 615–618.
Rothenberg, M., and Hanson, M. R. (1987a). *Mol. Gen. Genet.* **209,** 21–27.
Rothenberg, M., and Hanson, M. R. (1987b). *Curr. Genet.* **12,** 235–240.
Rothenberg, M., and Hanson, M. R. (1988). *Genetics* **118,** 155–161.
Rottman, W. H., Brears, T., Hodge, T. P., and Lonsdale, D. M. (1987). *Embo J.* **6,** 1541–1546.
Sangare, A., Weil, J.-H., Grienenberger, J.-M., Fauron, C., and Lonsdale, D. (1990). *Mol. Gen. Genet.* **223,** 224–232.
Schardl, C. L., Lonsdale, D. M., Pring, D. R., and Rose, K. R. (1984). *Nature (London)* **31,** 292–296.
Schardl, C. L., Pring, D. R., and Lonsdale, D. M. (1985). *Cell (Cambridge, Mass.)* **43,** 361–368.
Schuster, W., and Brennicke, A. (1986). *Mol. Gen. Genet.* **204,** 29–35.
Schuster, W., and Brennicke, A. (1987a). *EMBO J.* **6,** 2857–2863.
Schuster, W., and Brennicke, A. (1987b). *Mol. Gen. Genet.* **210,** 44–51.
Schuster, W., and Brennicke, A. (1987c). *Nucleic Acids Res.* **14,** 5943–5954.
Schuster, W., and A. Brennicke (1990). *FEBS Lett.* **268,** 252–256.
Schuster, W., Hiesel, R., Isaac, P. G., Leaver, C. J., and Brennicke, A. (1986). *Nucleic Acids Res.* **14,** 5943–5955.
Schuster, W., Hiesel, R., Manna, E., Schoble, W., Wissinger, B., and Brennicke, A. (1987). *Plant Physiol. Biochem.* **25,** 259–264.
Schuster, W., Hiesel, R., Wissinger, B., and Brennicke, A. (1990a). *Mol. Cell. Biol.* **10,** 2428–2431.
Schuster, W., Unseld, M., Wissinger, B., and Brennicke, A. (1990b). *Nucleic Acids Res.* **18,** 229–233.

Schuster, W., Wissinger, B., Unseld, M., and Brennicke, A. (1990c). *EMBO J.* **9**, 263–269.
Schuster, W., Wissinger, B., Marchfelder, A., Binder, S., Unseld, M., Gerold, E., Hiesel, R., Schobel, W., Scheike, R., Knoop, V., Gronger, P., Ternes, R., and Brennicke, A. (1991). *Physiol. Plant* **81**, 437–445.
Sederoff, R. R., Ronald, P., Bedinger, P., Rivin, C., Walbot, V., Bland, M., and Levings, C. S., III (1986). *Genetics* **113**, 469–482.
Shirzadegan, M., Christey, M., Earle, E. D., and Palmer, J. D. (1989). *Theor. Appl. Genet.* **77**, 17–25.
Shirzadegan, M., Christey, M., Earle, E. D., and Palmer, J. D. (1991). *Plant Mol. Biol.* **16**, 21–37.
Shumway, L. K., and Bauman, L. F. (1967). *Genetics* **55**, 33–38.
Siculella, L., and Palmer, J. D. (1988). *Nucleic Acids Res.* **16**, 3787–3799.
Small, I. D., Earle, E. D., Escote-Carlson, L. J., Gabay-Laughnan, S. G., Laughnan, J. R., and Leaver, C. J. (1988). *Theor. Appl. Genet.* **76**, 609–618.
Small, I. D., Isaac, P. G., and Leaver, C. J. (1987). *EMBO J.* **6**, 865–869.
Small, I., Suffolk, R., and Leaver, C. J. (1989). *Cell (Cambridge, Mass.)* **58**, 69–76.
Smith, A. G., and Pring, D. R. (1987). *Curr. Genet.* **12**, 617–623.
Sparks, R. B., and Dale, R. M. K. (1980). *Mol. Gen. Genet.* **180**, 351–355.
Stamper, S. E., Dewey, R. D., Bland, M. M., and Levings, C. S., III (1987). *Curr. Genet.* **12**, 457–463.
Stern, D. B., Bang, A. G., and Thompson, W. F. (1986). *Curr. Genet.* **10**, 857–869.
Stern, D. B., and Gruissem, W. (1987). *Cell (Cambridge, Mass.)* **51**, 1145–1157.
Stern, D. B., Jones, H., and Gruissem, W. (1989). *J. Biol. Chem.* **264**, 18,742–18,750.
Stern, D. B., and Lonsdale, D. M. (1982). *Nature (London)* **229**, 698–702.
Stern, D. B., and Newton, K. J. (1985). *Curr. Genet.* **9**, 396–405.
Stern, D. B., and Palmer, J. D. (1984a). *Proc. Natl. Acad. Sci. U.S.A.* **81**, 1946–1950.
Stern, D. B., and Palmer, J. D. (1984b). *Nucleic Acids Res.* **12**, 6141–5157.
Stern, D. B., and Palmer, J. D. (1986). *Nucleic Acids Res.* **14**, 5651–5666.
Stevens, B. (1981). *In* "The Molecular Biology of the Yeast *Saccharomyces:* Life Cycle and Inheritance" (J. N. Strathern, E. W. Jones, and J. R. Broach, eds.), pp. 471–504. Cold Spring Harbor Laboratory, Cold Spring Harbor, NY.
Sutton, C. A., Conklin, P. L., Pruit, K. D., Hanson, M. R. (1991). *Mol. Cell Biol.* **11**, 4274–4277.
Synenki, R. M., Levings, C. S., III, and Shah, D. M. (1978). *Plant Physiol.* **61**, 460–464.
Turano, F. J., DeBonte, L. R., Wilson, K. G., and Matthews, B. F. (1987). *Plant Physiol.* **84**, 1074–1079.
Turpen, T., Garger, S. J., Marks, M. D., and Grill, L. K. (1987). *Mol. Gen. Genet.* **209**, 227–233.
Van den Bogaart, P., Samallo, J., and Agsteribbe, E. (1982). *Nature (London)* **298**, 187–189.
Wahleithner, J. A., Macfarlane, J. L., and Wolstenholme, D. R. (1990). *Proc. Natl. Acad. Sci. U.S.A.* **87**, 548–552.
Wahleithner, J. A., and Wolstenholme, D. R. (1987). *Curr. Genet.* **12**, 55–67.
Wahleithner, J. A., and Wolstenholme, D. R. (1988a). *Curr. Genet.* **14**, 163–170.
Wahleithner, J. A., and Wolstenholme, D. R. (1988b). *Nucleic Acids Res.* **16**, 6897–6913.
Wallace, D. C., (1989). *Trends Genet.* **5**, 9–13.
Wallace, D. C., Zheng, X., Lott, M. T., Shoffner, J. M., Hodge, J. A., Kelley, R. I., Epstein, C. M., and Hopkins, L. C. (1988). *Cell (Cambridge, Mass.)* **55**, 601–610.
Ward, B. L., Anderson, R. S., and Bendich, A. J. (1981). *Cell (Cambridge, Mass.)* **25**, 793–803.
Warmke, H. E., and Lee, S. L. J. (1978). *Science* **200**, 561–563.
Weber, F., Dietrich, A., Weil, J.-H., and Marechal-Drouard, L. (1990). *Nucleic Acids Res.* **18**, 5027–5030.

Wells, R., and Ingle, J. (1970). *Plant Physiol.* **46,** 178–179.

Wintz, H., Grienenberger, J. R., Weie, H. J., and Lonsdale, D. R. (1988a). *Curr. Genet.* **13,** 247–254.

Wintz, H., Chen, H.-C., and Pillay, D. N. (1988b). *Curr. Genet.* **13,** 255–260.

Wintz, H., Chen, H.-C., and Pillay, D.T.N. (1989). *Curr. Genet.* **15,** 155–160.

Wintz, H., and Hanson, M. R. (1990). *Curr. Genet.* **19,** 61–64.

Wise, R. P., Fliss, A. E., Pring, D. R., and Gengenbach, B. G. (1987a). *Plant Mol. Biol.* **9,** 121–126.

Wise, R. P. Pring, D. R., and Gengenbach, B. G. (1987b). *Proc. Natl. Acad. Sci. U.S.A.* **84,** 2858–2862.

Wissinger, B., Hiesel, R., Schuster, W., and Brennicke, A. (1988). *Mol. Gen. Genet.* **212,** 56–65.

Wissinger, B., Schuster, W., and Brennicke, A. (1990). *Mol. Gen. Genet.* **224,** 389–395.

Wissinger, B., Schuster, W., and Brennicke, A. (1991). *Cell (Cambridge, Mass.)* **65,** 473–482.

Yang, A. J., and Mulligan, R. M. (1991). *Mol. Cell. Biol.* **11,** 4278–4281.

Young, E. G., and Hanson, M. R. (1987). *Cell (Cambridge, Mass.)* **50,** 41–49.

Young, E. G., Hanson, M. R., and Dierks, P. M. (1986). *Nucleic Acids Res.* **14,** 7995–8006.

Zabala, G., O'Brien-Vedder, C., Walbot, V. (1987). *Proc. Natl. Acad. Sci. U.S.A.* **84,** 7861–7865.

Zabala, G., and Walbot, V. (1988). *Mol. Gen. Genet.* **211,** 386–392.

Animal Mitochondrial DNA: Structure and Evolution

David R. Wolstenholme
Department of Biology, University of Utah, Salt Lake City, Utah 84112

I. Introduction

Mitochondria of multicellular animals (metazoa), like the mitochondria of other kinds of eukaryotes, contain their own genome. Acceptance of this notion followed the demonstration that circular DNA molecules of uniform contour length could be isolated from mitochondrial (mt-) fractions of cells of chicken and mouse (Van Bruggen *et al.*, 1966; Sinclair and Stevens, 1966; Nass, 1966). Within a short time thereafter circular DNA molecules were isolated from mitochondria of other vertebrates and some invertebrates (Wolstenholme *et al.*, 1971). With only one exception, all of the mt-genomes of metazoa examined to date, which range from sea anemones to man, are in the form of a single circular DNA molecule. The exception is the mt-genome of the cnidarian *Hydra attenuata,* which comprises two unique 8-kb linear DNA molecules (Warrior and Gall, 1985). The circular metazoan mt-genomes range in size from 14 kb (the nematode, *Caenorhabditis elegans;* Okimoto *et al.,* 1992a) to 42 kb (the scallop, *Placeopecten megellanicus;* LaRoche *et al.,* 1990).

Circular metazoan mtDNA molecules are covalently closed structures (Kroon *et al.*, 1966; Dawid and Wolstenholme, 1967), and a varying fraction occur as catenated dimers, or higher oligomers (see Wolstenholme *et al.,* 1973, for references). More rarely, and usually associated with cellular abnormalities and tissue culture, circular molecules comprising two copies of the mt-genome arranged head to tail (circular dimers) are found (Clayton and Vinograd, 1967, 1969; Wolstenholme *et al.,* 1973). The mtDNA molecule of human was the first mt-genome from any organism to be completely sequenced, analyzed, and interpreted in regard to gene content. The published sequence was, in fact, a mosaic. It was derived mainly from mtDNA of a placenta (Drouijn, 1980), but contained segments derived from HeLa cell mtDNA (Anderson *et al.*, 1981, 1982b).

This genome was shown to contain the genes for two RNAs homologous to the 16S and 23S ribosomal RNAs (rRNAs) of *Escherichia coli*, and for 22 transfer RNAs (tRNAs), and 13 open reading frames. Six of these open reading frames were identified as the genes for enzymes or components of enzymes involved in oxidative-phosphorylation: cytochrome *b* (Cyt b), subunits I–III of cytochrome *c* oxidase (COI–III), and subunits 6 and 8 of the F_0 ATPase complex (ATPase6 and ATPase8). The remaining seven open reading frames, designated URF1-6 and 4L, were later shown to encode subunits of the respiratory chain NADH dehydrogenase complex and have since been referred to as ND1-6 and ND4L (Chomyn *et al.*, 1985, 1986).

With few exceptions, mtDNA molecules from a broad range of vertebrates and invertebrates have proved to have the same gene content as the human mtDNA molecule, although there is a great deal of variation regarding the relative arrangements of the genes in different molecules. In all metazoan mtDNA molecules sequenced there are very few or no nucleotides between genes. However, there is a single noncoding region that in some vertebrates has been shown to contain sequences essential for the initiation of transcription and replication (see Clayton, 1984), and therefore has been designated the control region. Some of the size variation among metazoan mtDNAs results from differences in gene lengths. However, the greatest size variations are attributable to differences in the length of the control region, some of which contain repeated sequences. Mitochondrial DNA molecules that are of different size or that contain sequence differences are found in individuals of some species, a condition know as heteroplasmy (Solignac *et al.*, 1983; Hauswirth and Laipis, 1982; Rand and Harrison, 1989; Hoeh *et al.*, 1991).

In contrast to the relative uniformity of gene content, metazoan mtDNAs exhibit an abundance of genetic novelties. Metazoan mt-genetic codes are the most highly modified of all known genetic codes. Also, it appears that at least six unorthodox translation initiation codons are utilized in transcripts of metazoan mt-protein genes. Individual or bicistronic gene transcripts are precisely cleaved from primary transcripts of entire mtDNA strands and have no or few upstream and downstream nontranslated nucleotides. In some cases, gene transcripts end in U or UA that are polyadenylated to provide complete translation termination codons.

Metazoan mtDNAs encode tRNAs that have an unequaled variety of secondary structures. Also, the genes for both RNA components of the small and large mt-ribosomal subunits are reduced in size to greatly different degrees in different phylogenetic lines.

The complementary strands of mammalian and some other vertebrate and invertebrate mtDNA molecules differ sufficiently in guanine and thymine content that they can be separated in alkaline cesium chloride gradients. The complementary strands of these mtDNA molecules thus acquired the designations heavy (H) and light (L) strands that have been used as strand definitions in replication and transcription studies of mammalian mt-genomes (Clayton, 1982, 1984, 1991).

Replication of metazoan mtDNA has been studied in mammals and, to a lesser degree, in *Drosophila*. In both cases replication is unidirectional around the molecule and uniquely asymmetrical. In mouse and human tissue culture cell mtDNAs (the replication mechanism of which has been examined in greatest detail, reviewed in Clayton, 1982, 1991, 1992) synthesis of one (H) strand initiates at a specific location in the control region and proceeds for two-thirds of the distance around the molecule before synthesis of the second (L) strand is initiated in a noncoding sequence between two tRNA genes. A more symmetrical mechanism has been shown to be employed in mtDNA molecules from both normal and malignant rodent tissues (all liver derived): in this case L-strand synthesis appears to initiate at various locations within 5–66% of the H-strand origin (Wolstenholme *et al.*, 1974). In contrast, in *Drosophila* mtDNAs, synthesis of the first strand can be up to 99–100% complete before synthesis of the complementary strand is initiated, but again, replication of other molecules proceeds in a more nearly symmetrical manner (Goddard and Wolstenholme, 1978, 1980).

Mitochondrial DNA molecules of metazoa are maternally inherited (Dawid and Blacker, 1972; Hutchison *et al.*, 1974; Buzzo *et al.*, 1978; Fauron and Wolstenholme, 1980b), although a few cases of biparental inheritance have been reported (Kondo *et al.*, 1990; Hoeh *et al.*, 1991), and evidence for recombination between molecules is lacking (discussed in Brown, 1985; Moritz *et al.*, 1987). Consequently, mtDNA molecules from two female progeny of the same mother are inherited as separate clones and accumulate substitutions, small deletions/insertions, and sequence rearrangement. Further, in vertebrates, but questionably in invertebrates, nucleotide substitutions occur at a much higher rate in mtDNA than in nuclear DNA (Brown *et al.*, 1979, 1982; Brown, 1983, 1985; Moritz *et al.*, 1987; Vawter and Brown, 1986; Powell *et al.*, 1986; DeSalle *et al.*, 1987; Monnerot *et al.*, 1990). Differences in nucleotide sequences between mtDNA molecules of different populations of the same species, as well as mtDNA molecules of different species of a genus, have been greatly exploited in recent years to gain information regarding population structure and evolutionary histories of a wide variety of organisms (see for example Avis *et al.*, 1987; Avis, 1992).

The purpose of this chapter is to review what is known at the present time concerning the various structural features of metazoan mtDNA molecules and, from the distribution of these features among phylogenetically diverse groups, what we can deduce concerning the approximate time that each may have first appeared during the evolution of the metozoa. Previous reviews that contained information on the structure and evolution of metazoan mtDNAs include those of Wallace (1982), Brown (1983, 1985), Attardi (1985), Moritz *et al.* (1987), and Chomyn and Attardi (1987).

II. Genome Content and Organization

Complete nucleotide sequences and gene contents have been determined for five mammals, human, cow, mouse, rat, and fin whale (Anderson *et al.*, 1981, 1982a,b; Bibb *et al.*, 1981; Gadaleta *et al.*, 1989; Arnason *et al.*, 1991); a bird, *Gallus domesticus* (Desjardins and Morais, 1990); an amphibian, *Xenopus laevis* (Roe *et al.*, 1985); two echinoderms, *Strongylocentrotus purpuratus*, and *Paracentrotus lividus* (Jacobs *et al.*, 1988; Cantatore *et al.*, 1989); an insect, *Drosophila yakuba* (Clary and Wolstenholme, 1984, 1985a); three nematodes, *Ascaris suum*, *Caenorhabditis elegans*, and *Meloidogyne javanica* (Wolstenholme *et al.*, 1987; Okimoto *et al.*, 1991, 1992a,c); and a cnidarian, *Metridium senile* (Pont *et al.*, 1992a; Beagley *et al.*, 1992b). The exact size of each of these molecules is given in Table I. Also, sufficient nucleotide sequences have been obtained from mtDNA molecules of a fish, *Gadus morhua* (Johansen *et al.*, 1990), and a mollusk, *Mytilus edulis* (Hoffman *et al.*, 1992), to establish gene orders in these molecules.

Other partial mtDNA sequences that have been obtained include the following: chimpanzee, *Pan troglodytes*, pygmy chimpanzee, *P. paniscus*, lowland gorilla, *Gorilla gorilla* (Foran *et al.*, 1988); japanese monkey, *Macaca fucata* (Hayasaka *et al.*, 1991); two dolphins, *Cephalorhynchus commersonii* and *Delphinus delphis* (Southern *et al.*, 1988); a bird, japanese quail, *Coturnix japonica* (Desjardins and Morais, 1991); six species of salmonid fish (four of the genus *Orcorhynchus* and two of the genus *Salma*) (Thomas and Beckenbach, 1989); a sea urchin, *Arabacia lixula* (De Giorgi *et al.*, 1991); starfish and sea stars, *Pisaster ochraceus*, *Asterias forbessi*, *Asterias amureusis* (Smith *et al.*, 1989, 1990; Jacobs *et al.*, 1989), and *Asternia pectinifera* [Himeno *et al.*, 1987; the complete sequence of *A. pectinifora* mtDNA has been determined (Asakawa *et al.*, 1991), but not published]; the insects *Drosophila melanogaster* (de Bruijn, 1983; Garesse, 1988), *D. virilis* (Clary and Wolstenholme, 1987), and *D. teisieri* (Monnerot *et al.*, 1990) and honey bee, *Apis mellifera* (Crozier *et al.*, 1989), mosquito, *Aedes albopictus* (HsuChen *et al.*, 1984; Dubin *et al.*, 1986), and locust, *Locusta migratoria* (McCracken *et al.*, 1987; Uhlenbusch *et al.*, 1987; Hanke and Gellissen, 1988); the brine shrimp, *Artemia* sp. (Batuecas *et al.*, 1988); a platyhelminth, the liver fluke, *Fasciola hepat-*

TABLE I

Sizes, in Nucleotide Pairs, of the Metazoan, Circular mtDNA Molecules That Have Been Completely Sequenced

Organism	Total nucleotide pairs	Reference[a]
Mammalia		
Homo sapiens (human)	16,569	1
Mus musculus (mouse)	16,295	2
Bos taurus (cow)	16,338	3
Rattus norvegicus (rat)	16,298	4
Balaenoptera physalus (fin whale)	16,398	5
Aves		
Gallus domesticus (chicken)	16,775	6
Amphibia		
Xenopus laevis (South African clawed toad)	17,553[b]	7
Echinodermata		
Paracentrotus lividus (sea urchin)	15,697	8
Strongylocentrotus purpuratus (sea urchin)	15,650	9
Asterina pectinifera (starfish)	16,260	10
Athropoda		
Drosophila yakuba (fruit fly)	16,019	11
Nematoda		
Caenorhabditis elegans (soil nematode)	13,794	12
Ascaris suum (pig gut nematode)	14,284	12
Meloidogyne javanica (root knot nematode)	~20,500[c]	13
Cnidaria		
Metridium senile (sea anemone)	17,443	14

[a] 1, Anderson *et al.* (1981); 2, Bibb *et al.* (1981); 3, Anderson *et al.* (1982b); 4, Gadaleta *et al.* (1989); 5, Arnason *et al.* (1991); 6, Desjardins and Morais (1990); 7, Roe *et al.* (1985); 8, Cantatore *et al.* (1989); 9, Jacobs *et al.* (1988); 10, Asakawa *et al.* (1991); 11, Clary and Wolstenholme (1985a); 12, Okimoto *et al.* (1992a); 13, Okimoto *et al.* (1991, 1992c); 14, Beagley *et al.* (1992a).

[b] This value does not include the additional nucleotides found in the *X. laevis* mt-s-rRNA gene by Dunon-Bluteau and Brun (1986).

[c] The exact number of nucleotides contained in the series of 102-ntp direct repeats has not been determined (Okimoto *et al.*, 1991).

ica (Garey and Wolstenholme, 1989); two cnidarians, a coral, *Sarcophyton glaucum* (Pont *et al.*, 1992b), and *Hydra attenuata* (Pont *et al.*, 1992c). Each of the metazoan mtDNA molecules that has been completely sequenced has the same gene content (Fig. 1). with the following exceptions. An ATPase8 gene has not been located in the mtDNA molecules of the nematodes *C. elegans, A. suum*, and *M. javanica* (Wolstenholme *et al.*, 1987; Okimoto *et al.*, 1991; 1992a,c). The *M. javanica* mtDNA molecule contains an unidentified open reading frame of 116 codons downstream

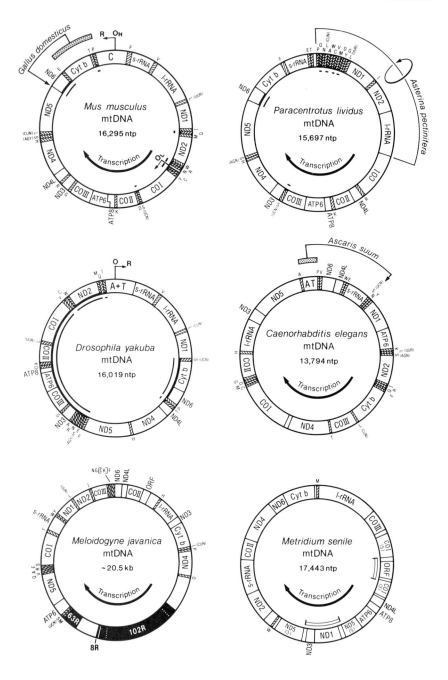

from the COII gene that would be transcribed in the same direction as all other protein, rRNA, and tRNA genes in this molecule (Fig. 1; Okimoto *et al.*, 1992c). In the mtDNA molecule of the sea anemone *Metridium senile* only two tRNA genes occur, for tryptophan and formyl-methionine.

FIG. 1   Gene maps of mtDNA molecules of a mammal (*Mus musculus:* Bibb *et al.*, 1981), an insect (*Drosophila yakuba;* Clary and Wolstenholme, 1985a), an echinoderm (*Paracentrotus lividus;* Cantatore *et al.,* 1989), two nematodes (*Caenorhabditis elegans* and *Meloidogyne javanica;* Okimoto *et al.,* 1991, 1992a,c), and a cnidarian (*Metridium senile,* Beagley *et al.,* 1992b) that have been completely sequenced. Each tRNA gene is identified by the one-letter amino acid code. Serine and leucine tRNA genes are also identified by the codon family that the corresponding tRNAs recognize. s-rRNA and l-rRNA identify the small and large rRNA genes, respectively. The 12 protein genes found in all of the molecules represented are COI, COII, and COIII (subunits, I, II, and III of cytochrome *c* oxidase); Cyt b (cytochrome *b*); ATP6 (subunit 6 of the F_0 ATPase complex); and ND1-ND6, and ND4L (components 1–6 and 4L of the respiratory chain NADH dehydrogenase). All sequenced mtDNA molecules other than those of nematodes also contain a gene (ATP8) for subunit 8 of the F_0ATPase complex. In the *C. elegans, M. javanica,* and *M. senile* mtDNA molecules, all genes are transcribed in a clockwise direction, as shown. In all other mtDNA molecules represented, some genes are also transcribed in a clockwise direction, and some, identified by bars inside the circular maps, are transcribed anticlockwise. The locations of the origins of replication in mouse [O_H in the control (C) region] and *Drosophila yakuba* (O in the adenine + thymine (A + T)-rich region) mtDNA molecules are shown (Goddard and Wolstenholme, 1980; Clayton, 1982). Also shown in the mouse mtDNA molecule is the origin of L-strand synthesis (O_L; Clayton, 1982). Arrows marked "R" indicate direction of replication around the molecule. The extent of two group I introns within the *M. senile* COI and ND5 genes are identified by open bars on the inside of the map. ORF in the *M. senile* map identifies the open reading frame within the COI gene intron. In the *M. javanica* map, designation of the tentative 5' end of the l-rRNA gene is shown by a dashed line; the open reading frame (ORF) upstream from the l-rRNA gene has not been identified; bracketed K and C indicate putative tRNAS that lack both DHU and TΨC arms; the blacked-in areas indicate the location of tandamly arranged, directly repeated sequences: 102R, approximately 36 copies of a 102-ntp sequence; 8R, five copies of an 8-ntp sequence; 63R, 11 copies of a 63 ntp sequence. The continuous sequence of the segment between the dotted lines in the 102R region has not been determined. Gene arrangement in four other sequenced mammalian mtDNA molecules (human, cow, rat, and fin whale; Anderson *et al.,* 1981, 1982b; Gadeleta *et al.,* 1989; Arnason *et al.,* 1991) and the amphibian *Xenopus laevis* (Roe *et al.,* 1985) are identical to that in mouse mtDNA. Gene arrangement in the mtDNA molecules of a bird (chicken, *Gallus domesticus*) differs from that of mouse only in the relative locations of the gene sets ND6 and tRNA[glu], and Cyt b, tRNA[thr], and tRNA[pro] (Desjardins and Morais, 1990) as indicated by the arrow outside the mouse map. The mtDNA molecule of the nematode *Ascaris suum* differs from that of the mtDNA molecule of *C. elegans* only in the location of the AT region (shown by the arrow; *Ascaris suum*) within the *C. elegans* map (Wolstenholme *et al.,* 1987; Okimoto *et al.,* 1992a). The mtDNA molecule of the sea urchin *Strongylocentroutus purpuratus* (Jacobs *et al.,* 1988) has an identical gene arrangement to that of the sea urchin *Paracentrotus lividus.* However, gene arrangement within the mtDNA molecule of the starfish *Asterina pectinifera* differs from that in the sea urchin mtDNA molecules in that a segment containing two protein genes (ND1 and ND2), the l-rRNA gene, and 14 tRNA genes has been inverted, as indicated (Asakawa *et al.,* 1991).

Also, in this molecule the COI and ND5 genes each contain a group I intron: the COI gene intron includes an open reading frame that, from predicted amino acid sequence motifs, may encode either an RNA splicase (maturase) or an endonuclease. The ND5 gene intron contains the molecule's only copies of the ND1 and ND3 genes (Beagley *et al.*, 1992a). From partial sequence data, it is also known that the mtDNA molecules of a cnidarian, *Sarcophyton glaucum*, contains an open reading frame that may encode an enzyme concerned with mismatch repair (Pont *et al.*, 1992b) and that the mtDNA molecule of the mollusk *Mytilus edulis* contains an extra tRNA^met gene that has a TAT anticodon (Hoffman *et al.*, 1992).

The component protein, rRNA, and tRNA genes have identical arrangements in fish, amphibian, and mammalian mtDNAs. An arrangement similar to that found in mammals and amphibia is also found in bird mtDNA, except that in this case the segments of the molecule containing the Cyt b, tRNA^pro, and tRNA^thr genes, and the ND6 and tRNA^glu genes have been transposed relative to each other (Desjardins and Morais, 1990, 1991) (Fig. 1). In comparison to vertebrate mtDNAs, some limited protein and rRNA gene rearrangements, and more extensive tRNA gene rearrangements, occur in mtDNA molecules of *D. yakuba*, sea urchin, and starfish (Fig. 1; Clary and Wolstenholme, 1985a; Jacobs *et al.*, 1988; Cantatore *et al.*, 1989; Asakawa *et al.*, 1991).

Differences in relative locations of from two to four tRNA genes (but not protein or rRNA genes) have been noted between the mtDNA molecules of *Drosophila yakuba* and the mosquito *Aedes albopictus* (Dubin *et al.*, 1986); *D. yakuba* and the honey bee, *Apis mellifera* (Crozier *et al.*, 1989); three marsupials (opossums, *Trichosurus vulpecular*, *Philander opossum andersoni*, *Marmosa germana rutteri*) and placental mammals (Pääbo *et al.*, 1991). These observations together with those concerning differences in gene order among vertebrate, echinoderm, and *Drosophila* mtDNAs clearly suggest that rearrangements involving tRNAs occur more frequently than rearrangements involving protein and rRNA genes.

At the 5' end of the ND5 gene of the echinoderm *Paracentrotus lividus* is a sequence encoding 24 amino acids that has no counterpart in other, nonechinoderm metazoan ND5 genes. From considerations of sequence similarities, it has been argued that this 5'-end proximal sequence was once a tRNA^leu(CUN) gene and that the present-day tRNA^leu(CUN) gene (found elsewhere in the mtDNA molecule, Fig. 1) originated as a duplication of the tRNA^leu(UUR) gene (Cantatore *et al.*, 1987). Conservation of these sequence arrangements have been found in the mtDNAs of two other echinoderms, *Arbacia lixula* and *Strongylocentrotus purpuratus* (De Giorgi *et al.*, 1991).

Gene arrangements in the *C. elegans* and *A. suum* mtDNA molecules are identical, except for the location of the AT region (Fig. 1). Otherwise, there is an extreme lack of conservation of gene arrangement among the sequenced mtDNA molecules of lower invertebrates (the root knot nematode, *M. javanica*, the mollusk, *M. edulis*, and the sea anemone, *M. senile*) and between each of these mtDNA molecules and the mtDNA molecules of *D. yakuba*, sea urchins, starfish, and vertebrates (Fig. 1).

III. Protein Genes

A. Identification and Size

Of the 13 open reading frames recognized in the human (and cow) mtDNA molecules, four (COI, COII, COIII, and Cyt b) were originally identified in regard to the proteins they encode, from similarities of their predicted amino acid sequences to known amino acid sequences of bovine proteins, and predicted amino acid sequences of yeast mt-protein genes. ATPase6 was identified from sequence similarities to the yeast ATPase6 gene (Anderson *et al.*, 1981, 1982b).

The protein encoded by URFA6L was later identified from predicted amino acid sequence similarities as the homolog of the yeast *aapl* gene product (a component of the ATP synthase complex, Macreadie *et al.*, 1983). It has since been clearly demonstrated that the protein encoded by bovine URFA6L is associated with the ATP synthase complex (Fearnley and Walker, 1986) and the gene for this protein is now called ATPase8. The protein products of the remaining seven open reading frames (URF1-6 and 4L) in human mtDNA were identified as components of the respiratory chain NADH dehydrogenase complex by precipitation with antibodies raised against bovine NADH ubiquinone oxidoreductase (Chomyn *et al.*, 1985, 1986). These open reading frames were therefore designated the ND1-6 and ND4L genes.

Identification of the 13 (or 12) protein genes in other metazoan mtDNAs have been based entirely on amino acid sequence similarities, or, particularly for the genes that are conserved to a lesser extent (ATPase6, ND4L, and ND6), amino acid sequence and hydropathic profile similarities to the corresponding genes of human mtDNA, or to genes that have been previously identified by comparisons to the human mt-protein genes (see, for example, Bibb *et al.*, 1981; Clary and Wolstenholme, 1985a; Okimoto *et al.*, 1992a).

The number of codons in the corresponding 12 or 13 protein genes of some phylogenetically diverse, completely sequenced metazoan mtDNAs are compared in Table II. Among the five mtDNAs represented, the genes COI, COII, and COIII and Cyt b vary in size within the range 2.8% (COIII)

TABLE II

Comparisons of the Numbers of Codons in the Protein Genes of Completely Sequenced mtDNAs of Various Metazoa

Gene	Mouse	P. lividus	D. yakuba	C. elegans	M. senile	Maximum percentage variation
COI	514	517	512	525	530	3.5
COII	227	229	228	231	248	9.2
COIII	261	260	262	255	262	2.8
Cyt b	381	380	378	370	393	6.2
ATPase6	226	232	224	199	229	16.6
ATPase 8	67	54	53	NF[a]	72	35.9
ND1	315	323	324	291	334	14.8
ND2	345	352	341	282	385	36.5
ND3	114	116	117	111	118	6.3
ND4	459	463	446	409	491	20.0
ND4L	97	97	96	77	99	28.6
ND5	607	638	573	628	600	11.3
ND6	172	160	174	145	202	39.3

Note. Data are taken from the following sources: M. musculus, Bibb et al. (1981); P. lividus, Cantatore et al. (1989); D. yakuba, Clary and Wolstenhome (1985a); C. elegans, Okimoto et al. (1992a); M. senile, Pont et al. (1992a) and Beagley et al. (1992b).
[a] NF, not found.

to 9.3% (COII). In contrast, the variations in size of the ATPase6 and ATPase8 genes are 16.6 and 35.9%, respectively, and the variations in size of the genes for NADH dehydrogenase subunits are within the range of 6.3% (ND3) to 39.3% (ND6).

All of the protein genes encoded in M. senile mtDNA, except the ND5 gene, are larger than in any other species. The smallest of each of the mt-protein genes, except for COI, COII, and ND5, are found in C. elegans mtDNA. All of the 12 or 13 proteins encoded by all metazoan mtDNAs are rich in hydrophobic amino acids, the most prevalent of which, in every case, is leucine (Table III).

Among mtDNAs of vertebrates and higher invertebrates (insects and echinoderms), there are genes that overlap. Some overlaps are between the 3' ends of two genes that are encoded in opposite strands of the molecule. However, there are two cases of overlap that involve genes encoded in the same strand: the ATPase8 gene overlaps the ATPase6 gene by between 2 and 46 ntp in vertebrate and higher invertebrate mtDNAs, and the ND4L gene overlaps the ND4 gene by 7 ntp in vertebrate mtDNAs.

TABLE III

Predicted Amino Acid Compositions of the Proteins Encoded in Some Completely Sequenced Metazoan mtDNAs

Amino acid	H. sapiens	M. musculus	G. domesticus	X. laevis	P. lividus	D. yakuba	C. elegans	M. senile
Phe	5.7	6.3	5.8	6.0	8.3	8.9	13.2	7.9
Leu	16.9	15.5	17.5	15.9	15.4	16.8	15.5	14.5
Ile	8.5	9.8	8.0	8.9	9.5^a	9.6	8.2	9.1
Met	5.5	6.5	4.4	5.2	2.7^a	5.7	5.2	3.5
Val	4.4	4.5	4.2	4.7	6.3	5.2	7.4	8.3
Ser	7.2	7.6	7.8	8.0	9.5	9.0	11.0	6.8
Pro	5.8	5.4	6.2	5.4	4.6	3.5	2.4	4.2
Thr	9.2	7.8	9.2	8.1	6.5	5.0	4.2	5.5
Ala	6.7	6.1	7.6	7.3	8.1	4.6	3.3	8.1
Tyr	3.6	3.3	2.8	3.1	2.9	4.6	5.1	4.2
His	2.6	2.6	3.0	2.5	2.0	2.1	1.7	2.0
Gln	2.4	2.2	2.3	2.6	2.2	1.9	1.3	2.0
Asn	4.3	4.4	3.3	4.0	4.8	5.5	4.4	2.9
Lys	2.5	2.7	2.4	2.3	1.4	2.3	3.2	2.6
Asp	1.7	1.9	1.7	2.0	1.7	1.7	1.8	2.5
Glu	2.3	2.5	2.5	2.6	2.7	2.2	2.3	2.6
Cys	0.6	0.8	0.7	0.8	0.8	1.1	1.4	0.8
Trp	2.8	2.7	2.9	3.1	2.8	2.7	2.1	2.5
Arg	1.7	1.7	1.9	1.8	2.0	1.6	0.9	2.5
Gly	5.6	5.6	5.7	5.8	5.9	6.0	5.5	7.6

Note: All mtDNAs represented contain the same 13 protein genes except that *C. elegans* mtDNA lacks a gene for ATPase8 (Fig. 1). Data are taken from the following sources: *H. sapiens*, Anderson *et al.* (1981); *M. musculus*, Bibb *et al.* (1981); *G. domesticus*, Desjardins and Morais (1990); *X. laevis*, Roe *et al.* (1985); *P. lividus*, Cantatore *et al.* (1989); *D. yakuba*, Clary and Wolstenholme (1985a); *C. elegans*, Okimoto *et al.* (1992a); *M. senile*, Pont *et al.* (1992a) and Beagley *et al.* (1992b).

[a] These frequencies are based on the interpretation that ATA specifies isoleucine in the *P. lividus* mt-genetic code (Cantatore *et al.*, 1989), as in the standard genetic code, rather than methionine as in all other currently analyzed metazoan mt-genetic codes.

Mature, bicistronic transcripts containing the ATPase8 and ATPase6 reading frames, and the ND4L and ND4 reading frames, have been isolated from HeLa cell mitochondria (Ojala *et al.*, 1981). Fearnley and Walker (1986) isolated two bovine proteins that correspond in size and sequence to the proteins predicted from the overlapping ATPase8 and ATPase6 reading frames. This clearly indicates that translation of the ATPase6 protein must involve ribosome binding and translation initiation within the ATPase8 reading frame.

B. Mitochondrial Genetic Codes

Evidence for an altered genetic code in metazoan mtDNA was first obtained from comparisons of the predicted amino acid sequences of the human COII gene with the directly determined protein sequences of bovine cytochrome c oxidase subunit II (Barrel et al., 1979). It was found that TGA codons occur internally in the human COII gene sequence and correspond in all cases with tryptophans in the bovine protein sequence. From correspondence in position of methionine in the bovine COII protein sequence with ATA codons in the human COII gene sequence, it was reasoned that ATA probably specifies methionine rather than isoleucine in the human mt-genetic code. Also, as AGA and AGG codons do not occur internally in any of the human mt-protein genes, and human and bovine mtDNAs do not contain a gene for a tRNA that could recognize AGA and AGG codons, it was concluded that neither of these codons specify an amino acid in the human mt-genetic code. However, as the human COI gene and the cow Cyt b gene appear to end in AGA, and the human ND6 gene appears to end in AGG, it was argued that AGA and AGG may act as termination codons in mammalian mtDNAs (Anderson et al., 1981, 1982b).

All other metazoa examined to date have been found to use modified mt-genetic codes. This has been deduced using data obtained from comparisons of nucleotide and predicted amino acid sequences of the more conserved mt-protein genes, and from some considerations of relative codon frequencies in homologous genes of different species (see, for example, Wolstenholme and Clary 1985; Okimoto et al., 1992a).

The distribution among metazoan mtDNAs of codons with unusual amino acid specificities is summarized in Table IV, and a scheme by which these codon specificities might have evolved is summarized in Fig. 2. Rather than being a stop codon, TGA appears to specify tryptophan in all mt-genetic codes, except those of plants (Fox, 1987). This change in specificity is, therefore, indicated to have occurred before the ancestral lines of fungi, protista, and metazoa diverged, assuming that the mitochondria of these groups do indeed share a common ancestry, and that the change in specificity of TGA occurred only one time.

In cnidarian mtDNA, AGA and AGG specify arginine, as in the standard genetic code. However, in all other invertebrates AGA and AGG specify serine with the exception that in *D. yakuba*, and *D. melanogaster* mtDNAs AGG does not occur (Clary and Wolstenholme, 1985a; Garesse, 1988). The latter observation may be a function of the very low G content of *Drosophila* mtDNAs. The 13 mt-protein genes of *D. yakuba* have an average G content of 12.2%, and only 2.9% of all codons of these genes end in G (Table VIII). Assuming that AGC and AGT specify serine in

TABLE IV

Metazoan Mitochondrial Genetic Code Modifications: Unusual Amino Acid Specifications for Five Codons among mtDNAs of Various Groups of Metazoa

Codon:	TGA	ATA	AGA	AGG	AAA
Mammalia[a]	Trp	Met	Stop or NF[i]	Stop or NF[i]	Lys
Aves[b]	Trp	Met	NF	Stop	Lys
Amphibia[c]	Trp	Met	Stop	NF	Lys
Echinodermata[d]	Trp	Ile	Ser	Ser	Asn
Insecta[e]	Trp	Met	Ser	NF	Lys
Nematoda[f]	Trp	Met	Ser	Ser	Lys
Platyhelminthes[g]	Trp	Met	Ser	Ser	Asn
Cnidaria[h]	Trp	Ile	Arg	Arg	Lys

[a] Human, mouse, cow, rat, fin whale (Anderson *et al.*, 1981, 1982b; Bibb *et al.*, 1981; Gadaleta *et al.*, 1989, Arnason *et al.*, 1991).

[b] *Gallus domesticus* (Desjardins and Morais, 1990).

[c] *Xenopus laevis* (Roe *et al.*, 1985).

[d] *Strongylocentrotus purpuratus, Paracentrotus lividus, Pisaster ochraceus, Asterina pectinifera* (Jacobs *et al.*, 1988; Cantatore *et al.*, 1989; Himeno *et al.*, 1987; Smith *et al.*, 1990; Asakawa *et al.*, 1991).

[e] *Drosophila yakuba, D. melanogaster* (Clary and Wolstenholme 1983a,b, 1985a; de Bruijn, 1983; Garesse, 1988).

[f] *Ascaris suum, Caenorhabditis elegans, Meloidogyne javanica* (Wolstenhome *et al.*, 1987; Okimoto *et al.*, 1992a,c).

[g] *Fasciola hepatica* (Garey and Wolstenholme, 1989; Ohama *et al.*, 1990).

[h] *Metridium senile, Sarcophyton glaucum* (Pont *et al.*, 1992 a,b; Beagley *et al.*, 1992b).

[i] NF, not found.

metazoan mtDNAs, as appears to be the case in all other known genetic codes, then some invertebrate mtDNAs are unique in that they contain two families, each of four codons (TCN and AGN) that specify serine. AGA and AGG codons do not occur internally in any vertebrate mt-protein genes. However, the *Balaenoptera phyalus* COI and Cyt b genes and the *Xenopus laevis* ND6 gene end in AGA, and the *G. domesticus* COI gene ends in AGG, which, together with the occurrence of these triplets at the ends of some human and bovine mt-protein genes noted above, and the lack of a gene for a tRNA expected to recognize AGA and AGG codons, support the view that in some instances AGA and AGG are termination codons in vertebrate mtDNAs.

From the above described observations, it appears, therefore, that the change in specificity of AGA and AGG codons from arginine to serine occurred following divergence of the cnidarian ancestral line, but before divergence of the platyhelminth ancestral line from the line to which all

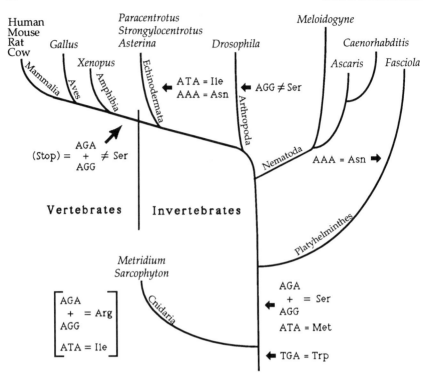

FIG. 2 Evolution of metazoan mitochondrial genetic code modifications. The earliest occurrence of each modification, based on information obtained from nucleotide and (predicted) amino acid sequence comparisons of mtDNAs of the organisms listed, is shown. Codon assignments for cnidarian mtDNAs, shown in brackets, are those that differ for other metazoan mtDNAs, but are equivalent to standard genetic code assignments. References to data on which this diagram is based are given in the text. The phylogenetic relationships shown follow those given by Wilson *et al.* (1978).

other invertebrate phyla can be rooted. The loss of amino acid coding specificity of AGA and AGG must have occurred sometime following divergence of the echinoderm ancestral line from the common vertebrate ancestral line.

In echinoderm mt-genetic codes, there appears to have been a reversion of AUA to specify isoleucine, and AAA specifies asparagine rather than lysine (Himeno *et al.*, 1987; Jacobs *et al.*, 1988; Cantatore *et al.*, 1989). The codon AAA has been interpreted as also specifying asparagine in the platyhelminth mt-genetic code (Ohama *et al.*, 1990). If substantiated this would indicate that during evolution there have been two independent changes in the specification of AAA from lysine to asparagine (Fig. 2).

Discussions of possible mechanisms by which codons in mt-protein genes may have changed specificity during metazoan evolution are contained in Osawa *et al.*, (1989a,b).

C. Translation Initiation Codons

That metazoan mt-protein gene transcripts employ unorthodox translation initiation codons (Table V) was first deduced from comparisons of human and bovine homologous gene sequences, and of human mt-protein gene sequences and the corresponding gene transcript sequences (Anderson *et al.*, 1981; Montoya *et al.*, 1981). It was concluded that AUA (methionine) and probably AUU, as well as AUG, were translation initiation codons. By directly sequencing amino-terminal regions of bovine and human mtDNA-encoded proteins, Fearnley and Walker (1987) showed that in the corresponding mtDNAs an initiation ATT, an initiation ATA, and internal ATAs all specify methionine, but internal ATTs specify isoleucine. From further DNA sequence comparisons it appears that among mt-protein genes of other mammals and echinoderms all four ATN triplets can act as initiation codons (Bibb *et al.*, 1981; Jacobs *et al.*, 1988; Cantatore *et al.*, 1989). Among *Drosophila yakuba* mt-protein genes, ATG, ATA, and ATT are used as translation initiation codons, and ATAA has been proposed as the translation initiation codon of *Drosophila* COI genes (Clary and Wolstenholme, 1983b; 1985a; de Bruijn, 1983). GTG appears to be used as the translation initiation codon of one gene in the mtDNAs of mouse (ND1, Clary and Wolstenholme, 1985a); rat (Gadaleta *et al.*, 1989); chicken (Desjardins and Morais, 1990); two echinoderms, *P. lividus* and *S. pupuratus* (ATPase8) (Jacobs *et al.*, 1988; Cantatore *et al.*, 1989); and *D. yakuba* (ND5) (Clary and Wolstenholme, 1985a).

Among the completely sequenced mtDNAs (Table I), only those of *X. laevis* and *M. senile* contain a full set of protein genes that all appear to begin with ATG (Roe *et al.*, 1985; Beagley *et al.*, 1992b). The mtDNAs of *C. elegans* and *A. suum* are the only ones that totally lack protein genes with an ATG initiation codon (Okimoto *et al.*, 1990). These mtDNAs are unusual in that 6 of the 12 mt-protein genes of *A. suum*, and 3 of the mt-protein genes of *C. elegans* appear to begin with TTG. Five of the remaining 6 *A. suum* mt-protein genes have a putative ATT initiation codon, and one, the COIII gene, has a putative GTT initiation codon. The translation initiation codons of the remaining *C. elegans* mt-protein genes appear to be ATT (6 genes) or ATA (3 genes). As homologous genes in mtDNAs of different species within a class or phylum sometimes begin with different putative translation initiation codons, it seems unlikely that the unorthodox codons play any role in regulating mt-protein gene expression, as has

TABLE V

Translation Initiation Codons and Complete and Partial Translation Termination Codons Predicted for the Protein Genes of Completely Sequenced Metazoan mtDNAs

Organism	Initiation codons	Termination codons	Reference
Mammalia			
Homo sapiens	ATG ATA ATT	TAA TAG TA T AGA AGG	1
Mus musculus	ATG ATA ATT ATC GTG[a]	TAA TA T	2
Bos taurus	ATG ATA	TAA TA T AGA	3
Rattus norvegicus	ATG ATA ATT GTG	TAA TA T	4
Balaenoptera phyalus	ATG ATA	TAA TAG TA T AGA	5
Aves			
Gallus domesticus	ATG GTG	TAA TA T AGG	6
Amphibia			
Xenopus laevis	ATG	TAA TA T AGA	7
Echinodermata			
Paracentrotus lividus	ATG ATA ATT ATC GTG	TAA TAG TA	8
Strongylocentrotus purpuratus	ATG ATA ATT ATC GTG	TAA TAG TA T	9
Arthropoda			
Drosophila yakuba	ATG ATA ATT (ATAA)	TAA TA T	10
Nematoda			
Caenorhabditis elegans	ATA ATT TTG	TAA TAG TA T	11
Ascaris suum	ATT TTG GTT	TAA TAG TA T	11
Cnidaria			
Metridium senile	ATG	TAA TAG	12

[a] See Clary and Wolstenholme, 1985a.

References: 1, Anderson et al. (1981); 2, Bibb et al. (1981); 3, Anderson et al. (1982b); 4, Gadaleta et al. (1989); 5, Arnason et al. (1991); 6, Desjardins and Morais (1990); 7, Roe et al. (1985); 8, Cantatore et al. (1989); 9, Jacobs et al. (1988); 10, Clary and Wolstenholme (1985a); 11, Okimoto et al. (1992a); 12, Beagley et al. (1992b).

been suggested for the TTG initiation codon of the *E. coli* adenylate cyclase gene (Reddy *et al.*, 1985).

D. Translation Termination

Some mammalian mt-protein genes end in either T or TA, rather than a complete translation termination codon (TAA or TAG; Table V), and in such cases the terminal nucleotide is immediately adjacent to the 5'-terminal nucleotide of the sense strand of a tRNA gene. Ojala *et al.* (1981) demonstrated that individual (mature) transcripts of such human mt-protein genes contain a complete termination codon created by post-transcriptional polyadenylation. It was argued that the primary transcription product of each strand of human mtDNA was a multicistronic RNA molecule, and that precise cleavage of this RNA between the 3'-terminal nucleotide of the protein gene transcript and the 5'-terminal nucleotide of the tRNA created the substrate for polyadenylation. It has also been argued that the precise cleavage is a function of the secondary structure of the tRNA (the tRNA punctuation model, Ojala *et al.*, 1980, 1981).

The observation that there are very few nucleotides between most pairs of genes in any of the metazoan mtDNAs that have been completely sequenced is consistent with a transcription mechanism that involves the generation of multicistronic primary transcripts. The finding that some protein genes in each of these mtDNAs, except that of *M. senile,* end in a T or TA (Table V) suggests that a cleavage–polyadenylation mechanism developed very early in the evolution of metazoan mitochondria.

IV. Ribosomal RNA Genes

All metazoan mtDNA molecules examined so far contain two genes that have been identified from sequence comparisons as the homologs of the 16S and 23S rRNA genes of *E. coli*. These mt-genes, s-rRNA and l-rRNA, encode the RNA components of the small and large subunits of mitochondrial ribosomes. A gene for an RNA equivalent of 5S RNA of the *E. coli* ribosomes is not found in metazoan mtDNAs, and a 5S rRNA homolog has not been isolated from any metazoan mt-ribosomes. The s-rRNA and l-rRNA genes are separated by a single tRNA gene in vertebrate and *Drosophila* mtDNAs. In sea star mtDNAs the two rRNA genes are separated from each other by only two tRNA genes and the short putative control region, but they are encoded in opposite strands of the molecule (Fig. 1). In all other invertebrates the two rRNA genes are contained in

TABLE VI

Comparisons of the Sizes, in Nucleotides, of the Mitochondrial s-rRNA (12S rRNA-like) Genes and l-rRNA (16S rRNA-like) Genes of Various Metazoa and of the Homologous 16S and 23S rRNA Genes of *Escherichia coli*[a]

Organism	s-rRNA	l-rRNA
Mus musculus	955	1582
Gallus domesticus	974	1621
Xenopus laevis	819	1640
Paracentrotus lividus	883	1549
Drosophila yakuba	789	1326
Caenorhabditis elegans	697	953
Metridium senile	1135	2259
Escherichia coli	1541	2904

[a] Data are taken from the following sources: *M. musculus*, Bibb *et al.* (1981); *G. domesticus*, Desjardins and Morais (1990); *X. laevis*, Roe *et al.* (1985); *P. lividus*, Cantatore *et al.* (1989); *D. yakuba*, Clary and Wolstenholme (1985a); *C. elegans*, Okimoto *et al.* (1992b); *M. senile*, Pont *et al.* (1992a) and Beagley *et al.* (1992b); *E. coli*, Brosius *et al.* (1978, 1980).

the same strand but they are separated by a varying number of protein genes (Fig. 1).

The mt-s-rRNAs and mt-l-rRNAs of mammals and insects are oligoadenylated at the 3' end (Dubin *et al.*, 1981, 1982; Kotin and Dubin, 1984; Van Etten *et al.*, 1983; Dubin and HsuChen, 1983) and contain methylated nucleotides, but to a lesser extent than their cytoplasmic counterparts (Attardi and Attardi, 1971; Dubin and Taylor, 1978; Dubin *et al.*, 1978, 1985; Baer and Dubin, 1981).

For the *E. coli* 16S and 23S rRNAs, secondary structure models have been developed, based on experimental data, and on comparisons of rRNAs of phylogenetically diverse organisms (see Noller *et al.*, 1986; Gutell and Fox, 1988). Within these models have been mapped specific nucleotides and secondary structural elements concerned with ribosomal subunit interactions, interactions between both rRNAs and tRNAs at the various stages of protein synthesis, and interactions between the s-rRNA and messenger RNAs (Noller *et al.*, 1986; Moazed and Noller, 1989, 1991; Dahlberg, 1989).

All known metazoan mt-s-rRNA and mt-l-rRNA genes are smaller than the corresponding rRNA genes of *E. coli* (Table VI). However, in each case, the mt-s-rRNA gene can be folded into a secondary structure that resemble the *E. coli* 16S rRNA model in regard to the relative locations of specific secondary structural elements, and the occurrence of highly conserved primary sequences (Gutell *et al.*, 1985; Dams *et al.*, 1988). Also, in each case, the smaller size results more from elimination of specific secondary structure elements that occur in the *E. coli* model than from deletion of single or small numbers of nucleotides throughout the structure (Zweib *et al.*, 1981; Glotz *et al.*, 1981; Steigler *et al.*, 1981; Clary and Wolstenholme, 1985b; Okimoto *et al.*, 1992b). Interestingly, however, even in the secondary structure model of the smallest of the mt-s-rRNAs represented in Table VI, that of *C. elegans* (Okimoto *et al.*, 1992b), there is clear retention of all of the sequence blocks that are universally conserved in all other s-rRNA secondary structure models (Steigler *et al.*, 1981; Gray *et al.*, 1984).

Also, the sequenced metazoan mt-l-rRNA genes can be folded into structures that resemble the *E. coli* l-rRNA in regard to conserved primary sequences and secondary structure elements (Gutell and Fox, 1988). However, for the mt-l-rRNA of *C. elegans* (and *A. suum*) this has proved possible only for the 3' 63% of the sequence. The remaining 5' portion of the nematode mt-l-rRNA sequences are extremely A + T-rich and have been severely truncated relative to other metazoan mt-l-rRNA genes (Okimoto *et al.*, 1992b).

Despite the large reduction in size of the *C. elegans* mt-rRNA genes, however, it is noteworthy that with few exceptions, the secondary structure models proposed for these mt-rRNA genes include the secondary structure element-forming sequences that in *E. coli* rRNAs contain nucleotides important for subunit interactions and for interactions with tRNAs (Okimoto *et al.*, 1992b).

All metazoan mt-s-rRNAs lack the 3'-end proximal sequence (anti-Shine-Dalgarno) that in *E. coli* pairs with a sequence (Shine-Dalgarno) 5' to the translation initiation codon of messenger RNAs to correctly position the latter for translation (Gutell and Fox, 1988). In fact, there is little information available regarding how mt-messenger RNAs interact with mt-ribosomes to achieve protein synthesis. It is known that metazoan messenger RNAs do not have a 5' 7-methyl-guanosine cap, but they are 3' polyadenylated (Clayton, 1984).

V. Transfer RNA Genes

A. Anticodon—Codon Interactions

The mtDNAs of almost all metazoa so far examined contain the genes for only 22 tRNAs, but these tRNAs are sufficient to decode the 12 or 13 metazoan mt-protein genes. This is because a single tRNA can apparently read all codons of a four-codon family (Barrell *et al.*, 1980; Anderson *et al.*, 1981). It is unclear whether the uridine (always unmodified) that occurs in the anticodon wobble (first) position of these tRNAs can bond with all four third position codon nucleotides, or whether in these cases two nucleotide pairs are sufficient to stabilize the anticodon–codon interactions (Lagervist, 1981).

In fungal mitochondria also, all members of each four-codon family are read by a single tRNA with an unmodified U in the wobble position (Heckman *et al.*, 1980; Bonitz *et al.*, 1980). It was first demonstrated for *Neurospora* mtDNA that two codon families ending in A or G are read by tRNAs that have a modified U in the wobble position, and it was suggested that the modified U prevents misreading of the complimentary two codon family ending in C or U (Heckman *et al.*, 1980). It has been proposed that a similar mechanism operates to maintain the accuracy of two codon family reading in metazoan mtDNAs. This is based on the finding of a modified U (see below) in the wobble position of mt-tRNAs that recognize two-codon families ending in G or A from rat, hamster, cow, and mosquito (Randerath *et al.*, 1981a,b; Roe *et al.*, 1982; Dubin *et al.*, 1986). Also, in tRNAs that recognize two-codon families ending in G or A (but not in other tRNAs) there is a consistent, specific modification of the nucleotide immediately following the anticodon (see below), suggesting that this nucleotide also plays a part in ensuring accurate anticodon–codon recognition (Roe *et al.*, 1982; Dubin *et al.*, 1986).

Other cases of unusual wobble are indicated to occur in metazoan mitochondria. The anticodon of all sequenced mt-tRNA[f-met] genes is CAT. Transfer RNAs seem to be imported into mitochondria only in Cnidaria (see below), and, except for the second methionine tRNA gene (anticodon TAT) identified in *Mytilus edulis* mtDNA (Hoffman *et al.*, 1992), tRNA[f-met] is the only tRNA for methionine encoded in metazoan mtDNAs; therefore, this tRNA must be able to recognize AUG and AUA codons when they occur internally, and, assuming all metazoan mt-protein genes begin with (formyl-)methionine, all four AUN codons together with TTG, GTG, and GTT (Table V) when they occur as initiation codons. This implies that the wobble position C in the tRNA[f-met] anticodon can pair with any codon third position nucleotide. Insect mt-tRNA[lys] genes each encode a tRNA with a CUU (rather than UUU) anticodon (Clary and Wolstenholme, 1983a; HsuChen *et al.*, 1983b). As both AAA and AAG codons appear to specify lysine in the insect mt-genetic code, then again there must be wobble C–A pairing. In mt-tRNA[f-met] of both cow and mosquito, and tRNA[lys] of mosquito, the C in the wobble position is unmodified (Roe *et al.*, 1982; HsuChen *et al.*, 1983b; Dubin and HsuChen, 1984).

In the mtDNA of vertebrates in which AGA and AGG are not used as amino acid-specifying codons, the anticodon of the tRNA gene that

recognizes AGT and AGC (serine) codons is GCT (GCU in the tRNA), as
expected (Arcari and Brownlee, 1980; de Bruijn *et al.*, 1980). However, a
GCT anticodon is also found in the single gene for the tRNA that recognizes the AGT, AGC, and AGA serine-specifying codons in *Drosophila
yakuba* mtDNA, and in the single gene for the tRNA that recognizes
all four serine-specifying AGN codons in echinoderm and platyhelminth
mtDNAs (Himeno *et al.*, 1987; Jacobs *et al.*, 1988; Cantatore *et al.*, 1989;
Garey and Wolstenholme, 1989). This implies that the G in the wobble
position of the GCU anticodons can pair with third position A in *D.
yakuba* AGA codons, and with third position A and G in echinoderm
and platyhelminth AGA and AGG codons. The possibility that only two
nucleotide pairs are necessary in each of the above-discussed anticodon–codon interactions is again an alternative to postulating unorthodox
anticodon first position–codon third position nucleotide pairings.

The mtDNA molecule of the sea anemone, *Metridium senile* (Cnidaria),
contains only two tRNA genes; for tRNAs expected to recognize trypto-
phan-specifying and methionine-specifying codons (Beagley *et al.*, 1992b).
As *M. senile* mtDNA also contains the genes for the two rRNAs and 13
proteins characteristic of other metazoan mtDNAs (Fig. 1), it seems likely
that *M. senile* mitochondria contain a functional protein synthesis system.
The remaining, necessary tRNAs (at least 20) are therefore indicated to
be transcribed from nuclear genes and imported. Within partial sequences
of mtDNAs of two other Cnidaria, *Hydra attenuata* and *Sarcophyton
glaucum* (an octocoral), genes for tRNA^{trp} and tRNA^{f-met}, but no other
tRNAs, have been identified (Pont *et al.*, 1992a,b). The occurrence in the
predicted cnidarian tRNA^{trp} of a UCA anticodon that could pair with both
UGA and UGG codons in transcripts of these organisms' mt-protein genes
suggests a rationale for the specific retention of this tRNA gene. The
cnidarian mt-tRNA^{f-met} gene (anticodon CAT) has characteristics more
closely resembling a prokaryotic initiator tRNA^{f-met} gene than a prokaryotic elongator tRNA^{f-met} gene. This suggests that retention of a tRNA^{f-met}
gene in *M. senile* mtDNA may be associated with the use of formyl-
methionine to initiate polypeptide synthesis in cnidarian mitochondria, as
is known to occur in mammalian mitochondria (Chomyn *et al.*, 1981). Like
cnidarian mtDNAs, protista mtDNAs and plant mtDNAs also appear to
encode only a fraction of the tRNAs used in mt-protein systhesis (Suyama,
1986; Boer and Gray, 1988; Pritchard *et al.*, 1990; Joyce *et al.*, 1988;
Marechal-Drouard *et al.*, 1988).

B. Variant Structural Forms of Mitochondrial tRNA Genes

The primary and secondary structures of tRNAs encoded in prokaryotic
DNAs, nuclear DNAs of protista, animals and plants, and chloroplast

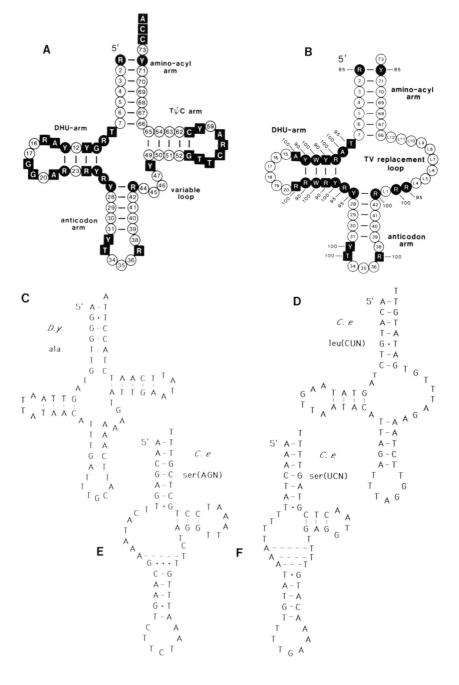

DNAs have been highly conserved (Fig. 3A) (Rich and RajBhandary, 1976; Dirheimer *et al.*, 1979; Singhall and Fallis, 1979; Sprinzl *et al.*, 1989). In contrast, among tRNAs encoded by metazoan mtDNAs there is a remarkable diversity of structure (Figs. 3B–3F).

Most vertebrate and invertebrate mt-tRNA genes can fold into configurations resembling the four-armed secondary structures of prokaryotic and eukaryotic nuclear-encoded tRNAs (standard tRNAs) (Fig. 3C). However, in most or all of the mt-tRNA genes of platyhelminthes, insects, echinoderms, and various vertebrates, there is variation in both the size and the sequence of the loops of the dihydrouridine (DHU) and TΨC arms (Fig. 3C; Table IV); Barrell *et al.*, 1979; Crews and Attardi, 1980; Van Etten *et al.*, 1980; Anderson *et al.*, 1981, 1982a; Clary *et al.*, 1982; HsuChen *et al.*, 1983a; Wong *et al.*, 1983; Himeno *et al.*, 1987; Garey and Wolstenholme, 1989). Only in three mt-tRNA genes of vertebrates [leu (UUR), ser (UCN), and gln] and echinoderms [leu (UUR), leu (CUN) and gln] have the primary sequences of both of these loops been completely conserved. The extent to which other nucleotides, which in standard tRNAs are considered to be either invariant or semi-invariant (Fig. 3A), are found among the sets of tRNA genes of completely sequenced mtDNAs is shown in Table VII. The least conserved of these nucleotides are G_{18}

FIG. 3   (A) The secondary structure model of a prokaryotic–eukaryotic nuclear-encoded (standard) tRNA gene. Letters in solid squares and solid circles identify nucleotide or nucleotide combinations that are considered invariant or semi-invariant, respectively. The 3′ terminal CCA nucleotides either are DNA-encoded (as shown; most prokaryotic tRNA genes) or are added post-transcriptionally (most eukaryotic tRNAs). The numbering system is the conventional numbering system used for yeast tRNA[phe] (Kim, 1979; Sprinzl *et al.*, 1989). (B) The consensus secondary structure of the TΨC arm–variable loop (TV)-replacement loop-containing mt-tRNA genes of *Caenorhabditis elegans* (from Okimoto *et al.*, 1992a). The number of nucleotides shown in the dihydrouridine (DHU) loop and the TV-replacement loop are the maximum numbers observed. Letters in solid squares identify nucleotides or nucleotide combinations that occur among *C. elegans* mt-tRNA genes with frequencies (as percentages) shown by the accompanying numbers, and are also considered invariant nucleotides in standard tRNAs. Letters in solid circles identify other nucleotide and nucleotide combinations that occur in *C. elegans* mt-tRNA genes with frequencies (as percentages) shown by the accompanying numbers. Ten of these nucleotides (1,9,10,13,22,25,26,27,43,72) are considered semi-invariant in standard tRNAs (Dirheimer *et al.*, 1979; Singhall and Fallis, 1979). L1 and L12 denote the maximum of 12 nt that occur in the TV-replacement loop of the *C. elegans* mt-tRNA genes. A, adenine; T, thymine; R, adenine or guanine; Y, cytosine or thymine; W, adenine or thymine. (C) The mt-tRNA[ala] gene of *Drosophila yakuba* that has short DHU and TΨC loops that lack most of the invariant nucleotides of standard tRNAs. (D) The TV-replacement loop-containing mt-tRNA [leu(CUN)] gene of *C. elegans*. (E,F) The DHU-replacement loop-containing mt-tRNA[ser(AGN)] and mt-tRNA [ser(UCN)] genes of *C. elegans*.

TABLE VII

The Frequencies of Occurrence among Sets of Various Metazoan mt-tRNA Genes of Nucleotides or Nucleotide Combinations That Are Highly Conserved (Invariant or Semi-invariant) in Prokaryotic and Eukaryotic Nuclear-Encoded tRNAs (Rich and RajBhandary, 1976; Singhal and Fallis, 1979).

Nucleotide or nucleotide combination[a]	H. sapiens	G. domesticus	X. laevis	P. lividus	D. yakuba	C. elegans
T8	71	81	76	86	81	95
R9	95	95	95	95	95	100
DHU arm						
G10	86	86	86	86	76	85
Y11[b]	90	95	95	100	100	100
A14	81	81	86	95	91	100
R15	76	76	95	81	52	100
G18[c]	14	19	24	14		
	(29)	(43)	(38)	(67)	(14)	(30)
G19[c]	14	19	24	14		
A21	71	71	71	86	76	100
R24[b]	84	95	95	100	100	100
Y25	86	95	86	100	95	100
R26	91	100	86	67	91	95
Anticodon loop						
Y32	100	96	96	100	96	100
T33	96	96	91	91	100	100
R37	100	100	100	100	100	100
Y48	86	91	82	76	77	d
TΨC arm						
G53	59	73	91	91	27	d
T54	55	64	73	86	41	d
T55	59	64	86	77	68	d
C56	32	46	41	27	5	d
R57	59	68	91	96	23	d
A58	68	82	77	96	23	d
Y60	68	82	82	91	77	d
C61	68	82	91	91	27	d
	Range of loop sizes (ntp)					
DHU	3–10	3–12	4–10	4–7	3–8	5–8
TΨC	5–9	5–9	6–7	6–7	3–8	3–6[e]

Note: Data are taken from the following sources: *H. sapiens,* Anderson *et al.* (1982b); *G. domesticus,* Desjardins and Morais (1990); *X. laevis,* Roe *et al.* (1985); *P. lividus,* Cantatore *et al.* (1989); *D. yakuba,* Clary and Wolstenholme (1985a); *C. elegans,* Okimoto *et al.* (1992a).

[a] The numbering system used is that given for yeast tRNA[phe] (Sprinzl *et al.,* 1989). A, adenine; T, thymine; R, adenine or guanine; Y, cytosine or thymine. Nucleotide positions in the dihydrouridine (DHU) and TΨC arms are based on the DHU stems and TΨC stems having 4 or 5 ntp, respectively, even though in some cases one or more mismatches occur in these stems. The only exceptions are as follows: *X. laevis* tRNA[cys] gene has a DHU arm with a stem of 3 ntp and a loop of 4 nt; the *D. yakuba* tRNA genes for Phe, Ile, Asn, Cys each have a TΨC arm with a stem of 4 ntp and a loop of 4 nt. As the tRNA[ser(AGY/A/N)] gene of all species and the tRNA[ser(UCN)] gene of *C. elegans* lack a DHU arm, all entries for T8, R9,

(*continued*)

TABLE VII (*continued*)

and R26 and all nucleotides within the DHU arm are based on 20 tRNA genes for *C. elegans* and 21 tRNA genes for all other species. All entries are based on the sets of predicted tRNA secondary structures given in the respective publications (see *Note*).

[b] Entries for Y11 and R24 are based on the 19 DHU-arm-containing tRNA genes (18 for *C. elegans*) other than the tRNA[trp] and tRNA[f-met] genes. In the latter two tRNA genes, nucleotides 11 and 24 are always R and Y, respectively, in metazoan mtDNAs.

[c] The entries for G18 and G19 are minimum values based on DHU loops that contain two adjacent guanines. The entries in parentheses are the maximum possible occurrences of G18 or G19 based on DHU loops that contain either one or two guanines.

[d] Twenty of the mt-tRNA genes of *C. elegans* lack a TΨC arm, and the TΨC arms of the remaining two mt-tRNA genes of each species (tRNA[ser(AGN)]; tRNA[ser(UCN)]) are unusually reduced in nucleotide number.

[e] Only tRNA[ser(AGN)] and tRNA[ser(UCN)] genes contain a TΨC arm in *C. elegans* mtDNA.

and G_{19} in the DHU loop, and C_{56} in the TΨC loop, indicating that the tertiary interactions G_{18}–Ψ_{55} and G_{19}–C_{56} that are essential for the folding of standard tRNAs into a functional tertiary configuration (Kim, 1979) are of minor or no importance in regard to how metazoan mt-tRNAs achieve their final folded form. Although T_{55} is conserved in the range 59 to 86% among species, a pseudouridine is not present at this position in mt-tRNAs (Randerath *et al.*, 1981a,b; Roe *et al.*, 1982; HsuChen *et al.*, 1983a,b; Dubin *et al.*, 1986). Purines at position 9 between the amino-acyl and DHU stems and at position 37 immediately following the anticodon, together with the purine$_{11}$–pyrimidine$_{24}$ pair in the DHU stem, and pyrimidines preceding the anticodon at positions 32 and 33 are the most highly conserved nucleotides among metazoan tRNA genes (Table VII). With minor exceptions, all other of the nucleotides that are highly conserved in standard tRNAs occur with frequencies in the range 70–100% among metazoan mt-tRNA genes. The 3' CCA, characteristic of all tRNAs sequenced to date, is not encoded in metazoan mt-tRNA genes, but is added post-transcriptionally (Randerath *et al.*, 1981a,b; Roe *et al.*, 1982; HsuChen *et al.*, 1983a,b).

In all metazoan mtDNAs sufficiently examined to date the gene for the tRNA that recognizes AGY, AGY/A, or AGN codons has the DHU arm replaced with a simple loop of between 2 and 11 nt (Garey and Wolstenholme, 1989). This tRNA gene was first identified in human and bovine mtDNAs. The corresponding mt-tRNA[ser(AGY)] has been isolated from cow and mosquito mitochondria, and sequenced. Also, the bovine tRNA[ser(AGY)] has been specifically charged with serine (Arcari and Brownlee, 1980; de Bruijn *et al.*, 1980; Dubin *et al.*, 1984). In mtDNAs of the nematodes *C. elegans, A. suum,* and *M. javanica,* but not of other metazoa, the tRNA[ser(UCN)] gene also contains a DHU arm-replacement loop (Fig. 3F; Okimoto and Wolstenholme, 1990; Okimoto *et al.*, 1992a,c). In the remaining 20 mt-tRNA genes of *C. elegans* and *A. suum* the TΨC arm and variable loop (TV) are together replaced with a loop of between 6 and 12 nt (Figs. 3B and 3D; Wolstenholme *et al.*, 1987; Okimoto and Wolstenholme, 1990; Okimoto *et al.*, 1992a). Eighteen such tRNA genes are found in *M. javanica* mtDNA, and the remaining 2 (the identification of which remains tentative) appear to lack sequences that can be folded into structures resembling either a DHU or a TΨC arm (Okimoto *et al.*, 1992c). Therefore, all of the mt-tRNA genes from *C. elegans*, *A. suum*, and *M. javanica* lack at least one of the arms found in standard tRNAs.

Interpretation of the structures described above as the mt-tRNA gene sets of *C. elegans*, *A. suum*, and *M. javanica* was supported by the following features. As in all other vertebrate and higher invertebrate (*D. yakuba* and echinoderm) mtDNA molecules, the number of putative tRNA genes found in each of these nematode mtDNA molecules was 22. The anticodons in each set of nematode putative mt-tRNA genes were the same and were compatible with codon usage in the *C. elegans* and *A. suum* mtDNAs. Also, these anticodons have a high correspondence with anticodons of vertebrate and higher invertebrate mt-tRNA genes. Except for G_{18} and G_{19} in the loop of the DHU arm [and nucleotides characteristic of the TΨC arm and variable loop], conservation of nucleotides that in standard tRNAs are considered invariant or semi-invariant is exceptionally high in the putative nematode mt-tRNA genes (Fig. 3A and Table VII).

Evidence was sought that tRNAs are in fact transcribed from the unusual *C. elegans* and *A. suum* mt-tRNA genes (Okimoto and Wolstenholme, 1990). Oligonucleotide probes complimentary to a portion of the sense strand of seven *C. elegans* and three *A. suum* TV-replacement loop-containing genes, the tRNA[ser(UCN)] genes of both nematodes and the tRNA[ser(AGN)] gene of *C. elegans*, were made and hybridized to a fraction of small (<150 nt) *C. elegans* and *A. suum* RNAs. Data from these experiments provided direct evidence that both kinds of unusual mt-tRNA genes are transcribed, and that the transcripts (tRNAs) are the same size as the respective tRNA genes to which 3 nt (presumably CCA) are added following transcription. The locations of the sequences within some of the mt-tRNA genes tested to which the probes hybridized specifically ruled out the possibility that tRNAs of the standard sort were created by any kind of *trans*-splicing or RNA editing.

On the basis of chemical probing data, de Bruijn and Klug (1983) built a tertiary structure model for the bovine DHU-replacement loop-containing mt-tRNA[ser(AGY)]. In this model, the overall shape of the tRNA resembles that of the crystal structure of yeast tRNA[phe] (Kim, 1979), but is smaller and includes a unique set of tertiary interactions. Because it is the case in all other known tRNAs, it seems likely that in TV-replacement loop-containing mt-tRNAs, tertiary interactions occur between nucleotides in the DHU loop and the TV-replacement loop. It also seems likely that the final tertiary configuration of these mt-tRNAs resembles that of the nematode DHU-replacement loop-containing tRNAs, as all of the mt-tRNAs must fit a common site in the ribosome for protein synthesis to occur (Okimoto and Wolstenholme, 1990).

The mt-tRNA$^{f\text{-met}}$ genes of the cnidaria *Metridium senile, Sarcophyton glaucum,* and *Hydra attenuata* and the *M. senile* mt-tRNA^{trp} gene can all fold into secondary structures closely resembling those of their prokaryotic counterparts, and in each case nucleotides G_{18} and G_{19} and the complete TΨC arm sequence (5′GTTCRANYC; Fig. 3A) characteristic of standard tRNA genes are present (Pont *et al.,* 1992a,b,c; Beagley *et al.,* 1992b). However, in the DHU arm of the *H. attenuata* tRNA^{trp} gene (Pont *et al.,* 1992c) there is an unusual pairing of a kind previously predicted only in some ciliated protozoan mt-tRNA genes (Morin and Cech, 1988).

C. Modified Bases in Mitochondrial tRNAs

A number of mt-tRNAs from various metazoa including rat, cow, hamster and mosquito have been directly sequenced and, using chromatography, shown to contain modified bases. Pseudouridine is the most abundant such base and occurs at different locations in many mt-tRNAs. In almost all mt-tRNAs, 1-methyladenosine or 1-methylguanosine occur at position 9. Also, in some mt-tRNAs N6-threoninocarboxyladenosine or 2-methlythio-N6-isopentenyladenosine occurs at position 37, the first nucleotide following the anticodon. Other modified bases that have been identified in a small number of mt-tRNAs include 5-methylcytidine and N2-methylguanosine (Randerath, 1981a,b; Roe *et al.,* 1982; HsuChen *et al.,* 1983a,b; Dubin *et al.,* 1984; Dubin and HsuChen, 1984).

D. Mitochondrial tRNA Gene Evolution

The evolution of different structural forms of tRNA genes among metazoan mtDNAs is summarized in Fig. 4. As the cnidarian mt-tRNA$^{f\text{-met}}$ and mt-tRNA^{trp} genes (except *Hydra* mt-tRNA^{trp} gene) have secondary structure potential very similar to that of standard tRNAs, the variously modified DHU and TΨC loops and arms of metazoan mt-tRNA genes are indicated to have arisen after divergence of the cnidarian ancestral line from the line leading to all other metazoa. From consideration of the sequences of the DHU and TΨC loops of individual homologous mt-tRNA genes in

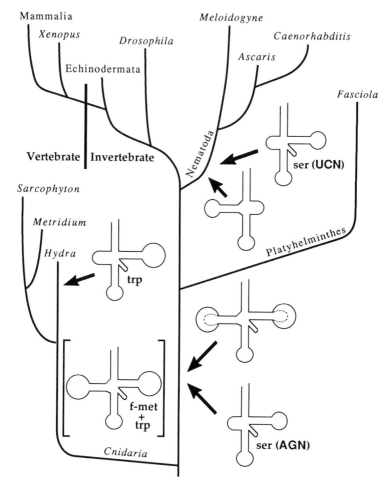

FIG. 4   Evolution of metazoan mt-tRNA gene structure. The earliest occurrence of each variant structure found in the mtDNAs of the individual organisms or groups of organisms listed is shown. The tRNA gene containing dotted dihydrouridine (DHU) and TΨC loops indicates the earliest deduced occurrence of sets of mt-tRNA genes in which the sequence of at least some contain DHU or TΨC loops that are changed from the standard and/or reduced. The brackets around the tRNA genes associated with the cnidarian ancestral line indicate that these tRNA genes have close structural similarity to the corresponding tRNA genes of prokaryotes. The phylogenetic relationships shown follow those given by Wilson *et al.* (1978).

organisms from different phyla, it seems likely that for most mt-tRNA genes, there have been independent changes in the DHU and TΨC loops in different ancestral lines.

The finding that all metazoan mt-tRNA$^{ser(AGN)}$ genes described so far have a DHU-replacement loop strongly suggests that this tRNA gene modification occurred early in metazoan evolution, at least prior to divergence of the platyhelminth line. As TV-replacement loop-containing mt-tRNA genes and the DHU-replacement loop-containing tRNA$^{ser(UCN)}$ gene have been found in all three nematodes examined to date, but not in any other metazoa, it seems likely that these mt-tRNA genes arose early in, or before, the establishment of the ancestral line that led to present-day nematodes. Determination of whether other groups of pseudocoelomates contain TV-replacement loop-containing mt-tRNA genes (20 or less) might be expected to elucidate both the time of origin of this kind of tRNA gene and the interrelationships of the various pseudocoelomate groups.

VI. The Control Region

Between the tRNA^{pro} and tRNA^{phe} genes of vertebrate mtDNAs is a noncoding region that varies in length between species; 1122 ntp in human, 879 ntp in mouse, 2134 ntp in *Xenopus laevis* (Anderson *et al.*, 1981; Bibb *et al.*, 1981; Wong *et al.*, 1983). Because in mammals and amphibia, this sequence has been shown to include the signals necessary for the initiation of H-strand synthesis (that is, the molecule's replication origin), and for both H-strand and L-strand transcription (Montoya *et al.*, 1982, 1983; Clayton, 1984), it has been designated the control region.

In variable portions of isolated vertebrate mtDNA molecules, a segment of the H-strand within the control region is displaced by a short segment of H-strand base-paired to the L-strand. The three-stranded structure thus formed is called the D-loop, and the short H-strand segment (520–700 nt in mouse) is repeatedly synthesized and degraded (Clayton, 1982). The 5' ends of the D-loop-forming H strands and the 5' ends of nascent, extended H strands have similar nucleotide map locations, but whether the extended H strands are produced by continuation of synthesis of the short, D-loop-forming H strands remains undetermined. Although the control region sequence is, in general, poorly conserved between species, close to the H-strand origin in mouse, rat, and human mtDNAs are three blocks of sequences that are very similar. These sequences are referred to as conserved sequence blocks (CSB) 1, 2, and 3 (Walberg and Clayton, 1981). The occurrences of the CSBs vary among different vertebrates. All three have been reported as present in other primates (Foran *et al.*, 1988;

Hayashi *et al.*, 1991) and a fish, *Acipenser transmontanus* (Buroker *et al.*, 1990). However, only CSB 2 is present in another fish, *Gadus morhua* (Johansen *et al.*, 1990), and this sequence has not been identified in chicken mtDNA (Desjardins and Morais, 1990). CSB 3 is absent from bovine mtDNA (Anderson *et al.*, 1982b). The H-strand and L-strand transcription promoters are located about 150 nt apart between the H-strand origin and the tRNA^{phe} gene of mouse and human mtDNAs (Clayton, 1992).

In *Drosophila* mtDNA molecules, there is an apparently noncoding region between the tRNA^{ile} and the s-rRNA genes that is extremely rich in adenine and thymine (the A + T-rich region) and varies in length from about 1 to 5 kb between species (Fauron and Wolstenholme, 1976, 1980a,b). As the origin of replication maps within this region, it is indicated to be the functional equivalent of the control region of vertebrate mtDNAs (Goddard and Wolstenholme, 1978, 1980; Goddard *et al.*, 1982).

In other invertebrate mtDNAs, there occur noncoding regions that vary in size from 121 ntp (sea urchin, Jacobs *et al.*, 1988, 1989) to over 20 kb (weevil, Boyce *et al.*, 1989) that have been designated putative control regions. The larger sizes of some control regions result from the inclusion within them of directly repeated sequences (described in more detail below). In all but the smaller putative control regions, there occur inverted repeat sequences that could fold into a variety of secondary structures (Attardi *et al.*, 1978; Buroker *et al.*, 1990; Okimoto *et al.*, 1992a). However, partly because such structures are not well conserved among species, it remains unclear how they might be associated with transcription and replication initiation.

In mammalian mtDNA molecules, there is a short sequence (mouse, 32 nt; human, 31 nt) between the tRNA^{asn} and the tRNA^{cys} genes that can fold into a stable hairpin structure with a T-rich loop, within which L-strand synthesis originates (see Clayton, 1982). A similar hairpin-forming loop has been found in the corresponding location in *Xenopus laevis* mtDNA, but not in chicken mtDNA (Wong *et al.*, 1983; Desjardins and Morais, 1990). However, in *Drosophila yakuba, D. virilis,* and *D. teisieri* mtDNAs, there is a sequence within the A + T-rich region that has been highly conserved in regard to its potential to form a hairpin structure with a T-rich loop (Clary and Wolstenholme, 1987; Monnerot *et al.*, 1990). As the location of this hairpin is within a region of the molecule that, from electron microscope data, seems to contain an origin of second-strand replication, it has been suggested as a possible functional equivalent of the vertebrate L-strand origin-containing sequence (Clary and Wolstenholme, 1987).

Nematode mtDNAs contain a noncoding segment of 109 ntp (*C. elegans*) and 117 ntp (*A. suum*) between the ND4 and COI genes that also includes a sequence with the potential to form a hairpin with a T-rich loop (Okimoto *et al.*, 1992a). Although information regarding nematode mtDNA replication is currently lacking, this latter observation raises the possibility that the association of L-strand synthesis initiation with a specific secondary structure may have arisen early in the evolution of the metazoa.

VII. Directly Repeated Sequences

In the mtDNA molecules of a number of metazoan species, directly repeated segments are found. The most common type comprises tandemly arranged repeated sequences in the (putative) control region. Such repeats have been reported in the mtDNA molecules of the root knot nematode *Meloidogyne javanica* (5 copies of an 8-ntp sequence, 11 copies of a 63-ntp sequence, and about 36 copies of a 102-ntp sequence; Okimoto *et al.*, 1991); the soil nematode *Caenorhabditis elegans* (6 copies of a 43-ntp sequence; Okimoto *et al.*, 1992a); some *Drosophila* species (between one and 5 copies of a 470-ntp sequence; Solignac *et al.*, 1986); a cricket, *Gryllus firmis* (between one and 7 copies of a 220-ntp sequence; Rand and Harrison, 1989); three weevils, *Pissodes* species (various numbers of copies of sequences of between 800 and 2000 ntp; Boyce *et al.*, 1989); various species of the lizard genus *Cnemidophorus* (between 3 and 9 copies of a 64-ntp sequence; Densmore *et al.*, 1985); three fishes, *Alosa sapidissima* (between one and 3 copies of a 1500-ntp sequence; Bentzen *et al.*, 1987), *Acipenser transmontanus* (between one and 4 copies of an 82-ntp sequence; Buroker *et al.*, 1990), and *Gadus morhua* (4 copies of a 40-ntp sequence; Steiner *et al.*, 1990); and rabbit, *Oryctolagus cuniculus* (various number of copies of a 20-ntp sequence and of a 153-ntp sequence; Mignotte *et al.*, 1990). Also, there are nontandemly arranged direct repeats in the control regions of mtDNA molecules of *Xenopus laevis* (2 copies of a 45-ntp sequence; Wong *et al.*, 1983) and chicken, *Gallus dometicus* (2 copies of a 29-ntp sequence; Desjardins and Morais, 1990). In mtDNA from the scallop, *Placopecten magellanicus,* there are between 2 and 8 copies of a 1442-ntp direct repeat, but the location of these repeated segments within the molecule has not been determined (Snyder *et al.*, 1987; LaRoche *et al.*, 1990). Variations in the number of copies of the repeats, as indicated above, are between individuals of a species and in some cases within individuals (heteroplasmy; Solignac *et al.*, 1983; Rand and Harrison, 1989).

It has been shown that mtDNA molecules of individuals of different *Cnemidophorus* species and of the gecko, *Heteronotia binoei,* include duplicated segments of heterogeneous lengths, most of which comprise various portions of the control region and adjacent rRNA, tRNA, and protein genes (Moritz and Brown, 1986, 1987; Moritz, 1991). Mitochondrial DNA molecules of the newt *Triturus cristatus* contain a single duplication of a sequence that includes the s-rRNA, 1-rRNA, ND1, and ND2 genes (Wallis, 1987). Also, a coding sequence-containing segment is both directly and inversely repeated in the mtDNA of a nematode (*Romanomermis culicivorax*) that is parasitic in mosquitos (Hyman *et al.*, 1988).

Of the above-mentioned mtDNA repeats, nucleotide sequence information has been obtained for those of *M. javanica*, *G. firmis*, *A. transmontanus*, *G. morhua*, *X. laevis*, *G. domesticus*, *O. cuniculus*, *P. megallanicus*, and *R. culicivorax*. A function for any of the control region-containing repeats has not been established, nor is it clear why repeats containing segments of coding sequences should be maintained in some metazoan mtDNAs.

VIII. Nucleotide Bias

In the protein genes of all metazoan mtDNAs there is bias regarding the overall occurrence of different nucleotides, the occurrence of different nucleotides between the complementary strands, and the occurrence of nucleotides in the third position of codons (Table VIII). In the sense strand of mt-protein genes of all vertebrates completely sequenced, there is a low frequency of guanine (range 11.4–12.7%). Among the four vertebrate mtDNAs represented in Table VIII, the most frequently used nucleotide is cytosine in humans and chicken, adenine in mouse, and thymine in *X. laevis*. Among the invertebrate mtDNAs represented, although guanine is again the least used nucleotide in *P. lividis*, in all of the remainder (*D. yakuba*, *A. suum*, *C., elegans*, and *M. senile*) cytosine is the least used nucleotide, and in all of these invertebrate mtDNAs thymine is the most used nucleotide (range 31.0–51.9%). Among vertebrate mt-protein genes, guanine is also the least used nucleotide in the third position of codons (range 3.2–5.1%, Table VIII). The most used nucleotide in the third position of codons in human, mouse, and chicken mt-protein genes is the same one as that most used overall. However, in both *X. laevis* and *P. lividus*, although thymine is the most used nucleotide overall, the most used nucleotide in the third position of codons is adenine. Among mt-protein genes of the remaining invertebrates, the most used third position nucleotide is thymine (range 40.4–62.7%), which is also the nucleotide that in all cases is the overall most used. In *A. suum*, *C. elegans*, and *M. senile*, the low use of cytosine in codon third positions (range 2.4–14.7%) is positively correlated with the overall low use of cytosine in the mt-protein genes in

TABLE VIII

Data Regarding Nucleotide Composition for the Protein Genes of Completely Sequenced mtDNAs of Various Metazoa[a]

Organism	Percentage nucleotide composition of the sense strand						Percentage of codons ending in						Percentage of codons that specify leucine	Percentage of leucine codons beginning with	
	T	C	A	G	A + T	G + T	T	C	A	G	A + T	G + T		T	C
H. sapiens	25.6	33.1	29.3	12.0	54.9	37.6	15.3	43.1	36.5	5.1	51.8	20.4	17.3	12.0	88.0
M. musculus	29.4	26.0	33.2	11.4	62.6	40.8	22.9	27.4	46.6	3.2	69.5	26.1	15.5	21.9	78.1
G. domesticus	24.0	34.8	29.1	12.1	53.1	36.1	12.4	44.8	39.0	3.9	51.4	16.3	17.7	9.2	90.8
X. laevis	31.7	24.5	31.1	12.7	62.8	44.4	29.3	23.2	43.5	4.0	72.8	33.4	16.1	37.8	62.2
P. lividus	31.0	23.6	28.6	16.8	59.6	47.8	25.5	22.5	39.9	12.1	65.4	37.6	15.4	28.7	71.3
D. yakuba	44.4	11.1	32.3	12.2	76.7	56.6	48.4	3.3	45.4	2.9	93.8	51.3	16.8	90.6	9.4
A. suum	51.9	8.0	18.5	21.5	70.4	73.4	62.7	2.4	11.5	23.4	74.2	86.1	15.0	86.9	13.1
C. elegans	46.0	9.3	29.5	15.3	75.5	61.3	49.7	4.7	36.6	9.0	86.3	58.7	15.5	82.5	17.5
M. senile	38.0	17.0	24.5	20.5	62.5	58.4	40.4	14.7	27.9	17.0	68.3	57.3	14.5	69.3	30.7

Note: Data for H. sapiens (Anderson et al., 1981), M. musculus (Bibb et al., 1981), G. domesticus (Desjardins and Morais, 1990), X. laevis (Roe et al., 1985), and P. lividus (Cantatore et al., 1989) mtDNAs are derived from the light (L) strand (Fig. 1), and both the nucleotide composition of the sense strand and the frequencies of nucleotides in the third positions of codons are complementary to those of the other 12 mt-protein genes of each species. Data for D. yakuba (Clary and Wolstenholme, 1985a), C. elegans, A. suum (Okimoto et al., 1992a), and M. senile mtDNAs (Beagley et al., 1992b) are derived from all protein genes. In D. yakuba mtDNA 9 protein genes are transcribed from one strand and 4 from the other (Fig. 1), but there is no clear distinction between the nucleotide compositions of the sense strands of genes transcribed from the two complementary strands. In C. elegans, A. suum, and M. senile mtDNAs, all protein genes are transcribed from the same strand (Fig. 1).

[a] Excluding nucleotides concerned with termination.

each of these species. In *Drosophila yakuba* mt-protein genes in which overall uses of cytosine and guanine are low but approximately the same (11.1 and 12.2%), the frequency of each of these nucleotides at third positions of codons is again low and similar (3.3 and 2.9%, respectively).

Both the overall high adenine + thymine content (76.7%) and the extreme bias in the use of adenine and thymine in the third position of codons (93.8%) in *D. yakuba* mtDNA are shared by *C. elegans* mt-protein genes: 75.5 and 86.3%, respectively. All codons that are not used in these genes end in C or G (*D. yakuba*) or C (*C. elegans*). Factors that may contribute to maintaining the nucleotide composition of an A + T-rich genome once it has been established are discussed in Wolstenholme and Clary (1985).

Leucine occurs with high frequency (range 15.0 - 16.9%) among the mt-proteins encoded in all completely sequenced metazoan mtDNAs (Table III). In all mt-DNAs except those of *X. laevis* and *P. lividus* the ratio of TTR to CTN triplets used as leucine-specifying codons is positively correlated with the differential use of T and C nucleotides in the third position of codons (Table VIII; Clary and Wolstenholme, 1985a; Garey and Wolstenholme, 1989; Okimoto *et al.*, 1992a). It has been argued that the lack of such a correlation among *X. laevis* and *P. lividus* mt-protein genes suggests that in both cases there is a difference in the constraints on synonymous T and C nucleotides in the first position and in the third position of codons (Okimoto *et al.*, 1992a).

In vertebrate mtDNAs, one gene (ND6) is encoded on the opposite strand (H strand) to the strand (L strand) that encodes all of the other 12 mt-protein genes (Fig. 1). The nucleotide bias (and codon endings) of the sense strand of the ND6 gene in all cases examined reflects that of the remainder of the strand containing the ND6 sense strand, rather than that of the complementary strand containing the sense strands of the remaining 12 mt-protein genes. In sea urchin mtDNAs, as in vertebrate mtDNAs, all protein genes except ND6 are encoded in the same strand (Fig. 1). However, a segment of the star fish (*Asterina pectinifera*) mtDNA molecule is inverted relative to the sea urchin mtDNA molecule (Fig. 1) and as a result, the ND1 and ND2 genes (and the ND6 gene) are encoded in the strand complementary to that in which all other genes are encoded. Again, overall nucleotide composition and frequencies of nucleotides in the third position of codons of these two genes reflect the overall nucleotide composition of the remainder of the strand (Asakawa *et al.*, 1991). These two observations suggest that in vertebrate mtDNAs and echinoderm mtDNAs, at least, nucleotide bias has resulted from differential selection of nucleotides in the strand as a whole. However, consideration of the differential nucleotide usage in the different positions of codons in nematode and human mt-protein genes (Okimoto *et al.*, 1992a) has led to the conclusion that differential selection for amino acids could have contributed to overall nucleotide bias between the sense strands of mt-protein genes of these species.

IX. Deletions and Nucleotide Substitutions Associated with Human Diseases

Because there is a highly compact arrangement of genes in metazoan mtDNA molecules, and because all of the mt-gene products are essential to the life of aerobic cells, it would be expected that deletions common to all mtDNA molecules of an organism would be limited to the short regions between genes, and to nonessential sequences of the control region.

It has been shown that mtDNAs containing large deletions of heterogeneous length (1.3–7.6 kb) are associated with a human neuromuscular disease (an encephalomyopathy) called Kearns–Sayre syndrome (Holt *et al.*, 1988; Shoffner *et al.*, 1989; Shoubridge *et al.*, 1990; Hayashi *et al.*, 1991). These mtDNA deletions are not inherited but apparently arise early in development, and through clonal events give rise to islands of muscle tissue that are deficient in mitochondrial function. Also, a low level of mtDNA molecules containing a deletion similar to Kearns–Sayre syndrome-type deletions was found to distinguish adult human cells from fetal human cells, suggesting that such deletions might be associated with aging (Cortopassi and Arnheim, 1990).

Some other human diseases apparently result from single nucleotide substitutions in coding regions of mtDNA molecules (Lander and Lodish, 1990; Hayashi *et al.*, 1991): mtDNA molecules from patients with Leber's hereditary optic neuropathy have a missense mutation in the ND4 gene (Wallace *et al.*, 1988); mtDNA molecules from patients with a neurological syndrome that includes retina pigmentosa and proximal muscle weakness have a missense mutation in the ATPase6 gene (Holt *et al.*, 1990); two other conditions, myoclonus epilepsy and ragged red-fiber disease (MERRF), and mitochondrial myopathy, encephalopathy, lactic acidosis, and stroke-like episodes (MELAS), are associated with point mutations in the mt-tRNA^{lys} and mt-tRNA$^{leu(UUR)}$ genes, respectively (Shoffner *et al.*, 1990; Goto *et al.*, 1990; Kobayashi *et al.*, 1990).

X. Conclusions and Perspectives

Throughout the evolution of metazoa, gene content of mt-genomes has been very highly conserved, as has the close packing of genes. Most of the occasional sequence expansions that have occurred, by way of either repeated or noncoding unique sequences, are found in the control or putative control region, rather than being dispersed between genes. The close packing of genes is likely related to the mechanism by which most metazoan mtDNAs appear to be transcribed: that is, the generation of primary transcripts of entire strands of the molecule from which mature mono- or bi-cistronic transcripts are produced by precise cleavage, possibly, in some cases, as a function of tRNA secondary structure.

Many of the peculiar unique features of metazoan mtDNAs seem to have arisen following divergence of the cnidarian line from the line that is ancestral to all other present-day invertebrates and vertebrates. Only for cnidarian mtDNAs is there cause for doubt that the above-mentioned transcription mechanism is operational: in *Metridium senile* (sea anemone) mtDNA there are longer sequences of noncoding nucleotides at some gene junctions; all protein genes end with a full termination codon; and there occur only two tRNA genes, neither of which immediately follows a protein gene. The only genetic code modification in cnidarian mtDNAs is that TGA specifies tryptophan rather than translation termination, a modification shared with fungal and protista mtDNAs. The observations that the two cnidarian mt-tRNA genes have, in most cases, primary and secondary structure potential close to that of standard tRNA genes suggest that the sequence modifications found among metazoan mt-tRNAs also arose after divergence of the cnidarian ancestral line.

The extent of size reduction within metazoan mt-tRNA gene sets strongly correlates with the degree to which the more variable secondary structure element-forming regions of mt-rRNA genes have been lost, but it is not clear why this should be so. However, in this regard, it is again interesting to note that of all metazoan mt-rRNA genes that have been sequenced to date, those of Cnidaria are closest in size and secondary structure potential to the rRNAs of *E. coli,* and that, as noted above, the two retained tRNA genes have in most cases near standard primary and secondary structures.

At this time, TV-replacement loop-containing tRNA genes and the DHU-replacement loop-containing tRNA$^{ser(UCN)}$ gene have been found only in nematode mtDNAs. It therefore seems plausible that determination of whether such mt-tRNA genes occur in various other groups of lower invertebrates will provide important clues regarding some poorly understood phylogenetic relationships. This approach is particularly promising because it seems unlikely that once sequences with specific secondary structure potential are lost from most members of a tRNA gene set, they would be regained. If such sequence losses are a function, in whole or in part, of the observed mt-rRNA size reductions, then the probability that the lost sequences would not be regained is further increased. Also, any changes in structure of mt-aminoacyl tRNA synthetases that accompanied tRNA size reduction might be expected to decrease the likelihood that the lost tRNA sequences would be regained. Therefore, organisms found to have sets of arm-deficient mt-tRNAs would be strongly indicated to be more closely related to nematodes than organisms that do not. The importance of utilizing specific sequence losses to gain insight into phylogenetic relationships is further enhanced by findings of substantial differences in average nucleotide substitution rates in mt-protein and mt-rRNA gene sets of organisms of different phyla, and of nucleotide substitution rates that vary greatly in homologous genes or organisms of different phylogenetic lines (see discussions in Brown, 1985; Okimoto *et al.,* 1992a,b).

Although cnidarian mt-genomes appear to share few of the unusual features of other metazoan mtDNAs, they have provided their own surprises: as mentioned above, they encode only two tRNA genes; group I introns occur in the COI and ND5 genes of sea anemones; a novel gene is present in octocoral mtDNA that has convincing sequence similarity to a component of bacterial mismatch repair systems. These latter two features, in particular, are interesting from an evolutionary standpoint.

Group I introns corresponding to those found in the COI and ND5 genes of *Metridium senile* are not widely distributed among cnidarian mtDNAs. In fact, to date such introns have only been detected in the COI and ND5 genes of another sea anemone, *Anthopleura elegantissima,* a member of the same order (Actiniaria) of hexacorals to which *M. senile* belongs (Beagley *et al.,* 1992a). As *A. elegantissima* contains endosymbiotic Zooxanthellae (dinoflagellates), this observation raises the interesting question whether the mt-protein gene-containing introns might have been acquired by a common ancestor of *M. senile* and *A. elegantissima,* through horizontal transfer from one of the three DNA genomes (mitochondria, chloroplasts, nuclear) of the endosymbiotic Zooxanthellae. The very large evolutionary distance predicted between homologous mt-protein and rRNA genes of Zooxanthellae and sea anemones should provide highly significant controls for judging sequence relatedness of any group I introns found in Zooxanthellae DNAs and those that occur in sea anemone mt-protein genes.

The novel gene (Orf 983) found in octocoral mtDNAs (Pont *et al.,* 1992b) appears to be the homolog of the *hexA* and *mutS* genes of Gram-positive and Gram-negative bacteria, respectively (Haver *et al.,* 1988; Priebe *et al.,* 1988). These latter genes are each a component of a multi protein complex that is responsible for the correction of nucleotide replication errors generated during replication and overlooked by the polymerase proofreading activity (Lahue *et al.,* 1989). Therefore, the occurrence of Orf 983 in octocoral mtDNAs suggests that it would be worthwhile to investigate whether a functional mismatch repair system operates in octocoral mitochondria. From considerations of the sizes of octocoral mtDNA molecules and currently available sequence information, it seems unlikely that any other protein components homologous to those of the bacterial mismatch repair systems are encoded in octocoral mtDNAs. Therefore, it seems reasonable to expect that other components are nuclear-encoded. Although mammalian mitochondria contain enzymes that could carry out base excision repair of DNA damaged by oxidation (Tomkinson *et al.*, 1990), evidence for other types of repair mechanisms in mammalian mitochondria has not been found (see Clayton *et al.*, 1974, and discussions in Brown, 1983). A deficiency of repair mechanisms has been suggested as the basis for the higher rate of (mutations) nucleotide substitution in mtDNA than in nuclear DNA in vertebrates (Brown, 1983; Cann *et al.*, 1984; Moritz *et al.*, 1987). In view of this, and the controversy concerning the relative rates of nucleotide substitutions in mtDNAs and nuclear DNAs of some invertebrates (Moritz *et al.*, 1987), it will be important to determine whether the occurrence of a mt-mismatch repair system in (some) Cnidaria can be correlated with a slower rate of mt-nucleotide substitution. If this turns out to be the case, then it will be interesting to ponder what selective forces might have operated to either retain or discard such a repair system during metazoan mtDNA evolution.

In view of the great diversity of structural features that have been found among the few metazoan mt-genomes examined to date, it seems clearly worthwhile to continue sequencing and related studies of the mtDNAs of the many remaining groups of invertebrates.

Acknowledgments

I am indebted to Ronald Okimoto, C. Timothy Beagley, Douglas O. Clary, Jane L. Macfarlane, Jill A. Wahleithner, Genevieve A. Pont, and Nori A. Okada for discussions and for permission to quote unpublished findings. The work from my laboratory reported in this article was supported by National Institutes of Health Grants GM 18375 and RR 07092, and U.S. Department of Agriculture Grant 86-CR-CR-1-1994.

References

Anderson, S., Bankier, A. T., Barrell, B. G., de Bruijn, M. H. L., Coulson, A. R., Drouin, J., Eperon, I. C., Nierlich, D. P., Roe, B. A., Sanger, F., Schreier, P. H., Smith, A. J. H., Staden, R., and Young, I. G. (1981). *Nature (London)* **290,** 457–465.
Anderson, S., Bankier, A. T., Barrell, B. G., de Bruijn, M. H. L., Coulson, A. R., Drouin, J., Eperson, I. C., Nierlich, D. P., Roe, B. A., Sanger, F., Scheier, P. H., Smith, A. J. H., Staden, R., and Young, I. G. (1982a). *In* "Mitochondrial Genes" (P. Slonimski,

P. Borst, and G. Attardi, eds.), pp. 5–43. Cold Spring Harbor Laboratory, Cold Spring Harbor, NY.

Anderson, S., de Bruijn, M. H. L., Coulson, A. R., Eperon, I. C., Sanger, F., and Young, I. G. (1982b). *J. Mol. Biol.* **156**, 683–717.

Arcari, P., and Brownlee, G. G. (1980). *Nucleic Acids Res.* **8**, 5207–5212.

Arnason, U., Gullberg, A., and Widegren, B. (1991). *J. Mol. Evol.* **33**, 556–568.

Asakawa, S., Kumazawa, Y., Araki, T., Himeno, H., Miura, K., and Watanabe, K. (1991). *J. Mol. Evol.* **32**, 511–520.

Attardi, B., and Attardi, G. (1971). *J. Mol. Biol.* **55**, 231–249.

Attardi, G. (1985). *Int. Rev. Cytol.* **93**, 93–145.

Attardi, G., Crews, S. T., Nishigushi, J., Ojala, D. K., and Posakony, J. W. (1978). *Cold Spring Harbor Symp. Quant. Biol.* **43**, 179–192.

Avis, J. C. (1992). *Oikos* **63**, 62–76.

Avis, J. C., Arnold, J., Bull, R. M., Jr. Bermingham, E., Lamb, T., Neigel, J. E., Reeb, C. A., and Saunders, N. C. (1987). *Annu. Rev. Ecol. Syst.* **18**, 489–522.

Baer, R. J., and Dubin, D. T. (1981). *Nucleic Acids Res.* **9**, 323–337.

Barrell, B. G., Anderson, S., Bankier, A. T., de Bruijn, M. H. L., Chen, E., Coulson, A. R., Drouin, J. Eperon, I. C., Nierlich, D. P., Roe, B. A., Sanger, F., Schreier, P. H., Smith, A. J. H., Staden, R., and Young, I. G. (1980). *Proc. Natl. Acad. Sci. U.S.A.* **77**, 3164–3166.

Barrell, B. G., Bankier, A. T., and Drouin, J. (1979). *Nature (London)* **282**, 189–194.

Batuecas, B., Garesse, R., Calleja, M., Valverde, J. R., and Marco, R. (1988). *Nucleic Acids Res.* **16**, 6515–6529.

Beagley, C. T., Okada, N. A., and Wolstenholme, D. R. (1992a), in preparation.

Beagley, C. T., Pont, G. A., Okada, N. A., and Wolstenholme, D. R. (1992b), in preparation.

Bentzen, P., Leggett, W. C., and Brown, G. G. (1987). *Genetics* **118**, 509–518.

Bibb, M. J., Van Etten, R. A., Wright, C. T., Walberg, M. W., and Clayton, D. A. (1981). *Cell (Cambridge, Mass.)* **26**, 167–180.

Boer, P. H., and Gray, M. W. (1988). *Curr. Genet.* **14**, 583–590.

Bonitz, S. G., Berlani, R., Coruzzi, G., Li, M., Macino, G., Nobrega, F. G., Nobrega, M. P., Thalenfeld, B. E., and Tzagoloff, A. (1980). *Proc. Natl. Acad. Sci. U.S.A.* **77**, 3167–3170.

Boyce, T. M., Zwick, M. E., and Aquadro, C. F. (1989). *Genetics* **123**, 825–836.

Brosius, J., Dull, T. J., and Noller, H. F. (1980). *Proc. Natl. Acad. Sci. U.S.A.* **77**, 201–204.

Brosius, J., Palmer, M. L., Kennedy, P. J., and Noller, H. F. (1978). *Proc. Natl. Acad. Sci. U.S.A.* **75**, 4801.

Brown, W. M. (1983). *In* "Evolution of Genes and Proteins" (M. Nei, and R. K., Koehn, eds.), pp. 62–88. Sinauer Associates, Sutherland, MA.

Brown, W. M., (1985). *In* "Molecular Evolutionary Genetics" (R. J. MacIntyre, eds.), pp. 95–130. Plenum, New York/London.

Brown, W. M., George, M., Jr., and Wilson, A. C. (1979). *Proc. Natl. Acad. Sci. U.S.A.* **76**, 1967–1971.

Brown, W. M., Prager, E. M., Wang, A., and Wilson, A. C. (1982). *J. Mol. Evol.* **18**, 225–239.

Buroker, N. E., Brown, J. R., Gilbert, T. A., O'Hara, P. J., Beckenbach, A. T., Thomas, W. K., and Smith, M. J. (1990). *Genetics* **124**, 157–163.

Buzzo, K., Fouts, D. L., and Wolstenholme, D. R. (1978). *Proc. Natl. Acad. Sci. U.S.A.* **75**, 909–913.

Cann, R. L., Brown, W. M., and Wilson, A. C. (1984). *Genetics* **106**, 479–499.

Cantatore, P., Gadaleta, M. N., Roberti, M., Saccone, C., and Wilson, A. C. (1987). *Nature (London)* **32**, 853–855.

Cantatore, P., Roberti, M., Rainaldi, G., Gadaleta, M. N., and Saccone, C. (1989). *J. Biol. Chem.* **264**, 10965–10975.

Chomyn, A., and Attardi, G. (1987). *Curr. Top. Bioenerg.* **15**, 295–329.

Chomyn, A., Cleeter, W. J., Ragan, C. I., Riley, M. Doolittle, R. F., and Attardi, G. (1986). *Science* **234**, 614–618.

Chomyn, A., Hunkapillar, M. W., and Attardi, G. (1981). *Nucleic Acids Res.* **9**, 867–877.

Chomyn, A., Mariottini, P., Cleeter, M. W. J., Ragan, C. I., Matsuno-Yagi, A., Hatefi, Y., Doolittle, R. F., and Attardi, G., (1985). *Nature* (*London*) **314**, 592–597.

Clary, D. O., and Wolstenholme, D. R. (1983a). *Nucleic Acids Res.* **11**, 4211–4227.

Clary, D. O., and Wolstenholme, D. R. (1983b). *Nucleic Acids Res.* **11**, 6859–6872.

Clary, D. O., and Wolstenholme, D. R. (1984). *In* "Oxford Surveys on Eukaryotic Genes" (N. Maclean, ed.), Vol. 1, pp. 1–35. Oxford Univ. Press, Oxford.

Clary, D. O., and Wolstenholme, D. R. (1985a). *J. Mol. Evol.* **22**, 252–271.

Clary, D. O., and Wolstenholme, D. R. (1985b). *Nucleic Acids Res.* **113**, 4029–4045.

Clary, D. O., and Wolstenholme, D. R. (1987). *J. Mol. Evol.* **25**, 116–125.

Clary, D. O., Goddard, J. M., Martin, S. C., Fauron, C. M. R., and Wolstenholme, D. R. (1982). *Nucleic Acids Res.* **10**, 6619–6637.

Clayton, D. A., (1982). *Cell* (*Cambridge, Mass.*) **28**, 693–705.

Clayton, D. A. (1984). *Annu. Rev. Biochem.* **53**, 573–594.

Clayton, D. A. (1991). *Annu. Rev. Cell Biol.* **7**, 453–478.

Clayton, D. A. (1992). *Int. Rev. Cytol.* **141**, 000–000.

Clayton, D. A., Doda, J. N., and Freidberg, E. C. (1974). *Proc. Natl. Acad. Sci. U.S.A.* **71**, 2777–2781.

Clayton, D. A., and Vinograd, J. (1967). *Nature* (*London*) **216**, 652–657.

Clayton, D. A., and Vinograd, J. (1969). *Proc. Natl. Acad. Sci. U.S.A.* **62**, 1077.

Cortopassi, G. A., and Arnheim, N. (1990). *Nucleic Acids Res.* **18**, 6927–6933.

Crews, S., and Attardi, G. (1980). *Cell* (*Cambridge, Mass.*) **19**, 775–784.

Crozier, R. H., Crozier, Y. C., and Mackinlay, A. G. (1989). *Mol. Biol. Evol.* **6**, 399–411.

Dahlberg, A. E. (1989). *Cell* (*Cambridge, Mass.*) **57**, 525–529.

Dams, E., Henricks, L, Van der Peer, Y., Neef, J. M., Smits, G., Vandenbempt, I., and Wachter, R-De (1988). *Nucleic Acids Res.* **16**, r87–r173.

Dawid, I. B., and Blacker, A. W. (1972). *Dev. Biol.* **29**, 152–161.

Dawid, I. B., and Wolstenholme, D. R. (1967). *J. Mol. Biol.* **28**, 233–245.

de Bruijn, M. H. L. (1983). *Nature* (*London*) **304**, 234–241.

de Bruijn, M. H. L., and Klug, A. (1983). *EMBO J.* **2**, 1309–1321.

de Bruijn, M. H. L., Schreier, P. H., Eperon, I. C., Barrell, B. G., Chen, E. Y., Armstrong, P. W., Wong, J. F. H., and Roe, B. A. (1980). *Nucleic Acids Res.* **8**, 5213–5222.

De Giorgi, C., Lanave, C., Musci, M. D., and Saccone, C. (1991). *Mol. Biol. Evol.* **8**, 519–529.

Densmore, L. D., Wright, J. W., and Brown, W. M. (1985). *Genetics* **110**, 689–707.

Desjardins, P., and Morais, R. (1990). *J. Mol. Biol.* **212**, 599–634.

Desjardins, P., and Morais, R. (1991). *J. Mol. Evol.* **32**, 153–161.

DeSalle, R., Freedman, T., Prager, E. M., and Wilson, A. C. (1987). *J. Mol. Evol.* **26**, 157–164.

Dirheimer, G., Keith, G., Sibler, A. P., and Martin, R. P. (1979). *In* "Transfer RNA: Structure, Properties, and Recognition" (P. R. Schimmel, D. Soll, and J. N. Abelson, eds.) pp. 19–41. Cold Spring Harbor Laboratory, Cold Spring Harbor, NY.

Drouijn, J. (1980). *J. Mol. Biol.* **140**, 15–34.

Dubin, D. T., and HsuChen, C. C. (1983). *Plasmid* **9**, 307–320.

Dubin, D. T., and HsuChen, C. C. (1984). *Nucleic Acids Res.* **12**, 4185–4189.

Dubin, D. T., HsuChen, C. C., Cleaves, G. R., and Timko, K. D. (1984). *J. Mol. Biol.* **176**, 251–260.

Dubin, D. T., HsuChen, C. C., and Tillotson, L. E. (1986). *Curr. Genet.* **10**, 701–707.

Dubin, D. T., Montoya, J., Timko, K. D., and Attardi, G. (1982). *J. Mol. Biol.* **157,** 1–19.

Dubin, D. T., Prince, D. L., and Kotin, R. M. (1985). *In* "Achievements and Perspectives of Mitochondrial Research" (E. Quagliariello, E. C. Slater, F. Palmieri, C. Saccone, and A. M. Kroon, eds.), Vol. II, pp. 165–174. Elsevier, Amsterdam.

Dubin, D. T., and Taylor, R. H. (1978). *J. Mol. Biol.* **121,** 523–540.

Dubin, D. T., Taylor, R. H., and Davenport, L. W. (1978). *Nucleic Acids Res.* **5,** 4385–4397.

Dubin, D. T., Timko, K. D., and Baer, R. J. (1981). *Cell (Cambridge, Mass.)* **23,** 271–278.

Dunon-Bluteau, D., and Brun, G. (1986). *FEBS Lett.* **198,** 333–337.

Fauron, C. M. R., and Wolstenholme, D. R. (1976). *Proc. Natl. Acad. Sci. U.S.A.* **73,** 3623–3627.

Fauron, C. M. R., and Wolstenholme, D. R. (1980a). *Nucleic Acids Res.* **8,** 2439–2452.

Fauron, C. M. R., and Wolstenholme, D. R. (1980b). *Nucleic Acids Res.* **8,** 5391–5410.

Fearnley, I. M., and Walker, J. E. (1986). *EMBO J.* **5,** 2003–2008.

Fearnley, I. M., and Walker, J. E. (1987). Biochemistry **26,** 8247–8251.

Foran, D. R., Hixson, J. E., and Brown, W. M. (1988). *Nucleic Acids Res.* **16,** 5841–5861.

Fox, T. D. (1987). *Annu. Rev. Genet.* **21,** 67–91.

Gadaleta, G., Pepe, G., Decandia, G., Quagliariello, C., Sbisa, E., and Saccone, C. (1989). *J. Mol. Evol.* **28,** 497–516.

Garey, J. R., and Wolstenholme, D. R. (1989). *J. Mol. Evol.* **28,** 374–387.

Garesse, R. (1988). *Genetics* **118,** 649–663.

Glotz, C., Zweib, C., and Brimacombe, R. (1981). *Nucleic Acids Res.* **9,** 3287–3306.

Goddard, J. M., Fauron, C. M. R., and Wolstenholme, D. R. (1982). *In* "Mitochondrial Genes" (P. Slonimsky, P. Borst, G. Attardi, eds.), pp. 99–103. Cold Spring Harbor Laboratory, Cold Spring Harbor, NY.

Goddard, J. M., and Wolstenholme, D. R. (1978). *Proc. Natl. Acad. Sci. U.S.A.* **75,** 3886–3890.

Goddard, J. M., and Wolstenholme, D. R. (1980). *Nucleic Acids Res.* **8,** 741–757.

Goto, Y. I., Nonaka, I., and Horais, S. (1990). *Nature (London)* **348,** 651–653.

Gray, M. W., Sankoff, D., and Cedergren, R. J. (1984). *Nucleic Acids Res.* **12,** 5837–5852.

Gutell, R., and Fox, G. E. (1988). *Nucleic Acids Res.* **16,** r175–r269.

Gutell, R., Weiser, B., Woese, C. R., and Noller, H. F. (1985). *Prog. Nucleic Acids Res. Mol. Biol.* **32,**155–216.

Haber, L. T., Pang, P. P., Sobell, D. I., Marikovich, J. A., and Walker, G. C. (1988). *J. Bacteriol.* **170,** 197–202.

Hancke, H. R., and Gellissen, G. (1988). *Curr. Genet.* **14,** 471–476.

Hauswirth, W. W., and Laipis, P. J. (1982). *Proc. Natl. Acad. Sci. U.S.A.* 79, 4686–4690.

Hayasaka, K., Ishida, T., and Horai, S. (1991). *Mol. Biol. Evol.* **8,** 399–415.

Hayashi, J.-L., Ohta, S., Kikuchi, A., Takemiten, M., Goto, Y. I., and Nonaka, I. (1991). *Proc. Natl. Acad. Sci. U.S.A.* **88,** 10614–10618.

Heckman, J. E., Sarnoff, J., Alzner-de Weerd, B., Yin, S., and RajBhandary, U. L. (1980). *Proc. Natl. Acad. Sci. U.S.A.* **77,** 3159–3163.

Himeno, H., Masaki, H., Kawai, T., Ohta, T., Kumagi, I., Miura, I., and Watanabe, K. (1987). *Gene* **56,** 219–230.

Hoeh, W. R., Blakley, K. H., and Brown, W. M. (1991). *Science* **251,** 1488–1490.

Hoffman, R. J., Boore, J. L., and Brown, W. M. (1992). *Genetics* **131,** 397–412.

Holt, I. J., Harding, A. E., and Morgan-Hughes, J. A. (1988). *Nature (London)* **331,** 717–719.

Holt, I. J., Harding, A. E., Petty, R. K. H., and Morgan-Hughes, J. A. (1990). *Am. J. Hum. Genet.* **46,** 428–433.

HsuChen, C. C., Cleaves, G. R., and Dubin, D. T. (1983a). *Plasmid* **10,** 55–65.

HsuChen, C. C., Cleaves, G. R., and Dubin, D. T. (1983b). *Nucleic Acids Res.* **11,** 8659–8662.

HsuChen, C.-C., Kotin, R. M., and Dubin, D. T. (1984). *Nucleic Acids Res.* **12,** 7771–7785.

Hutchison, C. A., III, Newbold, J. E., Potter, S. S., and Edgell, M. H. (1974). *Nature (London)* **251**, 536–538.

Hyman, B. C., Beck, J. L., and Weiss, K. C. (1988). *Genetics* **120**, 707–712.

Jacobs, H., Asakawa, S., Araki, T., Miura, K., Smith, M. J., and Watanabe, K. (1989). *Curr. Genet.* **15**, 193–206.

Jacobs, H. T., Elliot, D. J., Math, V. B., and Farquharson, A. (1988). *J. Mol. Biol.* **202**, 185–217.

Johansen, S., Guddal, P. H., and Johansen, T. (1990). *Nucleic Acids Res.* **18**, 411–419.

Joyce, P. B. M., Spencer, D. F., Bonen, L., and Gray, M. W. (1988). *Plant Mol. Biol.* **10**, 251–262.

Kim, S. H., (1979). *In* "Transfer RNA: Structure, Properties and Recognition" (P. R. Schimmel, D. Soll, and J. N. Abelson, eds.), pp. 83–100. Cold Spring Harbor Laboratory, Cold Spring Harbor, NY.

Kobayashi, Y., Momoi, M. Y., Tominaya, K., Momoi, T., Nihei, K., Yanagisawa, M., Kagawa, Y., and Ohta, S. (1990). *Biochem. Biophys. Res. Commun.* **173**, 816–822.

Kondo, R., Satta, Y., Matsuura, E. T., Ishiwa, H., Takahato, N., and Chigusa, S. I. (1990). *Genetics* **126**, 657–663.

Kotin, R. M., and Dubin, D. T. (1984). *Biochim. Biophys. Acta* **782**, 106–108.

Kroon, A. M., Borst, P., Van Bruggen, E. F. J., and Ruttenberg, G. J. C. M. (1966). *Proc. Natl. Acad. Sci. U.S.A.* **56**, 1836.

Lagerkvist, U. (1981). *Cell (Cambridge, Mass.)* **23**, 305–306.

Lahue, R. S., Au, K. G., and Modrich, P. (1989). *Science* **245**, 160–164.

Lander, E. S., and Lodish, H. (1990). *Cell (Cambridge, Mass.)* **61**, 925–926.

LaRoche, J., Snyder, M., Cook, D. I., Fuller, K., and Zouros, E. (1990). *Mol. Biol. Evol.* **7**, 45–64.

Marechal-Drouard, L., Weil, J.-H., and Guillemaut, P. (1988). *Nucleic Acids Res.* **16**, 4777–4788.

McCracken, A., Uhlenbusch, I., and Gellissen, G. (1987). *Curr. Genet.* **11**, 625–630.

Mcreadie, I. G., Novitski, C. E., Maxwell, R. J., John, U., Ooi, B. G., McMullen, G. L., Lukins, H.-B., Linnane, A. W., and Nagley, P. (1983). *Nucleic Acids Res.* **11**,4435–4451.

Mignotte, F., Gueride, M., Champagne, A. M., and Mounolou, J. C. (1990). *Eur. J. Biochem.* **194**, 561–571.

Moazed, D., and Noller, H. F. (1989). *Cell (Cambridge, Mass.)* **57**, 585–597.

Moazed, D., and Noller, H. F. (1991). *Proc. Natl. Acad. Sci. U.S.A.* **88**, 3725–3728.

Monnerot, M., Solignac, M., and Wolstenholme, D. R. (1990). *J. Mol. Evol.* **30**, 500–508.

Montoya, J., Christianson, T., Levens, D., Rabinowitz, M., and Attardi, G. (1982). *Proc. Natl. Acad. Sci. U.S.A.* **79**, 7195–7199.

Montoya, J., Gaines, G. L., and Attardi, G. (1983). *Cell (Cambridge, Mass.)* **34**, 151–159.

Montoya, J., Ojala, D., and Attardi, G. (1981). *Nature (London)* **290**, 465–470.

Morin, G. B., and Cech, T. R. (1988). *Nucleic Acids Res.* **16**, 327–346.

Moritz, C. (1991). *Genetics* **129**, 221–230.

Moritz, C. and Brown, W. M. (1986). *Science* **233**, 1425–1427.

Moritz, C., and Brown, W. M. (1987). *Proc. Natl. Acad. Sci. U.S.A.* **84**, 7183–7187.

Moritz, C., Dowling, T., and Brown, W. M. (1987). *Annu. Rev. Ecol. Syst.* **18**, 269–292.

Nass, M. M. K. (1966). *Proc. Natl. Acad. Sci. U.S.A.* **56**, 1215.

Noller, H. F., Asire, M., Barta, A., Douthwaite, S., Goldstein, T., Gutell, R. R., Moazed, D., Normanley, J., Prince, J. B., Stern, S., Triman, K., Turner, S., Van Stolk, B., Wheaton, V., Weiser, B., and Woese, C. R. (1986). *In* "Structure, Function, and Genetics of Ribosomes" (B. Hardesty and G. Kramer, eds.), pp. 141–163. Springer-Verlag, New York.

Ohama, T., Osawa, S., Watanabe, K., and Jukes, T. H. (1990). *J. Mol. Evol.* **30**, 329–332.

Ojala, D., Merkel, C., Gelfand, R., and Attardi, G. (1980). *Cell (Cambridge, Mass.)* **22**, 393–403.

Ojala, D., Montoya, J., and Attardi, G. (1981). *Nature (London)* **290**, 470–474.

Okimoto, R., and Wolstenholme, D. R. (1990). *EMBO J.* **9**, 3405–3411.

Okimoto, R., Macfarlane, J. L., and Wolstenholme, D. R. (1990). *Nucleic Acids Res.* **18**, 6113–6118.

Okimoto, R., Chamberlin, H. M., Macfarlane, J. L., and Wolstenholme, D. R. (1991). *Nucleic Acids Res.* **19**, 1619–1626.

Okimoto, R., Macfarlane, J. L., Clary, D. O., and Wolstenholme, D. R. (1992a). *Genetics* **130**, 471–498.

Okimoto, R., Macfarlane, J. L., and Wolstenholme, D. R. (1992b), in preparation.

Okimoto, R., Chamberlin, H. M., Macfarlane J. L., Okada, N. A., and Wolstenholme, D. R. (1992c), in preparation.

Osawa, S., Ohama, T., Jukes, T. H., and Watanabe, K. (1989a). *J. Mol. Evol.* **29**, 202–207.

Osawa, S., Ohama, T., Jukes, T. H., Watanabe, K., and Yokoyama, S. (1989b). *J. Mol. Evol.* **29**, 373–380.

Pääbo, S., Thomas, W. K., Whitfield, K. M., Kumazawa, Y., and Wilson, A. C. (1991). *J. Mol. Evol.* **33**, 426–430.

Pont, G. A., Beagley, C. T., Okimoto, R., and Wolstenholme, D. R. (1992a), in preparation.

Pont, G. A., Okada, N. A., Beagley, C. T., Cavalier-Smith, T., Clark-Walker, D., and Wolstenholme, D. R. (1992b), in preparation.

Pont, G. A., Vassort, C. G., and Wolstenholme, D. R. (1992c), in preparation.

Powell, J. R., Caccone, A., Amato, C. D., and Yoon, C. (1986). *Proc. Natl. Acad. Sci. U.S.A.* **83**, 9090–9093.

Priebe, S. D., Hadi, S. M., Greenberg, B., and Lacks, S. A. (1988). *J. Bacteriol.* **170**, 190–196.

Pritchard, A. E., Seilhammer, J. J., Mahalingam, R., Sable, C. L., Venuti, S. E., and Cummings, D. J. (1990). *Nucleic Acids Res.* **18**, 173–180.

Rand, D. M., and Harrison, R. G. (1989). *Genetics* **121**, 551–569.

Randerath, E., Agrawal, H. P., and Randerath, K. (1981a). *Biochem. Biophys. Res. Commun.* **103**, 739–744.

Randerath, K., Agrawal, H. P., and Randerath, E. (1981b). *Biochem. Biophys. Res. Commun.* **100**, 732–737.

Reddy, P., Peterkofsky, A., and McKenney, K. (1985). *Proc. Natl. Acad. Sci. U.S.A.* **82**, 5656–5660.

Rich, A., and RajBhandary, U. L. (1976). *Annu. Rev. Biochem.* **45**, 805.

Roe, B. A., Ma, D. P., Wilson, R. K., and Wong, J. F. H. (1985). *J. Biol. Chem.* **260**, 9759–9774.

Roe, B. A., Wong, J. F. H., Chen, E. Y., Armstrong, P. W., Stankiewica, A., Ma, D. P., and McDonough, J. (1982). *In* "Mitochondrial Genes" (P. Slonimski, P. Borst, and G. Attardi, eds.), pp. 45–49. Cold Spring Harbor Laboratory, Cold Spring Harbor, NY.

Shoffner, J. M., Lott, M. T., Leesa, A. M. S., Seibel, P., Ballinger, S. W., and Wallace, D. C. (1990). *Cell (Cambridge, Mass.)* **61**, 931–937.

Shoffner, J. M., Lott, M. T., Voljavec, A. S., Soueidan, S. A., Costigan, D. A., and Wallace, D. C. (1989). *Proc. Natl. Acad. Sci. U.S.A.* **86**, 7952–7956.

Shoubridge, E. A., Karpatti, G., and Hastage, K. E. M. (1990). *Cell (Cambridge, Mass.)* **62**, 43–49.

Sinclair, J. H., and Stevens, B. J. (1966). *Proc. Natl. Acad. Sci. U.S.A.* **56**, 508–514.

Singhal, R. P., and Fallis, P. A. M. (1979). *Prog. Nucleic Acids Res. Mol. Biol.* **23**, 227–290.

Smith, M. J., Banfield, D. K., Doteval, K., Gorski, S., and Kowbel, D. J. (1989). *Gene* **76**, 181–185.

Smith, M. J., Banfield, D. K., Doteval, K., Gorski, S., and Kowbel, D. J. (1990). *J. Mol. Evol.* **31**, 195–204.

Snyder, M., Fraser, A. R., LaRoche, J., Gartner-Kepkay, K. E., and Zouros, E. (1987). *Proc. Natl. Acad. Sci. U.S.A.* **84**, 7595–7599.

Solignac, M., Monnerot, M., and Mounolou, J. C. (1983). *Proc. Natl. Acad. Sci. U.S.A.* **80**, 6942–6946.

Solignac, M., Monnerot, M., and Mounolou, J. C. (1986). *J. Mol. Evol.* **24**, 53–60.

Southern, S. O., Southern, P. J., and Dizon, A. E. (1988). *J. Mol. Evol.* **28**, 32–42.

Sprinzl, M., Hartmann, T., Weber, J., Blank, J., and Zeidler, R. (1989). *Nucleic Acids Res.* **17**, r1–r172.

Steigler, P., Carbon, P., Ebel, J.-P., and Ehresmann, C. (1981). *Eur. J. Biochem.* **120**, 487–495.

Suyama, Y. (1986). *Curr. Genet.* **10**, 411–420.

Thomas, W. K., and Beckenbach, A. T. (1989). *J. Mol. Evol.* **29**, 233–245.

Tomkinson, A. E., Bonk, R. T., Kim, J., Bartfeld, N., and Linn, S. (1990). *Nucleic Acids Res.* **18**, 929–935.

Uhlenbusch, I, McCracken, A., and Gellissen, G. (1987). *Curr. Genet.* **11**, 631–638.

Van Bruggen, E. F. J., Borst, P., Ruttenberg, G. J. C. M., Grubber, M., and Kroon, A. M. (1966). *Biochim. Biophys. Acta.* **119**, 437–439.

Van Etten, R. A., Bird, R. A., and Clayton, D. A. (1983). *J. Biol. Chem.* **258**, 10104–10110.

Van Etten, R. A., Walberg, M. W., and Clayton, D. A. (1980). *Cell (Cambridge, Mass.)* **22**, 157–170.

Vawter, L., and Brown, W. M. (1986). *Science* **234**, 194–196.

Walberg, M. W., and Clayton, D. A. (1981). *Nucleic Acids Res.* **9**, 5411–5421.

Wallace, D. C. (1982). *Microbiol. Rev.* **46**, 208–240.

Wallace, D. C., Singh, G., Lott, M. T., Hodge, J. A., Schurr, T. G., Lezza, A. M. S., Elsas, L. J., II, and Nikroskelainen, E. K. (1988). *Science* **242**, 1427–1430.

Wallis, G. P. (1987). *Heredity* **58**, 229–238.

Warrior, R., and Gall, J. (1985). *Arch. Sci. Geneva* **38**, 439–445.

Wilson, E. O., Eisner, T., Briggs, W. R., Dickerson, R. E., Metzenberg, R. L., O'Brien, R. D., Susman, M., and Boggs, W. E. (1978). *In* "Life on Earth," 2nd ed. p. 617. Sinauer, Sunderland, MA.

Wolstenholme, D. R., and Clary, D. O. (1985). *Genetics* **109**, 725–744.

Wolstenholme, D. R., Koike, K., and Cochran-Fouts, P. (1974). *Cold Spring Harbor Symposium Quant. Biol.* **38**, 267–280.

Wolstenholme, D. R., Koike, K., and Renger, H. C. (1971). *In* "Oncology 1970. A. Cellular and Molecular Mechanisms of Carcinogenesis; B. Regulations of Gene Expression (Proc. Xth International Cancer Congress)" (D. R. L. Clark, R. W. Cumley, J. E. McCoy, and M. M. Copeland, eds.), 2nd ed. Vol. I, pp. 627–648. Yew Book Medical Publishers, Chicago.

Wolstenholme, D. R., Macfarlane, J. L., Okimoto, R., Clary, D. O., and Wahleithner, J. A. (1987). *Proc. Natl. Acad. Sci. U.S.A.* **84**, 1324–1328.

Wolstenholme, D. R., McLaren, J. D., Koike, K., and Jacobson, E. L. (1973). *J. Cell Biol.* **56**, 247–255.

Wong, J. F. H., Ma, D. P., Wilson, R. K., and Roe, B. A. (1983). *Nucleic Acids Res.* **11**, 4977–4995.

Zweib, C., Glotz, C., and Brimacombe, R. (1981). *Nucleic Acids Res.* **9**, 3621–3640.

Transcription and Replication of Animal Mitochondrial DNAs

David A. Clayton

Department of Developmental Biology, Stanford University School of Medicine, Stanford, California, 94305-5427

I. Introduction

The existence of animal mitochondrial DNA (mtDNA) as a separate and distinct eukaryotic genetic entity was firmly established over 25 years ago. Its physical properties and coding capacity are now well known and are discussed in depth by Wolstenholme (1992). Given its small size (~16 kb in most cases) and closed circular topology, it was recognized early after its discovery that it would be an attractive system in which to explore basic phenomena surrounding transcription and DNA replication. This initial optimism was warranted and, in particular, human and mouse mtDNAs have been exploited in studies aimed at understanding both the modes and the mechanisms of mtDNA transcription and replication. The overall issues of mitochondrial gene expression and organelle biogenesis are intimately connected with these processes as well. Earlier comprehensive reviews provide a background for transcription (Clayton, 1984; Attardi, 1985) and structure and replication (Clayton and Smith, 1975; Clayton, 1982) of mammalian mtDNA. For a more recent treatment of both topics, see Clayton (1991). The purpose of this chapter is to describe our current understanding of how animal mtDNA is expressed and replicated and to point to the potentially most fruitful future directions in this area of investigation.

II. Transcription of Mitochondrial DNA

A. Initiation, Elongation, and Termination

Since the first animal mitochondrial genomes were sequenced in their entireties over a decade ago, it has been clear that these DNAs contain only one significant stretch of nucleotide sequence that does not code for an RNA or a protein molecule. This region is the displacement-loop (D-loop) portion, which is a signature feature of at least vertebrate mtDNA. All available data are consistent with the conclusion that the essential major *cis*-acting elements necessary for transcriptional initiation are located within the confines of the D-loop region, which is defined explicitly as that sequence bounded by the genes for tRNA-phenylalanine and tRNA-proline. In addition, the D-loop region also contains all of the required template information that is presumed necessary for the initiation of nascent heavy (H)-strand DNA synthesis, which marks the beginning of a round of DNA replication in this system.

The initial breakthrough that permitted the definition of required promoter elements in animal mtDNA occurred in 1983 with the development of a faithful *in vitro* transcription system that was able to initiate transcription correctly by utilizing mitochondrial protein fractions (Clayton, 1991). This system was then exploited extensively at the level of template manipulation. By the now standard approaches of template deletion, substitution, and point-mutation analyses, an overall picture of mammalian mtDNA promoters has emerged. There are two major promoters in this system that are located usually within 150 bp or less of one another (Fig. 1). Each promoter is responsible for the synthesis of transcripts complementary to either the H strand (the H-strand promoter; HSP) or the light (L) strand (the L-strand promoter, LSP). Although shown here as unidirectional in function, it is known that mammalian mtDNA promoters are weakly bidirectional *in vitro* and *in vivo*. This bidirectionality feature is much more pronounced in the case of *Xenopus laevis* mtDNA (Bogenhagen and Romanelli, 1988) and, in fact, it may be that chicken mtDNA has a single promoter that is fully bidirectional in function (L'Abbé *et al.*, 1991).

Human mtDNA promoters include a short region encompassing the transcriptional start sites that has important sequence requirements, as evidenced by the fact that mutations in these start sites can be shown to

FIG. 1 Simplified schematic of the major mammalian mtDNA promoters. The H-strand promoter (HSP) and L-strand promoter (LSP) are labeled at the respective transcriptional start sites (bent arrows). The cross-hatched areas represent sequence required for correct transcriptional initiation. The black boxes depict the binding sites for the *trans*-activator mtTF1. The arrows pointing to the right signify the fact that these target sequences show the greatest homology when aligned with this polarity (Fisher *et al.*, 1987).

have serious consequences for promoter function (Hixson and Clayton, 1985). Efficient transcription requires the presence and action of the only currently identified *trans*-acting factor in vertebrate mitochondrial transcription. This factor, termed mtTF1 (Parisi and Clayton, 1991; Fisher *et al.*, 1992), functions by binding immediately upstream of the transcriptional start sites from positions −10 to −40 at each promoter. The necessity for proper interaction of mtTF1 with its target binding sequence has been shown by saturation mutagenesis experiments. These data have revealed an excellent correspondence between the loss of mtTF1 binding to particular promoter-DNA base changes and, in turn, the loss of function in transcriptional activation. The fact that mtTF1 appears to be functional in either orientation of the protein with respect to the polarity of transcription is suggestive of the possibility that its target site may represent a DNA sequence duplication event. At the current time, it is unknown whether mtTF1 exerts it action by the binding of one protein molecule per activation event or whether multimerization is somehow involved, at the level of either one target sequence (corresponding to the major, proximal upstream footprint) or, possibly, binding events at multiple sites. This latter prospect is worth consideration and future investigation given the propensity of the yeast homolog of mtTF1 (Diffley and Stillman, 1991; Fisher *et al.*, 1991, 1992; Xu and Clayton, 1992) to exhibit phased binding on DNA duplexes (Diffley and Stillman, 1992). Indeed, human mtTF1 can bind at multiple sites in the D-loop region of human mtDNA (Fisher *et al.*, 1987, 1992).

The minimum protein set that can initiate transcription on a mammalian mtDNA promoter is shown in Fig. 2. Core mitochondrial RNA (mtRNA) polymerase is shown binding at the transcriptional start site where it is presumed to interact. The *trans*-acting mtTF1 is placed on its target sequence and both of these proteins are actually operative on each of the

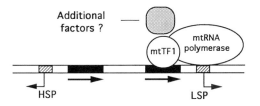

FIG. 2   Transcription proteins for mammalian mtDNA promoter function. Both mtTF1 and mtRNA core polymerase are shown next to their known and presumed, respectively, sites of template interaction. Core mtRNA polymerase has not been fully characterized from mammalian cells and it is unknown how many additional proteins constitute the functional holoenzyme. Other symbols are as in Fig. 1. In an attempt to simplify the nomenclature for mitochondrial transcription proteins, mtTF1 has been renamed mtTFA (Xu and Clayton, 1992).

two promoters, that shown for the LSP, as well as the HSP. At this point there is no documented difference in the nature of proteins involved in transcriptional initiation at each of these two promoters, although that possibility has not been ruled out. Because animal mtRNA polymerase has not been purified to homogeneity, nor has its gene been identified and sequenced in any system other than that of yeast (Masters *et al.,* 1987), it is unclear for the animal systems whether additional protein components are required for transcriptional initiation. Such putative proteins would most likely be ones that are fortuitously co-isolating with mtRNA polymerase under current purification strategies.

In contrast to the lack of success in purifying the core mtRNA polymerase, human mtTF1 was the first transcription protein to be unambiguously identified and characterized in a mitochondrial system. Its basic structural features are shown in Fig. 3. The mature protein, as isolated from human mitochondria, is 204 amino acids in size and largely comprises two sequence blocks that show similarity to sequences contained within nuclear high-mobility-group (HMG) proteins (Parisi and Clayton, 1991). These sequence blocks have been termed HMG box 1 and HMG box 2, and it is likely that they represent the most important structural domains of the protein relevant to DNA binding and stimulation of transcription. Although the exact mechanism by which mtTF1 enhances transcription is unknown, as noted above, DNA binding is critical. It has been established that mtTF1 has the capacity to wrap and bend duplex DNA upon binding (Fisher *et al.,* 1992). It is therefore expected that wrapping and bending are fundamental to the protein's action; this characteristic is shared with the yeast mtTF1 protein homolog as well (Diffley and Stillman, 1992; Fisher *et al.,* 1992).

Other than the HMG boxes, there are no obvious sequence features of mtTF1 that immediately suggest a particular functional consequence.

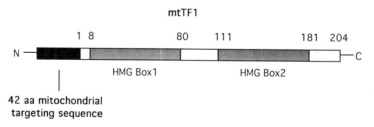

FIG. 3   Depiction of the mtTF1 protein. The sequence features of the human mtTF1 protein are shown as inferred from both protein purification studies and cDNA analysis. N and C denote the amino and carboxy terminus, respectively. The numbers on top denote the amino acid position in the mature protein (Parisi and Clayton, 1991).

Straightforward mutational analyses may permit assignment of the importance of each region of the protein. Sequence inspection of the mtTF1 cDNA reveals a putative 42 amino acid targeting sequence at the amino terminus (Fig. 3). This presumably represents the portion of mtTF1 that is processed and degraded after import of the protein into the mitochondrion.

The only other animal system for which functional DNA-binding data are available on mtTF1 is that of the mouse. In this case it is known that a protein of similar size and purification properties is able to interact with mouse mtDNA promoter elements and thereby stimulate correct transcriptional initiation in a manner virtually identical to that of human. Although there is a slight difference in the position of the transcriptional start sites relative to mtTF1 footprinting for one of the mouse promoters, the basic and essential phenomenon of upstream binding over a 30-bp region is conserved between species. A similar importance for the *cis*-acting sequences at the target site is also suggested by promoter deletional analyses. Finally, there is inherent flexibility in the manner in which the mtTF1 proteins interact with their target sequences in that the mouse and human mtTF1 proteins can be exchanged one for another. That is, mtTF1 can be swapped between human and mouse systems and correct transcriptional initiation, albeit at a reduced level, is achieved (Fisher *et al.*, 1989). The mtTF1 protein is currently the only one in the animal mitochondrial transcription system that has been shown to be capable of functioning across species boundaries.

With regard to elongation and production of transcripts around the genome after initiation, there has been little progress on this topic since it was summarized in earlier reviews of mitochondrial transcription (Attardi and Schatz, 1988; Clayton, 1991). One of the interesting questions to be asked about transcript elongation, and events surrounding subsequent maturation of transcripts, is whether these processes might be coupled. If so, this might explain the apparent inability to date of demonstrating any processing of primary mitochondrial transcripts. This is because most standard assays for such an activity would probably utilize substrates made *in vitro* with prokaryotic RNA polymerases in order to provide sufficient quantities and types of RNAs for reaction. Since *in vitro* transcription systems exist, it may be possible to test for processing in an actively transcribing system.

Some progress has been made in understanding the nature of transcription termination in mammalian mitochondria. It now seems clear that mammalian mtDNA maintains at least one site of legitimate transcription termination. This site is located at the 3′ terminus of the rRNA region and, in fact, encroaches upon the immediately adjacent gene for one of the leucyl-tRNAs. It has been known for several years that a 13-nt sequence, located at the 5′ end of the tRNA gene, is a necessary and sufficient template component to provide the *cis*-acting information required to effect transcription termination *in vitro* (Christianson and Clayton, 1988). This region and a few bases surrounding it have also been shown to be the target site for a protein or proteins that can footprint this sequence (Kruse *et al.*, 1989; Hess *et al.*, 1991). The protein fraction (~34 kDa, Hess *et al.*, 1991) has a high affinity for its target sequence and it is reasonable to suggest that the mechanism of transcription termination is simply the blockage of further elongation by mtRNA polymerase by the presence of the termination protein on the template. Interestingly, a point mutation in the target sequence, associated with a human disease, has been shown to alter the nature of binding of the termination protein and its effectiveness in supporting transcription termination (Hess *et al.*, 1991). This suggests that the mutation may be linked to proper RNA metabolism since it seriously affects the termination phenomenon, at least *in vitro*. The human cellular phenotype associated with this mtDNA point mutation has been examined with respect to mtRNA production and mitochondrial translational capacity (King *et al.*, 1992). A novel RNA species, encompassing the transcription termination site [16S rRNA–tRNA^{leu}–ND1], was observed. Since this RNA represents a readthrough across the termination template sequence, it may be a direct consequence of the mutation when present at high levels in the context of a living cell.

B. Processing of Primary Transcripts

As noted earlier, the very interesting situation surrounding the mechanism of processing of primary mitochondrial transcripts has remained unresolved. One would still logically infer that tRNA recognition and subsequent resection from primary transcripts should be the principal hallmark of processing animal mtRNAs. This would then invoke the activity of an RNase P enzyme whose full identification would be of great interest. The low abundance of critical enzymatic activities such as RNase P has hampered purification efforts.

III. Replication of Mitochondrial DNA

A. Initiation at the Origins

Most of the work on animal mtDNA replication has utilized mammalian systems. Mammalian mtDNAs have two separate origins of replication, one for each strand (Clayton, 1982). Based on conservation of DNA sequence information, it is likely that this is generally true for most vertebrate mtDNAs; even primitive vertebrates may have this feature (Johansen *et al.*, 1990). The origin of H-strand synthesis (O_H) is located within the D-loop region of the genome and the origin of L-strand synthesis (O_L) is surrounded by a cluster of five tRNA genes well distanced from the D loop. Genomic mtDNA replication begins by an initiation of H-strand synthesis that results in strand elongation that eventually produces a full-sized strand. Initiation of L-strand synthesis only occurs after O_L is displaced as a single-stranded template, and replication has never been found to begin at this origin. Thus, in this system, the H-strand (leading strand) origin is dominant with respect to replication initiation, whereas the L-strand (lagging-strand) origin is a secondary, but required, element. The O_L sequence appears to have been lost (or was never present) in chicken mtDNA (Desjardins and Morais, 1990). But the organization of the five tRNA genes has been conserved in this region in chicken mtDNA, and it has been suggested that some feature of these tRNA gene sequences may substitute for O_L (Desjardins and Morais, 1990). Alternative possibilities include a bidirectional origin in the D-loop region or other sequences in mtDNA, such as tRNA genes, that might have structural properties recognizable by a mitochondrial priming system.

There are several sites of transition from RNA synthesis to DNA synthesis at the leading-strand origin (Clayton, 1991). For both human and mouse species it is possible, in the majority of cases, to match RNA 3′ ends with DNA 5′ ends, suggesting a relationship between them in the replication process. The clearest example of this, in the case of mouse, is the existence of a species consisting of RNA at its 5′ end and DNA throughout its distal portion (Chang *et al.*, 1985a); the priming event occurred at the promoter for transcription of this strand of mtDNA. It was concluded that replication priming and transcription initiation begin at the same site in both human and mouse mtDNAs and, as a consequence, replication priming for O_H is an event dependent upon transcriptional promoter function. There are no known differences at the step of actual initiation of transcription at this promoter depending upon whether replication priming or transcription of genes on this strand will occur. Since all genes on this strand are downstream of O_H, one transcriptional initiation event could, in principle, serve *both* for replication priming and for transcription of the genes on this strand.

The transitions from RNA synthesis to DNA synthesis at O_H occur over a region of three short conserved sequence elements [termed conserved sequence blocks (CSBs) I, II, and III]. These sequences had been cited previously by Walberg and Clayton (1981) based on inspection of available mtDNA sequence information. In that light they stood out as conserved sequences in a region of mtDNA that was otherwise highly divergent. For this reason it was suggested that they would prove to have some regulatory role in mtDNA replication or expression and, indeed, that is likely the case. A search for an enzymatic activity that might recognize these CSBs in order to process RNA complementary to this origin region proved successful. The enzyme is a ribonucleoprotein named RNase MRP (for mitochondrial RNA processing) and it has now been characterized from mouse (Chang and Clayton, 1987a,b), human (Topper and Clayton, 1990), bovine (Dairaghi and Clayton, 1992), frog (Bennett *et al.*, 1992), and yeast (Stohl and Clayton, 1992) cellular sources. In the cases tested (mouse, human, bovine, and yeast), the RNase MRPs cleave RNA *in vitro* either exactly or within several nucleotides of a mapped RNA end or 5' DNA terminus. Heterologous cleavage reactions, where the species source of enzyme and RNA substrate are different, are also accurate in that the cleavage pattern reflects the substrate source. That is, the same RNA substrate is processed in a similar manner by RNase MRP independently of the source of enzyme. This result argues that the features of the substrate important for recognition and cleavage have been highly conserved. However, RNA processing appears more complex in the case of *Xenopus* (Bogenhagen and Morvillo, 1990) and activities other than RNase MRP may play roles in generating RNA and DNA D-loop species.

By utilizing an *in vitro* run-off replication assay, Wong and Clayton (1985a,b) found that human mitochondria contain a mtDNA primase with the apparent capacity to recognize O_L and to initiate priming and DNA synthesis. Daughter L-strand synthesis is primed by RNA which is largely complementary to the T-rich loop structure of O_L; the transition from RNA synthesis to DNA synthesis occurs about 15 bases downstream near the base of the stem. Interestingly, the site of transition from RNA synthesis to DNA synthesis is also part of a tRNA gene. Thus, as with transcription termination, there is a dual functional constraint on a mtDNA sequence. And in each of these cases one of the functions is coding for a tRNA. The development of an *in vitro* system for replication at O_H has so far been limited to studies with isolated organelles (Jui and Wong, 1991).

B. Elongation and Maturation of Progeny Strands

Mitochondrial DNA polymerase, commonly termed DNA polymerase-γ, has received limited attention over the years, and only recently has information become available on its size and subunit composition and some of its functional features. There is a consistent association of a 3' to 5' exonuclease activity with the catalytic polymerase; it may represent a proofreading ability consistent with studies of the error rate of the holoenzyme (Kunkel and Soni, 1988; Insdorf and Bogenhagen, 1989a,b;

Kaguni and Olson, 1989; Kunkel and Mosbaugh, 1989). The catalytic subunit of the holoenzyme appears to be a single polypeptide of over 100 kDa and the exonuclease function might be within a smaller subunit that copurifies in the most highly purified preparations. One would like to produce these proteins by recombinant technology, thus circumventing the normal problems of low abundance of specific mitochondrial enzymatic activities and thereby providing a beginning for the ultimate development of an *in vitro* replication system with defined components.

Single-stranded DNA binding proteins have been identified from mitochondrial preparations (Mignotte *et al.*, 1988, Pavco and Van Tuyle, 1985; Van Tuyle and Pavco, 1985). These are likely orthodox single-stranded DNA binding activities, and their molecular sequence and properties resemble those of prokaryotes (Hoke *et al.*, 1990; Ghrir *et al.*, 1991). Although there is not yet a strict functional assignment for these proteins in either replication or transcription of animal mtDNA, they are likely playing important roles in these processes (Mignotte *et al.*, 1988; Barat-Gueride *et al.*, 1989). Other protein candidates for regulation by specific targeting to mtDNA sequences associated with transcription and replication have been identified from *Xenopus* (Cordonnier *et al.*, 1987), sea urchin (Roberti *et al.*, 1991), and chicken (D'Agostino and Nass, 1992).

Mammalian mitochondria do not appear to have a system for the removal of pyrimidine dimers, nor do they exhibit anything in the way of a normal rate of recombination activity (Clayton, 1982). However, certain types of repair activities have been found in mitochondrial fractions (Tomkinson *et al.*, 1990). Given the high level of interest and activity surrounding genetic changes in mtDNA and human disease, it would be worthwhile to study mtDNA repair capacity in depth. Our knowledge of related enzymatic activities, such as those of a mtDNA ligase and topoisomerases, has also remained sparse.

Other animal mtDNA systems that are revealing detailed information on the nature of transcription and replication control sequences include bovine, chicken, *Drosophila*, rat, sea urchin, and *Xenopus*. These works have been important in establishing the central conserved features of mtDNA control region organization, as well as some divergent features. For a recent review of these findings, see Clayton (1991).

IV. Potential Points of Regulation

A. Controlling Transcription and Replication

The origin of L-strand synthesis, O_L, does not operate until a decision to replicate has already occurred and synthesis of the leading H strand is two-thirds complete. Although lack of function at this origin could result in the segregation of one duplex circle and one single-stranded circle that might be degraded, in principle it is unlikely that O_L plays any critical role in determining whether mtDNA begins to be replicated. This decision is taken at the D loop, where a relationship has been established between the processes of replication and transcription. At present, there is no distinguishing the fate of initiating transcription with respect to events at the transcriptional start site itself; as noted, both transcription and replication priming could be supported by one transcriptional event given the fact that the genes on this strand are all downstream of the leading-strand replication origin.

If only the mtRNA and mtDNA polymerases were present, but no auxiliary proteins, neither efficient mtDNA replication nor transcription would ensue. The functional presence of mtTF1 at lower concentration levels should result in LSP activation and transcription of genes encoded on the L strand as template. Raising the levels of mtTF1 should eventually activate the HSP (Fisher and Clayton, 1988), thereby producing the major-ity of mtRNAs, which are rRNAs and the gene products copied from the H strand as template. This simplified view neglects the possibility of post-transcriptional differences in the rate at which messages are processed to their mature form and turned over, as well as regulation at the level of translation (Loguercio Polosa and Attardi, 1991), either or both of which may be important. In addition, transcription termination may be involved in regulating the stoichiometry of rRNA and mRNA production (Christian-son and Clayton, 1986, 1988; Kruse *et al.*, 1989; Hess *et al.*, 1991).

Replication priming should follow a transcriptional event at the LSP, provided the necessary RNA processing and replication activities are in place. There is a possible role for mtTF1 in RNA primer formation, since it binds to mtDNA at sites near RNase MRP cleavages (Fisher *et al.*, 1987, 1992). After D-loop DNA strand synthesis, some mechanism must determine whether D-loop strands terminate to form the D loop or continue their elongation to replicate one strand of the genome (Clayton, 1982). This point has not been established; it is unknown whether mtDNA lead-ing-strand synthesis proceeds by elongation of D-loop strands, or whether a separate cycle of initiation and synthesis through the D-loop region occurs.

B. Mitochondrial Gene Expression and Nuclear Genes

Mitochondrial biogenesis involves, in part, the interaction of mtDNA and nuclear gene products that are present in the mitochondrion. An important point is that all known critical enzymatic activities and accessory proteins for transcription and replication are nuclear gene products; none is known to be encoded by animal mtDNA (but see discussion of octocoral mtDNA in Wolstenholme, 1992). As mentioned above, gene expression begins by transcriptional activation at one or both mtDNA promoters. Given the very low abundance of primary transcripts in all cases studied to date, it is likely that RNA processing, in general, is not rate-limiting. Of course, this need not be true in special cases, such as diseased tissues, and it is possible that defects in the RNA processing machinery might contribute to improper mitochondrial gene expression. In the case of deletion mutations of mtDNA (e.g., Hayashi *et al.,* 1991), it is plausible that newly created RNA molecules (that cross mtDNA deletion endpoints) might possess structure that could inhibit or sequester the RNA processing system. If so, the observed phenotype of lower levels of normal mtDNA expression could follow.

It is perhaps obvious to predict that deletions of mtDNA that removed critical regulatory regions for replication and expression would be lethal events for the molecule (Table I). This would include loss of the origins of replication and transcriptional promoters. In particular, loss of LSP function would be expected to result in the inability of that mtDNA molecule to replicate due to the absence of leading-strand replication priming. In cases of normal LSP function, but defects in the leading-strand origin, one would also predict a selective loss of such molecules.

Loss of HSP function would have no obvious effect on the ability of mtDNA to replicate. However, such a genome would be seriously impaired

TABLE I

Predicted Effects of Some mtDNA Defects That Would Cause Loss of Specific Functions

Loss of function of:	Consequence
LSP	No expression of ND6, 8 tRNA genes; no mtDNA replication
O_H, CSBs	No mtDNA replication
HSP	No expression of rRNA, 12 protein-coding and 14 tRNA genes
O_L	mtDNA replication yields no net increase in progeny unless alternative priming
TERM sequence	rRNA/mRNA amount and form

in that both rRNA and most mitochondrial mRNAs would not be produced. The general phenotype would be lack of mitochondrial gene expression (28 of the 37 genes would be silent, including the rRNA genes and
most of the tRNA genes) in a situation where a normal amount of mtDNA
might be present.

Loss of the origin of L-strand replication could result in a situation
where such molecules, when replicated, would yield only one progeny
circle; that is, replication would not result in a net increase in the total
number of molecules. However, in principle, a segregated progeny single-
stranded circle (see Clayton, 1982), if stable, might be primed for duplex
synthesis by any available complementary RNA species, including partially degraded mtRNAs. This would not necessarily require RNA complementary to a particular region of mtDNA. Furthermore, in the absence of
the wild-type origin, mtDNA primase might recognize other DNA template
elements that mimic, at least in part, the missing origin.

Of course, not all problems of mitochondrial genetic expression are
limited to changes in mtDNA itself. Loss of function of the nuclear genes
for mtDNA polymerase or mtRNA polymerase would be expected to
result in complete loss of mtDNA. This should be a lethal condition for an
animal cell (Table II). Mutations in the nuclear gene for the mitochondrial
transcriptional activator mtTF1 would be expected to affect, but not necessarily abolish, transcription (and mtDNA replication). It is possible that
other proteins can substitute for mtTF1, or that mtRNA polymerase alone
(or with another accessory protein) can initiate specific transcription at a
low level. This might be sufficient for maintaining mitochondrial gene
expression in a fully differentiated cell and could perhaps provide sufficient
primers for mtDNA replication under these conditions. More serious consequences would be expected during development, when demands on
organelle biogenesis are highest.

Another activity thought to be involved in mtDNA replication is the
site-specific endonuclease RNase MRP. RNase MRP has been implicated
in primer RNA metabolism for leading-strand DNA synthesis and its
absence would be expected to affect the ability of mtDNA to replicate. At
present, there is no role for this activity in mitochondrial transcription.

If the nucleus-encoded protein involved in transcription termination
(mtTERM) was not present in functional form or appropriate levels, then
the amount and form of H-strand transcripts would likely be altered. It is
not known whether there are alternative mechanisms to produce the required amounts of rRNA under these conditions. This issue may be resolved by further studies with possible termination-defective mtDNA mutations and the eventual precise identification of the mtTERM protein(s)
and its gene(s).

TABLE II

Predicted Effects of Loss of Correct Type or Amount of Some Nuclear
Gene Products Known or Proposed to Play Roles in Mitochondrial
Nucleic Acid Synthesis

Nuclear gene(s) encoding:	Consequence of loss of nuclear gene product
mtDNA polymerase	Loss of mtDNA
mtRNA polymerase	Loss of transcription; loss of mtDNA
mtTF1	Minimal transcription and replication
RNase MRP	Lack of proper primer RNA metabolism at leading-strand origin (O_H)
Primase	Lack of initiation at origin of light-strand synthesis (O_L)
mtTERM	Amount and form of transcription products

V. Summary and Conclusions

The development of *in vitro* transcription and replication systems has
allowed the identification of promoter sequences and origins of replication
for several animal mtDNAs. As a consequence, the necessary reagents
and basic information are available to permit the characterization of *trans*-
acting factors that are required for transcription and replication. All of the
animal *trans*-acting species purified at this time are known or reasoned to
be nuclear gene products. There is now the opportunity to learn how these
nuclear genes are regulated and the mechanisms that are utilized for the
import of their products into the organelle. With regard to import, the
human transcription factor mtTF1 appears to have an amino-terminal
sequence characteristic of other imported mitochondrial proteins (Parisi
and Clayton, 1991).

An interesting issue is the degree to which fundamental features of
mtDNA replication and transcription are conserved between species. With

regard to animal mtDNAs, there is very little in the way of conservation of DNA sequence at the promoters and origins of replication. The exceptions to this are the presence of a characteristic stem–loop L-strand origin of replication sequence in vertebrates [except for chicken mtDNA (Desjardins and Morais, 1990)] and the general presence of CSBs II and III (and to a lesser extent CSB I) in most higher animal mtDNAs.

Because mtDNA promoters are not highly conserved, it is perhaps not surprising that general cross-species transcription does not occur, except for very limited examples of closely related species and sequences (Chang *et al.*, 1985b). Using crude mtRNA polymerase holoenzyme preparations, there is no specific transcriptional initiation when proteins from human mitochondria are used with mouse mtDNA promoter templates, and vice versa. However, in contrast to this overall observation, purified fractions of human or mouse mtTF1 can be exchanged and shown to function across species boundaries (Fisher *et al.*, 1989). The ability of mitochondrial mtTF1-type proteins to operate across even greater evolutionary distances was suggested by the ability of human and yeast proteins to recognize some mitochondrial promoter sequences in common (Fisher *et al.*, 1992). More recent studies suggest that human mtTF1 can substitute for its yeast homolog *in vivo,* and thereby perform at least the most critical functions required to maintain yeast mtDNA in the cell (M.A. Parisi, B. Xu, and D.A. Clayton, submitted for publication).

The other sites of conserved macromolecular interactions are related to the two origins of DNA replication. There is conservation of O_L sequence and potential secondary structure in vertebrates, and some evidence of cross-species, but less than fully accurate, primase function (Wong and Clayton, 1985b). More general is the conservation of CSBs, in particular the guanosine-rich CSB II-type of sequence. This element is present even in *Saccharomyces cerevisiae* mtDNA and its locations therein are at putative origins of DNA replication (Baldacci *et al.,* 1984; de Zamaroczy *et al.,* 1984). It is intriguing that yeast cells contain an RNase MRP-type of activity (Stohl and Clayton, 1992) that cleaves a yeast mtRNA substrate exactly at this sequence which, as in vertebrates, is also near a promoter element (Schinkel and Tabak, 1989). In addition, RNase MRPs cleave across species boundaries in a site-specific manner that reflects the RNA substrate sequence being assayed (Chang and Clayton, 1987a; Dairaghi and Clayton, 1992; Stohl and Clayton, 1992). Thus it seems apparent that RNase MRP has been conserved in at least some basic features from yeast to humans. To what degree this reflects a demand for a yet-to-be described function in the nucleus, where a majority of the enzyme is located, or its proposed role in processing mtRNA, is not known. However, the prospect of manipulating the yeast nuclear genes for components of the RNase MRP enzyme sets the stage for learning its overall roles in cellular metabolism. This system should also provide a definitive answer as to any requirement for RNase MRP in maintaining wild-type yeast mtDNA.

The recent discovery that yeast mitochondria contain a protein homolog of mtTF1 may suggest a closer relationship in general fundamental aspects of transcription than previously imagined. Although an exact role of the yeast protein in transcription has yet to be established, it is known that it is essential for proper yeast mtDNA expression and maintenance (Diffley and Stillman, 1991) and very recent evidence demonstrates that it can enhance transcriptional initiation *in vitro* (Parisi *et al.*, submitted for publication). It may be anticipated that future investigations of mtDNA replication and transcription in these and other systems should reveal the full extent of basic features of these fundamental processes that are held in common.

Acknowledgments

Research from this laboratory has been supported by Grant R37-GM33088 from the National Institute of General Medical Sciences. I am grateful to current and former students and fellows who have been responsible for the work from our laboratory.

References

Attardi, G. (1985). *Int. Rev. Cytol.* **93**, 93–145.
Attardi, G., and Schatz, G. (1988). *Annu. Rev. Cell Biol.* **4**, 289–333.
Baldacci, G, Chérif-Zahar, B., and Bernardi, G. (1984). *EMBO J.* **3**, 2115–2120.
Barat-Gueride, M., Dufresne, C., and Rickwood, D. (1989). *Eur. J. Biochem.* **183**, 297–302.
Bennett, J. L., Jeong-Yu, S., and Clayton, D. A. (1992). *J. Biol. Chem.*, in press.
Bogenhagen, D. F., and Morvillo, M. V. (1990). *Nucleic Acids Res.* **18**, 6377–6383.
Bogenhagen, D. F., and Romanelli, M. F. (1988). *Mol. Cell. Biol.* **8**, 2917–2924.
Chang, D. D., and Clayton, D. A. (1987a). *EMBO J.* **6**, 409–417.
Chang, D. D., and Clayton, D. A. (1987b). *Science* **235**, 1178–1184.
Chang, D. D., Hauswirth, W. W., and Clayton, D. A. (1985a). *EMBO J.* **4**, 1559–1567.
Chang, D. D., Wong, T. W., Hixson, J. E., and Clayton, D. A. (1985b). *In* "Achievements and Perspectives of Mitochondrial Research," Vol. 2, "Biogenesis" (E. Quagliariello, E. C. Slater, R. Palmieri, C. Saccone, and Kroon, A. M., eds.), pp. 135–144. Elsevier, Amsterdam.
Christianson, T. W., and Clayton, D. A. (1986). *Proc. Natl. Acad. Sci. U.S.A.* **83**, 6277–6281.
Christianson, T. W., and Clayton, D. A. (1988). *Mol. Cell. Biol.* **8**, 4502–4509.
Clayton, D. A. (1984). *Annu. Rev. Biochem.* **53**, 573–594.
Clayton, D. A. (1982). *Cell (Cambridge, Mass.)* **28**, 693–705.
Clayton, D. A. (1991). *Annu. Rev. Cell Biol.* **7**, 453–478.
Clayton, D. A., and Smith, C. A. (1975). *Int. Rev. Exp. Pathol.* **14**, 1–67.
Cordonnier, A. M., Dunon-Bluteau, D., and Brun, G. (1987). *Nucleic Acids Res.* **15**, 477–490.
D'Agostino, M. A., and Nass, M. M. K. (1992). *Exp. Cell Res.* **199**, 191–205.
Dairaghi, D. J., and Clayton, D. A. (1992). *J. Mol. Evol.*, in press.
Desjardins, P., and Morais, R. (1990). *J. Mol. Biol.* **212**, 599–634.

de Zamaroczy, M., Faugeron-Fonty, G., Baldacci, G., Goursot, R., and Bernardi, G. (1984). *Gene* **32**, 439–457.

Diffley, J. F. X., and Stillman, B. (1991). *Proc. Natl. Acad. Sci. U.S.A.* **88**, 7864–7868.

Diffley, J. F. X., and Stillman, B. (1992). *J. Biol. Chem.* **88**, 3368–3374.

Fisher, R. P., and Clayton, D. A. (1988). *Mol. Cell. Biol.* **8**, 3496–3509.

Fisher, R. P., Lisowsky, T., Breen, G. A. M., and Clayton, D. A. (1991). *J. Biol. Chem.* **266**, 9153–9160.

Fisher, R. P., Lisowsky, T., Parisi, M. A., and Clayton, D. A. (1992). *J. Biol. Chem.* **267**, 3358–3367.

Fisher, R. P., Parisi, M. A., and Clayton, D. A. (1989). *Genes Dev.* **3**, 2202–2217.

Fisher, R. P., Topper, J. N., and Clayton, D. A. (1987). *Cell (Cambridge, Mass.)* **50**, 247–258.

Ghrir, R., Lacaer, J.-P., Dufresne, C., and Gueride, M. (1991). *Arch. Biochem. Biophys.* **291**, 395–400.

Hayashi, J.-I., Ohta, S., Kikuchi, A., Takemitsu, M., Goto, Y.-i., and Nonaka, I. (1991). *Proc. Natl. Acad. Sci. U.S.A.* **88**, 10,614–10,618.

Hess, J. F., Parisi, M. A., Bennett, J. L., and Clayton, D. A. (1991). *Nature (London)* **351**, 236–239.

Hixson, J. E., and Clayton, D. A. (1985). *Proc. Natl. Acad. Sci. U.S.A.* **82**, 2660–2664.

Hoke, G. D., Pavco, P. A., Ledwith, B. J., and Van Tuyle, G. C. (1990). *Arch. Biochem. Biophys.* **282**, 116–124.

Insdorf, N. F., and Bogenhagen, D. F. (1989a). *J. Biol. Chem.* **264**, 21,491–21,497.

Insdorf, N. F., and Bogenhagen, D. F. (1989b). *J. Biol. Chem.* **264**, 21,498–21,503.

Johansen, S., Guddal, P. H., and Johansen, T. (1990). *Nucleic Acids Res.* **18**, 411–419.

Jui, H. Y., and Wong, T. W. (1991). *Nucleic Acids Res.* **19**, 905–911.

Kaguni, L. S., and Olson, M. W. (1989). *Proc. Natl. Acad. Sci. U.S.A.* **86**, 6469–6473.

King, M. P., Koga, Y., Davidson, M., and Schon, E. A. (1992). *Mol. Cell. Biol.* **12**, 480–490.

Kruse, B., Narasimhan, N., and Attardi, G. (1989). *Cell (Cambridge, Mass.)* **58**, 391–397.

Kunkel, T. A., and Mosbaugh, D. W. (1989). *Biochemistry* **28**, 988–995.

Kunkel, T. A., and Soni, A. (1988). *J. Biol. Chem.* **263**, 4450–4459.

L'Abbé, D., Duhaime, J.-F., Lang, B. F., and Morais, R. (1991). *J. Biol. Chem.* **266**, 10,844–10,850.

Loguercio Polosa, P., and Attardi, G. (1991). *J. Biol. Chem.* **266**, 10,011–10,017.

Masters, B. S., Stohl, L. L., and Clayton, D. A. (1987). *Cell (Cambridge, Mass.)* **51**, 89–99.

Mignotte, B., Marsault, J., and Barat-Gueride, M. (1988). *Eur. J. Biochem.* **174**, 479–484.

Parisi, M. A., and Clayton, D. A. (1991). *Science* **252**, 965–969.

Pavco, P. A., and Van Tuyle, G. C. (1985). *J. Cell Biol.* **100**, 258–264.

Roberti, M., Mustich, A., Gadaleta, M. N., and Cantatore, P. (1991). *Nucleic Acids Res.* **19**, 6249–6254.

Schinkel, A. H., and Tabak, H. F. (1989). *Trends Genet.* **5**, 149–154.

Stohl, L. L., and Clayton, D. A. (1992). *Mol. Cell. Biol.* **12**, 2561–2569.

Tomkinson, A. E., Bonk, R. T., Kim, J., Bartfelt, N., and Linn, S. (1990). *Nucleic Acids Res.* **18**, 929–935.

Topper, J. N., and Clayton, D. A. (1990). *Nucleic Acids Res.* **18**, 793–799.

Van Tuyle, G. C., and Pavco, P. A. (1985). *J. Cell Biol.* **100**, 251–257.

Walberg, M. W., and Clayton, D. A. (1981). *Nucleic Acids Res.* **9**, 5411–5421.

Wolstenholme, D. R. (1992). In "International Review of Cytology," Vol. 141, "Mitochondrial Genomes" (D. R. Wolstenholme and K. W. Jeon, eds.), pp. 173–216. Academic Press, San Diego.

Wong, T. W., and Clayton, D. A. (1985a). *J. Biol. Chem.* **260**, 11,530–11,535.

Wong, T. W., and Clayton, D. A. (1985b). *Cell (Cambridge, Mass.)* **42**, 951–958.

Xu, B., and Clayton, D. A. (1992). *Nucleic Acids Res.* **20**, 1053–1059.

The Endosymbiont Hypothesis Revisited

Michael W. Gray

Department of Biochemistry, Dalhousie University, Halifax, Nova Scotia, Canada
B3H 4H7

I. Introduction

A. Scope of the Review

In the biological world, there are two distinctly different cell types, prokaryotic and eukaryotic (Chatton, 1938; Stanier and van Niel, 1962; Stanier, 1970). Eukaryotic cells are internally more complex than prokaryotic
cells: not only is their genetic material partitioned into a membrane-de-
limited *nucleus* (but see Fuerst and Webb, 1991), but they also contain
other functionally specialized structures (*organelles*), prominent among
which are *mitochondria* (present in most, but not all, eukaryotes) and
plastids (found in photosynthetic eukaryotes—land plants and algae).
How mitochondria and plastids arose has been a subject of continuing
interest and considerable speculation ever since they were first recognized
over a hundred years ago. Did these organelles, in concert with the nucleus, originate through a process of intracellular compartmentalization
and functional specialization, within a single kind of cell (*autogenous
origin*, or, origin from within; see, e.g., Raff and Mahler, 1972, 1975) or are
they the exolutionary remnants of (bacterial) endosymbionts (*xenogenous
origin*, or, origin from outside; see Margulis, 1970, 1975)? [For an illuminat-
ing historical overview of the central debates concerning the role of symbiosis in cell evolution, see Sapp (1990).] The discovery that genetic information in eukaryotic cells is also compartmentalized (the plastid and
mitochondrion each possessing its own unique genome; see Gray, 1989a;
Palmer, 1991) has raised a parallel question of origin: do nuclear, mitochondrial, and plastid genomes derive from a single progenitor genome within
the same cell; or, are the latter two genomes the direct evolutionary
descendants of the genomes of (bacterial) endosymbionts?

In 1982, W. F. Doolittle and I published a review entitled "Has the
Endosymbiont Hypothesis Been Proven?" (Gray and Doolittle, 1982). In

posing this question, we sought to bring together molecular data bearing on the evolutionary origin of the nucleus, mitochondrion, and plastid of eukaryotic cells, and to provide a critical analysis, in the light of such data, of competing autogenous and xenogenous theories of organelle origin (see Fig. 1). The former type of theory is often referred to as *direct filiation* and the latter is more commonly known as the *endosymbiont hypothesis.* As relevant evidence, we considered primarily "data derived from protein and nucleic acid sequence analyses and comparative studies of nuclear and organellar genomic organization and molecular biology." We asserted that it was upon these sorts of data "that a decision on the validity of the endosymbiont hypothesis must ultimately rest." We based this assertion on the expectation that the mitochondrial and plastid genomes of eukaryotic cells would still retain vestiges of their evolutionary ancestries: molecular clues that could be deciphered through the appropriate studies. We concluded at the time that there was already strong molecular support for an endosymbiotic origin of plastids from oxygenic–photosynthetic eubacteria (cyanobacteria), but that the case for an endosymbiotic origin of mitochondria was more problematic and considerably less certain, for reasons discussed later in the present review. Only in the case of plant mitochondria, which appeared to contain a relatively conservative organellar genome, were relics of a eubacterial origin clearly apparent (Bonen *et*

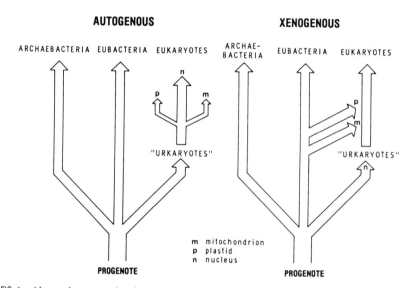

FIG. 1 Alternative scenarios for the evolutionary origin of mitochondria and plastids. Reprinted from Gray (1983), with permission.

al., 1977), and we suggested that further analyses of plant mitochondrial genomes offered the best chance for fulfillment of the sorts of proof of organelle origin that we had enunciated (see Section I,B).

The present review is intended to update and comment upon the main conclusions and predictions of the 1982 review, which in the interim has been supplemented by a number of other reviews dealing with various aspects of organelle origins and evolution (Gray, 1982, 1983, 1988, 1989a,b, 1991). For the most part, I will not reiterate the data presented in the 1982 review, which will form the essential background for the current commentary. Moreover, the presentation here will depart somewhat in form and content from the 1982 review, in that it will not attempt to provide as comprehensive a review of genome organization and expression. The sheer volume of relevant data appearing in the last decade makes that an impossible task within the confines of a single book chapter, and might well detract from the central theme of this review. I will be rather more selective than we were in 1982 in describing the molecular biology of archaebacterial, nuclear and organellar genomes, citing wherever possible more specific reviews that have appeared in recent years on various aspects of organism, organelle, and genome evolution. The contributions of the other authors to this special volume on mitochondrial genomes provide a comprehensive account of the current state of mitochondrial molecular biology.

Finally, I will not deal here with the events that preceded the origin of organelles, i.e., the progression from the precellular evolutionary world to the *progenote* to the *most recent common ancestor* of prokaryotes and eukaryotes. [The term "progenote" (Woese and Fox, 1977) was introduced to designate "entities that were in various stages of evolving a translation apparatus, whose linkage between genotype and phenotype was not yet as precise as that seen in modern cells" (Woese, 1990). Progenote should be distinguished from "most recent common ancestor," which may actually have been relatively advanced; it may be appropriate to use the term "protogenote" to designate the immediate precursor of prokaryotes and eukaryotes (Benner and Ellington, 1990), the modern cell being considered a "genote" (Woese, 1990).] Much of the discussion surrounding these issues has centered on the concept of a primordial "RNA world" (Gilbert, 1986). The reader should consult theoretical papers (Pace and Marsh, 1985; Darnell and Doolittle, 1986; Weiner and Maizels, 1987; Cedergren and Grosjean, 1987; Weiner, 1988; Benner *et al.*, 1989; Joyce, 1989; Wächtershäuser, 1988, 1990; Gibson and Lamond, 1990) and commentaries (Gilbert, 1986; Lewin, 1986; North, 1987; Benner and Ellington, 1987, 1991; Maizels and Weiner, 1987; De Duve, 1988; Pace, 1991) for a variety of viewpoints on this subject. An excellent

compendium of papers on these and related topics may be found in Vol.
52 (1987) of the *Cold Spring Harbor Symposia on Quantitative Biology*
("Evolution of Catalytic Function").

B. The Hypothesis Stated and Forms of Proof

Before discussing new evidence and insights that have emerged since 1982,
it would be useful to restate a number of the principles that guided our
earlier discussion of organelle origins. In the first place, we assumed "that
all contemporary genomes (including those of plastids and mitochondria)
ultimately derive from a single genome—the genome of a single, presumably cellular, entity which was the ancestor of all surviving forms of live."
Given this assumption (an eminently reasonable one in view of the great
similarity of metabolic pathways in all organisms and the near universality
of the genetic code):

> The question of organellar origins can then be restated simply: did the genomic
> lineages housed in contemporary nuclei, plastids, and mitochondria ever
> inhabit separate cellular lineages? If they did, then their present cohabitation
> can only be the result of symbioses, "endosymbioses" as classically de-
> scribed being the most easily imagined but not the only possible sort. (Mar-
> gulis, 1975, 1981)
>   The ease with which such a symbiotic origin can be proved depends on the
> antiquity of the original partitioning of genomes into separate cellular lineages
> and on the recentness of their subsequent reunion. It depends on the existence
> of contemporary free-living survivors of those separate lineages. It also de-
> pends on the rates of evolution within nuclear, organellar, and procaryotic
> lineages: these need not be, and demonstrably are not, the same.

In considering forms of proof, we stated that "the endosymbiont origin
of [a particular] organelle could be taken as proven" if (i) "the evolutionary
histories of nuclear genomes and one of the organellar genomes were
known with certainty, and were with certainty different"; (ii) "the nuclear
genome, although lacking modern free-living relatives, clearly descended
from a lineage *other* than that from which organellar genomes descended";
or (iii) "it could be shown that organelles of different major groups (as an
example, the plastids of rhodophytes [red algae] and chlorophytes [green
algae] derived from different lineages of procaryotes (in this case, the
cyanobacteria and the "prochlorophytes" [Lewin, 1981])."

In the 1982 review, we also discussed evidence which, unless integrated
with other data, cannot be considered to constitute proof of the endosymbiont hypothesis. In particular, we emphasized the distinction between *plesi-
omorphic* (primitive, ancient) and *apomorphic* (derived) traits: the former
being ones that were present in the common ancestor of eukaryotic and

prokaryotic cells, but selectively retained in only one of these groups following their separation from one another, and the latter being traits that evolved specifically in one of the groups after their separation from a common ancestor. Only derived traits are meaningful in evaluating organelle origin hypotheses [see Uzzell and Spolsky (1974, 1981) for specific discussion of this point]. We also noted that some "bacterial" traits of plastids and mitochondria are in fact determined by nuclear genes, whose presence in the nucleus is assumed to be the result of gene transfer from endosymbiont to host genome. We commented, "Although it is not unreasonable to invoke organellar–nuclear gene transfer after the establishment of endosymbioses, it is at least awkward to base arguments for the xenogenous origin of organellar genomes on traits not now encoded in these genomes." Recent evidence of evolutionary gene transfer from organellar to nuclear genomes (see IV,I and V,D) now make these arguments somewhat less awkward than in 1982. Finally, although symbiosis is rampant in the biological world, and bacterial endosymbionts within eukaryotic cells are legion (Margulis, 1981; Lee and Corliss, 1985; Reisser et al., 1985; Lee et al., 1985), that in itself cannot be considered part of the actual proof that mitochrondria and plastids arose in evolution as a result of such associations. On the other hand, the fact and frequency of symbiosis, by providing the basis for a xenogenous origin, cannot help but make such a scenario attractive. Indeed, an intriguing example of "endosymbiosis in real time" has recently been documented (Jeon, 1987).

The present review will not provide an exhaustive reevaluation of alternative hypotheses of organelle origin. In the light of the new data accumulated since 1982, that now seems rather pointless: the evidence increasingly compels acceptance of the endosymbiont hypothesis, at least in broad outline. Considering the forms of proof listed above, (ii) has now been clearly satisfied in the case of both plastid and mitochondrion, providing strong evidence of a separate, eubacterial, endosymbiotic origin of these two organelles, as we shall see.

II. The Lineages of Life

A. Phylogenetic Trees Based on Ribosomal RNA

Before dealing directly with the origin of mitochondria and plastids, it is essential to comment on the primary biological lineages, and on the role molecular data have played in defining these lineages and in elucidating phylogenetic relationships within and among them. In Fig. 1, alternative hypotheses of organelle origin are placed within the concept of three primary lineages (archaebacteria, eubacteria, and eukaryotes) diverging from a universal ancestor, the *progenote*. This fundamental division of life, and the recognition that *archaebacteria* and *eubacteria* represent two phylogenetically separate groups of prokaryotes, stems from the pioneering work of C. R. Woese and colleagues (Fox *et al.,* 1977, 1980; Woese and Fox, 1977; Woese *et al.,* 1978; Woese and Olsen, 1986). That the archaebacteria are a coherent assemblage of organisms, distinct from both eubacteria and eukaryotes, was initially based largely, although not exclusively (Woese 1981), on comparisons of ribosomal RNA (rRNA) sequences (Woese, 1987).

The role that rRNA sequence comparisons have played during the past decade in shaping our views of phylogeny and evolution cannot be overemphasized. The construction and interpretation of phylogenetic trees based on small subunit (SSU, or 16S-like) and large subunit (LSU, or 23S-like) rRNA sequences have proven especially informative in instances where morphological diversity tends to confound traditional methods of phylogenetic analysis [e.g., in deducing phylogenies of bacteria (Woese, 1987) and unicellular eukaryotes (protists) (Sogin *et al.,* 1989)]. As we shall see, rRNA sequence comparisons have been equally important in deciphering the pathways of organelle evolution. The features of SSU and LSU rRNAs that have prompted their designation as the "ultimate molecular chronometers" (Woese, 1987) have been discussed elsewhere (Woese, 1987; Gray, 1988). They include antiquity (descent of all extant rRNA homologs from a common ancestor); ubiquity (the presence of homologous rRNA genes in all genome types—archaebacterial, eubacterial, nuclear, mitochondrial, and plastid); functional equivalence and constancy, assuring reasonably good clocklike behavior (but see V,C); the presence in both molecules of slowly evolving and more rapidly evolving segments, facilitating the determination of relationships over a broad range of evolutionary distance; the ability of the slowly evolving segments to form a highly conserved core of secondary structure that facilitates accurate alignment of primary sequence (Gray *et al.,* 1984; Cedergren *et al.,* 1988); the availability of a large number of nucleotide positions (typically 500–1000), minimizing statistical fluctuations inherent in 5S rRNA and tRNA data sets; and relative ease of determination of complete or nearly complete rRNA sequences (Lane *et al.,* 1985; Sogin, 1990).

It is beyond the scope of this review to comment critically and in depth on the use of rRNA sequence information in phylogenetic tree construction, including the methodologies employed and the evaluation and interpretation of the resulting trees. Succinct discussions of some of the relevant issues may be found in Olsen (1987) and Felsenstein (1988), and wider ranging discussions of molecular phylogeny and treeing methods are presented in Doolittle (1990), Li and Graur (1991), and Sidow and Bowman (1991). Ribosomal RNA-based tree construction is not without its limitations (Woese, 1991); in some cases, inferences of phylogenetic relationship based on rRNA trees have proven quite contentious (see II,B). In this connection, a few caveats are worth noting. (i) A phylogenetic tree, whether based on rRNA or some other molecule, should be considered a *best approximation* of the true tree, to be confirmed, modified, or rejected as additional/better data and/or methodologies become available. To what extent the *inferred tree* recapitulates the *true tree* may depend critically on the nature and quality of the data set (number and phylogenetic distribution of compared taxa, number of homologous positions analyzed, alignment of sequences) and the particular method(s) used in tree construction. Consistency between trees generated by the same method with different data sets (Woese *et al.*, 1980), or by different methods applied to the same data set (Douglas *et al.*, 1991), obviously give one increased confidence in the resulting tree topologies. (ii) A phylogenetic tree based on a particular gene sequence (e.g., SSU rRNA) is precisely that: a tree of *that particular gene,* rather than a tree of the genome encoding that gene. To what extent the two coincide depends on a number of factors, chief among which is whether the gene in question has been subject to any sort of lateral transfer between genomes during its evolution, or whether the genome in question is actually a genetic mosaic (having acquired genes from different sources in the course of evolution). A basic assumption of rRNA trees is that lateral transfer of functional rRNA genes between genomes is not only much less likely to have occurred than lateral transfer of protein-coding genes, but that it has not, in fact, occurred. Whether there is now reason to question this assumption (Gray *et al.*, 1989) will be discussed later (V,C). (iii) Finally, lateral transfer of genes (including rRNA genes) as a result of lateral transfer of organelles and their contained genomes must now be considered a possible factor that may confound phylogenetic inference. As we shall see, transfers of this sort appear to have been prominent in plastid evolution (Douglas *et al.*, 1991).

Although rRNA sequence comparisons may not be "the Rosetta Stone of phylogenetics" (Rothschild *et al.*, 1986), they are still the most accessible and useful way to generate universal phylogenetic trees. Currently, they offer a database that is unparalleled in phylogenetic depth and breadth, and they have provided profound new insights into evolutionary relationships. For that reason, they are given special emphasis here. Obviously, phylogenies determined from comparisons of a single type of gene assume added significance when confirmed by analyses of other kinds of genes, and ultimately have to be evaluated in the light of all other available data.

B. Uniqueness of the Archaebacteria

In addition to rRNA sequence, there are a number of traits that characterize the archaebacteria as a distinct (albeit biochemically and genetically
diverse) group of organisms, separate from all other prokaryotes (eubacteria). These include [see Gray and Doolittle (1982) for earlier references
and Zillig (1991) for a more complete listing]: (i) cell walls that, when
present, lack peptidoglycan and muramic acid; (ii) the presence of lipids
containing phytanyl side groups in ether linkage; (iii) the presence of
RNA polymerases that are distinct from those of eubacteria in subunit
composition (Zillig *et al.*, 1985; Gropp *et al.*, 1986), response to RNA
synthesis inhibitors/stimulators (Schnabel *et al.*, 1982), immunological
reactivity (Huet *et al.*, 1983), and gene sequence (Berghöfer *et al.*, 1988);
(iv) unique patterns of rRNA and tRNA modification (Pang *et al.*, 1982;
Edmonds *et al.*, 1991); (v) transcriptional control signals unlike those of
eubacteria (Reiter *et al.*, 1990); and (v) elongation factors that (in contrast
to those of eubacteria) are sensitivite to diphtheria toxin-catalyzed ADP-
ribosylation. Two major divisions of archaebacteria have been recognized
(Woese and Olsen, 1986; Woese *et al.*, 1990): (i) sulfur-dependent, extreme
thermophiles; (ii) methane-producers (methanogens) and their relatives,
including the extreme halophiles. More detailed discussions of archaebacterial biochemistry and molecular biology can be found in reviews by
Woese (1981, 1987), Woese and Wolfe (1985), Dennis (1986), Zillig *et al.*
(1988), J. W. Brown *et al.* (1989), Garrett *et al.* (1991), and Zillig (1991).
A useful overview of the current state of "archaebacteriology" is provided
by the collection of papers representing the Proceedings of the Third
International Conference on "Molecular Biology of Archaebacteria—1988, " published as a special issue of the *Canadian Journal of
Microbiology* [Vol. 35, No. 1 (1989)].

   Although there is general agreement that the archaebacteria constitute
a group of organisms distinct from both eubacteria and eukaryotes, there
has been controversy about the phylogenetic coherence of the archaebacteria, i.e., whether they constitute a monophyletic grouping (Olsen, 1987;
Lake, 1987b, 1989; Olsen and Woese, 1989). The more widely accepted
view (Pace *et al.*, 1986) is that they do, and that the fundamental division
of life forms is into the three primary lineages (Eubacteria, Archaebacteria,
Eukaryotes) shown in the unrooted "archaebacterial" tree of Fig. 2. A
dissenting opinion has been vigorously voiced by Lake (1987b, 1988, 1989,
1990b, 1991), who has promoted an alternative tree (Fig. 3). Lake argues
both that the prokaryotes are not a proper phylogenetic group and that the
archaebacteria do not represent a monophyletic assemblage, but are in
fact paraphyletic (i.e., they are associated with more than one primary
lineage). Lake's "eocyte" tree (Fig. 3) places extremely thermophilic,

ARCHAEBACTERIA

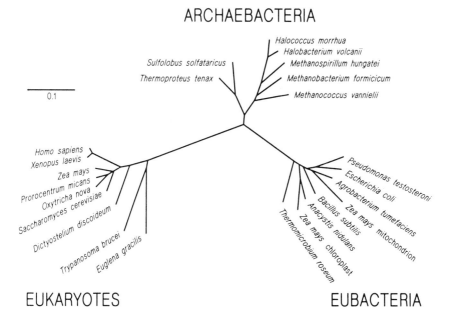

FIG. 2   The "archaebacterial tree" of Pace *et al.* (1986), based on comparison of SSU rRNA sequences. See text for discussion and Pace *et al.* (1986) and Olsen (1987) for details of the analysis. The length of each segment is proportional to the inferred number of fixed point mutations along that segment; the scale bar corresponds to 0.1 nucleotide substitutions per sequence position. The phylogenetic position of the common ancestor to the sequences is not identified; i.e., the tree is unrooted. Reprinted from Olsen (1987), with permission. Copyright 1987 Cold Spring Harbor Laboratory.

sulfur-dependent archaebacteria (*eocytes;* Lake *et al.*, 1984) together with eukaryotes (the two constituting the "karyotes"); puts halophilic archaebacteria (*photocytes;* Lake *et al.*, 1985) together with eubacteria; and leaves the *methanogens* as a sister group to the eubacteria/halobacteria clade (these three groups together comprising the "parkaryotes").

These alternative universal phylogenetic trees have sparked a heated debate in the scientific press about the validity of the various assumptions, methodologies, and conclusions (Lake, 1986a,b, 1990a; Zillig, 1986; Lederer, 1986; Woese *et al.*, 1986; Cavalier-Smith, 1986a; Penny, 1988; Guoy and Li, 1990). Although there is some independent support for the idea that archaebacteria may be paraphyletic (Wolters and Erdmann, 1986; but see Wolters and Erdmann, 1989), much other evidence favors the original view that archaebacteria are a monophyletic group. In addition to SSU rRNA data (which include both primary sequence and secondary structure "signature" features; Woese, 1987, 1991; Winker and Woese,

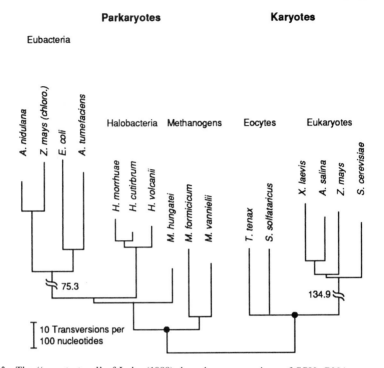

FIG. 3   The "eocyte tree" of Lake (1988), based on comparison of SSU rRNA sequences. This tree organizes all taxa into two monophyletic superkingdoms, the karyotes (eocytes and eukaryotes) and parkaryotes (eubacteria, halobacteria, and methanogens, including *Thermoplasma*). See text for discussion and Lake (1987b, 1988) for details of the analysis. Branch lengths indicate number of transversion differences rather than substitutions. The tree was rooted as described by Lake (1988). Reprinted from Lake (1987b), with permission. Copyright 1987 Cold Spring Harbor Laboratory.

1991), the evidence derives from comparisons of: (i) sequences of LSU rRNA (Leffers *et al.*, 1987; Guoy and Li, 1989), RNA polymerase subunits (Pühler *et al.*, 1989; Sidow and Wilson, 1990; Iwabe *et al.*, 1991; Zillig *et al.*, 1991), and ribosomal proteins (Wittmann-Liebold *et al.*, 1990; Matheson *et al.*, 1990); (ii) sequences flanking rRNA genes (Kjems and Garrett, 1990); (iii) RNA polymerase gene organization (Pühler *et al.*, 1989); (iv) antigenic determinants in elongation factor (EF)-Tu (Cammarano *et al.*, 1989); and (v) patterns of sensitivity/resistance to protein synthesis inhibitors (Amils *et al.*, 1989). Lake's proposal of archaebacterial paraphyly has centered on his use of a rate-independent technique for analysis of nucleic acid sequences (evolutionary parsimony; Lake, 1987a), but this particular technique has its own limitations (Jin and Nei, 1990), and Lake's application of evolutionary parsimony to the analysis of SSU rRNA sequence data has been challenged (Olsen and Woese, 1989). A recent study employing an extension of evolutionary parsimony (compositional statistics) did not, in fact, support paraphyly of the archaebacteria when applied to RNA polymerase sequence database (Sidow and Wilson, 1990).

On balance, and considering all of the available and relevant molecular data, there is no compelling reason at the present time to subdivide the archaebacteria at the highest taxonomic level. Thus, the currently prevailing view remains the original one: that there are three monophyletic lineages that represent the primary divisions of life. This view challenges the traditional prokaryote–eukaryote dichotomy (Stanier, 1970), as well as the more recent five-kingdom scheme (Whittaker, 1969; Whittaker and Margulis, 1978) that divides the living world into four eukaryotic kingdoms [Animalia, Plantae, Fungi, and Protista (or Protoctista; see Margulis and Schwartz, 1988)] and a single prokaryotic one (Monera). The five-kingdom scheme is perhaps the most familiar of current classifications; however, with the accumulation of molecular data, it has become apparent that at least two of these kingdoms (Protista, Monera) not only are of nonequivalent taxonomic rank, but are in fact artificial groupings (see Woese *et al.*, 1990). More recently, Cavalier-Smith (1987c, 1989b) has integrated molecular (i.e., sequence) data with ultrastructural and organismic properties in a phylogenetic scheme that comprises eight kingdoms within two "empires," Bacteria (kingdoms Eubacteria and Archaebacteria) and Eukaryota (kingdoms Archezoa, Protozoa, Chromista, Fungi, Plantae, and Animalia—all except Archezoa constituting the superkingdom Metakaryota).

In an effort to "bring formal taxonomy into line with the natural system emerging from molecular data," Woese *et al.* (1990) have proposed a new taxon (above the level of kingdom) called a *domain*. The biological world is seen as comprising three domains, the Bacteria (= eubacteria), the Archaea (= archaebacteria), and the Eucarya (= eukaryotes), each containing two or more kingdoms (see Fig. 4). For example, the Archaea are formally subdivided into two kingdoms, the Euryarchaeota (encompassing the methanogens and their phenotypically diverse relatives) and Crenarchaeota [comprising the relatively homogeneous group of extremely thermophilic archaebacteria (thermoacidophiles, or eocytes)]. The renaming of the Bacteria and Archaea is a deliberate attempt to avoid any suggestion that the eubacteria and archaebacteria are specifically related to one another. The domain Eucarya is seen as containing three previously recognized kingdoms (Animalia, Plantae, and Fungi) and a number of others (primarily unicellular ones) yet to be defined.

To what extent this revised taxonomy will be accepted remains to be seen; dealing as it does with the primary division of the biological world, it is certain to evoke controversy (Mayr, 1990; Margulis and Guerrero,

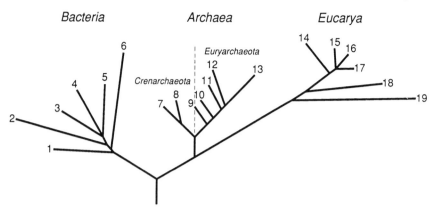

FIG. 4   Universal phylogenetic tree in rooted form, showing three domains (Bacteria, Archaea, Eucarya). Branching order and branch lengths are based upon SSU rRNA sequence comparisons (Woese, 1987). The position of the root was determined by comparing (the few known) sequences of pairs of paralogous (duplicated) genes that diverged from each other before the three primary lineages emerged from their common ancestral condition (Iwabe *et al.*, 1989; see Fig. 5). The numbers on the branches correspond to the following groups of organisms (Woese, 1987). **Bacteria:** 1, the Thermotogales; 2, the flavobacteria and relatives; 3, the cyanobacteria; 4, the purple bacteria; 5, the Gram-positive bacteria; and 6, the green nonsulfur bacteria. **Archaea:** the kingdom Crenarchaeota: 7, the genus *Pyrodictium;* and 8, the genus *Thermoproteus;* and the kingdom Euryarchaeota: 9, the Thermococcales; 10, the Methanococcales; 11, the Methanobacteriales; 12, the Methanomicrobiales; and 13, the extreme halophiles. **Eucarya:** 14, the animals; 15, the ciliates; 16, the green plants; 17, the fungi; 18, the flagellates; and 19, the microsporidia. Reprinted from Woese *et al.* (1990), with permission.

1991; Woese *et al.*, 1991; Cavalier-Smith, 1992). The new nomenclature, while making the distinction between eubacteria and archaebacteria more explicit, does potentially reintroduce an element of ambiguity into any discussion of the endosymbiont hypothesis, as the term "bacteria" has long been synonymous with "prokaryotes" (i.e., eubacteria *plus* archaebacteria). For that reason, although I accept the concept of three primary lineages and am sympathetic to the nomenclature proposed by Woese *et al.* (1990), I will continue in the present review to refer to "archaebacteria" (occasionally "archaeotes"), "eubacteria," and "eukaryotes," recognizing (as implied in Figs. 2 and 4) that eubacteria and archaebacteria, although both prokaryotes, are not specifically related at the highest taxonomic level.

C. Rooting the Universal Tree

A phylogenetic tree determined from rRNA sequence data alone cannot be rooted. For example, in the tree of Fig. 2 (assuming that a simultaneous three-way divergence is not realistic), there is no way to deduce whether the root leading to the common ancestor should be placed on the archaebacterial branch, the eubacterial branch, or the eukaryotic branch. However, the position of the root can be inferred from phylogenetic trees of duplicated genes (Iwabe *et al.*, 1989). In cases where pairs of gene duplicates are present in all three pirmary lineages, it is reasonable to suppose that this duplication occurred prior to the initial divergence of the lineages. If a composite phylogenetic tree is determined (a phylogenetic tree comprising two subtrees, one for each of the gene duplicates), the root can be placed at the point where the two genes diverged by gene duplication; this uniquely determines the evolutionary relationships among the universal ancestor and the three primary lineages. (This rooting strategy effectively uses one set of duplicated genes as an outgroup for the other.) Applying this approach, Iwabe *et al.* (1989) inferred composite phylogenetic trees for two pairs of duplicated genes, elongation factors Tu and G (Fig. 5) and the α and β subunits of ATPase. In both cases, the analysis showed that archaebacteria are specifically related to eukaryotes, to the exclusion of the eubacteria. A similar conclusion was reached by Gogarten *et al.* (1989) and Gogarten and Kibak (1992), also studying the evolution of ATPase subunits. This result allows one to root the rRNA tree of Fig. 2, as has been done in Fig. 4.

The rooting of the universal tree makes it possible to infer something about the nature of the universal common ancestor, by considering the distribution of shared characteristics among archaebacteria, eubacteria,

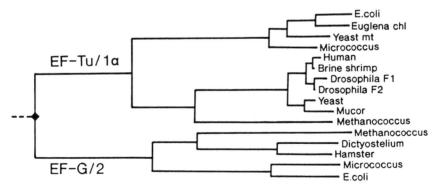

FIG. 5 Composite phylogenetic tree inferred from a simultaneous comparison of EF-Tu(1α) and EF-G(2) amino acid sequences from an archaebacterium (*Methanococcus*), eubacteria, and eukaryotes. The deepest root was arbitrarily chosen at a point between the two clusters corresponding to the different proteins. chl, plDNA-encoded gene product; mt, nDNA-encoded mitochondrial isozyme. Reprinted from Iwabe *et al.* (1989), with permission.

and eukaryotes. Features that are found in two of the three primary lineages to the exclusion of the third may (to a first approximation) be considered to be retained primitive traits. Thus, eubacteria and archaebacteria both have a prokaryotic type of cellular arrangement, and archaebacterial genes are in general organized into transcription units that are strikingly reminiscent of the homologous eubacterial operons. On the other hand, the cell membranes of eubacteria and eukaryotes (but not those of archaebacteria) have ester-linked lipids. From this, we may deduce that the universal common ancestor (protogenote) was likely a prokaryotic-type cell that possessed ester-linked membrane lipids. This conclusion is in accord with arguments (Cavalier-Smith, 1987a,c) that eubacteria preceded both eukaryotes and archaebacteria. It should be noted, however, that in unrooted rRNA phylogenetic trees (e.g., Fig. 2), the ends of the branches in the archaebacterial subtree appear closest to the trifurcation of the three primary lineages; the archaebacterial subtree also shows the smallest distance from this trifurcation point to the lowest branching point on the subtree. This may suggest that archaebacteria, and especially the Crenarchaeota, have diverged to a lesser extent from the protogenote than either eubacteria or eukaryotes. Thus, if the rooting of the universal phylogenetic tree in Fig. 4 is correct, eubacteria represent the earliest diverged group of organisms, having separated from other lifeforms well before archaebacteria and eukaryotes diverged from one another. However, archaebacteria (by virtue of a slower rate of evolution than either eubacteria and eukaryotes) may actually have remained closer in certain characteristics (in this case, rRNA sequence) to the protogenote.

This intermediate position of archaebacteria also provides a rationalization for the mix of eubacterial and eukaryotic features that characterizes the archaebacteria. For example, whereas archaebacterial genes tend to be clustered in operons, as in eubacteria, and arrayed within these operons in an order very similar to that of the homologous eubacterial operons, they are generally more similar in primary structure to their eukaryotic than their eubacterial homologs (Lechner and Böck, 1987; Auer *et al.*, 1989a,b; Köpke and Wittmann-Liebold, 1989; Pühler *et al.*, 1989; Arndt *et al.*, 1990; Wittmann-Liebold *et al.*, 1990; Matheson *et al.*, 1990). On the other hand, the *gln*A (glutamine synthetase) gene of the archaebacterium *Sulfolobus solfataricus* is strikingly similar in sequence to that of eubacteria, not eukaryotes (Sanangelantoni *et al.*, 1990). To Sanangelantoni *et al.* (1990), these results raised "the disturbing possibility that different relationships between kingdoms may be inferred depending upon which protein . . . is chosen as the phylogenetic probe."

III. Origin of the Nuclear Genome

A. Archaebacteria and the Nucleus

In the earlier review, we listed a number of major differences between eukaryotic nuclear and eubacterial molecular biologies [see p. 24 of Gray and Doolittle (1982)]. For the most part, these distinguishing characteristics remain, although new data have softened the differences in a number of areas.

1. Introns

Introns have now been found in the eubacterial lineage, first in certain protein-coding genes of bacteriophages of both Gram-negative and Gram-positive eubacteria (Chu *et al.*, 1986; Ehrenman *et al.*, 1986; Sjöberg *et al.*, 1986; Chu *et al.*, 1988; Shub *et al.*, 1988; Belfort, 1989; Quirk *et al.*, 1989; Goodrich-Blair *et al.*, 1990) and more recently in eubacterial tRNA genes (Xu *et al.*, 1990; Kuhsel *et al.*, 1990). However, the evolutionary relationship (if any) of these introns to those in eukaryotic protein-coding and tRNA genes remains to be determined.

2. Organization of rRNA Genes

Genes for 5S rRNA, physically and transcriptionally unlinked to 23S rRNA, have now been found in some eubacteria (Fukunaga and Mifuchi, 1989; Fukunaga *et al.*, 1990) and archaebacteria (Tu and Zillig, 1982; Larsen *et al.*, 1986; Reiter *et al.*, 1987; Kjems *et al.*, 1987; Culham and Nazar, 1988), whereas at least one example of very close physical linkage of eukaryotic 28S and 5S rRNA genes (potentially co-transcribed in that order) has been discovered (D. F. Spencer and M. W. Gray, unpublished observations).

3. Ribosome Size and rRNA Structure

The microsporidia, a group of obligately parasitic protists lacking mitochondria, are reported to contain prokaryotic-sized ribosomes and rRNAs (Curgy *et al.*, 1980; Ishihara and Hayashi, 1968). Moreover, *Vairimorpha nectarix*, a member of this group, has been shown to lack a separate 5.8S rRNA (Vossbrinck and Woese, 1986), an otherwise ubiquitous constituent of eukaryotic 80S ribosomes. As an eubacteria, a 5.8S-like sequence is covalently continuous with the rest of the *V. necatrix* LSU rRNA. Nevertheless, this organism trees as a eukaryote (albeit one diverging very early) in phylogenetic analyses based on rRNA sequence comparisons (Vossbrinck et al., 1987).

In contrast to the substantial differences between eukaryotes and eubacteria in traits related to the genome and its replication and expression, there are an increasing number of similarities that link archaebacterial and nuclear genomes (Searcy, 1987).

(i) With few exceptions, archaebacterial proteins resemble their eukaryotic homologs more closely in structure than their eubacterial ones. This has been seen with various ribosomal proteins (Auer et al., 1989a,b; Hatakeyama et al., 1989; Kimura et al., 1989; Köpke and Wittmann-Liebold, 1989; Arndt et al., 1990; Wittmann-Liebold et al., 1990; Matheson et al., 1990), α and β subunits of ATPase (Denda et al., 1988a,b; Gogarten et al., 1989; Sudhof et al., 1989), elongation factors EF-1α (= ET-Tu; Lechner and Böck, 1987; Cammarano et al., 1989) and EF-2 (= EF-G; Lechner et al., 1988; Itoh, 1989; Schröder and Klink, 1991), and DNA-dependent RNA polymerase (Schnabel et al., 1983; Berghöfer et al., 1988; Pühler et al., 1989).

(ii) Archaebacteria and eukaryotes share a few homologous ribosomal proteins that have no counterparts in eubacteria (Auer et al., 1989a,b; Kimura et al., 1989; Scholzen and Arndt, 1991).

(iii) The archaebacterium Methanothermus fervidus contains a DNA-binding protein that shows a close relationship to eukaryotic histones (Sandman et al., 1990).

(iv) A molecular chaperone from the thermophilic archaebacterium, Sulfolobus shibatae, displays high similarity to eukaryotic t-complex polypeptide-1 (TCP1), an essential protein that may play a part in mitotic spindle formation (Trent et al., 1991).

(v) Archaebacterial 5S rRNA (Hori and Osawa, 1987) and initiator methionine tRNA (Kuchino et al., 1982) are more similar to their eukaryotic than their eubacterial counterparts.

(vi) Archaebacterial promoters resemble those used by eukaryotic RNA polymerase II (Reiter et al., 1988, 1990; Hüdepohl et al., 1990) rather than eubacterial promoters.

(vii) Archaebacteria share with eukaryotes several distinctive functional characteristics not shown by eubacteria, such as stimulation of RNA polymerase by silybin (Schnabel et al., 1982) and sensitivity of EF-2 to ADP-ribosylation by diptheria toxin (Kessel and Klink, 1980). Notably, this toxin-mediated ADP-ribosylation occurs at a unique post-translationally modified histidine residue (diphthamide) (Van Ness et al., 1980), which is present in archaebacteria and eukaryotes but not in eubacteria

(Pappenheimer *et al.*, 1983). Another unusual amino acid, hypusine [N$^\varepsilon$-(4-amino-2-hydroxybutyl)lysine] (Shiba *et al.*, 1971), initially considered to be specific for eukaryotic initiation factor eIF-4D (Cooper *et al.*, 1983), has recently been shown to be present in cell protein from archaebacteria (Schümann and Klink, 1989; Bartig *et al.*, 1990) but not eubacteria (Gordon *et al.*, 1987; Schümann and Klink, 1989).

(viii) The 70S ribosomes of some archaebacteria appear to contain interaction sites for antibiotics such as anisomycin, which are characteristic of eukaryotic 80S, but not eubacterial 70S, ribosomes (Hummel and Böck, 1985).

Rooted universal trees (Figs. 4 and 5) imply that archaebacteria and eukaryotes are specific (albeit distant) relatives. Increasingly, then, it appears that a specific ancestor of the archaebacteria also contributed in a substantial way to the genetic make-up of the eukaryotic nuclear genome. The extent and nature of this contribution, and indeed the molecular characteristics of the Archaea–Eucarya common ancestor, remain to be clarified. It has been suggested that the last common ancestor of the archaebacteria was a thermophilic, anaerobic, and sulfur-metabolizing organism (Woese, 1991). Because the Crenarchaeota (Woese *et al.*, 1990) are thought to resemble most closely the archaeal ancestor, we would expect that crenarchaeotes (i.e., eocytes) should also resemble eukaryotes more closely than do members of the Euryarchaeota (euryarchaeotes, i.e., methanogens and extreme halophiles) (see Fig. 4). Such a relationship is also implicit in Lake's eocyte tree (Fig. 3) and seems to be supported by the fact that the crenarchaeotal translation apparatus shows a number of "eukaryote-like" characteristics not seen in euryarchaeotes save among some of the more deeply branching ones (Woese, 1991).

That the eukaryotic nuclear genome is to some degree a genetic mosaic is suggested by reports that certain of its genes are *more* (not less) similar to their eubacterial than their archaebacterial countrparts. A case in point concerns RNA polymerase (pol) subunits. The largest and second largest subunits of eukaryotic pol I, II, and III are homologs of one another and of the β' and β subunits, respectively, of the single eubacterial RNA polymerase. Homologous genes for the corresponding subunits of the archaebacterial RNA polymerase have also been identified. From comparisons of the β'-like sequences, Pühler *et al.* (1989) inferred different origins for eukaryotic pol I, II, and III: phylogenetic trees produced by these investigators show the archaebacteria as a coherent lineage related to the nuclear pol II and/or III lineage, whereas pol I appears to arise separately from a bifurcation with the eubacterial lineage (Jess *et al.*, 1990). From this, Pühler *et al.* (1989) suggested the possibility that pol I was acquired through a direct horizontal gene transfer from the eubacterial lineage or, more indirectly, from a eubacteria-like endosymbiont.

An even more radical interpretation sees the "eucyte" (the nuclear–cytoplasmic component of the eukaryotic cell) as a fusion of eubacteria-like (contributing pol I) and archaebacteria-like (contributing pol II and/or III) partners (Zillig *et al.*, 1989, 1991). Hartman (1984) and Sogin (1991), on the other hand, conjecture that the eukaryotic nucleus originated at least in part as (a) prokaryotic endosymbiont(s), but that the host organism was actually an RNA-based one, whose distinguishing feature was the possession of a cytoskeleton that conferred on it the ability to engulf other microorganisms. However, Forterre (1992) concludes, on the basis of DNA polymerase sequence comparisons, that the last common ancestor of eubacteria, archaebacteria, and eukaryotes was already a complex DNA-based organism that contained at least two DNA polymerases with proofreading (3' to 5' exonuclease) activity.

The results and interpretations of Pühler *et al.* and Zillig *et al.* (1989, 1991) have been challenged by Iwabe *et al.* (1991), whose comparisons of both β'-like and β-like RNA polymerase sequences led them to conclude that the three eukaryotic polymerases diverged from an ancestral progenitor *within* the eukaryotic lineage, after its separation from a common archaeael–eukaryotic ancestor. Another putative example of horizontal gene transfer from eubacteria to eukaryotes (that of a glyceraldehyde 3-phosphate dehydrogenase gene) (Martin and Cerff, 1986; Hensel *et al.*, 1989) has recently been reinterpreted as an example of a horizontal gene transfer in the *opposite* direction (i.e., from a eukaryote to a prokaryote) (Doolittle *et al.*, 1990). Kemmerer *et al.* (1991) have reported that the nuclear cytochrome *c* sequence of *Arabidopsis thaliana* (a dicotyledonous plant) resembles that of fungi more closely than it does the cytochrome *c* sequence from other plants, a situation that might be construed as a case of lateral gene transfer from the nuclear genome of one eukaryote to another. Finally, lateral transfer from an archaebacteria-like organism to a eubacterium has been inferred from the observation that *Thermus thermophilus*, which is classified as a eubacterium (Woese, 1987), has an ATPase that groups specifically with archaebacterial, not other eubacterial, ATPases (Gogarten and Kibak, 1992).

These few examples illustrate the present difficulties we face in drawing firm conclusions about the evolutionary origin of the nuclear genome, given the available data set. It is implicit in the endosymbiont hypothesis that there has been an evolutionary transfer of eubacteria-like genes from protomitochondrial and protoplastid genomes to the nuclear genome, and there is now good molecular evidence to support this assumption (see IV,I and V,D). Much additional and systematic study of selected protein-coding genes in archaebacterial, eubacterial, and nuclear genomes will be required before we can give a more informed assessment of how the nuclear genome originated. In this regard, knowledge of the molecular biology of the earliest diverging eukaryotes (see III,B) ought to be particularly enlightening. For the purposes of this review, it is sufficient that the uniqueness of the archaebacterial, eubacterial, and eukaryotic lineages is well enough established that we are able to infer their evolutionary affiliations (if any) with plastid and mitochondrial genomes.

B. Phylogenetic Diversity of the Eukaryotes

Over the past decade, phylogenetic analyses based on rRNA sequence comparisons, both SSU (McCarroll *et al.*, 1983; Sogin *et al.*, 1986, 1989; Gunderson *et al.*, 1987; Douglas *et al.*, 1991; Hendriks *et al.*, 1991; Sogin, 1991; Schlegel, 1991) and LSU (Baroin *et al.*, 1988; Qu *et al.*, 1988a,b; Perasso *et al.*, 1989, 1990; Lenaers *et al.*, 1991), have substantially enhanced our understanding of evolutionary relationships within the eukaryotes. Among the important new perspectives is recognition of the large evolutionary distances and deep divergences that exist within the eukaryotes, and in particular among unicellular ones (protists) (see Fig. 6). These distances and divergences are so great (Sogin *et al.*, 1989; Sogin, 1991) that the concept of a single "kingdom Protista (Protoctista)" (Whittaker, 1969; Whittaker and Margulis, 1978) seems increasingly untenable (Patterson and Sogin, 1992). At the same time, unexpected relationships among certain groups of protists, not readily discerned on the basis of traditional phenotypic, untrastructural or biochemical comparisons, have been proposed (Gunderson *et al.*, 1987; Gajadhar *et al.*, 1991). As well as testifying to the antiquity and diversity of the protists, rRNA sequence comparisons have suggested that the multicellular eukaryotes (animals, plants, fungi) represent a massive radiation that occurred relatively late in the evolution of this domain (Sogin *et al.*, 1986).

Several features of the eukaryotic nuclear tree (Fig. 6) are of particular relevance to the issue of organelle origins. The most deeply branching (earliest diverging) eukaryotes in this tree are *Vairimorpha necatrix* (a microsporidian) and *Giardia lamblia* (a diplomonad); these are members of a structurally simple group of eukaryotes placed in the kingdom Archezoa by Cavalier-Smith (1983). The Archezoa lack mitochondria and have a number of apparently primitive characters that suggest they are living relics of the earliest phases of eukaryote evolution (see Cavalier-Smith, 1987d; Brugerolle, 1991). Ribosomal RNA sequence comparisons support the proposed primitive position of the Archezoa and bolster the contention that these organisms lack mitochondria not because of some secondary loss, but because they never had them. If this is indeed the case, then a systematic investigation of the molecular biology of the Archezoa may well yield important information about retained primitive features in the

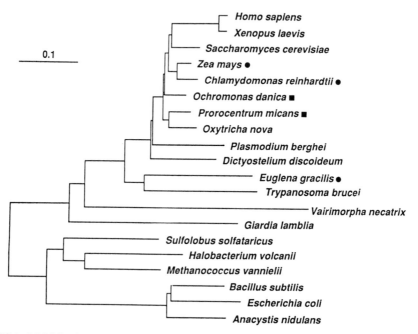

FIG. 6 Multikingdom tree inferred from comparison of SSU rRNA sequences. For details of the analysis, see Sogin *et al.* (1989). The tree is an unrooted one in which the horizontal component of separation represents the evolutionary distance between organisms. The distance that corresponds to 10 changes per 100 positions is indicated. (●) chl a/b algae and land plants; (■) chl a/c algae. Reprinted from Sogin *et al.* (1989), with permission. Copyright 1989 AAAS.

archezoan genome; this in turn could provide clues about the nature of the ancestral eukaryote (diplomonad?) (Cavalier-Smith, 1987d) that played host to the subsequent endosymbioses that led to mitochondria and plastids. At the same time, given their phylogenetic positions, the Archezoa would appear to be key organisms in any evolutionary linkage between archaebacteria and eukaryotes.

Among the earliest diverging eukaryotes that do possess mitochondria are the euglenoid and trypanosomatid protozoa (represented by *Euglena gracilis* and *Trypanosoma brucei,* respectively, in Fig. 6). The two groups appear to be (very distant) relatives of one another. These and other yet-to-be-identified representatives of the most primitive mitochondria-containing eukaryotes are likely to be pivotal in defining the nature of the proto-mitochondrial genome(s).

Finally, it should be noted that plastid-containing eukaryotes (plants and algae) are distributed on a number of branches throughout the eukaryotic tree (Fig. 6) (Perasso *et al.*, 1989, 1990). This intermingling implies either that plastids have originated multiple times in different eukaryotic phyla or that plastids have been lost secondarily numerous times in the couse of eukaryote evolution.

IV. Plastids

A. Introductory Comments

In concluding the section on Plastids in the 1982 review (Gray and Doolittle, 1982), we wrote that the data "leave no doubt that [free-living oxygenic-photosynthetic procaryotes] were ancestral to [plastids]." This assertion has been further bolstered by the data on plastid molecular biology that have been published since 1982; in addition, though, what is now starting to emerge are the outlines of the pathways and processes that led from the protoplastid to contemporary plastids. The application of more sophisticated methods of phylogenetic analysis to comprehensive protein and rRNA sequence databases has resulted in much more detailed and accurate phylogenies than were available a decade ago. As a result, we now have a much better appreciation of the phylogenetic positions and interrelationships of the various groups of photosynthetic eukaryotes (particularly protists) at the level of the nuclear as well as the plastid genome (reliable phylogenies of the host lineage being equally as important as reliable phylogenies of the plastid lineage itself).

In 1982, we also wrote that protein and rRNA sequence data "*all* indicate that plastids are of multiple (polyphyletic) origin." In view of the new phylogenetic insights since obtained, this statement requires some clarification. It is certainly true that there is an increasing awareness of the role that secondary symbioses (the acquisition of plastids from a eukaryotic rather than a prokaryotic endosymbiont) have played in plastid evolution (see Gibbs, 1981, 1990). As we shall see, plastids have undoubtedly originated through a series of separate endosymbioses in different branches of the eukaryotic tree, and from that perspective, plastids can be considered to have had multiple origins. However, a more fundamental issue is whether (within the context of a basically eubacterial origin) the plastids of all plants and algae trace their existence to the **same** primary endosymbiotic event involving a **single** eubacteria-like ancestor (*monophyletic origin*), or whether plastids arose separately in different eukaryotic lines as a result of independent associations with eubacteria-like progenitors (*polyphyletic origin*). Unfortunately, a clear distinction between *primary* and *secondary* endosymbiotic events has not always been made in discussions of plastid origin(s), even very recent ones (Penny and O'Kelly, 1991; Douglas and Gray, 1991). In this review, I use the terms "monophyletic" and "polyphyletic" strictly in the context of the primary endosymbiotic event(s) that generated plastids directly from (a) eubacterial ancestor(s). I use the term "multiple origins" to indicate that plastids in certain eukaryotes have been acquired through a series of separate, secondary endosymbioses. Although it is safe to say that plastids themselves have had mulitple origins, the question of whether the plastid genome is monophyletic or polyphyletic remains (in my view) unsettled.

I will not review in detail the data on plastid genes, genome organization and expression, and molecular evolution of plDNA that have appeared since 1982; these topics have been well covered in other reviews over the past ten years (Bohnert *et al.*, 1982; Whitfield and Bottomley, 1983; Palmer, 1985a,b, 1987, 1990; 1991, 1992; Cattolico, 1986; Weil, 1987; Hanley-Bowdoin and Chua, 1987; Gruissem, 1989a,b; Sugiura, 1989a,b; Clegg *et al.*, 1991; Shimada and Sugiura, 1991; Wolfe *et al.*, 1991). Rather, I will concentrate on the molecular data that are most directly relevant to the question of plastid genome origin. Material in this section that was first presented in Gray (1991) has been extensively revised and updated to reflect relevant new data and developments.

B. Phylogenetic Diversity and Distribution of Plastids

A variety of plastid types, distinguished by pigment composition and number of surrounding membranes, have been recognized (Table I). Although all plastids possess chlorophyll *a* as the primary photosynthetic pigment, they differ in their content of accessory pigments, and on this basis the algae containing them have been divided into three generally recognized groups (Christensen, 1964): (i) the red algae (Rhodophyta), which have phycobiliproteins (a/PB algae); (ii) the green algae (Chlorophyta and Gamophyta) and euglenoid algae (Euglenophyta), which contain chlorophyll *b* (a/b algae); and (iii) the chromophyte algae, a diverse assemblage of eight different phyla [using the classification of Margulis and Schwartz (1988)], all of which have chlorophyll *c* (a/c algae). Two of the chromophyte phyla, namely Cryptophyta (cryptomonads) and Eustigmatophyta, contain phycobiliproteins as well as chlorophyll *c* (a/c/PB algae). These grouping may be subdivided further according to the number of membranes enclosing the plastid (Table I). Like the chlorophyte algae, multicellular land plants (Tracheophyta and Bryophyta) possess chlorophylls *a* and *b*. An origin of land plants from within the Charophyceae, one of the three major groups of green algae, is suggested by much ultrastructural and biochemical data (Ragan and Chapman, 1978; Mattox and

TABLE I

Plastid Membrane and Pigment Diversity in Eukaryotic Algae[a]

	Number of bounding membranes		
Plastid type	Two	Three	Four
Green (chl a/b)	Chlorophyta Gamophyta	Euglenophyta	
Red (chl a/PB)	Rhodophyta		
Yellow-brown (chl a/c) ("chromophytes")		Dinoflagellata[b]	Chrysophyta Haptophyta Cryptophyta[c] Xanthophyta Eustigmatophyta[c] Bacillariophyta Phaeophyta

[a] Phyla as designated by Margulis and Schwartz (1988).
[b] Certain dinoflagellate groups have plastids with only two membranes.
[c] Also contain phycobiliproteins (PB).

Stewart, 1984), and is further supported by recent molecular evidence (Manhart and Palmer, 1990; Devereux *et al.*, 1990).

C. Alternative Hypotheses of Plastid Origin

Over the years, several authors have suggested that plastids of different pigment compositions developed from prokaryotes having the corresponding arrays of pigments, through a series of separate endosymbioses (polyphyletic origin) (Mereschkowsky, 1910; Raven, 1970; Whatley and Whatley, 1981; Mirabdullaev, 1985). Until recently, the only known oxygen-evolving eubacteria were the cyanobacteria, which (like rhodophyte plastids) have chlorophyll *a* and phycobiliproteins. Among prokaryotes, the phycobiliproteins are found only in the cyanobacteria, within which group they are universally present (Gray and Doolittle, 1982). A cyanobacteria-like (a/PB) progenitor of at least rhodophyte plastids therefore seemed likely from the outset, and indeed this idea is strongly supported by molecular data (Gray and Doolittle, 1982, and below). In the case of a/b and a/c plastids, potential eubacterial relatives having the appropriate pigment compositions have only recently been discovered: the a/b "prochlorophytes" (Lewin, 1976, 1977, 1981; Lewin and Withers, 1975; Burger-Wiersma *et al.*, 1986; Chisholm *et al.*, 1988) and the brownish photoheterotroph *Heliobacterium chlorum* (Gest and Favinger, 1983), respectively.

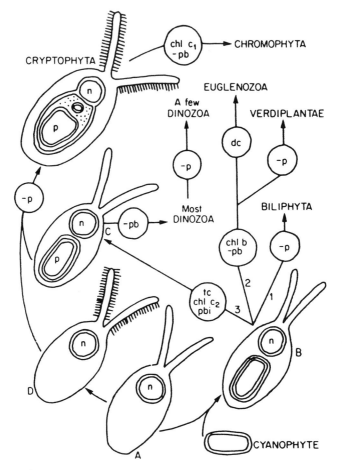

FIG. 7   A postulated phylogeny for the photosynthetic eukaryotes assuming only two endo-symbioses (one primary and one secondary). In this scheme, a nonphotosynthetic phago-trophic eukaryote (A) with a single nucleus (n), two anterior cilia bearing nontubular mastigo-nemes (lateral appendages), a gullet, and a pellicle engulfs a Gram-negative photosynthetic prokaryote (cyanophyte) with a plasma membrane and an outer membrane to produce an endosymbiotic cyanelle with three bounding membranes (B). The cyanelle rapidly becomes established in the resulting photosynthetic biciliated host and is stably passed on to its offspring at cell division. The conversion of the cyanelle into a plastid is assumed to involve many cell generations and therefore takes a different course in different descendants of the initial symbiotic complex (B). Three main lines (1, 2, and 3) evolve. In that leading to the Biliphyta (red algae), the phagosome membrane is lost (-p) but the thylakoids and pigments remain essentially unaltered; the cyanophyte starch-making machinery is lost but that of the host is retained. In the second line, phycobilins (pb) are replaced by chlorophyll b and the thylakoids consequently become stacked; this line diverges early by (on the one hand) loss of the phagosome membrane (-p) to produce the Viridiplantae (''Verdiplantae'' in the figure; green algae and land plants), and (on the other hand) loss of the starch-making machinery of

(continued)

The significance of these organisms with respect to plastid evolution is discussed below.

The alternative possibility, that diversity in pigment composition and other distinguishing features, such as membrane composition, are derived features that postdate a monophyletic origin of plastids from a cyanobacteria-like progenitor, has been argued by Cavalier-Smith (1982, 1987b) and Taylor (1979, 1987). In this case, evolutionary divergence occurring subsequent to the primary endosymbiotic event, rather than a polyphyletic origin, is assumed to have led to the various identified plastid types. Pigment composition differences might be readily rationalized in a monophyletic scenario by evolutionary modifications, *via* gene duplication/gain/loss, of a primary biochemical pathway. For example, differences between chlorophylls *a* and *b* (replacement of a side-chain methyl group by formyl) and *a* and *c* [acquisition of one (c_1) or two (c_2) double bonds] may be regarded as biochemically simple, secondary steps in the chlorophyll biosynthetic pathway. In these monophyletic schemes, membrane differences are in large part attributed to secondary symbiosis events. These issues of differences in pigment composition and membrane ultrastructure, and whether they occurred before or after the endosymbiosis(es) that generated the various plastid types, are addressed in detail in alternative monophyletic (Cavalier-Smith, 1982) and polyphyletic (J. M. Whatley, 1981, 1983; Whatley and Whatley, 1981) schemes of plastid origin.

A final issue is whether the plastids and mitochondria originated serially or simultaneously. The serial endosymbiosis theory (Taylor, 1974; Margulis, 1981) proposes a stepwise acquisition of organelles, with mitochondria (relatives of the nonsulfur purple bacteria) probably originating first and plastids (relatives of cyanobacteria) sometime later; this scenario

---

both host and symbiont to produce the Euglenozoa (euglenoid protozoa). Retention of the phagosome membrane in the euglenozoan line leads to the modification of the mitochondrial protein transport system and the origin of the disc-shaped cristae (dc). In the third line, the phycobilins are moved inside the thylakoids (pbi), which thus became able to stack, chlorophyll c_2 evolves, and the cyanophyte starch-making machinery is lost (C) but that of the host is retained, and the mitochondrial cristae become tubular (tc). This line also forks: one branch loses phycobilins (-pb) to yield the Dinozoa (dinoflagellates), a very few of which also lose the phagosome membrane (-p), whereas the other line loses the phagosome membrane before or during its incorporation by a nonphotosynthetic phagotroph (D) to form the Cryptophyta. The cryptophyte host (D), like the initial host (A), is assumed to have had a pellicle, trichocysts ("stinging organelles" underlying the surface), and a gullet, but to have differed from host (A) in having tubular rather than nontubular mastigonemes on its two anterior cilia. These major events are postulated (Cavalier-Smith, 1982) to have occurred very rapidly—perhaps even within 10 years of the primary endosymbiotic uptake of a cyanophyte by the first biciliated host (A). Reprinted from Cavalier-Smith (1982), with permission. Copyright 1982 Academic Press, Inc. (London), Ltd.

can readily accomodate multiple origins of each organelle in different eukaryotic lineages. An alternative view (Cavalier-Smith, 1987b) argues for a simultaneous origin of mitochondria and chloroplasts (and peroxisomes) within the same archezoan host cell. According to this scheme, the plastid and mitochondrion would not only have evolved simultaneously, but in fact synergistically, the two protoorganelles having begun their residence together within the same cellular environment.

D. Secondary Symbioses in Plastid Evolution

An integrated scheme that invokes a minimum of one primary (prokaryotic) and one secondary (eukaryotic) endosymbiosis is shown in Fig. 7 (see Cavalier-Smith, 1982, 1987b, for a more detailed discussion). Assuming that the original endosymbiont was a cyanobacteria-like eubacterium (cyanophyte), it presumably possessed a double lipoprotein membrane system with an interposed peptidoglycan layer. The double membrane surrounding the plastids of rhodophytes, chlorophytes, and metaphytes (land plants) is most easily rationalized as arising directly from the double lipoprotein membrane system of a cyanophyte, with loss of any peptidoglycan layer that may have existed between these two membranes (Cavalier-Smith, 1987b). This view is consistent with similarities in permeability properties displayed by the outer membranes of plastids (and mitochondria) on the one hand and eubacteria such as *Escherichia coli* on the other, as well as by the presence of porins in eubacterial and organellar membranes (Benz, 1985). That a peptidyloglycan layer initially existed is suggested by the presence of a rudimentary peptidoglycan wall in the *cyanelles* (plastids) of the photoautotrophic protist *Cyanophora paradoxa* (Aitken and Stanier, 1979). In the scheme of Fig. 7, establishment of the original endosymbiosis was followed by loss not only of the peptidoglycan layer of the endosymbiont (the *Cyanophora* cyanelle being an exception) but also of the engulfing phagosome membrane of the host. The extra membrane surrounding the plastids of some other algae (e.g., euglenophytes and most dinoflagellates) is attributed to retention of the phagosomal membrane of the original host. Finally, secondary endosymbioses are proposed to account for the four-membraned plastids of the chromophyte algae, in which the inner two membranes represent the membranes of the endosymbiont's plastid, the third membrane corresponds to the endosymbiont's plasma membrane, and the fourth (outermost) membrane is derived from the phagosome membrane of the new host. The latter two membranes comprise the "chloroplast endoplasmic reticulum" (CER) (Gibbs, 1962; Bouck, 1965).

Subsequent reduction of a eukaryotic endosymbiont is presumed to have taken a different course in different algae. In this regard, members of the Cryptophyta (cryptomonoad algae, Fig. 8) occupy a pivotal position. In these organisms, it is thought that vestiges of the endosymbiont's nucleocytoplasmic compartment have been retained as a periplastidal compartment (between the CER and the inner two plastid membranes) in which are located a distinctive body (the *nucleomorph*) (Greenwood *et al.*, 1977) and 80S-sized ribosomes. It has been postulated that the

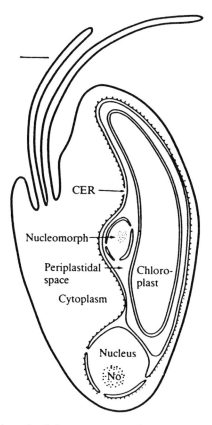

FIG. 8   Diagram showing subcellular compartments in a cryptomonad alga. At the anterior is a gullet with two emergent flagella. The main nucleus with its nucleolus (No) is at the posterior. Four membranes surround the chloroplast. An extension of the nuclear membranes, the chloroplast endoplasmic reticulum (CER), surrounds the chloroplast and associated compartment. Between the CER and the chloroplast envelopes is the periplastidal space, which contains 80S-sized ribosome-like particles and the nucleomorph. Bar, 1μm. Reprinted from McFadden (1990), with permission. Copyright 1990 Company of Biologists Ltd.

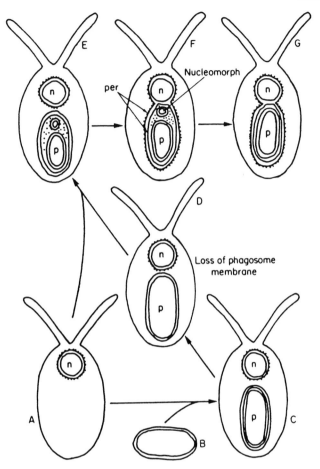

FIG. 9 Proposed origin of the characteristic topology of plastid envelopes in eukaryotes. A phagotrophic biciliate (A) with a single nucleus (n) engulfs a cyanophyte (B) to yield a photosynthetic biciliate (C) with three membranes surrounding its plastid (p), as in the euglenoids and most dinoflagellates (see also Fig. 7). The third (outermost) membrane is the phagosome membrane, the loss of which can produce a plastid with two bounding membranes (D) as in the red and green algae, land plants, and a very few dinoflagellates (Table I). Symbiotic uptake of such a eukaryote (D) by a nonphotosynthetic phagotroph (A) gives rise to (E), following which fusion of the phagosome membrane with the nuclear envelope produces a cryptophyte arrangement (F) with a plastid endoplasmic reticulum (per) (also called "CER"; see Fig. 8) surrounding the eukaryote endosymbiont's vestigial cytoplasm and nucleus (now represented by the nucleomorph). In this scheme, the outer membrane of the per/CER is derived by fusion of phagosome and outer nuclear envelope membranes and has ribosomes on it like the nuclear envelope; it is assumed that the three-membered envelope in the euglenoids and dinoflagellates (C) never fused with the nuclear envelope, so never acquired ribosome binding sites. The inner membrane of the per/CER is the former plasma membrane of the eukaryote endosymbiont (D). Loss of the contents of the nucleomorph-

(continued)

nucleomorph is the evolutionary remnant of the nucleus of the eukaryotic endosymbiont (Greenwood *et al.*, 1977; Whatley *et al.*, 1979; Gillott and Gibbs, 1980). Evidence supporting this idea is discussed below.

Cavalier-Smith (1982) has proposed that loss of the contents of the nucleomorph-containing compartment produced the arrangement of four concentric membranes seen in the other chromophyte algae (Fig. 9), i.e., that other chromophytes are descended directly from cryptophytes. Thus, the cryptophytes are central to a definition of the kingdom Chromista [those algae whose plastids are surrounded by more than three membranes (Table I; Cavalier-Smith, 1986b, 1987b)], and their phylogenetic relationship to other algae is of great interest (see IV,H).

Molecular evidence suggests that the scheme in Fig. 7, although representing a plausible outline of plastid evolution, is probably not correct in all of its details; indeed, alternative scenarios for various aspects of this scheme have been suggested by others. For example, John and the Whatleys (Whatley *et al.*, 1979; J. M. Whatley, 1981) consider that euglenophyte plastids evolved by the selective uptake of isolated plastids from other (a/b) eukaryotes, whereas Gibbs (1978, 1981) has argued that *Euglena* plastids were acquired by uptake of an entire a/b algae in a secondary symbiosis. According to Gibbs, the engulfed eukaryotic endosymbiont was subsequently reduced to the point where only its plastid and plasmalemma remained, the surrounding phagosome membrane of the host being lost. While still maintaining a primary monophyletic origin of plastids, Cavalier-Smith (1991) now concedes that the euglenoid plastid was most likely obtained from a chlorophyll *b*-containing eukaryotic symbiont.

The scheme of Fig. 7 proposes that the photosynthetic symbiont involved in the origin of the Chromophyta was a eukaryote with both phycobilins and chlorophyll *c:* an early dinozoan that had already evolved chlorophyll *c* but had not lost phycobilins (Cavalier-Smith, 1982, 1986b). Cavalier-Smith (1989b) notes that the recently discovered *Amphidinium wigrense* (Wilcox and Wedemayer, 1985) represents just such a dinozoan, having a plastid envelope of three membranes, like a typical dinoflagellate, but paired thylakoids containing chlorophyll *c* and enclosing soluble phycobilins, precisely as in cryptomonads. In Cavalier-Smith's view, this represents ". . . a superb example of a (formerly) missing link."

---

containing compartment is postulated to have produced the arrangement of four concentric membranes seen in the Chromophyta (G) (but see Douglas *et al.*, 1991). In this scenario, the outer membrane of the initial prokaryote endosymbiont is retained in all plastids, but the thin peptidoglycan layer (not shown here) lying between it and the prokaryotic plasma membrane is lost in all except *Cyanophora*. Reprinted from Cavalier-Smith (1982), with permission. Copyright 1982 Academic Press, Inc. (London), Ltd.

In assessing these alternative proposals of plastid origin and evolution, it must be recognized that they draw heavily on comparisons of morphological and ultrastructural characteristics. This once more raises the issue of how one distinguishes between primitive and derived character states, and how one determines whether shared characteristics reflect descent from a common ancestor (homology) or convergent evolution (analogy). For instance, assertions of "homology" between a particular plastid membrane and that of a symbiont or host have for the most part been made on the basis of little or no structural and/or functional data to support such inferences. New molecular data, relating to both plastid and host genomes and emanating from a variety of algal/plant species, are now providing rigorous tests of these evolutionary proposals.

E. Plastid Molecular Biology

1. Size, Gene, Content, and Genome Organization

The great majority of plastid genomes are between 120 and 160 kbp in size, with only a few approaching 200 kbp (Palmer, 1991). Greatest variation is seen within the green algae, with plDNA varying from 89 kbp in *Codium* to 400 kbp or larger in a few members of each of the three major classes of green algae. The highly reduced plDNAs of the nonphotosynthetic angiosperm *Epifagus virginiana* (de Pamphilis and Palmer, 1990) and the nonphotosynthetic euglenophyte *Astasia longa* (Siemeister and Hachtel, 1989; Siemeister *et al.*, 1990) are only about 70 kbp in size, being devoid of photosynthetic genes.

Over the past decade, two factors have greatly improved our knowledge of what genes are present in plDNA, how these genes are organized and expressed, and how plDNA as a genome evolves. First, limited physical and gene mapping of a large number (many hundreds) of plastid genomes and the more intensive characterization of a few (a dozen or so at latest count) have been carried out. These studies have culminated in the complete sequencing of three land plant chloroplast genomes: tobacco (Shinozaki *et al.*, 1986), liverwort (Ohyama *et al.*, 1986), and rice (Hiratsuka *et al.*, 1989; see also Ozeki *et al.*, 1989; Shimada and Sugiura, 1991). Second, although there was very little information in 1982 about plDNA in other than a/b plastids, data on the plastid genomes of a/PB a/c, and a/c/PB algae are starting to appear at an accelerating rate (Kowallik, 1989). As a result, we now have a better perspective on the range of genetic and organizational diversity in plDNA, leading to a more balanced view of plDNA evolution. An up-to-date and comprehensive discussion of the structure and evolution of plastid chromosomes may be found in Palmer (1991) and Wolfe *et al.* (1991).

What has emerged from these analyses is a picture of an organellar genome that is basically prokaryotic in gene organization. Tobacco plDNA (155,844 bp) contains genes specifying LSU (23S + 4.5S), SSU, and 5S rRNAs; 30 different tRNAs; 21 (Yokoi *et al.*, 1990) different ribosomal proteins; over 30 other homologs of eubacterial proteins (including RNA polymerase and photosynthetic genes); and more than two dozen unidentified open reading frames (ORFs). With few exceptions, the same genes are found in the same arrangement in the other two completely sequenced plDNAs. There are a number of examples of clustering and cotranscription of functionally related genes, and in several instances there is a striking similarity in gene order between corresponding plastid and eubacterial genes (Fig. 10), with counterparts of several well-characterized *E. coli* operons [e.g., *rrn* (rRNA genes), S10-*spc*-α and *str* (mostly ribosomal protein genes), *rif* (genes for subunits of RNA polymerase), and *unc* (genes for ATP synthase subunits)] readily identifiable (see Ozeki *et al.*, 1987). A particularly striking example is a cluster of 10 ribosomal protein genes that are in the same order as their homologs in the *E. coli* [S10]-[*spc*]-[α] operons (Fig. 10d). Similarly conserved linkage groups have been found not only in land plant plDNAs, but also in those of chlorophytes (Palmer, 1985b), euglenophytes (Christopher *et al.*, 1988), chromophytes (Janssen *et al.*, 1987; Markowicz *et al.*, 1988; Delaney and Cattolico, 1989; Evrard *et al.*, 1990; Douglas and Durnford, 1990; Michalowski *et al.*, 1990; Kraus *et al.*, 1990; Douglas, 1991) and rhodophytes (Maid and Zetsche, 1991). Considering these similarities in gene organization and the presence of many photosynthetic genes (including ones for phycobiliproteins in a/PB algae; Reith and Douglas, 1990) and other genes having eubacterial homologs (Ozeki *et al.*, 1987; Ohyama *et al.*, 1988), it is rather easy to view the plastid genome as a highly condensed (cyano)bacterial genome.

In evaluating the evolutionary implications of information on plastid gene arrangement, several points should be kept in mind:

(i) Because homologous and similarly arranged operons are present in both eubacterial and archaebacterial genomes [e.g., the spectinomycin (*spc*) operon; see Auer *et al.*, 1989a,b; Scholzen and Arndt, 1991] gene content and order may not seem, on first consideration, to be especially strong criteria for distinguishing between an archaebacterial and a eubacterial origin of the plastid genome. However, organizational details in homologous archaebacterial and eubacterial operons usually do allow such a distinction. For example, in the archaebacteria *Methanococcus vannielii* and *Halobacterium marismortui*, the *spc* operon contains additional ORFs (some of which are homologous to proteins of the eukaryotic 80S ribosome) that are not present in the same *E. coli* and plastid operons (Auer *et al.*, 1989a,b; Scholzen and Arndt, 1991). Moreover, as a result of apparent

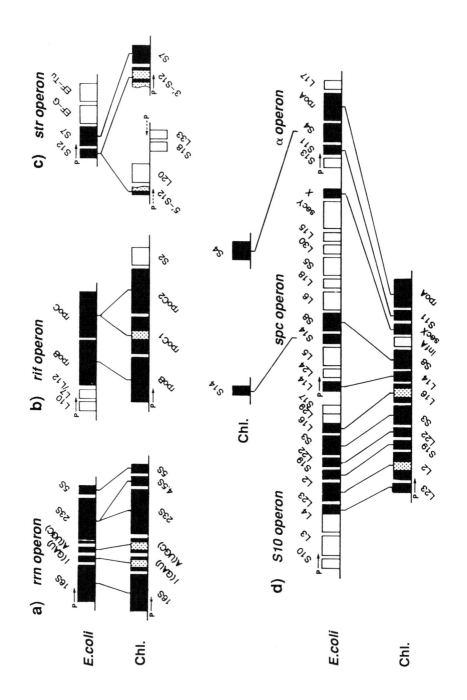

a) *rrn operon*

b) *rif operon*

c) *str operon*

d) *S10 operon*

insertion/deletion events, those archaebacterial genes that do not have eubacterial homologs are generally larger than the latter (and larger than the corresponding plastid ribosomal protein genes).

(ii) Although there are striking similarities in gene order in homologous eubacterial and plastid operons, there are also a number of differences that suggest considerable movement of genes in the course of evolution, with evidence of differential transfer of genes to the nucleus. For example, rbcS [the gene for the small subunit of ribulose-1,5-biphosphate carboxylase (Rubisco)] has been found in the plDNA of all non-a/b algae so far examined (Starnes et al., 1985; Reith and Cattolico, 1986; Douglas and Durnford, 1989; Hwang and Tabita, 1989; Valentin and Zetsche, 1989, 1990a; Kostrzewa et al., 1990; Shivji, 1991), where together with rbcL (the Rubisco large subunit gene) it constitutes a eubacteria-like operon (Valentin and Zetsche, 1990a; Kostrzewa et al., 1990; Assali et al., 1991). In a/b algae and plants, however, rbcS is a nuclear gene (Palmer, 1985a). In general, the plastid operons appear to have lost genes that are present in the corresponding eubacterial operons; this has occurred to different extents in different plDNAs (Michalowski et al., 1990). Certain genes (e.g., the homologs of rps14 and rps4 in the E. coli spc and α operons, respectively) are located elsewhere in the plDNAs of tobacco, rice, and liverwort, unlinked to other ribosomal protein genes. With the availability of molecular data on the plDNAs of a/b (e.g., Lemieux and Lemieux, 1985; Lemieux et al., 1985) as well as non-a/b algae, it is evident that there has been a great deal more evolutionary reordering of plDNAs, involving intra- and intergenomic movement of plastid genes, that had appeared to be the case when only land plant plDNAs were considered (Kowallik, 1989; Palmer, 1991; von Berg and Kowallik, 1992).

(iii) We know relatively little about the extent of variation in gene content and order in homologous operons within the eubacteria (and particularly within and among cyanobacteria); therefore, there is little basis for assessing whether there are aspects of plastid gene arrangement that we may consider specifically cyanobacterial. Separation of the atpB and atpE genes from the remainder of the genes in the unc operon is one such feature that appears to distinguish cyanobacterial and plastid genomes from those of other eubacteria (Cozens and Walker, 1987; Palmer, 1991). Likewise,

FIG. 10 Comparison of arrangement of homologous genes in E. coli and liverwort (Marchantia polymorpha) plastid genomes. (a) rRNA genes; (b) RNA polymerase genes; (c) and (d), ribosomal protein genes. P, promoter. Solid arrows indicate the direction of transcription starting from the first gene in each cluster; dotted arrows show the transcription starting further upstream. Corresponding genes are connected by thin lines. Reprinted from Ozeki et al. (1987), with permission. Copyright 1987 Cold Spring Harbor Laboratory.

the very similar organization of the *ndhC-psbG-orf157/159* operon (encoding NADH dehydrogenase subunits; Pilkington *et al.*, 1991b) in chloroplast and cyanobacterial (*Synechocystis* sp. PCC6803) DNA (Steinmüller *et al.*, 1989) contrasts with the presence of a lone *ndh* gene (the apparent *psbG* homolog; Nixon *et al.*, 1989) specifying the single flavoprotein component of respiratory NADH dehydrogenase in *E. coli* (Young *et al.*, 1981).

(iv) A number of plastid genes contain introns that until very recently had appeared to be uniformly absent from their eubacterial homologs. However, an intron has now been found in a leucine tRNA gene of several cyanobacterial species, in precisely the same position as in the homologous plastid gene (Xu *et al.*, 1990; Kuhsel *et al.*, 1990); even more striking is the fact that the cyanobacterial and plastid introns are homologous in primary sequence and secondary structure. This strongly indicates that the intron was already present in this particular leucine tRNA gene in the common ancestor of contemporary cyanobacteria and plastids (see Belfort, 1991). Although these observations would seem to provide another specific evolutionary link between plastid and cyanobacterial genomes, there are indications (Kuhsel *et al.*, 1990) that an intron may be present in the same leucine tRNA gene in other (nonphotosynthetic) eubacterial phyla.

2. Gene Expression

Many plastid genes are not only organized into eubacteria-like operons, but are expressed by way of eubacteria-like promoters (Hanley-Bowdoin and Chua, 1987; Gruissem, 1989a,b). Within the 5'-regions of many of the plastid transcription units are motifs that resemble the eubacterial "−10" and "−35" consensus promoter elements (Steinmetz *et al.*, 1983; Briat *et al.*, 1986; Hanley-Bowdoin and Chua, 1987), whose function in plDNA has been verified experimentally (reviewed in Gruissem, 1989a,b). Not surprisingly, *E. coli* RNA polymerase is able to transcribe plastid genes, initiating at discrete sites (Link, 1984; Gruissem, 1989b). Sequence analysis has revealed that land plant plDNAs possess ORFs homologous to the *E. coli rpoA, rpoB,* and *rpoC* loci (Ozeki *et al.*, 1987), which encode the α, β, and β' subunits, respectively, of the eubacterial RNA polymerase. This suggests the presence of a eubacteria-like RNA polymerase in plastids, although it remains to be demonstrated that such an enzyme is actually synthesized and functions in transcription in this organelle.

It should be noted that transcription in plastids also displays many distinct (noneubacterial) features. Plastid RNA polymerases may be distinguished from those in eubacteria (at least the *E. coli* one) by several biochemical criteria (e.g., insensitivity to rifampicin) (Gruissem, 1989b);

eubacteria-like promoter elements seem to be absent upstream of some plastid genes (Gruissem, 1989b); transcription appears to be terminated differently in land plant plastids than in eubacteria (Stern and Gruissem, 1987); homologous primary transcripts tend to be subjected to more extensive processing in land plant plastids than in eubacteria (Sugiura, 1989a); intron-containing plastid transcripts undergo *cis*-splicing (Koller and Delius, 1984); transcripts from certain split and rearranged plastid genes [e.g., *rps12* in liverwort (Fukuzawa *et al.*, 1986) and tobacco (Torazawa *et al.*, 1986; Koller *et al.*, 1987; Zaita *et al.*, 1987; Hildebrand *et al.*, 1988), *psaA* in *Chlamydomonas reinhardtii* (Kück *et al.*, 1987; Goldschmidt-Clermont *et al.*, 1991)] must be processed by a unique *trans*-splicing mechanism; and certain chloroplast mRNAs undergo a C-to-U type of RNA editing (Hoch *et al.*, 1991; Kudla *et al.*, 1992). However, it should also be stressed that (i) these differences do not detract from the very substantial similarities that are evident between plastid and eubacterial gene expression; (ii) similarities of this extent and magnitude are not evident when plastid gene expression mechanisms are compared with those in archaebacteria and the eukaryotic nucleus; and (iii) because most comparisons of gene expression have been between plastids and noncyanobacteria such as *E. coli*, it may be that some of these apparent differences will become less pronounced as we learn more about the mechanisms of gene expression in cyanobacteria.

The plastid translation system is strikingly eubacterial in its properties, and these particular similarities provided some of the earliest molecular evidence pointing to a bacterial origin of plastids (Margulis, 1970; Gillham and Boynton, 1981; Gray and Doolittle, 1982). The 70S ribosomes of plastids, like their prokaryotic counterparts, are composed of 50S and 30S subunits containing, respectively, high-molecular-weight RNA species 23S and 16S in size. There is both structural and immunological similarity between eubacterial and plastid ribosomal proteins (Schmidt *et al.*, 1984; Subramanian *et al.*, 1990). Functional similarity between plastid and eubacterial ribosomes is evidenced by interchangeability of ribosomal subunits and translation factors, use of *N*-formylmethionyl-tRNA as initiator of protein synthesis, and a similar spectrum of antibiotic sensitivity and resistance (discussed in Gray and Doolittle, 1982).

Plastids use the standard genetic code, and all 61 sense codons are found in plastid protein-coding genes. In plant chloroplasts, some tRNA species employ a "two out of three" or "U:N wobble" decoding mechanism during codon:anticodon interaction (Pfitzinger *et al.*, 1990). This means that 30 (in tobacco) or 31 (in Marchantia) plDNA-encoded tRNAs are apparently sufficient to read all 61 sense codons in plastid mRNAs.

3. Gene Structure

Numerous aspects of plastid genome organization and expression now testify to its essentially prokaryotic nature. However, I have emphasized that some of these features (e.g., similarities in the organization of certain operons, ribosome/rRNA size) are shared by both eubacteria and archaebacteria, and so are presumably primitive traits for these groups. Gene sequence, on the other hand, uniformly differentiates between eubacteria + plastids on the one hand and archaebacteria on the other. Morever, sequence information, both nucleotide and amino acid, clearly places the plastids within the cyanobacteria phylum of eubacteria.

Phylogenetic trees based on protein data (e.g., c-type cytochromes) show a specific clustering of plastids with cyanobacteria (Schwartz and Dayhoff, 1978, 1981; Hunt et al., 1985), to the exclusion of other types of eubacteria, including other photosynthetic species. Where clear homologs of plastid-encoded protein genes exist in other eubacteria as well as in cyanobacteria, the plastid genes display highest sequence similarity with their cyanobacterial counterparts (e.g., Cozens et al., 1986). Specific similarities between homologous plastid and cyanobacterial rRNAs are evident at the level of secondary structure as well as primary sequence (Delihas et al., 1982, 1985; Tomioka and Sugiura, 1983; Kumano et al., 1983; Douglas and Doolittle, 1984a,b; Delihas and Fox, 1987; Mannella et al., 1987). This is also the case with plastid-encoded tRNAs (Maréchal-Drouard et al., 1991). One notable exception to the cyanobacterial character of plastid gene sequences is discussed below (IV,F).

F. Plastid Phylogenies from Ribosomal RNA and Protein Sequence Comparisons

As pointed out earlier (II,A) rRNA sequence comparisons are particularly appealing because in principle they allow one to determine, simultaneously, phylogenies of nuclear, organellar, and prokaryotic genomes (Gray et al., 1984; Cedergren et al., 1988). Thus, not only do they provide a means to test alternative autogenous and xenogenous hypotheses of organelle origin, but they offer the prospect of being able to determine the number of primary endosymbiotic events and whether there as been any secondary movement of genes, genomes, or entire organelles in the course of evolution. An underlying assumption is that if a particular organelle has originated only once, and there has subsequently been no cross-organism movement of genetic material from that organelle, then phylogenies based on host nuclear and organellar genes should be congruent, because the two genomes will have coexisted and evolved in parallel (although not necessarily at the same rate) since the protoorganelle took up residence in the host. In such a case, we expect branching topologies in nuclear and organellar trees to be the same; in the event they are not, we ask whether there is a biological basis for the difference.

Over the past decade, and particulary within the last few years, there has been a veritable flood of rRNA sequence information, with hundreds of complete and many more hundreds of incomplete SSU rRNA sequences (and a lesser but still substantial number of complete LSU rRNA sequences) (Gutell *et al.*, 1990) now available for phylogenetic analysis. Sequences of plastid-encoded proteins are also appearing with increasing frequency, permitting the construction of independent, protein-based phylogenetic trees. The main conclusions that have been drawn about plastid evolution from these various sequence comparisons are the following:

(i) In global rRNA trees that define three separate and distinct lines of descent (archaebacteria, eubacteria, and eukaryotic nucleus), plastids fall clearly within the eubacterial lineage and cluster specifically with cyanobacteria (Gray *et al.*, 1984; Giovannoni *et al.*, 1988; Cedergren *et al.*, 1988; Turner *et al.*, 1989).

(ii) In rRNA trees, all plastid types (a/b, a/PB, a/c, a/c/PB) appear to be of cyanobacterial origin, and all are more closely related to one another than they are to a cyanobacterium such as *Anacystis nidulans* (Douglas and Turner, 1991; but see Markowicz and Loiseaux-de Goër, 1991). No specific affiliation between a/b eubacteria (prochlorophytes) and a/b plastids (see Lewin, 1981) is evident at the level of rRNA sequence. *Prochlorothrix hollandica*, a free-living prochlorophyte, branches deeply within the cyanobacteria but does not cluster with the coherent phylogenetic grouping of green plastids (Turner *et al.*, 1989; see Fig. 11). This conclusion is consistent with earlier T1 oligonucleotide cataloging studies of 5S (MacKay *et al.*, 1982) and 16S (Seewaldt and Stackerbrandt, 1982) rRNAs from the exosymbiotic prochlorophyte *Prochloron* (but see Van Valen, 1982). Although a common ancestry of prochlorophytes and a/b plastids was initially inferred from amino acid sequence comparisons of *psbA* (a photosystem protein encoded in plDNA) (Morden and Golden, 1989a), this conclusion was subsequently reassessed (Morden and Golden, 1989b, 1991; Kishino *et al.*, 1990). The separate branching of prochlorophytes and green plastids implies that the ability to synthesize chlorophyll *b* arose independently in the ancestors of these two groups, or was acquired by lateral transfer (Turner *et al.*, 1989). Other data indicating that prochlorophytes are polyphyletic within the cyanobacterial radiation, and that none of the known species is specifically related to chloroplasts, have been published (Palenik and Haselkorn, 1992; Urbach *et al.*, 1992).

(iii) Likewise, no specific relationship is apparent between the plastid 16S rRNA of the a/c chromophyte *Ochromonus danica* (phylum Chrysophyta) and that of the brownish photoheterotrophic eubacterium *Heliobacterium chlorum* (Witt and Stackebrandt, 1988). This result fails to support the proposal (Margulis and Obar, 1985) of an endosymbiotic

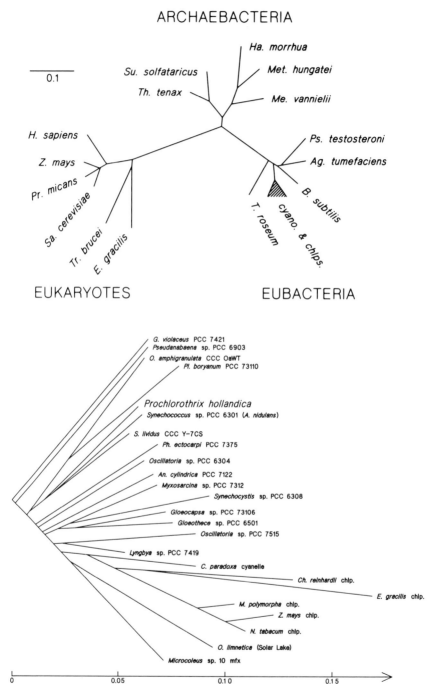

origin of the chrysophyte plastid from an ancestor that also gave rise to
H. chlorum.

(iv) A surprising discrepancy between rRNA and protein trees has
emerged from recent analyses of *rbcS* and *rbcL* genes. In the a/PB, a/c,
and a/PB/c algae, these genes appear specifically related to their homologs
in β-purple bacteria, not cyanobacteria (Boczar *et al.,* 1989; Assali *et al.,*
1990; Douglas *et al.,* 1990; Valentin and Zetsche, 1990b; Assali *et al.,*
1991). Biochemical similarities between the Rubiscos from chromophyte
and rhodophyte algae, and dissimilarities with those of green algae and
land plants, have also been reported (Newman *et al.,* 1989). Although it
has been suggested that this discordance between rRNA and protein data
supports a polyphyletic origin of plastids (Douglas *et al.,* 1990; Valentin
and Zetsche, 1990a,b), a noncyanobacterial origin of rhodophyte plastids
is incompatible with their strong resemblance to cyanobacteria in other
respects, notably pigment composition and photosynthetic membrane
structure. It would seem more likely that in rhodophyte and chromophyte
algae, the *rbcLS* operon as been acquired from a β-purple bacterium
through a process of lateral gene transfer, in which case the plastid genome
in these algae should be considered an evolutionary mosaic. Cavalier-
Smith (1989b) has suggested, in the context of a simultaneous symbiotic
origin of mitochondrion and plastid, that the chromophyte *rbcLS* operon
was transferred from the protomitochondrion purple bacterial symbiont
to the protoplastid cyanobacterial symbiont within the same host cell.
However, this postulate is not consistent with available molecular evidence that clearly traces the ancestry of mitochondria to the α-purple, not
β-purple, eubacteria (V,C).

FIG. 11   Rooted phylogenetic tree depicting evolutionary relationships among SSU rRNAs
from *Prochlorothrix hollandica* (a prochlorophyte), cyanobacteria, green plastids (chloroplasts), and a cyanelle. SSU rRNA sequences from the outgroup organisms *Agrobacterium
tumefaciens, Bacillus subtilis,* and *Pseudomonas testosteroni* were used to locate the root
(Giovannoni *et al.,* 1988). In this representation, evolutionary distances are proportional
only to the horizontal component of the segment lengths. The scale is in units of fixed point
mutations per sequence position. See Turner *et al.* (1989) for details of the analysis. Inset:
unrooted phylogenetic tree depicting the three primary kindgoms ("domains") and the
relationship of the cyanobacteria and chloroplasts to some other eubacteria. Segment lengths
are proportional to evolutionary distances. The scale bar represents 0.1 fixed point mutations
per sequence position. Abbreviations: A., *Anacystis;* Ag., *Agrobacterium;* An., *Anabaena;*
B., *Bacillus;* C., *Cyanophora;* CCC, Castenholz Culture Collection; Ch., *Chlamydomonas;*
chlp(s)., chloroplast(s); cyano., cyanobacteria; E., *Euglena;* G., *Gloeobacter;* H., *Homo;*
Ha., *Halococcus;* M., *Marchantia;* Me., *Methanococcus;* Met., *Methanospirillum;* N., *Nicotiana;* O., *Oscillatoria;* PCC, Pasteur Culture Collection; Ph., *Phormidium;* Pl., *Plectonema;*
Pr., *Prorocentrum;* Ps., *Pseudomonas;* S., *Synechococcus;* Sa., *Saccharomyces;* Su., *Sulfolobus;* T., *Thermomicrobium;* Th., *Thermoproteus;* Tr., *Trypanosoma;* Z., *Zea.* Reprinted
from Turner *et al.* (1989), with permission. Copyright 1989 Macmillan Magazines Ltd.

(v) In studies by Giovannoni *et al.*, (1988) and Turner *et al.*, (1989), plastids of the a/b type were seen to form a monophyletic group contained within the cyanobacterial radiation of eubacteria (Fig. 11). The plastid clade included representative metaphytes (maize, tobacco, and liverwort), a chlorophyte (*C. reinhardtii*), and a euglenophyte (*E. gracilis*). This result implies a common evolutionary ancestry of chlorophyte and euglenophyte plastids, despite some notable differences (see Palmer, 1991) in plastid genome organization in the two algal groups. A more comprehensive rRNA tree, containing additional plastid taxa, suggests a common origin of all plastids from a cyanophyte ancestor, but an early split between metaphytes/chlorophytes on the one hand and *Cyanophora paradoxa*/ rhodophytes/chromophytes on the other (Douglas and Turner, 1991). This and other recent rRNA trees (Markowicz *et al.*, 1988; Witt and Stacke-brandt, 1988; Markowicz and Loiseaux-de Goër, 1991) place eugleno-phytes together with chromophyte algae, rather than with the chlorophyte/ metaphyte cluster. This division is consistent with the structure of the plastid 16S–23S rDNA spacer region, which in *Euglena gracilis*, *Cyanophora paradoxa*, chromophyte, and rhodophyte algae is "small-sized," compared to the more complex structure of the same region in green algae and land plants (see Maid and Zetsche, 1991). If further analyses substantiate this separation, the results would argue that chlorophyll *b* arose on at least *three* separate occasions: once within the prochlorophyte eubacteria, once within the specific ancestor of green algae and land plants, and once within the specific ancestor of euglenophyte algae. Some additional observations [e.g., the fact that reserve carbohydrate is stored as cytoplasmic β-1,3-glucan in both euglenophyte and chromophyte algae but as plastid α-1,4-glucan in chlorophyte algae and metaphytes (Whatley and Whatley, 1981)] support the phylogenetic separation of euglenophyte and chlorophyte algae; however, other data (*rbc* trees (Douglas et al., 1990) and plastid SSU rRNA methylation patterns (M.N. Schnare and M.W. Gray, unpublished results)) do not. Markowicz and Loiseaux-de Goër (1991) have argued that the plastid genome of *Euglena*, like those of rhodophyte and chromophyte algae, also had a composite phylogenetic origin.

(vi) In the rRNA trees of Giovannoni *et al.* (1988) and Turner *et al.* (1989), the a/PB cyanelle of *Cyanophora paradoxa* represents an early diverging line within the plastid branch, splitting off before the diversifica-tion of the a/b plastids. In the tree of Douglas and Turner (1991), the cyanelle represents the earliest branching in the rhodophyte/chro-mophyte/euglenophyte cluster. However, a more recent study (Lockhart *et al.*, 1992a) suggest that a base compositional bias confounds our ability to correctly infer cyanelle origins from sequence comparisons. Thus, avail-able data offer conflicting evidence regarding the origin of the cyanelle of

Cyanophora paradoxa, its phylogenetic relationship to other plastid types, and the question of whether cyanelles as a group are monophyletic. Nevertheless, the results do support the contention (Cavalier-Smith, 1982) that the cyanelle is not of recent endosymbiotic origin, as had been proposed by Hall and Claus (1963). Consistent with this conclusion is the fact that although the cyanelle is structurally more conservative than other plastids in its retention of a rudimentary cell wall, its genome is typically plastid-like in size and organization (Herdmann and Stanier, 1977; Bohnert *et al.*, 1985; Lambert *et al.*, 1985).

G. Algal Phylogenies from Ribosomal RNA Sequence Comparisons

When rRNA sequences are used to trace the evolutionary descent of the host rather than the plastid genome, the resulting (nuclear) phylogenetic trees indicate that the eukaryotic algae do not form a coherent grouping to the exclusion of other eukaryotes: rather, there is considerable interspersion of photosynthetic and nonphotosynthetic types (Gunderson *et al.*, 1987; Bhattacharya and Druehl, 1988; Perasso *et al.*, 1989; Bhattacharya *et al.*, 1990) (see also Figs. 6 and 12). In these various trees, most of the algal types emerge later than a number of nonphotosynthetic eukaryotes. A notable exception is *E. gracilis*, which branches as a very early diverging protist that separated from the main eukaryotic line well before other algae (including other a/b ones) and whose closest relatives are the trypanosomatid protozoa.

Different interpretations follow from the observed distribution of plastid-containing and plastid-deficient eukaryotes in nuclear rRNA trees, depending on whether one favors a monophyletic or polyphyletic origin of plastids. In the former case, there would have had to have been a primary acquisition of a cyanophyte very early in the evolution of the eukaryotes, before the separation of *E. gracilis* from the trypanosomatid protozoa. Subsequently, plastids would had had to have been lost multiple times in different branches of the eukaryotic tree. There is, in fact, suggestive evidence for plastid loss in some eukaryotic lines (Nes *et al.*, 1990). On the other hand, if (as seems likely; see IV,H) the *Euglena* plastid was acquired by a nonphotosynthetic flagellate in a secondary endosymbiosis event, the primary endosymbiosis(es) that gave rise to plastids could have occurred considerably later in the evolution of the eukaryotes. However, the latter scenario effectively excludes a simultaneous origin of plastids and mitochondria in the same host cell. In a polyphyletic scheme, cyanophytes could have been acquired both early and late in different lineages,

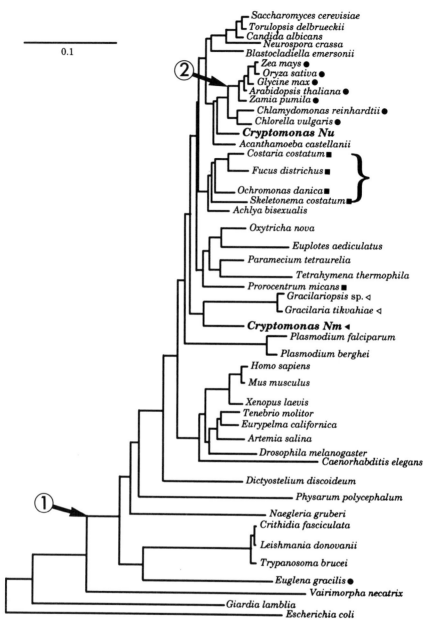

with consequently less need to invoke secondary loss of plastids in order to account for their present phylogenetic distribution. This interpretation would also be complicated by any occurrence of secondary endosymbioses. The bottom line is that it is still possible to maintain a **single primary origin** of algal plastids followed by subsequent loss in some lines, with secondary endosymbioses serving to redistribute plastids within the eukaryotic domain and repopulate plastid-less lineages.

It is premature at this time to make definitive statements about the relative branching positions of the various algal groups and their phylogenetic relationships with other (nonphotosynthetic) eukaryotes. Several individual branchings are quite robust, consistently appearing in trees produced by different investigators using different treeing methods; these branchings include an early-emerging clade of *E. gracilis* together with trypanosomatid protozoa; a late-emerging clade of chlorophytes together with metaphytes; and a clustering of *Prorocentrum micans* (dinoflagellate) with ciliates (cf. Figs. 6 and 12). The phylogenetic affiliations and positions of other algal groups are less certain; for example, in some trees (Perasso *et al.,* 1989; Bhattacharya *et al.,* 1990), red algae appear to be relatively advanced, diverging from the eukaryotic line of descent at approximately the same time or even later than other algal groups, an observation seemingly at odds with schemes (Cavalier-Smith, 1982) in which red algae are ancestral to all other algal types. In other trees (Douglas *et al.,* 1991; see also Fig. 12), red algae emerge relatively early, although still after *Euglena* and a number of nonphotosynthetic eukaryotes.

H. Molecular Evidence for Acquisition of Plastids from Eukaryote Endosymbionts

The early branching position of *E. gracilis,* whose divergence predates that of the rhodophytes, appears incompatible with Cavalier-Smith's (1982)

FIG. 12   Nuclear phylogenetic tree based on analysis of SSU rRNA sequences according to the neighbor-joining method of Saitou and Nei (1987). This tree is an expanded version of that presented in Douglas *et al.* (1991), which should be consulted for additional details of the analysis. Nu, nuclear; Nm, nucleomorph. Branch lengths on the horizontal axis represent the evolutionary distance between the nodes of the tree, with the horizontal bar representing 10 changes per 100 nucleotides. The bracket indicates the branching position of the chromophyte algae (see Table I). Photosynthetic eukaryotes are denoted by the following symbols: (●) chl a/b algae and land plants; (■) chl a/c algae; (◁) chl a/PB algae; (◀) chl a/c/PB alga. Circled numbers (1, 2) indicate possible endosymbiotic events leading to mitochondria (see text, V,C).

scheme in which euglenozoa are derived from red algae by loss of phycobilins and elaboration of chlorophyll *b*. The fact that the *Euglena* plastid contains chlorophylls *a* and *b* and is surrounded by three rather than two membranes has been taken as support for a secondary uptake of either a/b plastids (Whatley *et al.,*1979; J. M. Whatley, 1981) or an entire a/b eukaryote endosymbiont (Gibbs, 1978, 1981) by a flagellated host cell. Although a secondary symbiosis appears the most reasonable way to account for the *Euglena* plastid, it is not certain whether the endosymbiont was in fact an a/b-type alga, given the current conflicting data on the phylogenetic position of the *Euglena* plastid relative to other plastid types (IV,F).

The nucleomorph-containing cryptomonad algae (IV,D; Fig. 8) currently constitute the most convincing case of secondary acquisition of plastids from a eukaryotic algal endosymbiont. The cryptomonad nucleomorph, assumed to be the vestigial nucleus of such an endosymbiont, has been shown to contain DNA (Ludwig and Gibbs, 1985, 1987). More recently, McFadden (1990) demonstrated that eukaryotic nuclear-type rRNA genes are present in a nucleolus-like structure in the nucleomorph of the cryptomonad *Chroomonas caudata,* and Eschbach *et al.* (1991a) showed that the nucleomorph of another cryptomonad, *Pyrenomonas salina,* contains three linear chromosomes (195, 225, and 240 kbp), all of which hybridize with eukaryotic-type SSU and LSU rDNA probes. These results imply that nucleomorph DNA may encode the rRNA components of the periplastidal 80S ribosomes in cryptomonads. If this is the case, then phylogenetic analysis of cryptomonad nuclear and nucleomorph rRNA genes could provide a means both to reinforce the hypothesis of cryptomonad origins and to investigate the phylogenetic positions of host and endosymbiont.

Such data have recently been obtained by Douglas *et al.* (1991) who, working with the cryptonomad alga *Cryptomonas* Φ, isolated and completely sequenced two distinctly different eukaryotic-type 18S rRNA genes (Nu and Nm), both of which are expressed. The Nm (presumptive nucleomorph) sequence was found to cluster specifically with nuclear 18S rRNA sequences from red algae, whereas the nuclear (Nu) sequence clustered separately, in the assemblage containing chlorophytes and metaphytes (Fig. 12). These new data strongly support the idea that cryptomonad algae are evolutionary chimeras of two phylogenetically distinct eukaryotic cells, with the symbiont partner being a rhodophyte-type alga. This conclusion is in accord with other data that show a specific phylogenetic relationship between the plastids of cryptomonads and red algae (Douglas *et al.,* 1990; Valentin and Zetsche, 1990a; Douglas and Turner, 1991; Maerz *et al.,* 1992).

An additional feature of the tree shown in Fig. 12, and one that is consistently obtained with other treeing algorithms (Douglas *et al.,* 1991), is that *Cryptomonas* Φ (Nu) and chromophytes do not branch together. This observation [also made independently by Eschbach *et al.* (1991b) in the case of *P. salina*] is at variance with the hypothesis (Cavalier-Smith, 1982) that chromophyte algae evolved directly from a cryptomonad-like ancestor (Fig. 9), and it therefore calls into question the assumed monophyletic status of the proposed kingdom Chromista (Cavalier-Smith, 1989b).

I. Gene Transfer in Plastid Evolution

Assuming that the genome of the protoplastid endosymbiont had many more genes than currently are found in plDNA, an implicit assumption of the endosymbiont hypothesis is that massive loss and/or transfer of genetic information from the endosymbiont genome must have occurred in the course of plastid evolution. Such loss/transfer is necessary to account for the fact that in size and genetic information content, plastid genomes are at least an order of magnitude smaller than the average eubacterial genome (Cattolico, 1986). Moreover, the majority of the genes specifying plastid structure and function are, in fact, nuclear genes. Although some of latter may represent genes that were already present in the host genome and that happened to duplicate endosymbiont genes (in which case the redundant endosymbiont genes could have been lost), others were almost certainly transferred from protoplastid genome to host nuclear genome. Pathways and possible mechanisms for such transfer have been discussed (Bogorad, 1975, 1982; Weeden, 1981; Harington and Thornley, 1982; Obar and Green, 1985) and are postulated to have involved duplication of an endosymbiont gene in the host nuclear genome, acquisition of function by the nuclear copy (at which stage both nuclear and organellar genes might have been active), selection of the nuclear gene over the organellar gene for ultimate functioning, and eventual loss of the now silent organellar gene. In the case of plastids, such transfer has been invoked to account for the existence of nucleus-encoded, plastid-specific isozymes that are distinct from their nucleus-encoded, cytosolic counterparts, and that do not exist in animal cells (Weeden, 1981). This scenario is strongly supported by sequence analysis of nuclear genes encoding the cytosolic and plastid forms of glyceraldehyde 3-phosphate dehydrogenase (Shih *et al.,* 1986; Martin and Cerff, 1986; Brinkmann *et al.,* 1987; Liaud *et al.,* 1990).

Although the general conservation of plastid genome size and gene content suggests that most gene transfer occurred early in plDNA evolution, the fact remains that some genes (e.g., *rbcS*) are encoded in plDNA in some species but in nDNA in others. This indicates that such transfer must be an ongoing process, and strong evidence to this effect has been published. Baldauf and Palmer (1990) isolated a nucleus-encoded *tufA* gene [specifying chloroplast protein synthesis elongation factor Tu] from the land plant *Arabidopsis thaliana* and showed that this gene trees with cyanobacteria-like, plastid-encoded *tufA* genes of green algae. This result clearly argues for a plastid-to-nucleus *tufA* gene transfer in the metaphyte line, after its separation from a common ancestor with chlorophytes (Baldauf and Palmer, 1990). It was subsequently found (Baldauf *et al.*, 1990) that whereas some charophycean algae (the closest green algal relatives of land plants) appear to lack a plastid-encoded *tufA* gene, one charophyte, *Coleochaete orbicularis*, contains an intact but highly divergent plastid-encoded *tufA* homolog, whose product is unlikely to act as a functional EF-Tu. These results prompted Baldauf *et al.* (1990) to propose that "a copy of the *tufA* gene was functionally transferred from the chloroplast to the nucleus early in the evolution of the Charophyceae, with chloroplast copies of varying function being retained in some but not all of the subsequently diverging lineages." Other molecular and phylogenetic studies support a somewhat later plastid-to-nucleus relocation of ribosomal protein genes *rpl21* (Martin *et al.*, 1990; Smooker *et al.*, 1990) and *rpl22* (Grantt *et al.*, 1991) in a common ancestor of flowering plants. In the case of L21, it has been suggested (on the basis of amino acid sequence comparisons) that the nuclear gene for this plastid protein actually originated from the mitochondrial, not the plastid, genome (Martin *et al.*, 1990). This possibility, although certainly intriguing, needs to be substantiated by additional data.

J. Concluding Remarks

Over the past decade, new molecular data have (i) clarified the phylogenetic relationships among photosynthetic and nonphotosynthetic eukaryotes at the level of the nuclear genome; (ii) emphasized the diversity and antiquity of the unicellular eukaryotes (protists), among which are found the majority of photosynthetic eukaryotes; (iii) affirmed beyond reasonable doubt the eubacterial (specifically cyanobacterial), endosymbiotic origin of plastids; (iv) failed so far to provide compelling evidence of a polyphyletic, primary origin of plastids; (v) at the same time provided clear evidence of a multiple origin of plastids through secondary symbioses with eukaryotic rather than prokaryotic endosymbionts in different lines of algae; (vi) suggested that some plastid genomes are evolutionary mosaics, having acquired certain of their genes from outside the cyanobacterial lineage, presumably as a result of horizontal gene transfer; and (vii) verified the evolutionary transfer of genetic information from protoplastid to nuclear genome, while demonstrating that such transfer is an ongoing process.

Although many aspects of plastid evolution have been clarified by recent molecular data, the story is far from complete. Controversy still surrounds the question of plastid origins, with contradictory trees being generated from different sequence, biochemical, and ultrastructural data sets (Lockhart *et al.*, 1992b). Proposals for a polyphyletic origin of plastids continue to be made (Scherer *et al.*, 1991; Somerville *et al.*, 1992), while at the same time problems with current methods of tree construction continue to be of concern (Lockhart *et al.*, 1992b). Still, there is every reason to remain confident that more data and better refined methodology will provide continuing insights, particularly into the unresolved question of whether plastids had a single primary origin or more than one. As a more systematic and comprehensive molecular examination of photosynthetic eukaryotes is carried out, one that recognizes the full extent of the evolutionary diversity of plastids and their hosts, we should be able to reconstruct pathways of plastid evolution in some detail. From comparison of homologous plastid operons, for example, carried out within the framework of phylogenetic trees relating the various plastid genomes in question, we should be able to infer patterns of gene movement, both within the plastid genome and to the nucleus. At the moment, the database is rather spotty, with a heavy emphasis still on a/b-type plDNAs (primarily in land plants). A detailed investigation of non-a/b plDNAs is essential to give us a more balanced perspective on plastid evolution. Early indications are that some of the more "primitive" plDNAs (particularly rhodophyte and cryptophyte) have retained genes that are missing in plDNAs of more highly evolved algae, and have operons that display minimal deviation in gene content and organization from the homologous cyanobacterial operons (Bryant and Stirewalt, 1990; Bryant *et al.*, 1991; Neumann-Spallart *et al.*, 1990, 1991; Michalowski *et al.*, 1991a,b; Pancic *et al.*, 1991; Reith and Munholland, 1991; Wang and Liu, 1991; Douglas, 1991, 1992; Kessler *et al.*, 1992; Reith, 1992; Scaramuzzi *et al.*, 1992).

At the same time, additional comparative data will allow us to infer other specific changes that have occurred in different plDNAs since their divergence from a common ancestor, suggesting molecular mechanisms by which such changes have occurred. In land plants, for example, documented changes include loss of one copy of the inverted repeat that characterizes most plDNAs, with concomitant genome rearrangement, in several species of legume (Palmer and Thompson, 1982) and in conifers (Strauss *et al.*, 1988; Lidholm *et al.*, 1988); expansion of the inverted repeat into single-copy regions (Palmer *et al.*, 1987; Hiratsuka *et al.*, 1989); and a variety of additions, deletions, and inversions (Howe, 1985; Michalowski *et al.*, 1987; Palmer, 1985b). Even greater change is evident among algal plDNAs (Palmer, 1991, 1992), and the evolutionary basis of these differences will only be able to be discerned by judicious comparative analysis.

V. Mitochondria

A. Introductory Comments

When one considers the impact that molecular data have had on opposing theories of plastid and mitochondrial origins, two quite different pictures emerge. Over the past two decades, the growing body of knowledge about plastid molecular biology has increasingly supported the hypothesis of an endosymbiotic, cyanobacterial (i.e., xenogenous) origin of plastids. At the level of gene sequence, gene organization, and gene expression, similarities between plastid and eubacterial (particularly cyanobacterial) genomes are now so numerous and so pronounced that alternative (i.e., autogenous) hypotheses of plastid origin no longer demand serious consideration. The flood of evidence compelling a final affirmation of the endosymbiotic origin of plastids continues unabated: the question no longer is "whether?," but "how many times?," "when?," and "by what process(es) and pathway(s)?"

During the same two decades, the emerging mitochondrial molecular biology has not provided comparably overwhelming support for the long-standing proposal of an endosymbiotic, eubacterial origin of mitochondria; quite the contrary, much of the data has tended to confuse rather than enlighten. Early biochemical observations suggesting that mitochondria contain a bacteria-like translation system (Huang *et al.*, 1966; Lamb *et al.*, 1968; Smith and Marcker, 1968) were quickly followed by indications of physical similarities between mitochondrial and bacterial ribosomes, as well as their component subunits and rRNAs (Küntzel and Noll, 1967). However, as additional data appeared, it quickly became evident that the inferred structural "similarities" between mitochondrial and bacterial ribosomes were more apparent than real (discussed in Gray and Doolittle, 1982), and that mitochondrial ribosomes are actually extremely variable in their properties, and are in many respects unique (Curgy, 1985; Benne and Sloof, 1987). As the literature expanded, this theme of structural variability came to dominate our perception of mtDNA. Patterns of mitochondrial genome organization and gene expression have proven to be bewilderingly diverse (Gray, 1989a), and in very few instances are these features obviously eubacterial: indeed, molecular biology as practiced in mitochondria is in many of its specifics unlike anything seen in eubacteria, archaebacteria, or the eukaryotic nucleus. For that reason, it has proven much more difficult than with plastids to build a broadly based case for an endosymbiotic origin of mitochondria, one that is supported by a variety of molecular biological data.

Fortunately, gene sequence (particularly rRNA sequence) has strongly supported the initial biochemical evidence that suggested a eubacterial, endosymbiotic origin of mitochondria and has even pinpointed the closest contemporary eubacterial relatives of mitochondria (Gray, 1988). As with plastids, a central unsettled question is whether mitochondria are monophyletic or polyphyletic, but superimposed on this is the perplexing issue of what evolutionary forces and mechanisms have led to the extreme diversity in mitochondrial genome organization that we see today.

Not only does mitochondrial genome diversity greatly complicate reconstruction of the evolutionary history of mitochondria, but the task is made even more difficult by the novel (indeed sometimes bizarre) findings that have emerged from sutdies of mitochondrial molecular biology over the past decade. These unusual features include deviations from the universal genetic code (Fox, 1987; Jukes and Osawa, 1990; Bessho *et al.*, 1992); unusual tRNA structures (de Bruijn *et al.*, 1980; Wolstenholme *et al.*, 1987; Okimoto and Wolstenholme, 1990) and codon recognition patterns (Dirheimer and Martin, 1990); distinctive modes of gene organization (Attardi, 1985; Gray and Boer, 1988) and gene expression (Clayton, 1984, 1991; Schinkel and Tabak, 1989), including the involvement of *maturases* (Lazowska *et al.*, 1980) and other *trans*-acting protein factors in RNA processing (see Cech, 1990; Lambowitz and Perlman, 1990); promiscuous DNA (Stern and Lonsdale, 1982; Stern and Palmer, 1984); mosaic genes (Dewey *et al.*, 1986); RNA editing (see Cattaneo, 1991); scrabled rRNA genes (Boer and Gray, 1988a); RNA import (Nagley, 1989); and, most recently, *trans*-splicing (Chapdelaine and Bonen, 1991; Wissinger *et al.*, 1991; Conklin *et al.*, 1991). All of these molecular oddities must ultimately be accommodated in any full description of the origin and evolution of the mitochondrial genome.

B. Mitochondrial Molecular Biology

In this review, I will not attempt a comprehensive and detailed treatment of the data on mitochondrial genome structure, organization, expression, and evolution that have appeared since 1982. Rather, as in the preceding section on plastids, I will concentrate on those aspects (particularly gene sequence) that most directly bear on the question of the evolutionary origin of the mitochondrion and its genome. However, in order to put this material into the proper perspective, some discussion of the salient features of mtDNA structure and expression is necessary. More complete treatments of the genetics, molecular biology, and biogenesis of mitochondria can be found in Tzagoloff and Myers (1986), Chomyn and Attardi (1987), Attardi and Schatz (1988), and Gray (1989a), as well as in the other chapters of this special volume.

1. Mitochondrial Genome Diversity vs Constant Genetic Function

Several recent reviews have drawn attention to the extreme structural diversity found among characterized mtDNAs (Gray, 1982, 1989a; Wallace, 1982; Sederoff, 1984; Clark-Walker, 1985; Bonen, 1991). Mitochondrial genomes vary in size over more than a 150-fold range, from as little as 13.8 kbp in the nematode worm, *Caenorhabditis elegans* (Okimoto *et al.*, 1992), to as much as 2400 kbp in muskmelon, *Cucumis melo* (Ward *et al.*, 1981). Although most mtDNAs are considerably smaller than most plDNAs (i.e., <200 kbp in size), some plant mtDNAs may approach the size of a bacterial genome (Ward *et al.*, 1981). Most mtDNAs are either circular as isolated or map as circular genomes; however, linear mtDNAs have been identified in the coelenterates *Hydra attenuata* and *Hydra littoralis* (Warrior and Gall, 1985), the fungi *Hansenula mrakii* (Wesolowski and Fukuhara, 1981) and *Candida rhagii* (Kováč *et al.*, 1984), the ciliates *Paramecium aurelia* (Pritchard *et al.*, 1986) and *Tetrahymena pyriformis* (Suyama *et al.*, 1985), the chlorophyte alga *Chlamydomonas reinhardtii* (Grant and Chiang, 1980), and the chrysophyte alga *Ochromonas danica* (Coleman *et al.*, 1991).

Despite the two orders of magnitude size difference between the smallest and the largest mitochondrial genomes, there is no indication of a corresponding difference in gene content. Tables II and III list the genes that have been identified to date in a range of mtDNAs. Except in the case of angiosperms, these particular mitochondrial genomes have been completely or almost completely sequenced, so their coding functions are known with considerable certainty. It is clear from this compilation that throughout the range of mitochondria-containing eukaryotes, mtDNA has the same fundamental role: it specifies certain components of a mitochondrial protein-synthesizing system whose purpose is to translate a limited number of mitochondrially transcribed mRNAs that encode polypeptide components of the mitochondrial electron transport system (Chomyn and Attardi, 1987; Attardi and Schatz, 1988). The suite of respiratory chain genes is remarkably similar among structurally diverse mtDNAs, although there are some notable differences: in particular, the mitochondrial genome of certain ascomycete fungi (represented in Table II by *Saccharomyces cerevisiae* and *Schizosaccharomyces pombe*) appears to lack completely any NADH dehydrogenase subunit (*nad*) genes. Translation apparatus genes are even more variable in distribution, with ribosomal protein genes completely or almost completely lacking in animal and fungal mtDNAs, but more frequent in the mitochondrial genomes of land plants and some protists (Table III). Plant mtDNA is distinguished by having a 5S rRNA gene, which to date has not been identified in the mitochondrial

TABLE II

Respiratory Chain Genes Encoded by Mitochondrial DNA[a]

Organism	nad										cob	cox			atp				Other[e]
	1	2	3	4	4L	5	6	7[b]	"psbG"[c]	"ORF169"[d]		1	2	3	A	6	8	9	
Animals (except nematodes)[f]	+	+	+	+	+	+	+	−	−	−	+	+	+	+	−	+	+	−	−
Nematodes[g]	+	+	+	+	+	+	+	−	−	−	+	+	+	+	−	+	−	−	−
Land plants																			
Angiosperms[h]	+	+	+	+	+	+	−	+	−		+	+	+	+	+	+		+	orf25[i], orfB[j]
Bryophyte[k]	+	+	+	+	+	+	+	(+)	−		+	+	+	+	+	+		+	[+]
Fungi																			
Aspergillus nidulans[l,m]	+	+	+	+	+	+	+	−	−	−	+	+	+	+	−	+		+[n]	urfA3[m]
Neurospora crassa[l,o]	+	+	+	+	+	+	+	−	−	−	+	+	+	+	−	+		+[n]	[+]
Podospora anserina[p]	+	+	+	+	+	+	+	−	−	−	+	+	+	+	−	+		−	[+]
Saccharomyces cerevisiae[l]	−	−	−	−	−	−	−	−	−	−	+	+	+	+	−	+	+	+	[+]
Schizosaccharomyces pombe[l,q]	−	−	−	−	−	−	−	−	−	−	+	+	+	+	−	+	+	+	urfa
Allomyces macrogynus[q]	+	+	+	+	+	+	+	−	−	−	+	+	+	+	−	+		+	[+]
Protists																			
Phytophthora infestans[q,r]	+	+	+	+	+	+	+	+	−	+	+	+	+	+	+	+	−	+	[+]
Prototheca wickerhamii[s]	+	+	−	+	+	+	+	+	−	+	+	+	+	+	+	+	?	+	[+]
Chlamydomonas reinhardtii[t]	+	+	−	+	−	+	+	−	−		+	+	−	−	−	−	−	−	rtl[u]
Paramecium aurelia[v]	+	+	+	+	−	+	−	+	+	+	+	+	+	+	−	−	−	+	[+]
Trypanosoma brucei[w]	+	+	+	+	+	+	+	+	−	+	+	+	+	+	−	−	−	+	MURF1,2[x]

[a] Gene designations: nad, subunit of NADH–ubiquinol oxidoreductase (NADH dehydrogenase; Complex I); cob, apocytochrome b component of ubiquinol-cytochrome c oxidoreductase (cytochrome b/c_1, Complex III); cox, subunit of cytochrome c oxidase (cytochrome aa_3; Complex IV); atp, subunit of ATP synthase (H^+-ATPase; Complex V). Symbols: +, gene present; −, gene absent; [+], additional unassigned ORFs present. Where no symbol appears, insufficient data are available to decide. Intron-encoded maturases and transposition factors (Tzagoloff and Myers, 1986) are not listed here.

TABLE II (*Continued*)

[b] Homolog of a bovine nuclear gene encoding a 49-kDa NADH dehydrogenase subunit, and of the chloroplast genes orf393 (tobacco) and orf392 (liverwort) (Fearnley *et al.*, 1989).

[c] Probably not a photosystem 2 gene, but rather an additional subunit of NADH dehydrogenase (Nixon *et al.*, 1989; Mayes *et al.*, 1990; Arizmendi *et al.*, 1992).

[d] Homolog of ORFs in liverwort (orf169) and tobacco (orf159) chloroplast DNA; also present in the mtDNAs of *Dictyostelium discoideum* (ORF209; Tanaka *et al.*, 1990) and *Paramecium aurelia* (gene called "Pl"; Mahalingam *et al.*, 1986). This gene may encode another *nad* subunit, as it is part of the chloroplast *ndhC-psbG-orf159* operon.

[e] Genes unique (so far) to a particular organism or group of organisms, and whose function (in electron transport or coupled oxidative phosphorylation, or otherwise) has not yet been elucidated.

[f] Human (Anderson *et al.*, 1981), mouse (Bibb *et al.*, 1981), bovine (Anderson *et al.*, 1982), rat (Gadaleta *et al.*, 1989), chicken (Desjardins and Morais, 1990), frog (*Xenopus laevis*) (Roe *et al.*, 1985), fruitfly (*Drosophila yakuba*) (Clary and Wolstenholme, 1985), sea urchin (*Strongylocentrotus purpuratus*) (Jacobs *et al.*, 1988), and *Paracentrotus lividus* (Cantatore *et al.*, 1989).

[g] *Ascaris suum* and *Caenorhabditis elegans* (Wolstenholme *et al.*, 1987; Okimoto *et al.*, 1992).

[h] Genes identified in the mtDNA of at least one angiosperm. All except *nad2* (found in *Beta vulgaris* (sugarbeet) mtDNA; Xue *et al.*, 1990) and *orfB* have been identified in *Triticum aestivum* (wheat), a monocotyledon (Gray, 1990; L. Bonen, personal communication), and all except *nad2* and *nad7* have been found in *Petunia hybrida*, a dicotyledon (Conklin *et al.*, 1991; M. Hanson, personal communication).

[i] Stamper *et al.* (1987); Bonen *et al.* (1990).

[j] Hiesel *et al.* (1987); Quagliariello *et al.* (1990).

[k] Liverwort (*Marchantia polymorpha*) (Oda *et al.*, 1992). *nad7* appears to be a pseudogene, and a *cob* pseudogene is present in addition to the normal gene. Genome also contains homologs of *orf25* and *orfB* in angiosperm mtDNA, as well as 28 unassigned ORFs of >60 amino acid residues and 11 ORFs with similarity to yeast mitochondrial RNA maturases (10 of which are in introns).

[l] Wolf and Del Giudice (1988).

[m] Dyson *et al.* (1989); T. A. Brown *et al.* (1989).

[n] Apparently a silent gene; functional gene is located in the nucleus (van den Boogart *et al.*, 1982; Brown *et al.*, 1984).

[o] Collins (1990).

[p] Cummings *et al.* (1990).

[q] B. F. Lang, personal communication. *A. macrogynus* mtDNA also contains several unassigned ORFs (209, 212, 414) and an intron ORF having similarity to reverse transcriptase.

[r] Also contains a homolog of nuclear genes encoding largest subunit of the NADH dehydrogenase complex of bovine (75 kDa; Runswick *et al.*, 1989) and *Neurospora* (78 kDa; Prais *et al.*, 1991) mitochondria, as well as five unassigned ORFs, (32, 79, 100, 142, and 248), one of which (*orf248*) is homologous to *orf294* in liverwort mtDNA.

[s] G. Wolff, U. Kück, G. Burger and B. F. Lang, personal communication. Contains, in addition to homologs of *orf25* and *orfB* in angiosperm and liverwort mtDNAs, three unassigned ORFs (174, 183, 304) and two intronic ORFs having high similarity to those in fungal mtDNA.

[t] Gray and Boer (1988); Boer and Gray (1991).

[u] Boer and Gray (1988b).

[v] Pritchard *et al.* (1990a). Genome contains 16 additional unassigned ORFs.

[w] Benne (1990); also *Leishmania tarentole* and *Crithidia fasciculata* (Simpson and Shaw, 1989).

[x] The identities of MURF1 and MURF2 remain to be determined. MURF3 and MURF4 have recently been identified as, respectively, *nad7* and *atp6*, whose transcripts undergo extensive RNA editing (Koslowsky *et al.*, 1990; Bhat *et al.*, 1990).

285

TABLE III

Translation Apparatus Genes Encoded by Mitochondrial DNA[a]

	rrn			trn	rps												rpl					Other
	SSU	LSU	5S		1	2	3	4	7	8	10	11	12	13	14	19	2	5	6	14	16	
Animals[b] (including nematodes[c])	+	+	–	22[d]	–	–	–	–	–	–	–	–	–	–	–	–	–	–	–	–	–	–
Land plants																						
Angiosperms[m]	+[e]	+[e]	+[e]	<20[f,g]	+	+	+[h]	+	+	+	+	+	+[i]	+[j]	+[k]	+[l]	+	+	+	–	+[h]	–
Bryophyte[m]	+	+	+	27	+	+	+	+	+	+	+	+	+	+	+	+	+	+	+	–	+	
Fungi																						
Aspergillus nidulans[n,o]	+	+	–	28[d]	–	–	–	–	–	–	–	–	–	–	–	–	–	–	–	–	–	–
Neurospora crassa[n,p]	+	+	–	27[d]	–	–	–	–	–	–	–	–	–	–	–	–	–	–	–	–	–	S-5[q]
Podospora anserina[r]	+	+	–	27[d]	–	–	–	–	–	–	–	–	–	–	–	–	–	–	–	–	–	
Saccharomyces cerevisiae[n]	+	+	–	24[d]	–	–	–	–	–	–	–	–	–	–	–	–	–	–	–	–	–	
Schizosaccharomyces pombe[n,t]	+	+	–	25[d]	–	–	+	–	–	–	–	–	–	–	–	–	–	–	–	–	–	var-1[u], tsl[s]
Allomyces macrogynus[t]	+	+	–	25(+?)	–	–	–	–	–	–	–	–	–	–	–	–	–	–	–	–	–	–
Protists																						
Phytophthora infestans[t]	+	+	–	25(+?)	–	?	?	+	+	+	+	–	+	+	+	+	–	+	+	+	+	+
Prototheca wickerhamii[u]	+	+	–	25(+?)	–	+	–	?	+	–	+	–	+	+	+	–	+	+	+	–	+	+
Chlamydomonas reinhardtii[v]	+	+	–	3[f,v]	–	–	–	–	+	–	–	–	–	–	–	–	–	–	–	+	–	–
Paramecium aurelia[w]	+	+	–	4[f,x]	–	–	–	–	–	–	–	–	+	–	–	–	+	–	–	+	–	
Trypanosoma brucei[y]	+	+	–	0[f,z]	–	–	–	–	–	–	–	–	–	–	–	–	–	–	–	–	–	–

[a] Gene designations: rrn, ribosomal RNA (SSU, small subunit; LSU, large subunit); trn, transfer RNA; rps, SSU ribosomal protein; rpl, LSU ribosomal protein. Symbols: +, gene present; –, gene absent. Where no symbol appears, insufficient data are available to decide. Additional unassigned ORFs have been found in some of the larger mtDNAs (see Table II, footnote a).

[b] See Table III (footnote f) for primary references.

[c] Ascaris suum and Caenorhabditis elegans (Wolstenholme et al., 1987; Okimoto and Wolstenholme, 1990; Okimoto et al., 1992).

[d] Mitochondrial genome encodes a sufficient number of tRNAs to support mitochondrial protein synthesis.

[e] Bonen and Gray (1980); Huh and Gray (1982).

[f] Mitochondrial genome does not encode a sufficient number of tRNAs to support mitochondrial protein synthesis; import of nucleus-encoded tRNAs from cytosol demonstrated or suspected (Nagley, 1989).

[g] Sixteen distinct tRNA genes (10 "native," 6 "chloroplast-like") have been identified in the mitochondrial genomes of wheat (Joyce and Gray, 1989a) and maize (Sangaré et al., 1990), whereas potato mtDNA appears to encode 15 native and 5 chloroplast-like tRNA species (Maréchal-Drouard et al., 1990).

[h] Maize (Hunt and Newton, 1991).

[i] Wheat and maize (Gualberto et al., 1988).

[j] *Daucus carota* (carrot) and *Oenothera berteriana* (evening primrose) (Wissinger et al., 1990); *Nicotiana tobacum* (tobacco) (Bland et al., 1986); wheat (Bonen, 1987). Wheat mitochondrial *rps13* is apparently a silent gene (Bonen, 1987).

[k] *Vicia faba* (broad bean) (Wahleithner and Wolstenholme, 1988).

[l] *Petunia hybrida* (Conklin and Hanson, 1991).

[m] Liverwort (*Marchantia polymorpha*) (Oda et al., 1992).

[n] Wolf and Del Giudice (1988).

[o] Dyson et al., (1989).

[p] Collins (1990).

[q] *N. crassa* S-5 and *S. cerevisiae var-1* encode SSU-associated proteins that share limited sequence similarity with one another, but neither bears any resemblance to known bacterial ribosomal proteins (Benne and Sloof, 1987).

[r] Cummings et al. (1990).

[s] tRNA synthesis locus; encodes the RNA component of mitochondrial RNase P (Miller and Martin, 1983).

[t] B. F. Lang, personal communication. Genome also encodes *rpl12*. Putative *rps2* and *rps3* show only remote similarity to the corresponding *E. coli* genes.

[u] G. Wolff, U. Kück, G. Burger, and B. F. Lang, personal communication. Unassigned *orf511* shows similarity to *E. coli rps4*.

[v] Gray and Boer (1988); Boer and Gray (1988c); Boer and Gray (1991).

[w] Pritchard et al. (1990a).

[x] In addition to three typical tRNA genes identified in the *P. aurelia* mitochondrial genome (Pritchard et al., 1990a), the presence of a gene encoding an atypical tRNA[Met] has also been inferred (Heinonen et al., 1987).

[y] Benne (1990).

[z] Hancock and Hajduk (1990).

genome of other eukaryotes. Thus, although the overall role of mtDNA is highly conserved, there are actually only four mtDNA-encoded genes that are universally present in the mitochondrial genomes so far characterized: genes encoding the LSU and SSU rRNA components of the mitochondrial ribosome, and the protein-coding genes *coxl* and *cob* (Gray, 1989a).

2. Genome Organization

Within the four traditional eukaryotic kingdoms (Animalia, Plantae, Fungi, and Protoctista (Margulis and Schwartz, 1988), a sufficient number of mtDNAs have now been sequenced that it is possible to discern distinctive patterns of gene organization, expression, and evolution that characterize at least the mitochondrial genomes of the multicellular eukaryotes. The most striking manifestation of mtDNA variability is the contrast between the smallest (animal) and the largest (plant) mtDNAs, which in almost all of their distinguishing features are about as different as it is possible for two functionally equivalent genomes to be (Table IV).

a. Animal Mitochondrial DNA   For references see Attardi, 1985; Brown, 1985; Wilson *et al.*, 1985; and Cantatore and Saccone, 1987. The hallmark of the "parsimonious" animal mitochondrial genome (current size range 13.8 to 39.3 kbp; Gray, 1989a) is its exceptionally high information density, with coding regions directly abutting (and in some cases even overlapping; Cantatore *et al.*, 1989), and with no or few noncoding nucleotides separating them (Attardi, 1985). This extremely tight packing of genetic information (with over 90% of the genome having a coding function) was first described in the 16.5-kbp mtDNA of humans (Anderson *et al.*, 1981), but is in fact a distinctive feature of all characterized animal mtDNAs. The noncoding control region, which in vertebrate animals is between 1 and 2 kbp in size (Desjardins and Morais, 1990), is reduced to just over 100 bp in two sea urchin species (Jacobs *et al.*, 1988; Cantatore *et al.*, 1989). Further size reduction in *Caenorhabditis* and *Ascaris* mtDNAs (which at 13.8 and 14.3 kbp are the smallest animal mtDNAs identified so far) reflects the absence of the *atp8* locus (Okimoto *et al.*, 1992), whereas expansion of some other animal mtDNAs [up to about 40 kbp in the case of sea scallop, *Placopecten magellanicus* (Snyder *et al.*, 1987] is attributable to localized sequence amplification, resulting in variably sized direct tandem duplications of both coding and noncoding regions of the genome (Okimoto *et al.*, 1991). Until recently, it had been thought that animal mtDNA was devoid of intervening sequences; however, introns have now been found in the mtDNA of *Metridium*, a sea anemone (phylum Cnidaria) (D. R. Wolstenholme, personal communication).

TABLE IV

Contrasting Patterns of Mitochondrial Genome Organization and Expression

	Vertebrates	*C. reinhardtii*	Angiosperms
Genome size	Small (16–18 kbp)	Small (16 kbp)	Large (200–2400 kbp)
Coding sequence	>90%	>80%	<10% (est).
Intergenic sequence	Virtually absent	Very limited	Abundant and extensive
Introns	Absent	Absent[a]	Present
Gene linkage	Extensive	Extensive	Minimal
Transcriptional linkage	Extensive	Extensive	Minimal
Promoters	Single (1 on each strand)	Not known	Multiple
Transcript processing	Endonucleolytic, extensive[b]	Endonucleolytic, extensive[b]	Endonucleolytic (?)[b]
Oligo(A) added to mRNA?	Yes	No	No
5'-leader in mRNA	Absent	Absent	Present
3'-trailer in mRNA	Absent	Present	Present
UGA codon specifies:	Trp	Termination	Termination
Import of tRNA	No	(Yes)[c]	Yes

[a] An intron is present in the *cob* gene of the interfertile species *C. smithii* (Boynton *et al.*, 1987; Colleaux *et al.*, 1990).

[b] In mammalian (Attardi, 1985) and *C. reinhardtii* (Gray and Boer, 1988) mitochondria, precise endonucleolytic cleavages separate butt-joined coding regions contained in long cotranscripts; such extensive pre-RNA processing is not seen in plant mitochondria, although some endonucleolytic processing of precursor transcripts does occur (see Newton, 1988).

[c] Only three tRNA genes are encoded in *C. reinhardtii* mtDNA (Boer and Gray, 1988c); therefore, tRNA import is assumed.

Gene order is relatively constant within individual animal phyla, but varies between phyla. The mtDNAs of five vertebrate species [human, mouse, rat, bovine, and from (*Xenopus laevis*); phylum Chordata] have the same gene order, whereas a sixth species (chicken) shows a translocation relative to mammals and *Xenopus* (Desjardins and Morais, 1990). The order of rRNA and protein-coding genes in sea urchin (*Strongylocentrotus purpuratus* and *Paracentrotus lividus;* phylum Echinodermata) and *Drosophila yakuba* (phylum Arthropoda) can be related to the common vertebrate pattern by two apparent translocations and three inversions, respectively (Jacobs *et al.*, 1988), whereas there is no evident relationship between any of these patterns and that seen in the mtDNAs of *C. elegans* and *A. suum* (phylum Nematoda) (Wolstenholme *et al.*, 1987). Between different phyla, there is little or no similarity in the distribution of tRNA genes, suggesting there has been much greater movement of these genes than of rRNA and protein-coding genes during evolution of the animal mitochondrial genome.

Although mammalian mtDNA displays an invariant gene order, mammalian mitochondrial genes diverge in sequence at an extremely rapid rate (5–10 times the rate at which single-copy nuclear DNA diverges in the same species; Wilson *et al.*, 1985). Thus, animal mtDNA is generally considered to be evolving slowly in structure but rapidly in sequence, with mammalian mtDNA representing the most rapidly evolving cellular genome yet discovered.

b. Fungal Mitochondrial DNA   For references see Clark-Walker, 1985; Brown *et al.*, 1985; Breitenberger and RajBhandary, 1985; de Zamaroczy and Bernardi, 1986; Grossman and Hudspeth, 1985; Brown, 1987; Wolf, 1987a,b; and Wolf and Del Giudice, 1988. Fungal mitochondrial genomes, which range between 17.6 kbp [in the ascomycete *Schizosaccharomyces pombe* EF (Zimmer *et al.*, 1984)] and 176 kbp [in the basidiomycete *Agaricus bitorquis* (Hintz *et al.*, 1985)], nevertheless show few differences in gene content [the most notable being the complete absence of *nad* genes in at least some of the ascomycetes (Table II)]. Most of the size differences are accounted for by two factors: expansion/contraction of AT-rich intergenic spacer regions, and presence or absence of introns. The *coxI* gene in *Podospora anserina* race A is a spectacular example of intron extravagance: this gene spans 24.5 kbp (1.5 times the size of the entire human mitochondrial genome), comprises some 26% of the 94.2-kbp *Podospora* mtDNA, and contains 16 introns that together account for 93% of the total *coxI* sequence (Cummings *et al.*, 1990). A number of ORFs have been delineated on the basis of sequence analysis in some of the larger fungal mtDNAs [e.g., *S. cerevisiae* (de Zamaroczy and Bernardi, 1986) and *P. anserina* (Cummings *et al.*, 1990)], but whether any of these correspond to active genes remains to be determined. Two additional, intron-associated classes of genes are found in some of the larger fungal mtDNAs: (i) intron-encoded maturases that are required for intron excision during post-transcriptional processing of precursor RNAs; and (ii) intron-encoded site-specific endonucleases that mediate intron transposition (see Lambowitz, 1989; Perlman and Butow, 1989; Lambowitz and Perlman, 1990; Cech, 1990).

Given the extreme size variation evident in fungal mitochondrial genomes, it is not surprising that only limited conservation of gene order is evident when mtDNAs from distantly related fungi are compared (Clark-

Walker *et al.*, 1983). At shallower phylogenetic depths, common or closely related patterns of fungal mtDNA gene organization begin to emerge (Hoeben and Clark-Walker, 1986; Bruns *et al.*, 1988; Bruns and Palmer, 1989; Tian *et al.*, 1991a,b), and such data have begun to reveal mechanisms and pathways of fungal mtDNA evolution. Feature of yeast mtDNA that appear to contribute to evolutionary rearrangement include (i) a high rate of mtDNA recombination; (ii) circularity of the mitochondrial genome; (iii) a high proportion of intergenic, noncoding spacer sequences; and (iv) the existence within spacers of short repeated sequences, dispersed throughout the genome (Clark-Walker and Miklos, 1974). A model based on illegitimate recombination between short repetitive elements has been proposed to account for the evolutionary rearrangement of fungal mtDNAs having these characteristics (Evans and Clark-Walker, 1985; Clark-Walker *et al.*, 1985b).

c. Plant Mitochondrial DNA   For references see Leaver and Gray, 1982; Lonsdale, 1984, 1989; Palmer, 1985b, 1990; Pring and Lonsdale, 1985; Newton, 1988; Levings and Brown, 1989; Gray, 1990; and André *et al.*, 1992. Virtually the only plant mtDNAs that have been studied in any detail are those from the phylum Angiospermophyta (flowering plants), which is only one of 10 recognized phyla within the plant kingdom (Margulis and Schwartz, 1988). Our concept of the "plant" mitochondrial genome is therefore likely to prove as incomplete as our view of the "animal" mitochondrial genome was before information became available for species other than the chordates [which constitute but one of the 33 animal phyla listed by Margulis and Schwartz (1988)].

Apart from its large size (200–2400 kbp in the range of angiosperms examined to date), the most distinctive feature of the "profligate" plant mitochondrial genome is its propensity for rapid structural change in the face of an extremely slow rate of sequence divergence (Palmer and Herbon, 1988): precisely the opposite of what is observed in mammalian mtDNA. In angiosperms, mtDNA evolves about 10-fold less rapidly in sequence than single-copy nDNA, and about 3- to 4-fold less rapidly than plDNA, which means that the plant mitochondrial genome is the most slowly evolving cellular genome so far characterized (Wolfe *et al.*, 1987). In contrast, overall gene organization is exceptionally fluid in plant mtDNA, and marked differences in restriction patterns and gene arrangement are often evident between closely related genera (Coulthart *et al.*, 1990) [and even between different cultivars of the same species (Ichikawa *et al.*, 1989)]. Only very limited conservation of gene linkage is apparent in plant mtDNA, the most notable being the close juxtaposition of 18S and 5S rRNA genes in both monocotyledons and dicotyledons (Huh and Gray, 1982).

This organizational fluidity seems to be related to the existence of an active recombination system that mediates frequent rearrangement of the plant mitochondrial genome (Lonsdale, 1984, 1989). Plant mtDNA contains repeated blocks of sequence, some several kilobase pairs long, present in both direct and inverted orientation, and many of these appear to be recombinationally active (André et al., 1992). Recombination between direct repeats will resolve a "master circle" (constituting the entire genetic information complement) into two subgenomic circles, each containing one copy of the repeat, and having a size corresponding to the distance between the repeats in the master circle. Such a tripartite model, with two subgenomic circles in dynamic equilibrium with a master circle, is thought to describe the physical arrangement of the simplest plant mtDNAs, those of *Brassica* sp., most of which contain a single pair of recombinationally active direct repeats (Palmer and Shields, 1984). A more complex multipartite arrangement is postulated for larger plant mtDNAs containing a greater number of repeats (Lonsdale et al., 1984). An attractive model for the evolutionary restructuring of plant mtDNA, invoking rare recombination events between small direct repeats, has been proposed (Small et al., 1989; Fauron et al., 1990). Complex rearrangement events are also implicated in the generation of *mosaic genes* that encode novel open reading frames, and whose existence in mtDNA has been correlated with the pehnomenon of cytoplasmic male sterility (Dewey et al., 1986; Young and Hanson, 1987).

Despite the fact that even the smallest plant mtDNAs are larger in size than plant plDNAs, the former appear to contain about 10-fold fewer genes than the latter; indeed, the evidence so far is that the plant mitochondrial genome encodes only a few novel genes (e.g., 5S rRNA) in addition to the basic set of respiratory chain and translation apparatus genes found in its much smaller counterparts in other eukaryotes (Tables II and III). Moreover, despite its spaciousness, plant mtDNA (unlike animal mtDNA) does not appear to encode the minimal set of tRNA genes required to support mitochondrial protein synthesis: in three plants (wheat, maize, and potato), there is now strong evidence that the mtDNA contains less than 20 distinct tRNA genes (Joyce and Gray, 1989a; Sangaré et al., 1990; Maréchal-Drouard et al., 1990), and there is evidence that plant mitochondria import nucleus-encoded tRNAs from the cytosol to function in mitochondrial protein synthesis (Maréchal-Drouard et al., 1988, 1990; Joyce and Gray, 1989a). Perhaps because of this very low gene density (<10% of the mtDNA having a coding function) in the face of an active recombination system, plant mtDNA is readily able to incorporate and accommodate "foreign" DNA. Sequences originating from plDNA (Stern and Lonsdale, 1982; Stern and Palmer, 1984) [but also occasionally from nDNA (Schuster and Brennicke, 1987)] are rather frequent in plant mtDNA, and although most of this *promiscuous DNA* appears to be largely nonfunctional, there is evidence that some plastid-derived tRNA genes are actively transcribed and processed to mature tRNAs (see Joyce and Gray, 1989a). Thus, it appears that the plant mitochondrial genome has been able to recruit genes from other sources in the course of its evolution.

Part of the size difference between animal and plant mtDNAs is accounted for by the presence in the latter of repeated sequences (some >10 kbp long), novel genes, introns, and promiscuous DNA. Overall, however, these features account for only a small part of this large size differential. In fact, introns are relatively scarce in plant (at least angiosperm) mitochondrial genes. In contrast to their fungal mitochondria homologs angiosperm mitochondrial *cob, cox1,* and LSU rRNA genes do not contain introns; so far, these have only been reported in *cox2* [which in many although not all plants contains a single group II intron; Pruitt and Hanson (1991)], *nad5* (Wissinger *et al.,* 1988), and *nad1* (one of whose introns contains a maturase-like ORF) (Wahleithner *et al.,* 1990; Chapdelaine and Bonen, 1991; Wissinger *et al.,* 1991; Conklin *et al.,* 1991). Mostly, plant mtDNA appears to consist of noncoding intergenic (spacer) DNA, which is fairly nondescript in the sense that it displays no distinctive nucleotide or sequence bias (in contrast, e.g., to the very AT-rich spacer DNA found in many fungal mtDNAs; Wolf and Del Giudice, 1988).

An intriguingly different picture has emerged from the recent determination of the complete sequence of the 186,608-bp mtDNA of *Marchantia polymorpha* (liverwort; phylum Bryophyta) (Oda *et al.,* 1992). This work has identified the standard set of respiratory chain genes, including *nad4L* and *nad6,* which so far have not been found in angiosperm mtDNA (Table II); a substantial number of ribosomal protein genes (Table III); and a total of 32 ORFs ranging in size from 60 to 732 codons. In contrast to angiosperm mtDNA, a large number of introns (32) are present, in rRNA and tRNA genes as well as in most protein-coding genes (only *nad1, nad6,* and *atp6* do not contain introns). Ten of the introns have ORFs that are thought to be translated by read-through from the preceding exons, as in the case of intron-encoded maturases in fungal mtDNA (Lambowitz and Perlman, 1990). Liverwort mtDNA encodes 27 different tRNA species (including separate initiator and elongator tRNAs[Met]), which together could potentially translate almost all codons [only genes for tRNAs specific for AUU and AUC (Ile) and ACA and ACG (Thr) codons have not been identified in *Marchantia* mtDNA]. Although many repeated sequences (up to 800 bp long) are present, these do not appear to be involved in homologous recombination. In overall structure and organization, liverwort mtDNA looks more like a large fungal mtDNA than a conventional angiosperm mtDNA, even though sequence similarity, the presence of a 5S rRNA gene, and the conserved linkage of 18S and 5S rRNA genes all argue strongly that bryophyte and angiosperm mtDNAs are specific evolutionary relatives. Interestingly, the mitochondrial genomes of two other nonflowering land plants, *Onoclea sensibilis* (sensitive fern) and *Equisetum arvense* (common horsetail), appear to have the large size and complex structure characteristic of angiosperm mtDNA (Palmer *et al.*, 1992).

d. Protistan Mitochondrial DNA   Protistan mtDNAs, including those green algal mtDNAs that have been examined so far, vary between about 15 and 60 kbp in size (Gray, 1989a). Considering that the unicellular eukaryotes are an ancient and phylogenetically disparate group (Margulis and Schwartz, 1988; Sogin *et al.*, 1989; Patterson and Sogin, 1992), it is hardly surprising that no common pattern of genome organization is seen among the relatively few protist mtDNAs that have been studied in detail. Those protist mitochondrial genomes for which most information is available are briefly highlighted here, to emphasize some of the structural novelties these genomes exemplify.

   a. Phylum Zoomastigina (Genera Crithidia, Leishmania, Trypanosoma)   For references see Benne, 1985, 1989, 1990; Simpson, 1986, 1987, 1990; Simpson and Shaw, 1989; Stuart *et al.*, 1989; Feagin, 1990; and Stuart, 1991. The trypanosomatid (kinetoplastid) protozoa possess a single, structurally complex mitochondrion containing kinetoplast DNA (kDNA), an unusual genome consisting of a network of concatenated circular DNA molecules of two types, minicircles (465 to 2500 bp) and maxicircles (23–36 kbp) (Simpson, 1990; Borst, 1991). Maxicircle DNA, the functional equivalent of mtDNA in other eukaryotes, is a compact genome containing many of the recognized respiratory and translation genes (Tables II and III), although tRNA genes appear to be totally absent (Simpson *et al.*, 1989; Hancock and Hajduk, 1990). A number of conventional mitochondrial genes were originally not recognized in maxicircle DNA; however, several previously unidentified "MURFs" are now known to be cryptic genes (*cryptogenes*) whose transcripts undergo RNA editing, a U addition–deletion process that generates a translatable mRNA coding for a normal respiratory chain protein (Benne, 1989; Simpson and Shaw, 1989; Stuart, 1991; see Table III). Small RNAs (*guide RNAs*) implicated in providing the information for editing are encoded in both maxicircle and minicircle DNAs (Blum *et al.*, 1990; Sturm and Simpson, 1990; Blum and Simpson, 1990; Pollard *et al.*, 1990; van der Spek *et al.*, 1991), providing at last a genetic *raison d'etre* for the long-enigmatic minicircle (Borst, 1991). Consistent with the deep branching of the trypanosomatid flagellates in the eukaryotic tree (see Figs. 6 and 12), trypanosomatid mitochondrial genes are less similar in sequence to the homologous yeast or human genes than the latter are to each other (de la Cruz *et al.*, 1984). Trypanosomatid 9S (SSU) and 12S (LSU) rRNAs are the most highly diverged rRNAs known, displaying only the barest resemblance to typical rRNAs (Eperon *et al.*, 1983; Sloof *et al.*, 1985).

   b. Phylum Ciliophora (Genera Paramecium, Tetrahymena)   The complete sequence of the 40.5-kbp *Paramecium aurelia* mitochondrial genome has been published (Pritchard *et al.*, 1990a). This work revealed a conventional set of respiratory chain plus rRNA genes, although neither *cox3* nor *atp9* was identified (Tables II and III). Only three conventional tRNA genes appear in the entire genome, which may also encode an unconventional tRNA^{Met} (Heinonen et al., 1987). This small number of tRNA genes indicates a need for tRNA import, as is also the case for *Tetrahymena pyriformis*, whose mtDNA appears to specify no more than about 10 tRNAs (Suyama, 1986). On the other hand, several ribosomal protein genes as well as the *nad7* and *"psbG"* genes, which are not found in animal or fungal mtDNAs, are present in *P. aurelia* mtDNA. The latter also contains a number of unidentified ORFs, all of which have an associated transcript (Pritchard *et al.*, 1990a). In *T. pyriformis*, the mitochondrial rRNA genes have an unusual arrangement (Schnare *et al.*, 1986; Heinonen *et al.*, 1987), with each giving rise to a discontinuous rRNA composed of two "subunits": α [corresponding to the 5'-terminal 208 (SSU) or 280 (LSU) nt] and β (corresponding to the rest of the rRNA). LSUα and -β coding modules are rearranged at the genome level, so that LSUα is downstream of LSUβ and separated from it by a tRNA^{Leu} gene, all three cistrons being in the same transcriptional orientation (Heinonen *et al.*, 1987). A similar situation has been described in *Paramecium* mitochondria (Seilhamer *et al.*, 1984a,b), except that in this case there is some controversy about the position of the discontinuity in the SSU rRNA (see Schnare *et al.*, 1986), and LSUα and LSUβ coding modules are in the conventional order [but still, apparently, separated by a tRNA gene (Heinonen *et al.*, 1987)]. Single SSU and LSU rRNA genes are present in *P. aurelia* mtDNA (Pritchard *et al.*, 1990a), but *T. pyriformis* mtDNA has duplicate, functional (Heinonen *et al.*, 1987) LSU rRNA genes located in inverted subterminal repeats (Goldbach *et al.*, 1978). The two copies of the *T. pyriformis* LSU rRNA gene have been completely sequenced and, surprisingly, turn out to be heterogeneous at several positions (Heinonen *et al.*, 1990).

   c. Phylum Chlorophyta (Genera Chlamydomonas, Chlorella)   For references see Gray and Boer, 1988; Michaelis *et al.*, 1990; and Boer and Gray, 1991. The complete sequence of the linear 15.8-kbp mtDNA of *Chlamydomonas reinhardtii* has now been determined (referenced in Michaelis *et al.*, 1990; Boer and Gray, 1991) and its genetic function defined (Tables II and III). Several features of this compactly organized genome are noteworthy: (i) extensive physical linkage of genes (Gray and Boer, 1988; Boer and Gray, 1991); (ii) absence of certain otherwise ubiquitous protein-coding genes, notably *cox2* and any *atp* genes; (iii) presence of a novel gene (*rtl*) specifying a reverse-transcriptase-like polypeptide (Boer and Gray, 1988b); (iv) absence of introns; (v) adherence to the universal genetic code (Kück and Neuhaus, 1986), but with a pronounced codon distribution bias such that eight codons are completely absent from all protein-coding genes, whereas an additional four are used exclusively and only sparingly in the *rtl* gene (Boer and Gray, 1988c; Michaelis *et al.*, 1990); (vi) presence of only three tRNA genes, indicating a requirement for tRNA import to support mitochondrial translation (Boer and Gray, 1988c); and (vii) a highly unusual arrangement of rRNA genes, in which SSU and LSU cistrons are broken up into a number of separate modules that are rearranged from the normal order of transcription and interspersed with tRNA and protein-coding genes throughout a 6-kbp region of the mtDNA (Gray and Boer, 1988). These rRNA coding modules give rise to high abundance, small RNAs that have the potential to interact noncovalently (by way of specific base-pairing) to form the structural equivalents of continuous SSU and LSU rRNAs in other systems (Boer and Gray, 1988a).

Chlamydomonas smithii (a species interfertile with *C. reinhardtii*) has a mtDNA that is colinear with that of *C. reinhardtii* except that its *cob* gene contains a mobile intron (Boynton *et al.*, 1987; Colleaux *et al.*, 1990). A linear 20-kbp mtDNA has been reported in *Pandorina morum* (Moore and Coleman, 1989) which, like *Chlamydomonas* sp., is a member of the Volvocales class of green algae. Somewhat surprisingly, new data indicate that some other *Chlamydomonas* species have mtDNA patterns that deviate substantially from the *C. reinhardtii* pattern; e.g., *C. eugametos* and *C. moewusii* (which are interfertile with each other but not with the *C. reinhardtii/C. smithii* pair) have circular rather than linear mtDNAs. These mitochondrial genomes are colinear but are highly rearranged compared with *C. reinhardtii* mtDNA (Lee *et al.*, 1991; Denovan-Wright and Lee, 1992). The *C. eugametos* mitochondrial genome contains a number of introns that appear to account for its increased size (24 kbp) relative to *C. reinhardtii* mtDNA; gene content, on the other hand, appears to be the same in the two mtDNAs (Denovan-Wright and Lee, 1992, and unpublished observations). These differences may reflect deep evolutionary divergences within the genus "Chlamydomonas," whose phylogenetic coherence has recently been questioned (Buchheim *et al.*, 1990).

Although data on mtDNA size and organization in other classes of green algae are sparse, the available information suggests that the patterns are very different than those seen so far in the volvocacean algae. Within the Chlorococcales, a 76-kbp mtDNA has been described in an exosymbiotic *Chlorella*-like alga (Waddle *et al.*, 1990); this genome appears to contain several genes (*cox2, atpA, atp6, atp9*) absent from *C. reinhardtii* mtDNA. Mitochondrial rRNA genes have been characterized in *Prototheca wick-*

erhamii (a nonphotosynthetic relative of *Chlorella*) (Wolff and Kück, 1990) and in *Scenedesmus obliquus* (Kück *et al.,* 1990); it appears that these genes encode typical, continuous rRNAs rather than the small, unspliced rRNA pieces found in *C. reinhardtii* mitochondria. These structural differences in mtDNA are striking in view of the close phylogenetic affiliation of the volvocacean and chlorococcacean algae at the level of the nuclear genome (see, e.g., Fig. 12).

The above survey, although admittedly incomplete, is intended not only to give some indication of the nature and extent of mitochondrial genome structural diversity, but also to emphasize that the various patterns described to date differ markedly from patterns that typify both prokaryotic and eukaryotic nuclear genomes. Whereas eubacteria-like gene clusters are readily identifiable in plDNA, and provide a strong evolutionary link between plastids and eubacteria (Fig. 10), eubacteria-like operons are virtually absent from sequenced mtDNAs. Ribosomal RNA gene arrangement provides a particularly good illustration of this point. In eubacteria such as *E. coli* and *Anacystis nidulans,* rRNA and some tRNA genes are physically and transcriptionally linked, in the order (5′) SSU rRNA–spacer–tRNA^{Ile}-spacer–tRNA^{Ala}–spacer–LSU rRNA–spacer–5S rRNA (3′), and an identical arrangement (but see Yamada and Shimaji, 1987) is found in almost all plDNAs, although sometimes with introns in the tRNA genes and introns/additional spacers in the LSU rRNA gene (Palmer, 1985a,b, 1991, 1992; Turmel *et al.,* 1991) (see Fig. 10). In contrast, a variety of patterns, all of them unique, have been observed in mitochondria. In vertebrate animals as well as in *Drosophila,* mitochondrial SSU and LSU rRNA genes are separated only by a tRNA^{Val} gene, and in vertebrates the three genes are cotranscribed from a single promoter in the order (5′) SSU rRNA–tRNA^{Val}–LSU rRNA (3′). However, in sea urchin and nematode mtDNA, SSU and LSU rRNA genes are separated by a number of protein-coding and tRNA genes. In the fungus *S. cerevisiae,* SSU and LSU rRNA genes are widely separated in the mitochondrial genome and transcribed from independent promoters; moreover, the LSU rRNA gene contains an intervening sequence, so that pre-rRNA processing involves intron excision and exon splicing. In plant mitochondria, SSU and 5S rRNA genes are closely linked and cotranscribed, but the pair is physically and transcriptionally separated from the LSU rRNA gene (Bonen and Gray, 1980; Maloney and Walbot, 1990), even though these genes, like their plastid counterparts, are obviously eubacterial in primary sequence (see V,B,5,d). In trypanosomatid mitochondria, rRNA genes are adjacent, but are transcribed in the order LSU to SSU. Finally, in *T. pyriformis* and *C. reinhardtii* mtDNA, rRNA genes are split into separate coding modules that are rearranged in the genome, as summarized above. The upshot, then, it that patterns of gene organization in mtDNA provide little or no basis for discriminating between opposing autogenous and xenogenous hypotheses of mitochondrial origin: a marked contrast to what organizational patterns in plDNA tell us about the origin of plastids. To date, only one clear structural similarity between mitochondrial and prokaryotic genomes has been found: in liverwort mtDNA, 10 ribosomal protein genes are clustered together and arrayed in the same order as their counterparts in the *E. coli* [S10]-[*spc*]-[α] operons (Oda *et al.*, 1992).

3. Gene Expression: Transcription and Post-transcriptional Processing

Given the structural variability discussed above, it is not particularly surprising that diverse modes and mechanisms of gene expression have been found to operate in different mitochondrial systems. With respect to the biochemistry of mitochondrial transcription, the most detailed work has been carried out in animal (particularly human, mouse, and *Xenopus*) and fungal (particularly *S. cerevisiae* and *Neurospora crassa*) systems.

a. Animal Mitochondria   For references see Clayton, 1984, 1991; and Chang *et al.*, 1987. In vertebrate animals transcription control signals have been localized to the 1- to 2-kbp D-loop region (DLR), the only extensive noncoding stretch in vertebrate mtDNA. In mammals, transcription of mtDNA is initiated from a single major promoter on each of the two complementary (H and L) strands, with each strand being completely transcribed. Long cotranscripts are processed by discrete endonucleolytic cleavages, and the location of tRNA genes between rRNA and protein-coding genes suggests a possible role for tRNA sequences in precursor RNA processing. It appears, however, that such a mechanism cannot operate in the sea urchin mitochondrial system where tRNA genes are highly clustered and many protein-coding genes directly abut one another (Jacobs *et al.*, 1988; Cantatore *et al.*, 1989). In mammals, there is evidence (Christianson and Clayton, 1986, 1988) that transcription attenuation, mediated by a site-specific DNA-binding factor (Kruse *et al.*, 1989), operates just downstream of the mitochondrial LSU rRNA gene, allowing enhanced production of mature rRNA species relative to mature mRNAs. Newly processed mRNA transcripts usually lack any 5'-untranslated leader, with the AUG initiation codon located directly at the 5'-terminus (Attardi, 1985). Also absent are complete termination codons, with many nascent mRNAs ending in -U or -UA; 3'-polyadenylation creates a UAA termination codon in these cases (Attardi, 1985).

Intensive biochemical characterization, utilizing *in vitro* extracts that faithfully and efficiently transcribe mtDNA templates, has made possible a precise delineation of the nature of promoters and other regulatory sequences in the DLRs of human (Chang and Clayton, 1984; Hixson and Clayton, 1985; Topper and Clayton, 1989), mouse (Chang and Clayton, 1986a,b), and *X. laevis* (Bogenhagen *et al.*, 1986; Bogenhagen and Yoza, 1986) mtDNA, as well as the identification of individual components of the transcriptional machinery (Fisher and Clayton, 1985, 1988). In human and mouse mtDNA, the L- and H-strand promoters (LSP and HSP, respectively) are situated about 150 bp apart and are functionally non-overlapping, whereas in *X.laevis* mtDNA, overlapping HSP and LSP pairs are situated at two different sites (Bogenhagen and Yoza, 1986). In all three cases, transcription is initiated within each promoter element. Within each species, there is considerable sequence similarity between LSP and HSP; however, between species there is no obvious similarity in the promoter consensus sequences, nor are these obviously related to identified fungal mitochondrial or eubacterial promoter elements (Schinkel and Tabak, 1989).

At least two, single-polypeptide protein factors interact to direct transcription in mammalian mitochondria: a relatively nonspecific, core RNA polymerase (mtRNAP) and a specificity factor (transcriptional activator, mtTF1). The latter is a 25-kDa protein able to unwind and bend DNA (Fisher *et al.*, 1987; 1992), and capable of sequestering promoter-containing DNA templates in preinitiation complexes. The mtTF1 interacts with a stretch of sequence immediately upstream of the functionally defined basal promoter element, and in combination with mtRNAP is essential for accurate and efficient transcription initiation. Surprisingly, although the mtTF1 binding sites are very different in sequence in mouse and human mtDNAs, mtTF1 is rather flexible in its recognition of the rapidly evolving transcriptional control elements in these two species: thus, mouse mtTF1 is able to interact productively with human mtRNAP to initiate transcription from a human mtDNA template, and *vice versa* (Fisher *et al.*, 1989). These results suggest that the mtRNAP itself, or some as-yet-unrecognized component of partially purified mtRNAP preparations, must be responsible for the strict species specificity shown by transcription of mammalian mtDNA (such that mouse mitochondrial extracts will not transcribe human mtDNA, and *vice versa*). The mtTF1 has been purified to homogeneity from human mitochondria (Fisher and Clayton, 1988; Fisher *et al.*, 1991), and the nuclear gene encoding this protein has recently been cloned and sequenced (Parisi and Clayton, 1991), revealing that mtTF1 shares sequence similarity with nuclear high-mobility-group (HMG) proteins and with a pol I (nuclear rDNA) transcription factor.

To date, few studies have examined mitochondrial transcription and post-transcriptional processing in nonvertebrate animals. In sea urchin (*S. purpuratus*), it appears that precursor mitochondrial transcripts containing butt-joined coding sequences are processed *via* precise endonucleolytic cleavages (Elliott and Jacobs, 1989), as in vertebrate mitochondria. However, nothing is known yet about number, location, and structure of promoter elements in sea urchin or other invertebrate mtDNAs.

So far, only one animal mitochondrial RNA processing activity (meeting the criteria of copurification with mitochondria and ability to faithfully recognize and/or process *bona fide* mitochondrial sequences) has been described. RNase MRP (for *M*itochondrial *R*NA *P*rocessing), a site-specific endoribonuclease (Chang and Clayton, 1987a; Bennett and Clayton, 1990), has been identified in mouse (Chang and Clayton, 1987a) and human (Gold *et al.,* 1989; Topper and Clayton, 1990) mitochondria and has been characterized as a ribonucleoprotein consisting of nucleus-encoded protein and RNA components, both of which must be imported into mitochondria (Chang and Clayton, 1987b, 1989; Topper and Clayton, 1990). RNase MRP cleaves nascent L-strand transcripts between two closely spaced *C*onserved *S*equences *B*oxes, CSB II and CSB III (located downstream of LSP), thereby mediating a switch from RNA transcription to DNA replication. A mitochondrial RNase P activity (the enzyme responsible for creating the mature 5'-end of tRNAs) has been isolated from vertebrate mitochondria (Doersen *et al.,* 1985; Manam and Van Tuyle, 1987), but has not yet been shown to be active on the homologous (vertebrate mitochondrial) tRNA-containing transcripts.

b. Fungal Mitochondria   For references see Tabak *et al.,* 1983; Tzagoloff and Myers, 1986; and Schinkel and Tabak, 1989. In contrast to animal mtDNA, multiple transcription units served by separate promoters are present in many of the fungal mtDNAs that have been studied in detail. In the ascomycete yeasts *S. cerevisiae, Kluyveromyces lactis,* and *Torulopsis glabrata,* a highly conserved, nonanucleotide consensus sequence has been identified immediately upstream of proven or putative transcription initiation sites (Osinga *et al.,* 1982, 1984; Christianson and Rabinowitz, 1983; Clark-Walker *et al.,* 1985a). More recently, a less-well-conserved promoter motif was identified in the mtDNA of the filamentous fungus *Neurospora crassa* (Kennell and Lambowitz, 1989; Kubelik *et al.,* 1990). Little sequence similarity is evident between the *N. crassa* and the *S. cerevisiae* consensus blocs.

In yeast as in animal mitochondria, the transcriptional machinery seems to comprise a core, nonselective mtRNAP and a separate specificity factor, both of which are required for accurate initiation at mitochondrial promoters (Schinkel and Tabak, 1989). However, the yeast specificity factor by itself, in contrast to mammalian mtTF1, does not appear to interact with DNA (Schinkel and Tabak, 1989), a feature it apparently shares with the *X. laevis* specificity factor (Bogenhagen and Romanelli, 1988). In *S.*

cerevisiae, the specificity factor has been characterized as a 43-kDa protein (Schinkel *et al.,* 1987), the product of the *MTF1* gene (Lisowsky and Michaelis, 1988; Jang and Jaehning, 1991). The combination of mtRNA polymerase and MTF1 protein is necessary and sufficient for accurate transcription from yeast mitochondrial promoters (Xu and Clayton, 1992). Interestingly, a procedure for preparing highly purified human mtTF1 yields, when applied to yeast mitochondria, a protein having properties (including specific DNA-binding characteristics), very similar to those of the 25-kDa mammalian mtTF1, but with a molecular mass of only 19 kDa (Fisher *et al.,* 1991). This protein (ABF2) displays structural homology to human mtTF1 (Diffley and Stillman, 1991), but does not enable yeast mtRNA polymerase to recognize a yeast mtDNA promoter (Xu and Clayton, 1992).

In our 1982 review, Doolittle and I noted that mitochondrial RNA polymerases "may all contain but a single polypeptide of modest molecular weight" [e.g., the 145-kDa core polypeptide of the *S. cerevisiae* mtRNAP (Schinkel *et al.,* 1988)], and that these enzymes "resemble the RNA polymerases of phages T3 and T7 . . . much more than they do the high-molecular-weight, multisubunit RNA polymerases of either eubacteria or nuclei, a fact that must distress adherents of all reasonable evolutionary hypotheses" (Gray and Doolittle, 1982). Subsequent determination of the nucleotide sequence of the genetic locus for yeast mtRNAP (the nuclear gene *RPO41* (Greenleaf *et al.,* 1986)) revealed a 1351-codon ORF (predicting a 153-kDa polypeptide) sharing evident similarity with the DNA-directed RNA polymerases of bacteriophages T3 and T7, especially in those regions most highly conserved between the T3 and the T7 enzymes themselves (Masters *et al.,* 1987). The evolutionary significance of this intriguing finding is discussed below.

The expression of yeast mitochondrial genes is regulated primarily at a post-transcriptional level (Conrad-Webb *et al.,* 1990), and processing of mitochondrial transcripts is in general more complex in fungi than in animals, partly reflecting the excision/splicing events that occur during maturation of the transcripts of intron-containing genes in fungal mitochondria (Tzagoloff and Myers, 1986). This processing requires an array of *trans*-acting protein factors, including mtDNA-encoded maturases (Lazowska *et al.,* 1980) and nucleus-encoded proteins; certain of the latter have been identified as mitochondrial tRNA synthetases (Lambowitz and Perlman, 1990).

Even among the few fungal mitochondrial systems that have been analyzed in detail, there appear to be substantial differences in transcriptional patterns. For example, in *S. cerevisiae* mitochondria about 20 primary transcripts have been identified, each encoding one or rarely two proteins and a variable number of tRNAs (Tzagoloff and Myers, 1986). In contrast, mapping results suggest that transcription of the *Aspergillus nidulans* mitochondrial genome proceeds via a very limited number of primary transcripts, with mature RNAs being produced by extensive processing events, including tRNA excision (Dyson *et al.*, 1989). The latter situation resembles the transcription and processing events that take place in metazoan mitochondria. As in vertebrate mitochondria, tRNA sequences have been suggested as possible post-transcriptional processing signals in both *Aspergillus* (Dyson *et al.*, 1989) and *Neurospora* (Breitenberger *et al.*, 1985; Burger *et al.*, 1985) mitochondria.

Purification and characterization of individual fungal mitochondrial RNA processing activities has largely been limited to *S. cerevisiae*, and these studies have focused on tRNA processing enzymes. An RNase P has been isolated from yeast mitochondria and has been shown to be a ribonucleoprotein containing essential RNA and protein components (Hollingsworth and Martin, 1986; Morales *et al.*, 1989), as in the RNase P enzymes from both eubacteria and eukaryotes (Hanic-Joyce and Gray, 1990). The RNA component of the yeast mitochondrial enzyme is encoded in mtDNA (the *t*RNA *s*ynthesis *l*ocus, *tsl*) (Miller and Martin, 1983), whereas the protein component is nDNA-encoded (Hollingsworth and Martin, 1986).

By themselves, eubacterial RNase P RNAs are catalytic (Guerrier-Takada *et al.*, 1983) and display limited but compelling primary and secondary structural similarities (James *et al.*, 1988; Brown and Pace, 1992). In contrast, the RNA components of eukaryotic nuclear and mitochondrial RNase P enzymes are not catalytic under any of the *in vitro* conditions tested to date (Forster and Altman, 1990). Nevertheless, very similar core secondary structures have been proposed for the RNA components of all RNase P enzymes (Forster and Altman, 1990). Because yeast mitochondrial RNase P RNA is exceedingly AU-rich (about 87%) (Miller and Martin, 1983), it is not possible to tell whether this RNA is more closely related to its eubacterial or its eukaryotic homologs. Intriguingly, the RNase P RNA core secondary structure can also be assumed by the RNase MRP RNA (discussed above), implying that the RNA components of all RNase P and RNase MRP enzymes share a common ancestor and function. This is a particularly provocative suggestion in view of the demonstration that human RNase P and RNase MRP share common antigenic determinants (Gold *et al.*, 1989).

c. Plant Mitochondria   For references see Mulligan and Walbot, 1986; Newton, 1988; Lonsdale, 1989; Levings and Brown, 1989; Gray, 1990; and Gray *et al.*, 1992. Compared to what is known about transcription and post-transcriptional processing in animal and fungal mitochondria, our understanding of these processes in plant mitochondria is rather limited.

Gene mapping studies suggest a fairly even dispersion of genes throughout the master circle physical maps that have been constructed for a number of plant mtDNAs (Dawson *et al.*, 1986; Stern and Palmer, 1986; Makaroff and Palmer, 1987; Brears and Lonsdale, 1988; Fauron and Havlik, 1988; Fauron *et al.*, 1989; Conklin *et al.*, 1991). In keeping with the spaciousness of the plant mitochondrial genome, there is little evidence of close physical or transcriptional linkage of genes, although there are a few documented cases of tightly clustered genes [e.g., the tRNA^{fMet}–(1 bp)–18S rRNA–(114 bp)–5S rRNA gene trio in wheat mtDNA (Gray and Spencer, 1983)] and of cotranscription (Bland *et al.*, 1986; Rottmann *et al.*, 1987; Gualberto *et al.*, 1988; Wissinger *et al.*, 1988; Rasmussen and Hanson, 1989; Maloney and Walbot, 1990; Quagliarello *et al.*, 1990). How much of the (largely noncoding) plant mitochondrial genome is transcribed is not yet clear: there is some indication of extensive transcription of noncoding regions, although transcripts from such regions do not accumulate (Finnegan and Brown, 1990).

Transcriptional regulatory sequences in plant mtDNA have remained elusive, primarily because the low degree of structural conservation (a consequence of the rapid evolutionary rearrangement of plant mitochondrial genomes) makes a search for such sequences difficult. In comparing homologous genes in different plant mtDNAs, it is commonly found that primary sequence diverges abruptly immediately upstream and downstream of otherwise highly conserved coding regions. Thus, sequence comparisons have been of little value in uncovering conserved sequence motifs (potential regulatory elements) in the regions flanking identified genes. Recently, through analysis of the 5′-ends of primary transcripts, it has been possible to identify multiple transcription initiation sites in the mtDNAs of maize (Mulligan *et al.*, 1988a,b, 1991) and wheat (Covello and Gray, 1991). Such studies have defined a loosely conserved sequence motif immediately upstream of these sites; this consensus bears some similarity to the yeast mitochondrial promoter consensus, but it has yet to be established whether the plant mitochondrial motifs actually function as promoter elements (but see Rapp and Stern, 1992). The data, though limited, clearly point to important differences in the location and structure of putative promoters in the mtDNAs of wheat and maize (both monocotyledons and members of the same family), and even more substantial differences between monocotyledon and dicotyledon (Brown *et al.*, 1991) transcription initiation sites. The recent development of an *in vitro* transcription system from plant mitochondria should permit the precise functional delineation of promoters in plant mtDNA (Hanic-Joyce and Gray, 1991). Putative transcription termination signals (potential secondary structures reminiscent of bacterial terminators) have been identified immediately upstream of the 3′ termini of some plant mitochondrial mRNAs (Schuster *et al.,* 1986); however, it still has to be directly demonstrated that such sequences actually function in transcription termination and/or processing.

Extensive transcription of plant mtDNA, if it indeed occurs, implies extensive posttranscriptional processing involving specific endo- and exonucleases. Activities capable of processing tRNA-containing precursors to the mature species have been described in mitochondrial extracts from wheat (Hanic-Joyce and Gray, 1990; Hanic-Joyce *et al.,* 1990) and *Oenothera* (Marchfelder *et al.,* 1990). In the former case, it was shown that processed tRNAs contain a 3' -CCA terminus that is not encoded in the mtDNA and therefore must be added by a mitochondrial nucleotidyltransferase (Hanic-Joyce and Gray, 1990).

Conventional *cis*-splicing of intron-containing pre-mRNAs occurs in plant mitochondria, but more recently a novel form of intron excision/ exon splicing, involving a *trans* mechanism, was discovered in the mitochondria of three different plants: wheat (Chapdelaine and Bonen, 1991), *Oenothera* (Wissinger *et al.,* 1991), and *Petunia* (Conklin *et al.,* 1991). In these three cases, the mitochondrial *nad1* gene is fragmented into four or five coding segments that are scattered throughout the genome. The *nad1* exons in these segments are flanked by sequences having feature of group II introns, and formation of a functional, continuous *nad1* mRNA is postulated to involve both *cis*-and *trans*-splicing events. This is the first such example of *trans*-splicing in mitochondria, and in proposed mechanism it closely resembles the *trans*-splicing previously reported in chloroplasts (see Kück *et al.,* 1987).

Post-transcriptional processing in plant mitochondria recently acquired a new dimension with the discovery of an RNA editing system that introduces numerous C-to-U changes in the nascent mRNAs of most protein-coding genes (Covello and Gray, 1989; Gualberto *et al.,* 1989; Hiesel *et al.,* 1989; Mulligan, 1991; Walbot, 1991; Gray *et al.,* 1992), thereby altering the amino acid sequence of the encoded protein relative to that predicted by the gene sequence. In some cases, a new termination codon is also generated (Bégu *et al.,* 1990; Wintz and Hanson, 1991). Because most of the internal C-to-U edits occur at first or second positions of codons, most result in an amino acid replacement. The extent of editing can be substantial, with between 4.6 and 6.5% of amino acids changed by editing in the *cox2* mRNAs of wheat, maize, and pea (Covello and Gray, 1990a). Based on amino acid sequence comparisons between homologous plant and nonplant proteins, it would appear that most if not all of the observed editing is directed toward producing functional proteins (Covello and Gray, 1990b). The discovery of RNA editing in plant mitochondria eliminated what had appeared to be a deviation from the universal genetic code. It is now evident that in this system, CGG does not code for Trp, as originally proposed (Fox and Leaver, 1981), but instead is edited to UGG (the normal Trp codon) at selected positions in plant mitochondrial mRNAs. To date, plants are the only multicellular eukaryotes in which mitochondrial RNA editing has been shown to take place, although it apparently does not occur in bryophyte (liverwort) mitochondria (Oda *et al.*, 1992).

d. Protistan Mitochondria Although transcripts have been mapped onto various protistan mtDNAs (Simpson *et al.*, 1982; Hoeijmakers *et al.*, 1982; Stuart and Gelvin, 1982; Pritchard *et al.*, 1990a), the details of mitochondrial transcription and post-transcriptional processing have been little studied in this group of eukaryotes. There is evidence of extensive cotranscription of closely linked genes in *C. reinhardtii* mitochondria, with long precursor RNAs being processed by discrete endonucleolytic scissions to yield mRNAs having little or no 5'-untranslated leader (Boer and Gray, 1986, 1988b; Gray and Boer, 1988; P. H. Boer and M. W. Gray, manuscripts in preparation). This pattern is reminiscent of post-transcriptional processing in animal mitochondria, except that *C. reinhardtii* mitochondrial transcripts possess termination codons and 3'-untranslated trailers that are encoded by the mtDNA.

The phenomenon of RNA editing in protistan mitochondria has received considerable attention recently, with documentation of U insertion/deletion events in trypanosomatid mitochondria (Benne, 1990; Simpson, 1990; Feagin, 1990; Stuart, 1991), and C insertion events in the mitochondria of the slime mold *Physarum polycephalum* (Mahendran *et al.*, 1991). In *T. brucei*, the transcripts of several maxicircle genes undergo very extensive editing, such that upwards of 50% of the final coding sequence is provided by U residues added in the course of the editing process (Feagin *et al*, 1988; Bhat *et al.*, 1990; Koslowsky *et al.*, 1990). Although there is still considerable debate about the mechanism of RNA editing in trypanosome mitochondria (Decker and Sollner-Webb, 1990; Weiner and Maizels, 1990; Cech, 1991), the involvement of maxicircle- and minicircle-encoded guide RNAs (see V,B,2,d) in determining the specificity of editing seems beyond doubt, and enzymatic activities that may be involved in editing have been identified (Bakalara *et al.*, 1989). The discovery of RNA editing raises a major evolutionary question, namely whether this is a primitive relic of the postulated RNA world or a newly independently derived trait within different mitochondrial lineages. At this point, the RNA editing systems in trypanosome, slime mold, and angiosperm mitochondria seem unrelated; furthermore, the apparent absence of RNA editing in liverwort mitochondria (Oda *et al.*, 1992) would seem to favor the view that at least the C-to-U type of editing is a trait recently acquired by angiosperm mitochondria.

As in the case of gene structure, there is little in the various mitochondrial gene expression patterns summarized above to link mitochondrial genomes with those of either prokaryotes or the eukaryotic nucleus. In contrast to plastids, mitochondria do not appear to use eubacteria-type promoters, and there is little indication that the transcriptional apparatus in mitochondria closely parallels that in eubacteria. There is also not much to suggest that mitochondrial transcriptional mechanisms in different eukaryotes have a common evolutionary origin, although some tentative biochemical similarities do seem to be emerging (Schinkel and Tabak, 1989; Clayton, 1991).

The demonstration of obvious sequence similarity between the yeast mitochondrial and the bacteriophage T3/T7 RNA polymerases (Masters *et al.*, 1987) has some tantalizing evolutionary implications. I have previously suggested (Gray, 1989b) that this homology "raises the possibility that some genetic information for mitochondrial biogenesis and function has been acquired not from a eubacteria-like endosymbiont, but from phage-like entities that accompanied either an original or subsequent endosymbiont": a possibility independently considered by Schinkel and Tabak (1989). In this regard, it may be significant that linear plasmid-like DNAs have been identified in both fungal and plant mitochondria (Esser *et al.*, 1986; Samac and Leong, 1989). Of particular interest are the RNA polymerase-like ORFs encoded by the S-2 plasmid of maize mitochondria (Kuzmin *et al.*, 1988) and the *kalilo* plasmid of *Neurospora intermedia* mitochondria (Chan *et al.*, 1991), which share sequence similarity with one another and with the yeast mitochondrial and T-odd phage RNA polymerases.

4. Gene Expression: Translation and the Genetic Code

As noted earlier, certain features of organellar protein synthesis provided some of the earliest indications that mitochondria and plastids possess a bacteria-like translation apparatus. These similarities are much more striking for plastids than for mitochondria, and we acknowledged this previously (Gray and Doolittle, 1982) by commenting, "In their attack on [the endosymbiont] hypothesis, Raff and Mahler [1975] pointed out, quite correctly we think, that these [presumed structural and functional] homologies [between mitochondrial and eubacterial ribosomes] are not compelling." We went on to discuss the many ways in which mitochondria and (eu)bacterial translation systems differ, concluding that the two major aspects of similarity are the use of N-formylmethionyl-tRNA^{Met} as initiator and sensitivity to a number of the same antibiotics (e.g., chloramphenicol).

Ten years later, the situation is largely unchanged. There are a few additional indications of the prokaryotic nature of the mitochondrial translation system [e.g., further characterization of eubacterial-type initiation

(Denslow *et al.*, 1988) and elongation (Schwartzbach and Spremulli, 1989) factors]; however, we also have many additional data that emphasize the differences between eubacterial and mitochondrial translation systems. Mitochondrial ribosomes show a structural diversity as great as the organizational diversity of the mitochondrial genes encoding their component RNAs (see Curgy, 1985; Benne and Sloof, 1987); homologous mitochondrial rRNAs vary markedly in length and base composition (Gray, 1988); and mitochondrial tRNAs run the gamut from the perfectly normal to the exceedingly odd (Dirheimer and Martin, 1990). Mitochondrial protein synthesis has many unique characteristics that are not phylogenetically informative, and that vary with the particular mitochondrial system being considered. Thus, only limited support for a eubacterial, endosymbiotic origin of mitochondria is provided by comparative analysis of organismal and organellar translation systems (Benne and Sloof, 1987).

Limited though these similarities may be, they are nevertheless significant. In particular, the combination of initiator tRNA type and antibiotic sensitivity does discriminate among the translation systems in eubacteria, archaebacteria, and the eukaryotic cytosol (archaebacteria and eukaryotes using an unformylated Met-tRNA^{Met} as initiator, and protein synthesis in the eukaryotic cytosol being inhibited by cycloheximide but not chloramphenicol) (Woese, 1981; Hilpert *et al.*, 1981; Amils *et al.*, 1989). On this basis, mitochondria fit squarely in the eubacterial camp. Many of the determinants of these functional differences among the various translation systems are now known to reside in the constituent rRNA molecules (Dahlberg, 1989; Cundliffe, 1990), and as such they represent a set of fundamental, lineage-specific structural characters that can be exploited in phylogenetic analysis. This point is pursued in V,B,5.

Mitochondria provided the first examples of deviations from the standard genetic code, with codon reassignments of various sorts and use of UGA to specify Trp rather than termination (Fox 1987; Jukes, 1990; Jukes and Osawa, 1990; Bessho *et al.*, 1992). The latter variation is widespread among mitochondria, with only those of plants (Schuster and Brennicke, 1985) and *Chlamydomonas reinhardtii* (Boer and Gray, 1988c) using UGA as a termination codon. Animal and fungal mitochondria exhibit expanded codon recognition patterns, whereby a single tRNA is able to read all codons in a four-codon family specifying a single amino acid (Barrell *et al.*, 1980; Bonitz *et al.*, 1980; Heckman *et al.*, 1980). In this regard, and in the use of UGA to specify Trp, the translation systems of animal and fungal mitochondria are remarkably similar to that of the eubacterium *Mycoplasma capricolum* (Andachi *et al.*, 1989; Muto *et al.*, 1990). Because mycoplasms inhabit eukaryotic cells, these similarities have prompted the speculation that "mycoplasma-like organisms may have been the actual eubacterial ancestors of mitochondria" (Andachi *et al.*, 1989). However, an origin of mitochondria from within this particular eubacterial group is not supported by rRNA sequence comparisons (see V,B,4 and V,C), and it is much more likely that genetic code variations in mitochondria (and elsewhere) are derived, not primitive, characteristics. Archaebacteria, most eubacteria, most eukaryotes, plastids, and at least plant mitochondria all use the universal code, which argues strongly that this code was already extant in the last common ancestor of all these lineages, and that the aberrations we see today have been independently and secondarily derived. Indeed, Lang *et al.* (1987) contend that "variation in the mitochondrial genetic code is so great as to be useless as a taxonomic marker." Evolutionary pressures and mechanisms that may account for observed differences in the genetic code have been discussed by Jukes and Osawa (1990) and Andersson and Kurland (1991).

5. Gene Structure

There are several limitations in the use of gene structure data to establish phylogenetic relationships within the mitochondrial lineage and between that lineage and others. First, the low and somewhat variable gene content of mtDNA (Tables II and III) means that we have very many fewer candidate sequences on which to base such analyses than we do in the case of plastids. As noted earlier in this review, only two protein-coding genes (*coxI* and *cob*) and two rRNA genes (SSU and LSU) have proven to be universally present in characterized mtDNAs. Second, there are striking differences (in the extreme, several orders of magnitude) in the rates at which homologous genes diverge in sequence in different mitochondrial genomes. In fact, the mitochondrial lineage contains both the most rapidly evolving (mammalian) and the most slowly evolving (angiosperm) genomes known. Although these rate differences are particularly pronounced at the level of nucleotide sequence, they are also evident when comparisons are made at the amino acid level. Both of these factors considerably constrain the construction and interpretation of global, sequence-based phylogenetic trees. Despite this, gene structure data have provided our best insights so far into the origin of mitochondria.

a. Respiratory Chain Proteins   Fifteen years ago, John and Whatley (1975a,b) pointed out that *Paracoccus denitrificans,* a member of the α-subdivision of the nonsulfur purple eubacteria (Woese *et al.,* 1984), "resembles a mitochondrion more closely than do other bacteria, in that it effectively assembles in a single organism those features of the mitochondrial respiratory chain and oxidative phosphorylation which are otherwise distributed at random among most other aerobic bacteria." Since that time, genes for individual protein components of the *Paracoccus* respiratory chain have been cloned and sequenced, providing strong support for the thesis that *P. dentrificans* and mitochondria are specific evolutionary relatives. Initially, a two-subunit (45 and 28 kDa) cytochrome *c* oxidase was isolated from *P. dentrificans* (Ludwig and Schatz, 1980), and direct protein sequencing (Steffens *et al.*, 1983) followed by cloning and gene sequencing (Raitio *et al.*, 1987) demonstrated a striking degree of similarity between these eubacterial proteins and the two largest subunits of mito-chondrial cytochrome oxidase (COX1 and COX2), which are endoced by mtDNA (Table II). These studies revealed an additional *Paracoccus* gene with strong similarity to mitochondrial *cox3* (Saraste *et al.*, 1986; Raitio *et al.*, 1987), suggesting loss of a COX3 subunit during isolation of the *P. dentrificans* cytochrome oxidase complex. *Paracoccus denitrificans* therefore contains evident homologs of the three mtDNA-encoded cyto-chrome oxidase subunits. Strong functional and structural similarities be-tween other mitochondrial and purple bacterial respiratory components, including NADH dehydrogenase (Complex I) (Pilkington *et al.*, 1991a), the cytochrome b/c_1 complex (Complex III) (Gabellini *et al.*, 1985) and ATP synthase (F_1F_0 ATPase, Complex V) (Walker *et al.*, 1982a,b), have also been reported. Relatively few studies of the respiratory chain in aerobic archaebacteria have been reported yet, but these already suggest substantial differences between eubacterial and archaebacterial respira-tory complexes (Anemüller and Schäfer, 1989; Hildebrant *et al.*, 1991; Denda *et al.*, 1991; Lübben *et al.*, 1992).

Proteins of the mitochondrial respiratory complexes cannot provide the basis of a rigorous test of autogenous vs xenogenous mitochondrial origin hypotheses because electron transport and oxidative phosphorylation (un-like protein synthesis) take place only in the mitochondrion, and the functional genes for respiratory proteins are present in either the mitochon-drial genome or the nuclear genome, but not both. Mitochondrial *cox1* and *cob* sequences are now sufficiently abundant that phylogenetic trees based on these protein-coding genes could in principle be used to distinguish between a eubacterial and an archaebacterial origin of mitochondria, ex-cept that at the moment very few of the required homologous sequences from aerobic prokaryotes have been determined (but see Shapleigh and Gennis, 1992, and references therein). Comparison of *cox1* and/or *cob* nucleotide or amino acid sequences could also be used to investigate evolutionary relationships among different mitochondria; however, no systematic investigations along these lines have been reported. Phylo-genetic trees based on *atp9* (nucleotide) sequence comparisons have been reported (Recipon *et al.*, 1992); these show all mitochondrial/nuclear *atp9* sequences clustering specifically with that of the α-purple bacterium *Rho-dospirillum rubrum*.

Although sequence comparisons at the amino acid rather than nucleotide level might be expected to diminish the effects of such complicating factors as RNA editing, codon usage, base composition bias, and variable (and in some cases quite high) rates of mutation fixation at third positions of codons, differences in apparent rates of mitochondrial protein sequence evolution still persist. For example, although trypanosomatid mitochondrial protein sequences are less similar to the homologous yeast or human sequences than the latter are to each other (de la Cruz *et al.*, 1984), ciliate mtDNA-encoded protein sequences are even more divergent (Pritchard *et al.*, 1990b), despite the fact that ciliates appear to have diverged from other eukaryotes considerably more recently than trypanosomes (see Figs. 6 and 12).

Other potentially useful, phylogenetically informative proteins include nucleus-encoded ones that interact with mitochondrion-encoded proteins to form the various respiratory complexes. Implicit in the endosymbiont hypothesis is the assumption that (for the most part) the genes for these components were transferred from the protomitochondrial to the nuclear genome. The varied location of certain genes encoding mitochondrial proteins lends credence to this idea (e.g., the gene for the α subunit of the F_1 ATPase is in the mtDNA in some eukaryotes but in the nDNA in others; see Table II). If the foregoing assumption is correct, we would anticipate that at least some of these transferred genes would have retained vestiges of their evolutionary ancestry, even though their change in genomic venue may well have subjected them to different evolutionary pressures than those experienced by genes remaining in the mitochondrial genome. Support for this postulate of gene relocation comes from the fact that mitochondrial cytochrome c's are not only eubacterial in character (Schwartz and Dayhoff, 1978), but they share greater structural and functional similarities with their *Paracoccus* homolog (cytochrome c_{550}) than they do with other bacterial c-type cytochromes (Timkovich and Dickerson, 1976); thus, evolution of this **nDNA-encoded** constituent of the mitochondrial respiratory chain closely parallels evolution of its **mtDNA-encoded** partner proteins (e.g., COX1, COX2, and COX3). The wide availability of cytochrome c sequences made this protein an early choice for phylogenetic analyses, which supported an origin of the mitochondrion from within the purple photosynthetic bacteria (Schwartz and Dayhoff, 1978; Dayhoff and Schwartz, 1981). An interesting feature of these cytochrome c trees is that the branching topology appears to follow closely that expected on the basis of sequence comparisons of mtDNA-encoded respiratory proteins: i.e., *Tetrahymena* (ciliate) cytochrome c is the earliest diverging one (Dayhoff and Schwartz, 1981). Given the numerous additional respiratory protein sequences that have accumulated since these early trees appeared, and the subsequent innovations in phylogenetic tree construction, it would be instructive to determine parallel trees based on representative nDNA-encoded (cytochrome c) and mtDNA-encoded (e.g., COX1) proteins. Assessment of the degree of congruence between such trees, and agreement with trees based on different databases (e.g., mitochondrial rRNA sequences; see V,C), could provide some important insights into the gene transfer events that we assume have shaped mitochondrial evolution (see also V,D).

b. Translation Apparatus Proteins   The current database of ribosomal protein sequences is not extensive enough to permit the definitive phylogenetic placement of mtDNA-encoded ribosomal protein genes (found mostly in land plants; Table III). Comprehensive comparisons require that there be at least one sequenced homolog from all three primary lineages, in addition to the organellar one, a condition that so far has not been met for any ribosomal protein. Homologous ribosomal proteins encoded by mtDNA and plDNA are about equally divergent from their *E. coli* homologs and from each other (Wahleithner and Wolstenholme, 1988), indicating that (i) the mitochondrial versions are eubacterial in origin and (ii) the mitochondrial and plastid genes have had separate origins (i.e., the mitochondrial version is not a promiscuous plastid gene). Ribosomal protein genes have recently been found in a number of protistan mtDNAs, as well as in liverwort mtDNA (Table III). The sequences of most of these genes have not yet been published; when they appear, they will prove to be a rich new source of phylogenetic information.

Genes for most of the proteins of the mitochondrial transcription and translation complexes are encoded in the nucleus, and those that could be of particular use in phylogenetic analysis include genes encoding ribosomal proteins, initiation and elongation factors, aminoacyl-tRNA synthetases, and perhaps rRNA and tRNA modifying enzymes. Much of the current information about protein components of the mitochondrial translation apparatus has come from yeast (*S. cerevisiae*). Nucleus-encoded mitochondrial components reported to have significant sequence similarity with *E. coli* or other bacterial proteins include ribosomal proteins (Myers *et al.*, 1987; Fearon and Mason, 1988; Kitakawa *et al.*, 1990; Dang and Ellis, 1990), elongation factors Tu (Nagata *et al.*, 1983) and G (Vambutas *et al.*, 1991), initiation factor 2 (Vambutas *et al.*, 1991), and aminoacyl-tRNA synthetases (Myers and Tzagoloff, 1985; Pape *et al.*, 1985; Koerner *et al.*, 1987; Herbert *et al.*, 1988; Tzagoloff *et al.*, 1989). Nucleus-encoded mitochondrial proteins having apparent *E. coli* homologs have also been characterized in *N. crassa;* these include ribosomal proteins (Kuiper *et al.*, 1988; Kreader *et al.*, 1989) and tyrosyl-tRNA synthetase (Akins and Lambowitz, 1987).

From the perspective of rigorous phylogenetic analysis, there are a number of problems in using ''similarity'' data derived from analyses of these particular nucleus-encoded mitochondrial proteins. Considerations that may limit or compromise evolutionary inferences drawn from such analyses include the following:

(i) Although evident homologies between nucleus-encoded mitochondrial proteins and corresponding eubacterial (e.g., *E. coli*) proteins have been demonstrated in a number of instances, it is difficult to assess the phylogenetic significance of these homologies, i.e, whether the level of sequence similarity is significantly greater with the eubacterial homolog than with representative sequences (if such exist) from the other two primary lineages. This difficulty may simply reflect the fact that the appropriate selection of sequences is not available to permit such a comparison to be made at the present time; alternatively, the overall level of sequence similarity may be too low to permit meaningful conclusions to be drawn. For example, yeast mitochondrial lysyl-tRNA synthetase appears equally distant from its yeast cytoplasmic and *E. coli* homologs, which are in fact more similar to one another than either is to the mitochondrial protein (Gatti and Tzagoloff, 1991).

(ii) Although separate nuclear genes exist for many proteins that perform the same function in the mitochondrion and cytosol, there are cases where a single nuclear gene encodes isozymes that are individually targeted to the two compartments (Surguchov, 1987). Examples include a number of aminoacyl-tRNA synthetases (Natsoulis *et al.*, 1986; Chatton *et al.*, 1988) and tRNA modification enzymes (Ellis *et al.*, 1986; Dihanich *et al.*, 1987). The evolutionary origin of these genes (transferred from protomitochondrial to nuclear genome?; already present in the nuclear genome?) is unclear.

(iii) If a particular protein and its gene were already present in the host, or were recruited from elsewhere, the present functional localization of that protein (i.e., in the mitochondrion) may not be directly relevant to the question of the origin of this organelle and its genome, and may even prove misleading. A case in point may be the mitochondrial pyruvate dehydrogenase complex, which much more closely resembles that from Gram-positive than Gram-negative eubacteria (Henderson *et al.*, 1979; Keha *et al.*, 1982).

(iv) Genes being compared may be paralogous (related through gene duplication) rather than homologous (related by descent from a single ancestral gene). Present-day prokaryotic and eukaryotic aminoacyl-tRNA synthetases are thought to comprise two distinct subgroups of synthetases, each of whose members was probably derived from a single ancestral synthetase by way of gene duplications (Gatti and Tzagoloff, 1991). Thus, yeast mitochondrial lysyl-, aspartyl-, and asparaginyl-tRNA synthetases are paralogs of one another, and also show substantial similarity within the carboxy-termini domain with the same region in *E. coli* ammonia-dependent asparagine synthetase (Gatti and Tzagoloff, 1991).

(v) There is some evidence that nuclear genes for mitochondrial ribosomal proteins may evolve at a substantially faster rate than nuclear genes for the homologous cytosol proteins (Pietromonaco *et al.*, 1986); i.e., the molecular clock may not run at the same speed **even within the same genome.**

(vi) Nucleus-encoded mitochondrial proteins have been described that as yet have no obvious homologs, either prokaryotic or eukaryotic; such is the case with a number of yeast mitochondrial ribosomal proteins (Myers *et al.*, 1987; Grohmann *et al.*, 1989; Matsushita *et al.*, 1989; Partaledis and Mason, 1988; Kitakawa *et al.*, 1990). Obviously, we have no basis on which to assess the evolutionary origin of such proteins.

(vii) Finally, similarity is no guarantee of homology. Proteins that appear functionally the same in mitochrondia and eubacteria may be unrelated structurally, indicating an independent origin of the corresponding genes (Williamson *et al.*, 1989).

Given these potential pitfalls, we should be properly circumspect about evolutionary conclusions that are based on alleged similarities between eubacterial proteins and **nucleus-encoded** mitochondrial proteins. Only certain nucleus-encoded proteins active in mitochondrial transcription and translation are likely to prove generally useful as phylogenetic indicators. Among these are protein synthesis factors such as elongation factor Tu, which display many of the characteristics of an ideal phylogenetic probe, including antiquity, ubiquity, and high conservation of sequence (recall II,A). A phylogenetic tree based on EF-Tu-like sequences (Baldauf and Palmer, 1990) places the nDNA-encoded yeast mitochondrial EF-Tu (Nagata *et al.*,1983) in the eubacterial lineage, distinct from the grouping of archaebacterial and eukaryotic EF-Tu homologs that includes the yeast cytosol EF-1α (also nDNA-encoded). A potential limitation in using EF-Tu-like sequences for phylogenetic analysis may be the difficulty, in the case of photosynthetic eukaryotes, of distinguishing clearly between different eubacteria-like, nuclear *tufA* genes that separately encode plastid and mitochondrial versions of this protein.

c. Transfer RNA   In 1982, mitochondrial tRNA sequences were only available from a limited number of mammalian and fungal species. In reviewing the data at that time, we commented on unique structural features that distinguish individual animal and fungal mitochondrial tRNAs from both their eubacterial and their eukaryotic counterparts, and we suggested that rapid (and perhaps variable) evolution of mtDNA could account for much of the diversity in sequence among mitochondrial tRNAs. We felt this could explain the fact that individual animal and fungal mitochondrial tRNAs could not be clearly related to either prokaryotic or eukaryotic tRNAs (Gray and Doolittle, 1982). We went on to say, "In this regard, clear-cut primary sequence homologies that *are* found with either class are likely to be particularly significant." Our overall conclusion at the time was that because of their highly divergent primary and secondary structures, animal mitochondrial tRNA sequences provided no strong indications of their evolutionary ancestry; however, at least some of the known fungal mitochondrial tRNAs [which in general are considerably more "normal" in structure than animal mitochondrial tRNAs (Dirheimer and Martin, 1990)] did display structural and functional properties characteristic of their eubacterial counterparts (Gray and Doolittle, 1982).

In the intervening decade, this rather fuzzy picture has not been clarified much by additional data from animal and fungal mitochondrial systems; indeed, mitochondrial tRNAs from some invertebrate animals have proven to be even more atypical than their vertebrate counterparts (Okimoto and Wolstenholme, 1990). Nor has much new evolutionary insight been provided by sequences of a number of protistan mtDNA-encoded tRNA genes and/or tRNAs, including ones from the ciliates *T. pyriformis* (Schnare *et al.*, 1985; Suyama, 1985; Heinonen *et al.*, 1987; Suyama *et al.*, 1987; Suyama and Jenney, 1989), *T. thermophila* (Morin and Cech, 1988; Hekele and Beier, 1991), and *P. aurelia* (Pritchard *et al.*, 1990a). In contrast to animal mitochondrial tRNAs, characterized ciliate mitochondrial tRNAs generally display few or no deviations from the standard (Singhal and Fallis, 1979) tRNA structure (but see Heinonen *et al.*, 1987). Despite this, phylogenetic comparisons involving these particular tRNAs have not been very informative. Suyama (1985) claimed "an extraordinary extent of homology" between *Tetrahymena* mitochondrial tRNA^{His} and its *E. coli* counterpart on the one hand, and between *Tetrahymena* mitochondrial tRNA^{Phe} and the corresponding mammalian nDNA-encoded tRNA on the other. However, in comprehensive database searches that I have carried out, these tRNA sequences show comparably high levels of sequence similarity with both eubacterial and eukaryotic tRNAs, both cognate and noncognate; thus, no particularly persuasive conclusions can be drawn about the evolutionary origin of ciliate mitochondrial tRNA genes (Schnare *et al.*, 1985). Similarly, of the three (quite conventional) tRNAs encoded by *C. reinhardtii* mtDNA, only the tRNA^{Gln} shows an especially high level of sequence similarity with eubacterial (as opposed to archaebacterial or eukaryotic nucleus-encoded) tRNAs^{Gln} (Boer and Gray, 1988c).

Particularly persuasive evidence of eubacterial ancestry has come only from plant mtDNA-encoded tRNA sequences, which have turned out to be evolutionarily much more conservative than their fungal, protistan, and animal mtDNA-encoded counterparts, and therefore much easier to place phylogenetically. The first published plant mitochondrial tRNA sequence clearly showed that plant mitochondria encode a eubacteria-like initiator tRNA^{Met} (Gray and Spencer, 1983). Since that time, a great many plant mitochondrial tRNA sequences, both monocotyledon and dicotyledon, have been determined (for references, see Joyce *et al.*, 1988; Joyce and Gray, 1989ab; Sangaré *et al.*, 1990; Maréchal-Drouard *et al.*, 1990; Binder *et al.*, 1990). These studies have defined two classes of plant mtDNA-encoded tRNA sequences: "native," displaying 65–80% sequence identity with the homologous eubacterial and chloroplast tRNAs; and "chloroplast-like," showing a specifically high level of identity (>90%) with their chloroplast counterparts (see Maréchal *et al.*, 1985, 1987; Wintz *et al.*, 1988; Sangaré *et al.*, 1989; Joyce and Gray, 1989a). Native plant mitochondrial tRNAs and their bona fide chloroplast counterparts are about equally divergent in sequence from their *E. coli* homologs, attesting to the eubacterial character of mitochondrial sequences; the genes for these tRNAs are assumed to have been contributed by the eubacterial endosymbiont that gave rise to plant mitochondria. Chloroplast-like tRNA genes in plant mitochondria appear to have been recruited directly and relatively recently from plDNA, as a component of promiscuous DNA; they either are identical to the authentic chloroplast tRNA genes or differ by only one or a few nucleotides (Joyce and Gray, 1989a,b).

An especially interesting plant mitochondrial tRNA, whose gene was originally isolated from maize mtDNA and thought to encode an elongator tRNA^{Met} [predicted anticodon CAU (Parks *et al.*, 1984)], actually appears to have Ile rather than Met aminoacylation specificity, by virtue of a posttranscriptional modification at the anticodon position (Weber *et al.*, 1990). This modification converts the encoded C to a derivative that is very similar to the novel nucleoside lysidine (L) (Muramatsu *et al.*, 1988a), found at the same position in an AUA (Ile)-decoding tRNA from *E. coli* (Muramatsu *et al.*, 1988b) and *Mycoplasma capricolum* (Andachi *et al.*, 1989). This particular modification may also occur in the analogous chloroplast tRNA (Francis and Dudock, 1982), whereas the eukaryotic cytosol system translates AUA using either IAU or UAU anticodons (Jukes and Osawa, 1990). These observations may be taken as another specific indication of a eubacterial ancestry of plant mitochondria, in that it is highly unlikely that the enzymatic machinery for the biosynthesis of lysidine-like nucleosides arose independently in eubacteria and plant mitochondria, after their divergence from a remote common ancestor. It is, however, somewhat surprising that the tRNA^{Ile} (anticodon LAU) of maize (Parks *et al.*, 1984) and potato (Weber *et al.*, 1990) shares only 52–53% sequence identity with the analogous *E. coli* tRNA^{Ile}.

There are a number of potential limitations attending the use of tRNA sequence data for determination of phylogenies, including their small size (and consequent paucity of phylogenetically informative sites) and the possibility of switches in aminoacylation specificity (Cedergren *et al.*, 1990). To overcome the length limitation, Cedergren and Lang (1985) have proposed that all tRNA gene families in a given genome be considered as a single, continuous sequence, and Cedergren *et al.* (1990) have compiled aligned tRNA databases on that principle. Such an approach still has several limitation in the case of mitochondrial tRNA sequences, in that (i) many plant and protistan mtDNAs do not encode a full set of tRNA genes; (ii) mtDNA-encoded tRNA genes may have different evolutionary origins (e.g., the "native" and "chloroplast-like" classes in angiosperms); and (iii) the alignment of certain structurally atypical mitochondrial tRNA sequences (e.g., those in animals) is so problematic that they must be excluded from such a database. Despite these difficulties, tRNA phylogenetic trees have confirmed the plastid origin of plant mitochondrial chloroplast-like tRNAs and (under certain conditions) show fungal and native plant mitochondrial tRNAs as specific relatives of their purple bacterial counterparts (Cedergren *et al.*, 1990).

d. Ribosomal RNA   Some of the most compelling evidence supporting a eubacterial, endosymbiotic origin of mitochondria has come from studies of mitochondrial rRNA sequences, particularly those in land plants. Because most of the relevant information has been presented and discussed at length elsewhere (Gray, 1988), I will only briefly highlight it here.

As in the case of tRNA sequences, mitochondrial rRNA sequences from ciliates, fungi, and particularly animals have many unique primary and secondary structural features (Gutell *et al.*, 1985, 1990) that make it very difficult to assess their degree of similarity with one another and with their nonmitochondrial homologs (Gray, 1988). In striking contrast, not only do plant mitochondrial SSU and LSU rRNAs show a remarkably high degree of primary sequence and secondary structural correspondence with their prokaryotic, and specifically eubacterial, counterparts (Cunningham *et al.*, 1976b; Bonen *et al.*, 1977; Schnare and Gray, 1982; Spencer *et al.*, 1984; Chao *et al.*, 1984; Dale *et al.*, 1984, 1985; Brennicke *et al.*, 1985; Grabau *et al.*, 1985; Manna and Brennicke, 1985; Falconet *et al.*, 1988), but wheat mitochondrial SSU rRNA has been shown to contain eubacteria-specific, post-transcriptional modifications that have so far not been found in its mitochondrial homolog from other eukaryotes (Cunningham *et al.*, 1976b; Schnare and Gray, 1982; Gray, 1988). Moreover, among eukaryotes, plants stand alone (so far) in containing a unique mitochondrial 5S rRNA (Leaver and Harmey, 1976; Cunningham *et al.*, 1976a; Spencer *et al.*, 1981), encoded by the plant mitochondrial genome (Bonen and Gray, 1980; Huh and Gray, 1982). Plant mitochondrial ribosomes resemble prokaryotic 70S rather than eukaryotic 80S ribosomes in containing a 5S but not a 5.8S rRNA component (Leaver and Harmey, 1976), despite the fact that plant mitochondrial rRNAs appear "eukaryotic" (i.e., "26S" and "18S") in size. [The size difference between homologous plant mitochondrial and eubacterial rRNAs is attributable to the insertion of additional blocs of sequence into a few variable regions in the former (Spencer *et al.*, 1984; Gutell *et al.*, 1990)].

The availability of relatively conservative mitochondrial sequences has allowed their unambiguous placement in phylogenetic trees based on rRNA sequence comparisons (see V,C). Not only has it been possible to show from such analyses that plant mitochondria and eubacteria are specific evolutionary relatives (Gray *et al.*, 1984), but it has also proven possible to identify which particular eubacteria are the closest contemporary relatives of mitochondria (Yang *et al.*, 1985; Cedergren *et al.*, 1988; Gray *et al.*, 1989; Van de Peer *et al.*, 1990). These turn out to be members of the α-subdivision of the purple bacteria (Fig. 13): precisely the group of eubacteria singled out much earlier, on biochemical grounds, as the likely source of a protomitochondrial endosymbiont (John and Whatley,

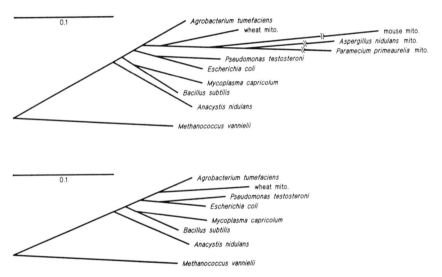

FIG. 13 Phylogenetic trees of eubacterial and mitochondrial SSU rRNA sequences. The evolutionary distances (estimated number of mutational events per sequence position) between the nodes of the trees are reflected in their horizontal separation. In the upper tree (which includes mitochondrial sequences representing all of the eukaryotic kingdoms), branches to the mouse, *Aspergillus nidulans,* and *Paramecium primaurelia* mitochondrial sequences have been shortened (note the line breaks) by 0.4, 0.2, and 0.2 unit, respectively. The lower tree was calculated without the nonplant mitochondrial sequences to eliminate the distortion introduced by the faster-evolving sequences and, thereby, to illustrate more accurately the affiliation of the *A. tumefaciens* and mitochondrial lineages. Reprinted from Yang *et al.* (1985), with permission.

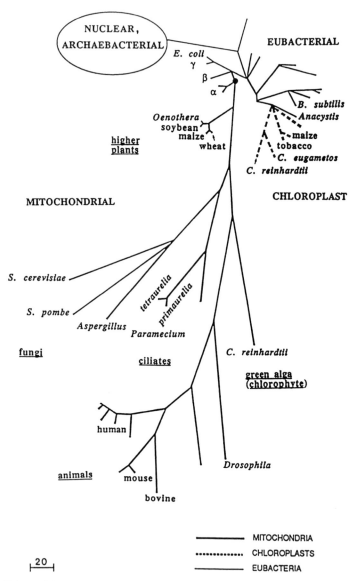

⊢―20―⊣

FIG. 14   Eubacterial–chloroplast–mitochondrial portion of an unrooted phylogenetic tree inferred from a SSU rRNA sequence database (Cedergren *et al.,* 1988) comprising a total of
76 sequences (9 archaebacterial, 15 eubacterial, 26 eukaryotic nuclear, 5 chloroplast, and 21
mitochondrial). The length of each branch is proportional to the number of fixed mutational
events represented by its two endpoints, as indicated by the scale. The identity of all
organisms at terminal nodes and bootstrap (Felsenstein, 1985) values indicating the ro-
bustness of each internal branch are given in Cedergren *et al.* (1988). Trypanosome mitochon-

(continued)

1975a,b; see V,B,5,a). Although the phylogenetic affiliation of the plant mitochondrial 5S rRNA sequence was initially uncertain (Spencer *et al.*, 1981; Gray and Spencer, 1981), its origin, too, was eventually traced to the α-subdivision of the purple bacteria (Villanueva *et al.*, 1985).

C. Mitochondrial Phylogenies from Ribosomal RNA Sequence Comparisons

As discussed in the preceding section, different mitochondrial components have been useful to varying degrees in providing molecular data to address the question of the origin of the mitochondrion and its genome. For reconstructing a detailed evolutionary history of the mitochondrial lineage, the most important molecules have unquestionably been SSU and LSU rRNAs. The reasons outlined earlier (II,A) for using rRNA sequences as a basis for construction of phylogenetic trees apply with even greater force to the mitochondrial lineage. In particular, comparisons of rRNA secondary structure provide the only objective basis for selecting and accurately aligning regions of homologous sequence from structurally diverse mitochondrial SSU or LSU rRNAs. Another consideration is that no other molecular compilation currently contains the number and diversity of organellar and non-organellar sequences as that provided by LSU and (particularly) SSU rRNA databases.

The utility of rRNA sequence comparisons for probing mitochondrial origins was demonstrated by Küntzel and Köchel (1981), Köchel and Küntzel (1982), and Gray *et al.* (1984). Phylogenetic trees generated in these studies provided unambiguous support for an endosymbiotic origin of mitochondria (animal, fungal, and ciliate as well as plant) from within a group of aerobic eubacteria, later narrowed to the α-purple bacteria (Yang *et al.*, 1985; see Fig. 13). More comprehensive rRNA trees, both SSU and LSU (Cedergren *et al.*, 1988; Gray *et al.*, 1989), have confirmed these conclusions (Fig. 14).

Given that molecular and biochemical data together support an endosymbiotic, α-purple bacterial origin of mitochondria, the central question now at issue is whether such an event occurred only once, or more than once. A monophyletic origin of mitochondria, one simultaneous with that

---

drial SSU rRNA sequences have been excluded from this analysis because they do not contain the full "universal structural core" (Gray *et al.*, 1984) upon which the sequence alignment is based (Cedergren *et al.*, 1988). Reprinted from Gray *et al.* (1989), with permission.

of plastids, has been advocated by Cavalier-Smith (1987b). Perhaps the strongest support for a single mitochondrial origin are the data compiled in Tables II and III, which bear witness to a high degree of conservation of a very limited genetic function among diverse mtDNAs. If we assume that the protomitochondrial genome, like contemporary eubacterial ones, possessed many more genes than are currently found in mtDNA, then we must postulate that the progenitor genome(s) suffered a drastic reduction in genetic information, through gene loss and/or transfer. It is difficult to envisage what evolutionary forces could have operated on independently acquired eubacterial genomes in different eukaryotes to yield basically the same overall function, specified by a very similar (and very small) set of genes. Nevertheless, the possibility of a polyphyletic origin of mitochondria has been considered on both molecular (Dayhoff and Schwartz, 1981) and morphological (Stewart and Mattox, 1984) grounds, so it is pertinent to ask whether rRNA phylogenetic trees offer any insight into this question.

On this point, current rRNA trees are equivocal. Küntzel and Köchel (1981) and Köchel and Küntzel (1982) inferred an independent origin of fungal and mammalian mitochondria, because in their trees a common root could not be found for both groups of sequences. Gray *et al.* (1984) also considered the possibility of a biphyletic origin of mitochondria, but in this case rRNA trees showed animal and fungal mitochondria branching together but separately from plant mitochondria. In contrast, the tree of Yang *et al.* (1985) clustered fungal, animal, ciliate, and plant mitochondria together as specific relatives of the α-purple bacterium *Agrobacterium tumefaciens* (Fig. 13).

More recent rRNA trees, based on a larger number of eubacterial and mitochondrial sequences (Cedergren *et al.,* 1988; Gray *et al.,* 1989), also show all mitochondria as a single clade branching together with the α-purple bacteria (Fig. 14). However, a curious topological discrepancy has been noted between such trees and nuclear ones containing the same organisms. In nuclear SSU rRNA trees, plants (angiosperms) and the green alga *C. reinhardtii* branch together as specific evolutionary relatives (Fig. 12), whereas in mitochondrial trees (Fig. 14), plants and *C. reinhardtii* branch separately. Moreover, in global SSU rRNA trees, plants occupy a seemingly anomalous branching position in the mitochondrial lineage, in that they are positioned very close to the root of the mitochondrial subtree, near its connection with the eubacteria (Fig. 14). This positioning is consistent with the strongly eubacterial character of plant mitochondrial rRNA sequences relative to their mitochondrial homologs from other eukaryotes, but it differs markedly from the branching position that plants occupy in the nuclear lineage (Fig. 12). In considering this noncongruence in the branching positions of angiosperms and *C. reinhardtii* in the nuclear and mitochondrial lineages, and the distinctive molecular biologies of angiosperm and *C. reinhardtii* mitochondria (Table IV), Gray *et al.* (1989) were led to suggest that the rRNA genes of plant mitochondria may have been acquired in a separate (and more recent) endosymbiotic event than the rRNA genes in other mitochondria.

In considering this proposal, a major issue is the validity of the mitochondrial trees on which such a suggestion is based. The studies in question (Cedergren *et al.*, 1988; Gray *et al.*, 1989) highlight the difficulty of inferring accurate branching patterns in a lineage in which the compared sequences are diverging at radically different rates. In Fig. 14, it can be seen that animals, fungi, ciliates, and *C. reinhardtii* are all on very long branches, whereas plant mitochondria, plastids, and eubacteria are on very short ones, even though the most conservative portions of SSU rRNA sequence are selected for comparison. This raises the question of whether the observed topology in these parsimony-based analyses is a methodological artifact, attributable to "long branches attracting" (Felsenstein, 1978), regardless of the true topology.

There are tests that one can apply to inferred topologies to evaluate the likelihood of such a methodological problem (see Cedergren *et al.*, 1988, and Gray *et al.*, 1989, for a fuller discussion). However, even when such tests are negative (Gray *et al.*, 1989), there is concern that the existing methodology may not be adequate to deal with the rate difference problem. In this regard, Van de Peer *et al.* (1990) have shown that two substantially different mitochondrial topologies are generated when two different treeing algorithms are applied to the same data set (Fig. 15). In one tree (Fig. 15b), plant mitochondria branch with the α-purple eubacteria, but all other mitochondria cluster outside of the three major lineages, with animal mitochondria separate from a clade of fungal, ciliate, and *C. reinhardtii* mitochondria. This branching topology is more or less in line with the *archigenetic hypothesis* of Mikelsaar (1987), who postulates that "mitochondria have arisen neither from pro- nor eukaryotes, but from free-living organisms (mitobionts) that were derivatives of primitive cells (protobionts)."

Recent analyses (D. F. Spencer and M. W. Gray, unpublished results) suggest that in these rRNA trees, the branching positions of nonplant mitochondria are strongly dependent on the combination of database and treeing algorithm, whereas the branching position of plant mitochondria is robust (not influenced by choice of database and methodology). When the database of Cedergren *et al.*, (1988) (489 aligned positions) was analyzed by the neighbor-joining method of Saitou and Nei (1987) [a distance method, like the method of Van de Peer *et al.* (1990)], the topology of Fig. 15b was reproduced in its essential features. However, when animal mitochondrial sequences were excluded from the analysis and an aligned database consisting of over twice as many nucleotide positions as that of

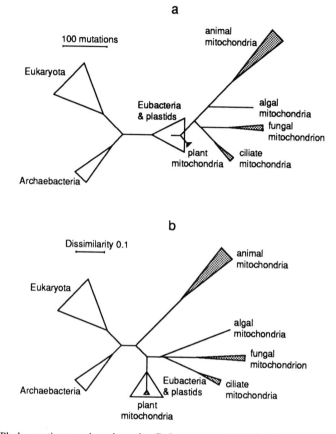

FIG. 15 Phylogenetic trees based on the Cedergren *et al.* (1988) alignment of SSU rRNA sequences (see Van de Peer *et al.*, 1990, for details of the analysis). **(a)** Outline of the tree published by Cedergren *et al.* (1988) (see also Fig. 14). As this is a parsimony tree, the distance scale measures net substitutions rather than dissimilarity. **(b)** Outline of the tree constructed by Van de Peer *et al.* (1990) using an algorithm to compute and optimize a dissimilarity matrix. See text for further discussion. Reprinted from Van de Peer *et al.* (1990), with permission. Copyright 1989 Springer-Verlag.

Cedergren *et al.* (1988) was compiled and analyzed, all mitochondria now branched together and with the α-purple bacteria, as in the topology of Fig. 15a. The latter topology is consistently generated with the parsimony-based algorithm of Cedergren *et al.* (1988), even using the database encompassing a more limited number of positions.

The volatility in branching position of nonplant mitochondrial sequences is a vexing problem in these analyses and complicates interpretation of the resulting trees. However, leaving aside the vagaries of mitochondrial phylogenetic trees, we still must explain the especially pronounced structural similarities between plant mitochondrial and eubacterial rRNA genes and rRNAs (Gray, 1988), and the substantial differences in mitochondrial genome organization and expression in angiosperms and *C. reinhardtii* (Table IV). As emphasized previously (Gray *et al.*, 1989), the available molecular data provide little indication that angiosperms and *C. reinhardtii* shared a common mitochondrial ancestor as recently as they shared a common nuclear or plastid ancestor. We have suggested (Gray *et al.*, 1989) that two alternative scenarios could account for the fact that the plant mitochondrion uniquely displays distinctive features of a eubacterial heritage (high degree of primary and secondary structural similarity in SSU and LSU rRNAs, eubacteria-specific post-transcriptional modifications in rRNA, presence of 5S rRNA): either (i) there features have been selectively retained in plant mitochondria but lost in the mitochondria of *C. reinhardtii* and other nonplant mitochondria, due to rapid evolutionary change in the latter; or (ii) the conservative features of plant mitochondrial rRNAs reflect a more recent acquisition by plant mitochondria of eubacteria-like rRNA genes, occurring after the separation of *C. reinhardtii* and plants from their last common ancestor.

In these alternative schemes [discussed in more depth by Gray *et al.* (1989)], the particular eubacteria-like features peculiar to plant mitochondrial rRNAs are assumed to represent retained traits of an α-purple bacterial endosymbiont, whether the endosymbiosis in question occurred early or late. If one invokes an early, monophyletic origin of mitochondria from such an ancestor (1, Fig. 12), then one also has to assume a number of independent losses of these particular eubacteria-like features in a number of different nonplant lineages (trypanosome, ciliate, fungal, animal) throughout the evolutionary history of the mitochondrial lineage. These conservative features would have had to have been present in the last common ancestor of angiosperms and *C. reinhardtii* mitochondria, with retention of them in the former but loss in the latter. The alternative view is that those eubacteria-like traits that are specific to plant mitochondria have been uniquely acquired in a separate, more recent endosymbiotic event (2, Fig. 12). In considering this possibility, it should be remembered that the recruitment of tRNA genes from plDNA (see V,B,2,c) provides a precedent for the relatively recent acquisition of new components of the plant mitochondrial translation system. Also, it is intriguing that Ochman and Wilson (1987), employing a temporal calibration of SSU rRNA trees, and with the wheat mitochondrial SSU rRNA sequence as the mitochondrial representative, inferred that mitochondria arose about 900 million years ago: in their words, "more recently than predicted by other authors," and considerably later than the estimated time of origin of chloroplasts, inferred from the same calibrated tree.

In presenting the scenarios outlined above, we stated (Gray *et al.*, 1989): "Validation or rejection of the secondary acquisition hypothesis . . . will obviously require further comparative data; in particular, additional mitochondrial genomes within the chlorophyte–metaphyte grouping must be examined. If the thesis developed here is correct, then we should expect to see an abrupt transition from a prototypical chlorophyte mitochondrial genome (if such can be defined) to a prototypical metaphyte mitochondrial genome somewhere within the green algal lineage leading directly to higher plants."

New information from *Prototheca wickerhamii*, a colorless *Chlorella*-like chlorophyte, indicates that its mitochondrial SSU rRNA is much more similar to that of higher plants than is *C. reinhardtii* mitochondrial SSU rRNA (Wolff and Kück, 1990). This is also the case with the mitochondrial SSU rRNA from *Acanthamoeba castellanii* (K. Lonergan and M. W. Gray, unpublished results), a nonphotosynthetic amoeboid protozoan that branches close to the chlorophyte–metaphyte line in nuclear rRNA trees (Fig. 12). These results suggest that the *C. reinhardtii* mitochondrial genome may not be representative of chlorophyte mtDNAs, and that it may well have diverged substantially from a pattern more characteristic of plant mtDNA. However, in none of the rRNA trees we have determined to date do *P. wickerhamii, A. castellanii,* or other as-yet-unpublished mitochondrial SSU rRNAs branch as **specific** evolutionary relatives of those from land plants (i.e., they do not emerge from a common node) (D. F. Spencer and M. W. Gray, unpublished results). Clearly, to critically evaluate the possibility of a secondary acquisition of plant mitochondria and/or mitochondrial rRNA genes, much additional comparative data will be required. Particularly crucial will be the search for other "conservative" mitochondrial genomes, especially within the green algae and among early diverging protists (e.g., Karlovsky and Fartmann, 1992). Such genomes, if they exist, could well provide the evolutionary link (currently missing) to the plant mitochondrial genome.

D. Gene Transfer in Mitochondrial Evolution

Although mitochondrion-to-nucleus gene transfer is implicit in the endosymbiont hypothesis, and is supported by the nuclear location of genes for certain mitochondrial proteins in some or all eukaryotes (see V,B,5), up to now there has been no clear example of recent evolutionary transfer of a gene from mtDNA to nDNA. This situation has changed with the recent discovery that a respiratory chain gene was transferred from the mitochondrion to the nucleus at some time during angiosperm evolution. Nugent and Palmer (1991) have found that *cox2*, normally encoded by mtDNA, is absent from the mitochondrial genome of two species of legume (mung bean and cowpea). In these plants, a copy of *cox2* has been transferred to and has become functional in the nucleus. A nuclear *cox2* copy is also active in soybean (Covello and Gray, 1992), although the mtDNA of this legume still contains a potentially functional *cox2* (Grabau, 1987); the latter, however, is apparently not expressed. Most legumes have *cox2*-related sequences in both the nucleus and the mitochondrion, although there is no evidence for simultaneous *cox2* expression in both compartments. Transfer of *cox2* is estimated to have occurred between 60 and 200 million years ago, whereas *cox2* loss from mtDNA has been much more recent (Nugent and Palmer, 1991). Because *cox2* mRNA is edited in both monocotyledon and dicotyledon mitochondria (Covello and Gray, 1990a), it is possible to infer something about the route by which *cox2* was transferred to the nucleus in legumes. In both cowpea (Nugent and Palmer, 1991) and soybean (Covello and Gray, 1992), the nuclear *cox2* gene more closely resembles an edited *cox2* transcript than it does the unedited sequence corresponding to a mitochondrial *cox2* gene. This provides strong evidence that the gene transfer event involved reverse transcription of an edited RNA intermediate.

Activation of a nuclear copy of a transferred mitochondrial gene necessitates the acquisition of nuclear transcriptional control elements, as well as information to target the protein back into the mitochondrion. Conforming to this expectation, the cowpea and soybean *cox2* genes encode a putative mitochondrial transit peptide. They also contain an intron at or near the junction of this sequence and the mature protein-coding sequence, and it is suggested (Nugent and Palmer, 1991) that this intron may have played a role in the recruitment of upstream expression and targeting signals. It is noteworthy that this evolutionary transfer of the *cox2* gene follows rather closely a model elaborated by Obar and Green (1985) for net transfer of nucleic acid sequences between nonhomologous genomes.

E. The Strange Case of *Plasmodium*

Some recent observations arising out of studies of *Plasmodium*, the malaria parasite, serve as a graphic illustration of some of the unexpected situations with which we must contend in studying the structure and evolutionary origin of organellar genomes. In several species of this genus, two nonnuclear, apparently unrelated, presumptive organellar DNAs have been described: a 6-kbp tandemly repeated element and a 35-kbp covalently closed, circular molecule (Feagin *et al.*, 1991; Wilson *et al.*, 1991). The 6-kbp DNA contains *cox1*, *cox3*, and *cob* as well as fragmented SSU and LSU rRNA coding segments (Vaidya *et al.*, 1989, 1990; Aldritt *et al.*,

1989; Feagin, 1992; Feagin *et al.*, 1992); the latter are reminiscent of the scrambled rRNA coding modules in *C. reinhardtii* mtDNA (Boer and Gray, 1988a,b). Based on the genes it contains and the fact that it copurifies with mitochondria (Wilson *et al.*, 1992), this 6-kbp element is presumably all or part of the *Plasmodium* mitochondrial genome; if all, it represents the smallest mitochondrial genome yet discovered.

The 35-kbp DNA has been identified in several *Plasmodium* species (Kilejian, 1975; Williamson *et al.*, 1985; Gardner *et al.*, 1988). It encodes conventional, uninterrupted SSU and LSU rRNA genes, duplicated within a 10.5-kbp inverted repeat (Gardner *et al.*, 1991a) and sharing only very limited sequence similarity with the homologous rRNA sequences encoded by the 6-kbp DNA. The 35-kbp DNA also contains tRNA and ribosomal protein genes and, intriguingly, two contiguous genes encoding polypeptides corresponding to the β and β' subunits of a eubacteria-like RNA polymerase (Gardner *et al.*, 1991b). The latter genes are commonly encoded in plDNA, but have never before been found in mtDNA. This raises the question of whether the 35-kbp DNA is a plDNA remnant and, if so, how it ended up on a nonphotosynthetic organism, where it is localized in the organism, and what its function is.

Because the 35-kbp DNA is very A+T-rich (>80%), it has been difficult to trace the origin of its encoded SSU and LSU rRNA sequences: rRNA trees suggest a plastid affiliation, but are not definitive (Gardner *et al.*, 1991a). Intriguingly, recent comparisons of nuclear SSU rRNA sequences (Johnson *et al.*, 1988; Gajadhar *et al.*, 1991) provide support for a postulated origin of the phylum Apicomplexa (of which *Plasmodium* is a member) from a common ancestor with marine dinoflagellates (Levine, 1985): i.e., *Plasmodium* may have evolved from a plDNA-containing progenitor. In subcellular fractionation studies, the 35-kbp DNA does not seem to copurify with mitochondria (Wilson *et al.*, 1992), in contrast to the 6-kbp molecule; this raises the possibility that it is in a distinct subcellular compartment. In this regard, it should be noted that the Apicomplexa are so named because certain stages contain a group of structures, the apical complex, at the anterior end (Levine, 1985). The origin and function of these structures are currently unknown.

Given the phylogenetic diversity of the eukaryotic domain, we may expect that other unusual situations such as that described above will surface during future investigations of organellar DNA, particularly among the protists.

F. Concluding Remarks

In summing up our 1982 review, we wrote (Gray and Doolittle, 1982): "The data so far seem too few to allow us to conclude that criteria for any of the three sorts of "proof" outlined at the beginning of this paper [see I,B] have been fulfilled for mitochondria. However, the data do give hope that the criteria ultimately will be fulfilled and indicate that the best chance for fulfillment lies in further analyses of plant mitochondrial genomes." Ten years later, I would assert that this prediction has come to pass (see particularly V,B,5,b; V,B,5,c; and V,C) It is true that phylogenetically informative molecular data are less plentiful and less compelling for mitochondria than for plastids. So much of mitochondrial molecular biology is unique, and does not show a clear affinity with the corresponding processes/mechanisms/structures in archaebacteria, eubacteria, or the eukaryotic nucleus/cytosol. However, where homologies are evident, they clearly tie mitochondria to the eubacterial lineage, and more specifically to the α-subdivision of the purple bacteria. Whereas plants have unquestionably led the way in solidifying this evolutionary connection, rRNA trees (Fig. 14) as well as other data (V,B,5) now affirm this origin for animal, fungal, ciliate, and algal (*C. reinhardtii*) mitochondria.

As in the case of plastids, interest is now focusing on the question of whether mitochondria are monophyletic or polyphyletic. Although an α-purple eubacterial ancestry of the mitochondrion now seems sure, it is still not certain whether the endosymbiotic event(s) that established this organelle happened once or more than once. Resolution of this issue is confounded by the highly divergent patterns of mtDNA organization, expression, and evolution in different eukaryotes, so that organisms that are obviously related at the level of nuclear and/or plastid genome may show little or no evidence of relatedness at the level of the mitochondrial genome (e.g., angiosperms and *C. reinhardtii;* Table IV). Interpretation of mitochondrial phylogenetic trees, which encompass slowly evolving and rapidly evolving rRNA sequences, is difficult: whereas branching order within major groups (animals, fungi, plants) is robust in both nuclear and mitochondrial trees, branching orders between and among these groups are not. For the most part, the deep evolutionary connections among major eukaryotic groups, and between multicellular and unicellular groups, remain elusive in both the nuclear and the mitochondrial lineages. Whether plant mitochondria (or at least the rRNA genes therein) have had a different evolutionary origin than their counterparts in other mitochondria (Gray *et al.,* 1989) remains to be seen. However, because plants are the only multicellular eukaryotic group showing a clear evolutionary affiliation with a unicellular group (green algae) at the nuclear level, further comprehensive studies of chlorophyte mtDNA will undoubtedly be crucial in validating or rejecting the idea of a secondary acquisition of plant mitochondrial rRNA genes.

Regardless of whether mitochondria arose once or more than once, the available data clearly suggest that they did so from organisms whose closest eubacterial contemporaries are members of the α-subdivision of the purple bacteria. As far as the origin of mitochondria is concerned, there seems to be something very special about this particular group of eubacteria. As Yang et al. (1985) pointed out, "The α subdivision of the purple bacteria contains several types of organisms whose common characteristic is intimate, usually intracellular, association with eukaryotic cells—the rhizobacteria, agrobacteria and . . . the rickettsias." In this regard, it is rather intriguing how often newly recognized pathogens/symbionts are turning out to be members of the α-purple bacteria, inhabiting eukaryotes as disparate as ciliates (Amann et al., 1991) and man (Relman et al., 1990).

Just as eukaryotic cells are evolutionary mosaics, so too are their genomes, to varying extents. There is a growing appreciation of the role of gene transfer in genome evolution (Maynard Smith et al., 1991; Heinemann, 1991; Sprague, 1991), and increasing evidence that genomes have acquired genetic information from phylogenetically distinct sources. In the case of mitochondria, much (most?) of the genetic information governing the biogenesis and functioning of this organelle appears to have been obtained from the genome of an α-purple-bacteria-like endosymbiont; however, contributions have also been made by plastids (Joyce and Gray, 1989a) and perhaps phage-like entities (Masters et al., 1987; Nargang et al., 1992). Mobile introns, many of which encode genetic functions (Perlman and Butow, 1989; Lambowitz, 1989), are another possible source of genetic information that may not have been present in the original protomitochondrial genome. It was in this context that I recently urged (Gray, 1989a) "a new perspective on mtDNA evolution," suggesting that "some of the genetic information in mtDNA (e.g., the genes encoding components of the respiratory chain) [may well] derive from a single endosymbiotic event, whereas information encoding other components (e.g., of the mitochondrial transcription, translation, and/or RNA processing machinery) has been more labile in the course of evolution," with mtDNA "having derived (and lost) genetic information through various processes of lateral gene transfer in the course of its evolutionary history."

Finally, as we discover smaller and smaller mitochondrial genomes (e.g., the 6-kbp DNA of Plasmodium falciparum), we may ask whether some organisms with mitochondrial function nevertheless lack an extrachromosomal mtDNA component. Recognition that genes may be successfully relocated from mtDNA to nDNA, both naturally (Nugent and Palmer, 1991; Covello and Gray, 1992) and artificially (Nagley, 1988; Thorsness and Fox, 1990), raises the prospect that in an evolutionary time frame, mtDNA may ultimately be dispensable. On the other hand, the (so far) universal occurrence of cox1 and cob in mtDNA strongly suggests that these genes may not be able to be transferred, in which case a minimal mitochondrial genome and the associated replication, transcription, and translation machinery would have to be maintained. Constraints related to cotranslational protein export across the endoplasmic reticulum (von Heijne, 1986) and an aggressive maintenance mechanism (Jacobs, 1991) have been suggested as possible reasons for the persistence of mtDNA. Time and additional data will tell whether there is a point in the reduction of mitochondrial genetic function beyond which Nature cannot go.

VI. Other Organelles of Endosymbiotic Origin?

If plastids and mitochondria are of eubacterial, endosymbiotic origin, is it possible that other eukaryotic organelles also arose in this way? Suggestions along these lines have been made by a number of authors, and are briefly summarized here. In none of these cases is there compelling molecular evidence of a xenogenous origin, and unless any of the organelles in question can be shown to possess DNA that encodes structural and functional components of that organelle, it may never be possible to adduce the sorts of "hard data" that have been instrumental in affirming the xenogenous origin of mitochondria and plastids.

A. Peroxisomes

A symbiotic origin of peroxisomes (including glyoxysomes in plants and glycosomes in trypanosomes) has been suggested on the basis that they multiply by growth and division and their matrix proteins are inserted post-translationally (De Duve, 1982; Borst and Swinkels; 1989; Opperdoes and Michels, 1989). A scenario for an origin of this single-membrane organelle from a "posibacterium" (Gram-positive eubacterium) or archaebacterium has been outlined by Cavalier-Smith (1987b, 1989c). Key features of this scheme include the provision of all or most of the peroxisomal matrix proteins by preexisting nuclear genes, and the transfer of all membrane protein genes (encoded by a putative protoperoxisomal genome) to the nucleus. Cavalier-Smith (1987b, 1989c) notes the absence of peroxisomes among members of the Archezoa and suggests that eukaryotes may have first acquired them as symbionts at the same time as mitochondria and plastids; indeed, he postulates a simultaneous conversion of three different symbionts into their respective organelles (mitochondrion, plastid, and peroxisome) within the same host cell. Because it is proposed that genetic information originally present in a protoperoxisome has been completely lost (peroxisomes having no DNA), Cavalier-Smith (1987b) admits that this organelle "would so effectively have covered its tracks that it may prove impossible to gain phylogenetic evidence for the above scenario . . . ."

B. Hydrogenosomes

The hydrogenosome (Lindmark and Müller, 1973; Müller, 1980) is an unusual, double-membrane organelle that plays a central role in pyruvate degradation in trichomonad flagellates [members of the archezoan phylum Parabasalia (Cavalier-Smith (1987a)] and some other anaerobic protozoa. Hydrogenosomes generate electrons leading to molecular hydrogen by reactions similar to those occurring in anaerobic eubacteria such as clostridia (Lindmark and Müller, 1973), and this has prompted the suggestion that the hydrogenosome has evolved from a free-living clostridia-like prokaryote (Whatley et al., 1979; F. R. Whatley, 1981). Čerkasovová et al. (1976) reported the presence of circular DNA in hydrogenosomes, but subsequent work by Turner and Müller (1983) failed to confirm this observation. Therefore, as in the case of peroxisomes, there is no easy way to critically evaluate the proposed xenogenous origin of hydrogenosomes, although Müller (1985) has suggested "sequencing of various hydrogenosomal proteins . . . to detect homologies with other organelles in order to be able to define the intriguing evolutionary position of hydrogenosomes." Cavalier-Smith (1987b) considers that hydrogenosomes may have evolved from preexisting mitochondria or peroxisomes, but Müller (1988) thinks this less likely than an endosymbiotic origin from anaerobic prokaryotes that already contained the characteristic pathway of pyruvate oxidation.

C. Undulipodia (Flagella)

An endosymbiotic, spirochete origin of the eukaryotic motility organelle (variously referred to as "undulipodium," "cilium," or "9+2 flagellum") and its underlying structure, the kinetosome (homologous to the centriole), has been proposed by Margulis and co-workers (Margulis et al., 1979, 1981; Bermudes et al., 1987). These investigators "infer the prokaryotic origin of the undulipodium from the biology of the organelle itself, its distribution and the occurrence of related and analogous structures" (Bermudes et al., 1987). As in the other two cases of proposed xenogenous origin discussed above, this hypothesis is based only on circumstantial evidence, and is not bolstered by any of the sorts of molecular data that we have considered essential in establishing the endosymbiotic origin of mitochondria and plastids. The recent report of basal body/centriolar

DNA in *C. reinhardtii* (Hall *et al.*, 1989) raised the exciting possibility of being able to rigorously test the proposed spirochete origin of eukaryotic flagella. However, this prospect has dimmed considerably with publication of recent evidence arguing against the existence of basal body DNA (Johnson and Rosenbaum, 1990; Johnson and Dutcher, 1991).

VII. Epilogue and Future Prospects

With respect to the evolution of the eukaryotic cell, current molecular data strongly support (i) an origin of the nuclear genome from a direct common ancestor with archaebacteria; (ii) an endosymbiotic origin of the plastid genome from a direct common ancestor with cyanobacteria; and (iii) an endosymbiotic origin of the mitochondrial genome from a direct common ancestor with members of the α-subdivision of the purple bacteria. In terms of the types of proof for the endosymbiotic origin of mitochondria and plastids outlined in I,B, the second form is now clearly satisfied: the nuclear genome, although lacking modern free-living relatives, clearly descended from a lineage *other* than that from which organellar genomes descended. Although it was previously possible to maintain an autogenous origin for the nuclear genome and at least one of the organellar genomes (from either a cyanobacteria-like or an α-purple-bacteria-like progenitor) (Doolittle, 1980), that is no longer the case. Although the evolutionary pathway to the nuclear genome is still obscure, it was evidently through the archaebacteria, not the eubacteria.

So, where do we go from here? Is there any life left in the endosymbiont hypothesis and the question of organelle evolution? I would suggest that, aside from the accumulation of additional data to bolster the endosymbiotic scenarios discussed in this review, there are still a number of important evolutionary questions (let alone questions of biochemistry and cell biology) that demand continuing intensive investigation of organellar genomes.

   1. Are mitochondria and plastids monophyletic or polyphyletic? In my view, this question is still very much up in the air, and only likely to be resolved by comprehensive molecular analyses involving a number of different mitochondrial genes in a wide variety of eukaryotes, especially protists.

   2. Did mitochondria and plastids arise simultaneously or serially? Cavalier-Smith (1987b) supposes that the reason cyanobacteria became plastids whereas purple bacteria became mitochondria is that they were taken up at the same time by the same host. He argues that because purple bacteria cannot photosynthesize in the presence of oxygen, the protomitochondrion lost photosynthesis; conversely, because cyanobacteria cannot control respiration separately from photosynthesis, the protoplastid lost respiration and in the dark came to rely on the more complete and efficient respiratory system of the purple bacterial symbiont. Accordingly, all extant mitochondrial and plastid types are assumed to be descendants of these single and essentially simultaneous symbioses. On the other hand, the *serial endosymbiosis theory* (Taylor, 1974; Margulis, 1981) proposes a stepwise acquisition of organelles, with mitochondria probably entering first and plastid sometime later; this scenario can readily accommodate multiple primary origins of each organelle.

Examination of nuclear phylogenetic trees (Figs. 6 and 12) reveals an interspersion of photosynthetic and nonphotosynthetic types, with a number of early diverging protists containing mitochondria but not plastids (*E. gracilis,* an early diverging photosynthetic eukaryote, almost certainly obtained its plastid secondarily; see IV,H). This implies, if mitochondria and plastids originated simultaneously, that plastid-less eukaryotes have lost plastids secondarily. There are insufficient data to argue persuasively one way or the other. The more cases of secondry plastid loss that can be documented, particularly among early diverging protists, the more tenable will the simultaneous origin hypothesis become. Conversely, clear evidence of polyphyly of plastids and/or mitochondria will certainly bolster the serial scenario.

3. To what extent has gene/genome migration played a role in shaping organelle exolution? Even if mitochondria and plastids are monophyletic, their evolution has undoubtedly been influenced by association with each other and with other symbionts. Our view of organelle evolution is rather simplistic if we maintain that nothing much happened after an initial endosymbiosis followed by an abrupt, massive transfer of genetic information to the nucleus. DNA transfer is demonstrably ongoing among the various genomes of the eukaryotic cell, and DNA transfer between these and the genomes of prokaryotic symbionts, either extra- or intracellular, is a distinct possibility. As well, now that we have good evidence of secondary acquisition of plastids, perhaps we should consider whether secondary acquisition of mitochondria may also have occurred in some instances. In cases where eukaryotic cells come together in various contexts to form new combinations (Goff and Coleman, 1987; Douglas *et al.,* 1991), we may well ask whose mitochondria are retained, and whose are lost. The pathways of eukaryotic evolution are becoming increasingly complex as we realize the roles that gene, genome, and organelle transfer play in this process.

4. What are the pathways and mechanisms by which organellar genomes evolve? As our understanding of phylogenetic relationships within the eukaryotic domain has expanded, and as we have moved to examine mitochondrial genomes other than those of mammals, yeast, and angiosperms, and plastid genomes other than those of land plants and green algae, it has become evident that we have just scratched the surface of organellar molecular biology, in all its novelty and diversity. Understanding how that diversity (in organellar genome organization, expression and evolution) has been generated will be the next major challenge, and will require that we first catalog its nature and extent, before we are able to understand the mechanisms that brought it about. Viewed from that perspective, proof of the endosymbiont hypothesis is merely the beginning of the story.

Whether acceptance of the endosymbiont hypothesis "will ever become truly universal" remains to be seen; even allowing that mitochondrion and plastid arose in this way, many of the details of this evolutionary process remain obscure. Judging from the record of the past decade, it is likely that a few more startling surprises await us before the evolutionary histories of the organelles are reconstructed to whatever extent is possible.

Acknowledgments

I am grateful to D. F. Spencer for the analysis presented in Fig. 14 and for preparation of that figure, and I thank those individuals who granted permission for reproduction of the other, previously published figures appearing in this review. I am indebted to all those who supplied unpublished information for inclusion here, in particular H. Bertrand, L. Bonen, R. J. Cedergren, R. A. Collins, S. Douglas, M. R. Hanson, B. F. Lang, R. W. Lee, P. Liu, K. Ohyama, M. Reith, R. J. M. Wilson, and D. R. Wolstenholme. In preparing this commentary, I benefited greatly from discussions with members of my laboratory group and with other local and external members of the Evolutionary Biology Program of the Canadian Institute for Advanced Research (CIAR). I sincerely acknowledge salary and travel support from CIAR, as well as assistance in the form of an operating grant (MT-4124) from the Medical Research Council of Canada, which has sustained my research on structure, function, and evolution of mitochondrial nucleic acids for the past two decades.

References

Aitken, A., and Stanier, R. Y. (1979). *J. Gen. Microbiol.* **112,** 219–223.
Akins, R. A., and Lambowitz, A. M. (1987). *Cell (Cambridge, Mass.)* **50,** 331–345.
Aldritt, S. M., Joseph, J. T., and Wirth, D. F. (1989). *Mol. Cell. Biol.* **9,** 3614–3620.
Amann, R., Springer, N., Ludwig, W., Görtz, H.-D., and Schleifer, K.-H. (1991). *Nature (London)* **351,** 161–164.
Amils, R., Ramírez, L., Sanz, J. L., Marín, I., Pisabarro, A. G., and Ureña, D. (1989). *Can. J. Mircrobiol.* **35,** 141–147.
Andachi, Y., Yamao, F., Muto, A., and Osawa, S. (1989). *J. Mol. Biol.* **209,** 37–54.
Anderson, S., Bankier, A. T., Barrell, B. G., de Bruijn, M. H. L., Coulson, A. R., Drouin, J., Eperon, I. C., Nierlich, D. P., Roe, B. A., Sanger, F., Schreier, P. H., Smith, A. J. H., Staden, R., and Young, I. G. (1981) *Nature (London)* **290,** 457–465.

Anderson, S., de Bruijn, M. H. L., Coulson, A. R., Eperon, I. C., Sanger, F., and Young, I. G. (1982). *J. Mol. Biol.* **156**, 683–717.

Andersson, S. G. E., and Kurland, C. G. (1991). *Mol. Biol. Evol.* **8**, 530–544.

André, C., Levy, A., and Walbot, V. (1992). *Trends Genet.* **8**, 128–132.

Anemüller, S., and Schäfer, G. (1989). *FEBS Lett.* **244**, 451–455.

Arizmendi, J. M., Runswick, M. J., Skehel, J. M., and Walker, J. E. (1992). *FEBS Lett.* **301**, 237–242.

Arndt, E., Krömer, W., and Hatakeyama, T. (1990). *J. Biol. Chem.* **265**, 3034–3039.

Assali, N.-E., Mache, R., and Loiseaux-de Göer, S. (1990). *Plant Mol. Biol.* **15**, 307–315.

Assali, N.-E., Martin, W. F., Sommerville, C. C., and Loiseaux-de Goër, S. (1991). *Plant Mol. Biol.* **17**, 853–863.

Attardi, G. (1985). *Int. Rev. Cytol.* **93**, 93–145.

Attardi, G., and Schatz, G. (1988). *Annu. Rev. Cell. Biol.* **4**, 289–333.

Auer, J., Lechner, K., and Böck, A. (1989a). *Can J. Microbiol.* **35**, 200–204.

Auer, J., Spicker, G., and Böck, A. (1989b). *J. Mol. Biol.* **209**, 21–36.

Bakalara, N. Simpson, A. M., and Simpson, L. (1989). *J. Biol. Chem.* **264**, 18,679–18,686.

Baldauf, S. L., Manhart, J. R., and Palmer, J. D. (1990). *Proc. Natl. Acad. Sci. U.S.A.* **87**, 5317–5321.

Baldauf, S. L., and Palmer, J. D. (1990). *Nature (London)* **344**, 262–265.

Baroin, A., Perasso, R., Qu, L.-H., Brugerolle, G., Bachellerie, J.-P., and Adoutte, A. (1988). *Proc. Natl. Acad. Sci. U.S.A.* **85**, 3474–3478.

Barrell, B. G., Anderson, S., Bankier, A. T., de Bruijn, M. H. L., Chen, E., Coulson, A. R., Drouin, J., Eperon, I. C., Nierlich, D. P., Roe, B. A., Sanger, F., Schreier, P. H., Smith, A. J. H., Staden, R., and Young, I. G. (1980). *Proc. Natl. Acad. Sci. U.S.A.* **77**, 3164–3166.

Bartig, D., Schümann, H., and Klink, F. (1990). *System. Appl. Microbiol.* **13**, 112–116.

Bégu, D., Graves, P.-V., Domec, C., Arselin, G., Litvak, S., and Araya, A. (1990). *Plant Cell* **2**, 1283–1290.

Belfort, M. (1989). *Trends Genet.* **5**, 209–213.

Belfort, M. (1991). *Cell (Cambridge, Mass.)* **64**, 9–11.

Benne, R. (1985). *Trends Genet.* **1**, 117–121.

Benne, R. (1989). *Biochim. Biophys. Acta* **1007**, 131–139.

Benne, R. (1990). *Trends Genet.* **6**, 177–181.

Benne, R., and Sloof, P. (1987). *BioSystems* **21**, 51–68.

Benner, S. A., and Ellington, A. D. (1987). *Nature (London)* **329**, 295–296.

Benner, S. A., and Ellington, A. D. (1990). *Science* **248**, 943–944.

Benner, S. A., and Ellington, A. D. (1991). *Science* **252**, 1232.

Benner, S. A., Ellington, A. D., and Tauer, A. (1989). *Proc. Natl. Acad. Sci. U.S.A.* **86**, 7054–7058.

Bennett, J. L., and Clayton, D. A. (1990). *Mol. Cell. Biol.* **10**, 2191–2201.

Benz, R. (1985). *CRC Crit. Rev. Biochem.* **19**, 145–190.

Berghöfer, B., Kröckel, L., Körtner, C., Truss, M., Schallenberg, J., and Klein, A. (1988). *Nucleic Acids Res.* **16**, 8113–8128.

Bermudes, D., Margulis, L., and Tzertzinis, G. (1987). *Ann. N.Y. Acad. Sci.* **503**, 187–197.

Bessho, Y., Ohama, T., and Osawa, S. (1992). *J. Mol. Evol.* **34**, 331–335.

Bhat, G. J., Koslowsky, D. J., Feagin, J. E., Smiley, B. L., and Stuart, K. (1990). *Cell (Cambridge, Mass.)* **61**, 885–894.

Bhattacharya, D., and Druehl, L. D. (1988). *J. Phycol.* **24**, 539–543.

Bhattacharya, D., Elwood, H. J., Goff, L. J., and Sogin, M. L. (1990). *J. Phycol.* **26**, 181–186.

Bibb, M. J., Van Etten, R. A., Wright, C. T., Walberg, M. W., and Clayton, D. A. (1981). *Cell (Cambridge, Mass.)* **26**, 167–180.

Binder, S., Schuster, W., Grienenberger, J.-M., Weil, J. H., and Brennicke, A. (1990). *Curr. Genet.* **17**, 353–358.

Bland, M. M., Levings, C. S., III, and Matzinger, D. F. (1986). *Mol. Gen. Genet.* **204**, 8–16.

Blum, B., Bakalara, N., and Simpson, L. (1990). *Cell (Cambridge, Mass.)* **60**, 189–198.

Blum, B., and Simpson, L. (1990). *Cell (Cambridge, Mass.)* **62**, 391–397.

Boczar, B. A., Delaney, T. P., and Cattolico, R. A. (1989). *Proc. Natl. Acad. Sci. U.S.A.* **86**, 4996–4999.

Boer, P. H., and Gray, M. W. (1986). *EMBO J.* **5**, 21–28.

Boer, P. H., and Gray, M. W. (1988a). *Cell (Cambridge, Mass.)* **55**, 399–411.

Boer, P. H., and Gray, M. W. (1988b). *EMBO J.* **7**, 3501–3508.

Boer, P. H., and Gray, M. W. (1988c). *Curr. Genet.* **14**, 583–590.

Boer, P. H., and Gray, M. W. (1991). *Curr. Genet.* **19**, 309–312.

Bogenhagen, D. F., and Romanelli, M. F. (1988). *Mol. Cell. Biol.* **8**, 2917–2924.

Bogenhagen, D. F., and Yoza, B. K. (1986). *Mol. Cell. Biol.* **6**, 2543–2550.

Bogenhagen, D. F., Yoza, B. K., and Cairns, S. S. (1986). *J. Biol. Chem.* **261**, 8488–8494.

Bogorad, L. (1975). *Science* **188**, 891–898.

Bogorad, L. (1982). *In* "On the Origins of Chloroplasts" (J. A. Schiff, ed.), pp. 277–295. Elsevier, North Holland, Amsterdam.

Bohnert, H. J., Crouse, E. J., and Schmitt, J. M. (1982). *In* "Nucleic Acids and Proteins in Plants: II. Structure, Biochemistry and Physiology of Nucleic Acids" (Vol. 14B of "Encyclopedia of Plant Physiology, New Series") (B. Parthier and D. Boulter, eds.), pp. 475–530. Springer-Verlag, Berlin/Heidelberg/New York.

Bohnert, H. J., Michalowski, C., Bevacqua, S., Mucke, H., and Löffelhardt, W. (1985). *Mol. Gen. Genet.* **201**, 565–574.

Bonen, L. (1987). *Nucleic Acids Res.* **15**, 10393–10404.

Bonen, L. (1991). *Curr. Opin. Genet. Dev.* **1**, 515–522.

Bonen, L., Bird, S., and Belanger, L. (1990). *Plant Mol. Biol.* **15**, 793–795.

Bonen, L., Cunningham, R. S., Gray, M. W., and Doolittle, W. F. (1977). *Nucleic Acids Res.* **4**, 663–671.

Bonen, L., and Gray, M. W. (1980). *Nucleic Acids Res.* **8**, 319–335.

Bonitz, S. G., Berlani, R., Coruzzi, G., Li, M., Macino, G., Nobrega, F. G., Nobrega, M. P., Thalenfeld, B. E., and Tzagoloff, A. (1980). *Proc. Natl. Acad. Sci. U.S.A.* **77**, 3167–3170.

Borst, P. (1991). *Trends Genet.* **7**, 139–141.

Borst, P., and Swinkels, B. W. (1989). *In* "Evolutionary Tinkering in Gene Expression" (M. Grunberg-Manago, B. F. C. Clark, and H. G. Zachau, eds.), pp. 163–174. Plenum, New York.

Bouck, G. B. (1965). *J. Cell Biol.* **26**, 523–537.

Boynton, J. E., Harris, E. H., Burkhart, B. D., Lamerson, P. R., and Gillham, N. W. (1987). *Proc. Natl. Acad. Sci. U.S.A.* **84**, 2391–2395.

Brears, T., and Lonsdale, D. M. (1988). *Mol. Gen. Genet.* **214**, 514–522.

Breitenberger, C. A., Browning, K. S., Alzner-De Weerd, B., and RajBhandary, U. L. (1985). *EMBO J.* **4**, 185–195.

Breitenberger, C. A., and RajBhandary, U. L. (1985). *Trends Biochem. Sci.* **10**, 478–483.

Brennicke, A., Möller, S., and Blanz, P. A. (1985). *Mol. Gen. Genet.* **198**, 404–410.

Briat, J. F., Lescure, A. M., and Mache, R. (1986). *Biochimie* **68**, 981–990.

Brinkmann, H., Martinez, P., Quigley, F., Martin, W., and Cerff, R. (1987). *J. Mol. Evol.* **26**, 320–328.

Brown, G. G., Auchincloss, A. H., Covello, P. S., Gray, M. W., Menassa, R., and Singh, M. (1991). *Mol. Gen. Genet.* **228**, 345–355.

Brown, J. W., Daniels, C. J., and Reeve, J. N. (1989). *CRC Crit, Rev. Microbiol.* **16**, 287–337.

Brown, J. W., and Pace, N. R. (1992). *Nucleic Acids Res.* **20**, 1451–1456.

Brown, T. A. (1987). *In* "Gene Structure in Eukaryotic Microbes" (Vol. 22, Special Publications of the Society for General Microbiology) (J. R. Kinghorn, ed.), pp. 141–162. IRL Press, Oxford/Washington.

Brown, T. A., Constable, A., Waring, R. B., Scazzocchio, C., and Davies, R. W. (1989). *Nucleic Acids Res.* **17**, 5838.

Brown, T. A., Ray, J. A., Waring, R. B., Scazzocchio, C., and Davies, R. W. (1984). *Curr. Genet.* **8**, 489–492.

Brown, T. A., Waring, R. B., Scazzocchio, C., and Davies, R. W. (1985). *Curr. Genet.* **9**, 113–117.

Brown, W. M. (1985). *In* "Molecular Evolutionary Genetics" (R. J. MacIntyre, ed.), pp. 95–130. Plenum, New York/London.

Brugerolle, G. (1991). *Protoplasma* **164**, 70–90.

Bruns, T. D., and Palmer, J. D. (1989). *J. Mol. Evol.* **28**, 349–362.

Bruns, T. D., Palmer, J. D., Shumard, D. S., Grossman, L. I., and Hudspeth, M. E. S. (1988). *Curr. Genet.* **13**, 49–56.

Bryant, D. A., Schluchter, W. M., and Stirewalt, V. L. (1991). *Gene* **98**, 169–175.

Bryant, D. A., and Stirewalt, V. L. (1990). *FEBS Lett.* **259**, 273–280.

Buchheim, M. A., Turmel, M., Zimmer, E. A., and Chapman, R. L. (1990). *J. Phycol.* **26**, 689–699.

Burger, G., Citerrich, M. H., Nelson, M. A., Werner, S., and Macino, G. (1985). *EMBO J.* **4**, 197–204.

Burger-Wiersma, T., Veenhuis, M., Korthals, H. J., Van de Wiel, C. C. M., and Mur, L. R. (1986). *Nature (London)* **320**, 262–264.

Cammarano, P., Tiboni, O., and Sanangelantoni, A. M. (1989). *Can. J. Microbiol.* **35**, 2–10.

Cantatore, P., Roberti, M., Rainaldi, G., Gadaleta, M. N., and Saccone, C. (1989). *J. Biol. Chem.* **264**, 10,965–10,975.

Cantatore, P., and Saccone, C. (1987). *Int. Rev. Cytol.* **108**, 149–208.

Cattaneo, R. (1991). *Annu. Rev. Genet.* **25**, 71–88.

Cattolico, R. A. (1986). *Trends Ecol. Evol.* **1**, 64–67.

Cavalier-Smith, T. (1982). *Biol. J. Linn. Soc.* **17**, 289–306.

Cavalier-Smith, T. (1983). *In* "Endocytobiology II" (W. Schwemmler and H. E. A. Schenk, eds.), pp. 265–279. de Gruyter, Berlin.

Cavalier-Smith, T. (1986a). *Nature (London)* **324**, 416–417.

Cavalier-Smith, T. (1986b). *In* "Progress in Phycological Research" (F. E. Round and D. J. Chapman, eds.), Vol. 4, pp. 309–347. Biopress Ltd., Bristol.

Cavalier-Smith, T. (1987a). *Ann. N.Y. Acad. Sci.* **503**, 17–54.

Cavalier-Smith, T. (1987b). *Ann. N.Y. Acad. Sci.* **503**, 55–71.

Cavalier-Smith, T. (1987c). *Cold Spring Harbor Symp. Quant. Biol.* **52**, 805–824.

Cavalier-Smith, T. (1987d). *Nature (London)* **326**, 332–333.

Cavalier-Smith, T. (1989a). *Nature (London)* **339**, 100–101.

Cavalier-Smith, (1989b). *In* "The Chromophyte Algae: Problems and Perspectives" (J. C. Green, B. S. C. Leadbeater, and W. L. Diver, eds.), pp. 381–407. Clarendon Press, Oxford.

Cavalier-Smith, T. (1989c). *In* "Endocytobiology IV" (P. Nardon, V. Gianinazzi-Pearson, A. M. Grenier, L. Margulis, and D. C. Smith, eds.), pp. 515–521. Institut National de la Recherche Agronomique, Paris.

Cavalier-Smith, T. (1991). *In* "Evolution of Life" (S. Osawa and T. Honjo, eds.), pp. 271–304. Springer-Verlag, New York.

Cavalier-Smith, T. (1992). *Nature (London)* **356**, 570.

Cech, T. R. (1990). *Annu. Rev. Biochem.* **59**, 543–568.

Cech, T. R. (1991). *Cell* **64**, 667–669.

Cedergren, R., Abel, Y., and Sankoff, D. (1991). *In* "Molecular Techniques in Taxonomy" (Vol. 57 in "Series H: Cell Biology, NATO ASI Series") (G. M. Hewitt, A. W. B. Johnston, and J. P. W. Young, eds.), pp. 87–99. Springer-Verlag, Berlin.

Cedergren, R., Gray, M. W., Abel, Y., and Sankoff, D. (1988). *J. Mol. Evol.* **28**, 98–112.

Cedergren, R., and Grosjean, H. (1987). *BioSystems* **20**, 175–180.

Cedergren, R., and Lang, B. F. (1985). *BioSystems* **18**, 263–267.

Čerkasovová, A., Cerkasov, J., Kulda, J., and Reischig, J. (1976). *Folia Parasitol.* **23**, 33–37.

Chan, B. S.-S., Court, D. A., Vierula, P. J., and Bertrand, H. (1991). *Curr. Genet.* **20**, 225–237.

Chang, D. D., and Clayton, D. A. (1984). *Cell (Cambridge, Mass.)* **36**, 635–643.

Chang, D. D., and Clayton, D. A. (1986a). *Mol. Cell. Biol.* **6**, 3253–3261.

Chang, D. D., and Clayton, D. A. (1986b). *Mol. Cell. Biol.* **6**, 3262–3267.

Chang, D. D., and Clayton, D. A. (1987a). *EMBO J.* **6**, 409–417.

Chang, D. D., and Clayton, D. A. (1987b). *Science* **235**, 1178–1184.

Chang, D. D., and Clayton, D. A. (1989). *Cell (Cambridge, Mass.)* **56**, 131–139.

Chang, D. D., Fisher, R. P., and Clayton, D. A. (1987). *Biochim. Biophys. Acta* **909**, 85–91.

Chao, S., Sederoff, R., and Levings, C. S., III (1984) *Nucleic Acids Res.* **12**, 6629–6644.

Chapdelaine, Y., and Bonen, L. (1991). *Cell (Cambridge, Mass.)* **65**, 465–472.

Chatton, B., Walker, P., Ebel, J.-P., Lacroute, F., and Fasiolo, F. (1988). *J. Biol. Chem.* **263**, 52–57.

Chatton, E. (1938), "Titres et Travous Scientifiques (1906-1937) de Edouard Chatton." E. Sottano, Sète, France.

Chisholm, S. W., Olson, R. J., Zettler, E. R., Goericke, R., Waterbury, J. B., and Welsch-meyer, N. A. (1988). *Nature (London)* **334**, 340–343.

Chomyn, A., and Attardi, G. (1987). *Curr. Top. Bioenerg.* **15**, 295–329.

Christensen, T. (1964). *In* "Algae and Man" (D. F. Jackson, ed.), pp. 59–64. Plenum, New York.

Christianson, T., and Rabinowitz, M. (1983). *J. Biol. Chem.* **258**, 14,025–14,033.

Christianson, T. W., and Clayton, D. A. (1986). *Proc. Natl. Acad. Sci. U.S.A.* **83**, 6277–6281.

Christianson, T. W., and Clayton, D. A. (1988). *Mol. Cell. Biol.* **8**, 4502–4509.

Christopher, D. A., Cushman, J. C., Price, C. A., and Hallick, R. B. (1988). *Curr. Genet.* **14**, 275–286.

Chu, F. K., Maley, G. F., and Maley, F. (1988). *FASEB J.* **2**, 216–223.

Chu, F. K., Maley, G. F., West, D. K., Belfort, M., and Maley, F. (1986). *Cell (Cambridge, Mass.)* **45**, 157–166.

Clark-Walker, G. D. (1985). *In* "The Evolution of Genome Size" (T. Cavalier-Smith, ed.), pp. 277–297. Wiley, New York.

Clark-Walker, G. D., and Miklos, G. L. G. (1974) *Genet. Res. Camb.* **24**, 43–57.

Clark-Walker, G. D., McArthur, C. R., and Sriprakash, K. S. (1983). *J. Mol. Evol.* **19**, 333–341.

Clark-Walker, G. D., McArthur, C. R., and Sriprakash, K. S. (1985a). *EMBO J.* **4**, 465–473.

Clark-Walker, G. D., Evans, R. J., Hoeben, P., and McArthur, C. R. (1985b). *In* "Achieve-ments and Perspectives of Mitochondrial Research," Vol. II, "Biogenesis" (E. Quagliar-iello, E. C. Slater, F. Palmieri, C. Saccone, and A. M. Kroon, eds.), pp. 71–78. Elsevier, Amsterdam.

Clary, D. O., and Wolstenholme, D. R. (1985). *J. Mol. Evol.* **22**, 252–271.

Clayton, D. A. (1984). *Annu. Rev. Biochem.* **53**, 573–594.

Clayton, D. A. (1991). *Trends Biochem. Sci.* **16**, 107–111.

Clegg, M. T., Learn, G. H., and Golenberg, E. M. (1991). *In* "Evolution at the Molecular Level" (R. K. Selander, A. G. Clark, and T. S. Whittam, eds.), pp. 135–149. Sinauer, Sunderland, MA.

Coleman, A. W., Thompson, W. F., and Goff, L. J. (1991). *J. Protozool.* **38**, 129–135.

Colleaux, L., Michel-Wolwertz, M.-R., Matagne, R. F., and Dujon, B. (1990). *Mol. Gen. Genet.* **223**, 288–296.

Collins, R. A. (1990). *In* "Genetic Maps" (S. J. O'Brien, ed.), Book 3, 5th ed., pp. 19–21. Cold Spring Harbor Laboratory, Cold Spring Harbor, NY.

Conklin, P. L., and Hanson, M. R. (1991). *Nucleic Acids Res.* **19**, 2701–2705.

Conklin, P. L., Wilson, R. K., and Hanson, M. R. (1991). *Genes Dev.* **5**, 1407–1415.

Conrad-Webb, H., Perlman, P. S., Zhu, H., and Butow, R. A. (1990). *Nucleic Acids Res.* **8**, 1369–1376.

Cooper, H. L., Park, M. H., Folk, J. E., Safer, B., and Braverman, R. (1983). *Proc. Natl. Acad. Sci. U.S.A.* **80**, 1854–1857.

Coulthart, M. B., Huh, G. S., and Gray, M. W. (1990). *Curr. Genet.* **17**, 339–346.

Covello, P. S., and Gray, M. W. (1989). *Nature (London)* **341**, 662–666.

Covello, P. S., and Gray, M. W. (1990a). *Nucleic Acids Res.* **18**, 5189–5196.

Covello, P. S., and Gray, M. W. (1990b). *FEBS Lett.* **268**, 5–7.

Covello, P. S., and Gray, M. W. (1991). *Curr. Genet.* **20**, 245–251.

Covello, P. S., and Gray, M. W. (1992). *EMBO J.*, in press.

Cozens, A. L., and Walker, J. E. (1987). *J. Mol. Biol.* **194**, 359–383.

Cozens, A. L., Walker, J. E., Phillips, A. L., Huttly, A. K., and Gray, J. C. (1986). *EMBO J.* **5**, 217–222.

Culham, D. E., and Nazar, R. N. (1988). *Mol. Gen. Genet.* **212**, 382–385.

Cummings, D. J., McNally, K. L., Domenico, J. M., and Matsuura, E. T. (1990). *Curr. Genet.* **17**, 375–402.

Cundliffe, E. (1990). *In* "The Ribosome. Structure, Function, and Evolution" (W. E. Hill, A. Dahlberg, R. A. Garrett, P. B. Moore, D. Schlessinger, and J. R. Warner, eds.), pp. 479–490. American Society for Microbiology, Washington.

Cunningham, R. S., Bonen, L., Doolittle, W. F., and Gray, M. W. (1976a). *FEBS Lett.* **69**, 116–122.

Cunningham, R. S., Gray, M. W., Doolittle, W. F., and Bonen, L. (1976b). *In* "Acides Nucléiques et Synthèse de Protéines Chex les Végétaux" (L. Bogorad and J. H. Weil, eds.), pp. 243–248. Centre National de la Recherche Scientifique, Paris.

Curgy, J.-J. (1985). *Biol. Cell.* **54**, 1–38.

Curgy, J. J., Vavra, J., and Vivares, C. (1980). *Biol. Cell* **38**, 49–52.

Dahlberg, A. E. (1989). *Cell (Cambridge, Mass.)* **57**, 525–529.

Dale, R. M. K., McClure, B. A., and Houchins, J. P. (1985). *Plasmid* **13**, 31–40.

Dale, R. M. K., Mendu, N., Ginsburg, H., and Kridl, J. C. (1984). *Plasmid* **11**, 141–150.

Dang, H., and Ellis, S. R. (1990). *Nucleic Acids Res.* **18**, 6895–6901.

Darnell, J. E., and Doolittle, W. F. (1986). *Proc. Natl. Acad. Sci. U.S.A.* **83**, 1271–1275.

Dawson, A. J., Hodge, T. P., Isaac, P. G., Leaver, C. J., and Lonsdale, D. M. (1986). *Curr. Genet.* **10**, 561–564.

Dayhoff, M. O., and Schwartz, R. M. (1981). *Ann. N. Y. Acad. Sci.* **361**, 92–103.

de Bruijn, M. H. L., Schreier, P. H., Eperon, I. C., Barrell, B. G., Chen, E. Y., Armstrong, P. W., Wong, J. F. H., and Roe, B. A. (1980). *Nucleic Acids Res.* **8**, 5213–5222.

Decker, C. J., and Sollner-Webb, B. (1990). *Cell (Cambridge, Mass.)* **61**, 1001–1011.

De Duve, C. (1982). *Ann. N.Y. Acad. Sci.* **386**, 1–4.

De Duve, C. (1988). *Nature (London)* **336**, 209–210.

de la Cruz, V. F., Neckelmann, N., and Simpson, L. (1984). *J. Biol. Chem.* **259**, 15,15–15,147.

Delaney, T. P., and Cattolico, R. A. (1989). *Curr. Genet.* **15**, 221–229.

Delihas, N., Anderson, J., and Berns, D. (1985). *J. Mol. Evol.* **21**, 334–337.

Delihas, N., Andresini, W., Andersen, J., and Berns, D. (1982). *J. Mol. Biol.* **162**, 721–727.

Delihas, N., and Fox, G. E. (1987). *Ann. N.Y. Acad. Sci.* **503**, 92–102.

Denda, K., Fujiwara, T., Seki, M., Yoshida, M., Fukumori, Y., and Yamanaka, T. (1991). *Biochem. Biophys. Res. Commun.* **181**, 316–322.

Denda, K., Konishi, J., Oshima, T., Date, T., and Yoshida, M. (1988a). *J. Biol. Chem.* **263**, 6012–6015.

Denda, K., Konishi, J., Oshima, T., Date, T., and Yoshida, M. (1988b). *J. Biol. Chem.* **263**, 17,251–17,254.

Dennis, P. P. (1986). *J. Bacteriol.* **168**, 471–478.

Denovan-Wright, E. M., and Lee, R. W. (1992). *Curr. Genet.* **21**, 197–202.

Denslow, N. D., LiCata, V. J., Gualerzi, C., and O'Brien, T. W. (1988). *Biochemistry* **27**, 3521–3527.

de Pamphilis, C. W., and Palmer, J. D. (1990). *Nature (London)* **348**, 337–339.

Desjardins, P., and Morais, R. (1990). *J. Mol. Biol.* **212**, 599–634.

Devereus, R., Loeblich, A. R., III, and Fox, G. E. (1990). *J. Mol. Evol.* **31**, 18–24.

Dewey, R. E., Levings, C. S., III, and Timothy, D. H. (1986). *Cell (Cambridge, Mass.)* **44**, 439–449.

de Zamaroczy, M., and Bernardi, G. (1986). *Gene* **47**, 155–177.

Diffley, J. F. X., and Stillman, B. (1991). *Proc. Natl. Acad. Sci. U.S.A.* **88**, 7864–7868.

Dihanich, M. E., Najarian, D., Clark, R., Gillman, E. C., Martin, N. C., and Hopper, A. K. (1987). *Mol. Cell. Biol.* **7**, 177–184.

Dirheimer, G., and Martin, R. P. (1990). *In* "Chromatography and Modification of Nucleosides. Part B: Biological Roles and Function of Modification" (Vol. 45b in "Journal of Chromatography Library") (C. W. Gehrke and K. C. T. Kuo, eds.), pp. B197–B264. Elsevier, Amsterdam.

Doersen, C.-J., Guerrier-Takada, C., Altman S., and Attardi, G. (1985). *J. Biol. Chem.* **260**, 5942–5949.

Doolittle, R. F. (ed.) (1990). "Molecular Evolution: Computer Analysis of Protein and Nucleic Acid Sequences" (Vol. 183 of "Methods in Enzymology"). Academic Press, San Diego.

Doolittle, R. F., Feng, D. F., Anderson, K. L., and Alberro, M. R. (1990). *J. Mol. Evol.* **31**, 383–388.

Doolittle, W. F. (1980). *Trends Biochem. Sci.* **5**, 146–149.

Douglas, S. E. (1991). *Curr. Genet.* **19**, 289–294.

Douglas, S. E. (1992). *FEBS Lett.* **298**, 93–96.

Douglas, S. E., and Doolittle, W. F. (1984a). *FEBS Lett.* **166**, 307–310.

Douglas, S. E., and Doolittle, W. F. (1984b). *Nucleic Acids Res.* **12**, 3373–3386.

Douglas, S. E., and Durnford, D. G. (1989). *Plant Mol. Biol.* **13**, 13–20.

Douglas, S. E., and Durnford, D. G. (1990). *DNA Sequence—J. DNA Sequencing Mapp.* **1**, 55–62.

Douglas, S. E., Durnford, D. G., and Morden, C. W. (1990). *J. Phycol.* **26**, 500–508.

Douglas, S. E., and Gray, M. W. (1991). *Nature (London)* **352**, 290.

Douglas, S. E., Murphy, C. A., Spencer, D. F., and Gray, M. W. (1991). *Nature (London)* **350**, 148–151.

Douglas, S. E., and Turner, S. (1991). *J. Mol. Evol.* **33**, 267–273.

Dyson, N. J., Brown, T. A., Ray, J. A., Waring, R. B., Scazzocchio, C., and Davies, R. W. (1989). *J. Mol. Biol.* **208**, 587–599.

Edmonds, C. G., Crain, P. F., Gupta, R., Hashizume, T., Hocart, C. H., Kowalak, J. A., Pomerantz, S. C., Stetter, K. O., and McCloskey, J. A. (1991). *J. Bacteriol.* **173**, 3138–3148.

Ehrenman, K., Pedersen-Lane, J., West, D., Herman, R., Maley, F., and Belfort, M. (1986). *Proc. Natl. Acad. Sci. U.S.A.* **83**, 5875–5879.

Elliott, D. J., and Jacobs, H. T. (1989). *Mol. Cell. Biol.* **9**, 1069–1082.

Ellis, S. R., Morales, M. J., Li, J.-M., Hopper, A. K., and Martin, N. C. (1986). *J. Biol. Chem.* **261**, 9703–9709.

Eperon, I. C., Janssen, J. W. G., Hoeijmakers, J. H. J., and Borst, P. (1983). *Nucleic Acids Res.* **11**, 105–125.

Eschbach, S., Hofmann, C. J. B., Maier, U.-G., Sitte, P., and Hansmann, P. (1991a). *Nucleic Acids Res.* **19**, 1779–1781.

Eschbach, S., Wolters, J., and Sitte, P. (1991b). *J. Mol. Evol.* **32**, 247–252.

Esser, K., Kück, U., Lang-Hinrichs, C., Lemke, P., Osiewacz, H. D., Stahl, U., and Tudzynski, P. (1986). "Plasmids of Eukaryotes, Fundamentals and Applications." Springer-Verlag, Berlin.

Evans, R. J., and Clark-Walker, G. D. (1985). *Genetics* **111**, 403–432.

Evrard, J. L., Kuntz, M., and Weil, J. H. (1990). *J. Mol. Evol.* **30**, 16–25.

Falconet, D., Sevignac, M., and Quétier, F. (1988). *Curr. Genet.* **13**, 75–82.

Fauron, C., Havlik, M., Lonsdale, D., and Nichols, L. (1989). *Mol. Gen. Genet.* **216**, 395–401.

Fauron, C. M.-R., and Havlik, M. (1988). *Nucleic Acids Res.* **16**, 10,395–10,396.

Fauron, C. M.-R., Havlik, M., and Brettell, R. I. S. (1990). *Genetics* **124**, 423–428.

Feagin, J. E. (1990). *J. Biol. Chem.* **265**, 19,373–19,376.

Feagin, J. E. (1992). *Mol. Biochem. Parasitol.* **52**, 145–148.

Feagin, J. E., Abraham, J. M., and Stuart, K. (1988). *Cell (Cambridge, Mass.)* **53**, 413–422.

Feagin, J. E., Gardner, M. J., Williamson, D. H., and Wilson, R. J. M. (1991). *J. Protozool.* **38**, 243–245.

Feagin, J. E., Werner, E., Gardner, M. J., Williamson, D. H., and Wilson, R. J. M. (1992). *Nucleic Acids Res.* **20**, 879–887.

Fearnley, I. M., Runswick, M. J., and Walker, J. E. (1989). *EMBO J.* **8**, 665–672.

Fearon, K., and Mason, T. L. (1988). *Mol. Cell. Biol.* **8**, 3636–3646.

Felsenstein, J. (1978). *Sys. Zool.* **27**, 401–410.

Felsenstein, J. (1985). *Evolution* **39**, 783–791.

Felsenstein, J. (1988). *Annu. Rev. Genet.* **22**, 521–565.

Finnegan, P. M., and Brown, G. G. (1990). *Plant Cell* **2**, 71–83.

Fisher, R. P., and Clayton, D. A. (1985). *J. Biol. Chem.* **260**, 11,330–11,338.

Fisher, R. P., and Clayton, D. A. (1988). *Mol. Cell. Biol.* **8**, 3496–3509.

Fisher, R. P., Lisowsky, T., Breen, G. A. M., and Clayton, D. A. (1991). *J. Biol. Chem.* **266**, 9153–9160.

Fisher, R. P., Lisowsky, T., Parisi, M. B., and Clayton, D. A. (1992). *J. Biol. Chem.* **267**, 3358–3367.

Fisher, R. P., Parisi, M. A., and Clayton, D. A. (1989). *Genes Dev.* **3**, 2202–2217.

Fisher, R. P., Topper, J. N., and Clayton, D. A. (1987) *Cell (Cambridge, Mass.)* **50**, 247–258.

Forster, A. C., and Altman, S. (1990). *Cell (Cambridge, Mass.)* **62**, 407–409.

Forterre, P. (1992). *Nucleic Acids Res.* **20**, 1811.

Fox, G. E., Magrum, L. J., Balch, W. E., Wolfe, R. S., and Woese, C. R. (1977). *Proc. Natl. Acad. Sci. U.S.A.* **74**, 4537–4541.

Fox, G. E., Stackebrandt, E., Hespell, R. B., Gibson, J., Maniloff, J., Dyer, T. A., Wolfe, R. S., Balch, W. E., Tanner, R. S., Magrum, L. J., Zablen, L. B., Blakemore, R., Gupta, R., Bonen, L., Lewis, B. J., Stahl, D. A., Luehrsen, K. R., Chen, K. N., and Woese, C. R. (1980). *Science* **209**, 457–463.

Fox, T. D. (1987). *Annu. Rev. Genet.* **21**, 67–91.

Fox, T. D., and Leaver, C. J. (1981). *Cell (Cambridge, Mass.)* **26**, 315–323.

Francis, M. A., and Dudock, B. S. (1982). *J. Biol. Chem.* **257**, 11,195–11,198.

Fuerst, J. A., and Webb, R. I. (1991). *Proc. Natl. Acad. Sci. U.S.A.* **88**, 8184–8188.

Fukunaga, M., Horie, I., and Mifuchi, I. (1990). *J. Bacteriol.* **172**, 3264–3268.

Fukunaga, M., and Mifuchi, I. (1989). *J. Bacteriol.* **171**, 5763–5767.

Fukuzawa, H., Kohchi, T., Shirai, H., Ohyama, K., Umesono, K., Inokuchi, H., and Ozeki, H. (1986). *FEBS Lett.* **198**, 11–15.

Gabellini, N., Harnisch, U., McCarthy, J. E. G., Hauska, G., and Sebald, W. (1985). *EMBO J.* **4**, 549–553.

Gadaleta, G., Pepe, G., De Candia, G., Quagliariello, C., Sbisà, E., and Saccone, C. (1989). *J. Mol. Evol.* **28**, 497–516.

Gajadhar, A. A., Marquardt, W. C., Hall, R., Gunderson, J., Ariztia-Carmona, E. V., and Sogin, M. L. (1991). *Mol. Biochem. Parasitol.* **45**, 147–154.

Gantt, J. S., Baldauf, S. L., Calie, P. J., Weeden, N. F., and Palmer, J. D. (1991). *EMBO J.* **10**, 3073–3078.

Gardner, M. J., Bates, P. A., Ling, I. T., Moore, D. J., McCready, S., Gunasekera, M. B. R., Wilson, R. J. M., and Williamson, D. H. (1988). *Mol. Biochem. Parasitol.* **31**, 11–18.

Gardner, M. J., Feagin, J. E., Moore, D. J., Spencer, D. F., Gray, M. W., Williamson, D. H., and Wilson, R. J. M. (1991a). *Mol. Biochem. Parasitol.* **48**, 77–88.

Gardner, M. J., Williamson, D. H., and Wilson, R. J. M. (1991b). *Mol. Biochem. Parasitol.* **44**, 115–124.

Garrett, R. A., Dalgaard, J., Larsen, N., Kjems, J., and Mankin, A. S. (1991). *Trends Biochem. Sci.* **16**, 22–26.

Gatti, D. L., and Tzagoloff, A. (1991). *J. Mol. Biol.* **218**, 557–568.

Gest, H., and Favinger, J. L. (1983). *Arch. Microbiol.* **136**, 11–16.

Gibbs, S. P. (1962). *J. Cell Biol.* **14**, 433–444.

Gibbs, S. P. (1978). *Can. J. Bot.* **56**, 2883–2889.

Gibbs, S. P. (1981). *Ann. N. Y. Acad. Sci.* **361**, 193–207.

Gibbs, S. P. (1990). *In* "Cell Walls and Surfaces, Reproduction, Photosynthesis" (Vol. 1 of "Experimental Phycology") (W. Wiessner, D. G. Robinson, and R. C. Starr, eds.), pp. 145–157. Springer-Verlag, Berlin.

Gibson, T. J., and Lamond, A. I. (1990). *J. Mol. Evol.* **30**, 7–15.

Gilbert, W. (1986). *Nature (London)* **319**, 618.

Gillham, N. W., and Boynton, J. E. (1981). *Ann. N.Y. Acad. Sci.* **361**, 20–40.

Gillott, M. A., and Gibbs, S. P. (1980). *J. Phycol.* **16**, 558–568.

Giovannoni, S. J., Turner, S., Olsen, G. J., Barns, S., Lane, D. J., and Pace, N. R. (1988). *J. Bacteriol.* **170**, 3584–3592.

Goff, L. J., and Coleman, A. W. (1987). *Ann. N.Y. Acad. Sci.* **503**, 402–423.

Gogarten, J. P., and Kibak, H. (1992). *In* "The Origin and Evolution of Prokaryotic and Eukaryotic Cells" (H. Hartman and K. Matsuno, eds.). World Scientific, New Jersey, in press.

Gogarten, J. P., Kibak, H., Dittrich, P., Taiz, L., Bowman, E. J., Bowan, B. J., Manolson, M. F., Poole, R. J., Date, T., Oshima, T., Konishi, J., Denda, K., and Yoshida, M. (1989). *Proc. Natl. Acad. Sci. U.S.A.* **86**, 6661–6665.

Gold, H. A., Topper, J. N., Clayton, D. A., and Craft, J. (1989). *Science* **245**, 1377–1380.

Goldbach, R. W., Borst, P., Bollen-De Boer, J. E., and van Bruggen, E. F. J. (1978). *Biochim. Biophys. Acta* **521**, 169–186.

Goldschmidt-Clermont, M., Choquet, Y., Girard-Bascou, J., Michel, F., Schirmer-Rahire, M., and Rochaix, J.-D. (1991). *Cell (Cambridge, Mass.)* **65**, 135–143.

Goodrich-Blair, H., Scarlato, V., Gott, J. J., Xu, M.-Q., and Shub, D. A. (1990) *Cell (Cambridge, Mass.)* **63**, 417–424.

Gordon, E. D., Mora R., Meredith, S. C., Lee, C., and Lindquist, S. L. (1987). *J. Biol. Chem.* **262**, 16,585–16,589.

Grabau, E. A. (1985). *Plant Mol. Biol.* **5,** 119–124.
Grabau, E. A. (1987). *Curr. Genet.* **11,** 287–293.
Grant, D., and Chiang, K.-S. (1980). *Plasmid* **4,** 82–96.
Gray, M. W. (1982). *Can. J. Biochem.* **60,** 157–171.
Gray, M. W. (1983). *BioScience* **33,** 693–699.
Gray, M. W. (1988). *Biochem. Cell Biol.* **66,** 325–348.
Gray, M. W. (1989a). *Annu. Rev. Cell Biol.* **5,** 25–50.
Gray, M. W. (1989b). *Trends Genet.* **5,** 294–299.
Gray, M. W. (1990). *In* "Plant Physiology, Biochemistry and Molecular Biology" (D. T. Dennis and D. H. Turpin, eds.), pp. 147–159. Longman Group, Harlow, UK.
Gray, M. W. (1991). *In* "The Molecular Biology of Plastids" (Vol. 7A in "Cell Culture and Somatic Cell Genetics of Plants") (L. Bogorad and I. K. Vasil, eds.), pp. 303–330. Academic Press, San Diego.
Gray, M. W., and Boer, P. H. (1988). *Phil. Trans. R. Soc. Lond. B* **319,** 135–147.
Gray, M. W., Cedergren, R., Abel, A., and Sankoff, D. (1989). *Proc. Natl. Acad. Sci. U.S.A.* **86,** 2267–2271.
Gray, M. W., and Doolittle, W. F. (1982). *Microbiol. Rev.* **46,** 1–42.
Gray, M. W., Hanic-Joyce, P. J., and Covello, P. S. (1992). *Annu. Rev. Plant Physiol. Plant Mol. Biol.* **43,** 145–175.
Gray, M. W., Sankoff, D., and Cedergren, R. J. (1984). *Nucleic Acids Res.* **12,** 5837–5852.
Gray, M. W., and Spencer, D. F. (1981). *Nucleic Acids Res.* **9,** 3523–3529.
Gray, M. W., and Spencer, D. F. (1983). *FEBS Lett.* **161,** 323–327.
Greenleaf, A. L., Kelly, J. L., and Lehman, I. R. (1986). *Proc. Natl. Acad. Sci. U.S.A.* **83,** 3391–3394.
Greenwood, A. D., Griffiths, H. B., and Santore, U. J. (1977). *Br. Phycol. J.* **12,** 119.
Grohmann, L., Graack, H.-R., and Kitakawa, M. (1989). *Eur. J. Biochem.* **183,** 155–160.
Gropp, F., Reiter, W. D., Sentenac, A., Zillig, W., Schnabel, R., Thomm, M., and Stetter, K. O. (1986). *System. Appl. Microbiol.* **7,** 95–101.
Grossman, L. I., and Hudspeth, M. E. S. (1985). *In* "Gene Manipulations in Fungi" (J. W. Bennett and L. L. Lasure, eds.), pp. 65–103. Academic Press, New York.
Gruissem, W. (1989a). *Cell (Cambridge, Mass.)* **56,** 161–170.
Gruissem, W. (1989b). *In* "Molecular Biology" (Vol. 15 of "The Biochemistry of Plants") (Marcus, A., ed.), pp. 151–191. Academic Press, San Diego.
Gualberto, J. M., Lamattina, L., Bonnard, G., Weil, J.-H., and Grienenberger, J.-M. (1989). *Nature (London)* **341,** 660–662.
Gualberto, J. M., Wintz, H., Weil, J.-H., and Grienenberger, J.-M. (1988). *Mol. Gen. Genet.* **215,** 118–127.
Guerrier-Takada, C., Gardiner, K., Marsh, T., Pace, N., and Altman, S. (1983). *Cell (Cambridge, Mass.)* **35,** 849–857.
Gunderson, J. H., Elwood, H., Ingold, A., Kindle, K., and Sogin, M. L. (1987). Proc. Natl. Acad. Sci. U.S.A. **84,** 5823–5827.
Gouy, M., and Li, W.-H. (1989). *Nature (London)* **339,** 145–147.
Gouy, M., and Li, W.-H. (1990). *Nature (London)* **343,** 419.
Gutell, R. R., Schnare, M. N., and Gray, M. W. (1990). *Nucleic Acids Res.* **18** (Suppl.), 2319–2330.
Gutell, R. R., Weiser, B., Woese, C. R., and Noller, H. F. (1985). *Progr. Nucleic Acid Res. Mol. Biol.* **32,** 155–216.
Hall, J. L., Ramanis, Z., and Luck, D. J. L. (1989). *Cell (Cambridge, Mass.)* **59,** 121–132.
Hall, W. T., and Claus, G. (1963). *J. Cell Biol.* **19,** 551–563.
Hancock, K., and Hajduk, S. L. (1990). *J. Biol. Chem.* **265,** 19,208–19,215.
Hanic-Joyce, P. J., and Gray, M. W. (1990). *J. Biol. Chem.* **265,** 13,782–13,791.

Hanic-Joyce, P. J., and Gray, M. W. (1991). *Mol. Cell. Biol.* **11,** 2035–2039.

Hanic-Joyce, P. J., Spencer, D. F., and Gray, M. W. (1990). *Plant Mol. Biol.* **15,** 551–559.

Hanley-Bowdoin, L., and Chua, N.-H. (1987). *Trends Biochem. Sci.* **12,** 67–70.

Harington, A., and Thornley, A. L. (1982). *J. Mol. Evol.* **18,** 287–292.

Hartman, H. (1984). *Specul. Sci. Technol.* **7,** 77–81.

Hatakeyama, T., Kaufmann, F., Schroeter, B., and Hatakeyama, T. (1989). *Eur. J. Biochem.* **185,** 685–693.

Heckman, J. E., Sarnoff, J., Alzner-DeWeerd, B., Yin, S., and RajBhandary, U. L. (1980). *Proc. Natl. Acad. Sci. U.S.A.* **77,** 3159–3163.

Heinemann, J. A. (1991) *Trends Genet.* **7,** 181–185.

Heinonen, T. Y. K. Schnare, M. N., and Gray, M. W. (1990). *J. Biol. Chem.* **265,** 22,336–22,341.

Heinonen, T. Y. K., Schnare, M. N., Young, P. G., and Gray, M. W. (1987). *J. Biol. Chem.* **262,** 2879–2887.

Hekele, A., and Beier, H. (1991). *Nucleic Acids Res.* **19,** 1941.

Henderson, C. E., Perham, R. N., and Finch, J. T. (1979). *Cell (Cambridge, Mass.)* **17,** 85–93.

Hendriks, L., De Baere, R., Van de Peer, Y., Neefs, J., Goris, A., and De Wachter, R. (1991). *J. Mol. Evol.* **32,** 167–177.

Hensel, R., Zwicki, P., Fabry, S., Lang, J., and Palm, P. (1989). *Can. J. Microbiol.* **35,** 81–85.

Herbert, C. J., Labouesse, M., Dujardin, G., and Slonimski, P. P. (1988). *EMBO J.* **7,** 473–483.

Herdman, M., and Stainer, R. Y. (1977). *FEMS Microbiol. Lett.* **1,** 7–12.

Hiesel, R., Schobel, W., Schuster, W., and Brennicke, A. (1987). *EMBO J.* **6,** 29–34.

Hiesel, R., Wissinger, B., Schuster, W., and Brennicke, A. (1989). *Science* **246,** 1632–1634.

Hildebrand, M., Hallick, R. B., Passavant, C. W., and Bourque, D. P. (1988). *Proc.. Natl. Acad. Sci. U.S.A* **85,** 372–376.

Hildebrandt, P., Heibel, G., Anemüller, S., and Schäfer, G. (1991). *FEBS Lett.* **283,** 131–134.

Hilpert, R., Winter, J., Hammes, W., and Kandler, O. (1981). *Zbl. Bakr. Hyg. I. Abt. Orig. C* **2,** 11–20.

Hintz, W. E., Mohan, M., Anderson, J. B., and Horgen, P. A. (1985). *Curr. Genet.* **9,** 127–132.

Hiratsuka, J., Shimada, H., Whittier, R., Ishibashi, T., Sakamoto, M., and Mori, M., Kondo, C., Honji, Y., Sun. C.-R., Meng, B.-Y., Li, Y.-Q., Kanno, A., Nishizawa, Y., Hirai, A., Shinozaki, K., and Sugiura, M. (1989). *Mol. Gen. Genet.* **217,** 185–194.

Hixson, J. E., and Clayton, D. A. (1985). *Proc. Natl. Acad. Sci. U.S.A.* **82,** 2660–2664.

Hoch, B., Maier, R. M., Appel, K., Igloi, G. L., and Kössel, H. (1991). *Nature (London)* **353,** 178–180.

Hoeben, P., and Clark-Walker, G. D. (1986). *Curr. Genet.* **10,** 371–379.

Hoeijmakers, J. H. J., Schoutsen, B., and Borst, P. (1982). *Plasmid* **7,** 199–209.

Hollingsworth, M. J., and Martin, N. C. (1986). *Mol. Cell. Biol.* **6,** 1058–1064.

Hori, H., and Osawa, S. (1987). *Mol. Biol. Evol.* **4,** 445–472.

Howe, C. J. (1985). *Curr. Genet.* **10,** 139–145.

Huang, M., Briggs, D. R., Clark-Walker, G. D., and Linnane, A. W. (1966). *Biochim. Biophys. Acta* **114,** 434–436.

Hüdepohl, U., Reiter, W.-D., and Zillig, W. (1990). *Proc. Natl. Acad. Sci. U.S.A.* **87,** 5851–5855.

Huet, J., Schnabel, R., Sentenac, A., and Zillig, W. (1983). *EMBO J.* **2,** 1291–1294.

Huh, T. Y., and Gray, M. W. (1982). *Plant Mol. Biol.* **1,** 245–249.

Hummel, H., and Böck, A. (1985). *Mol. Gen. Genet.* **198,** 529–533.

Hunt, L. T., George, D. G., and Barker, W. C. (1985). *Biosystems* **18**, 223–240.
Hunt, M. D., and Newton, K. J. (1991). *EMBO J.* **10**, 1045–1052.
Hwang, S.-R., and Tabita, F. R. (1989). *Plant Mol. Biol.* **13**, 69–79.
Ichikawa, H., Tanno-Suenaga, L., and Imamura, J. (1989). *Theor. Appl. Genet.* **77**, 39–43.
Ishihara, R., and Hayashi, Y. (1968). *J. Invertebr. Pathobiol.* **11**, 377–385.
Itoh, T. (1989). *Eur. J. Biochem.* **186**, 213–219.
Iwabe, N., Kuma, K.-i., Hasegawa, M., Osawa, S., and Miyata, T. (1989). *Proc. Natl. Acad. Sci. U.S.A.* **86**, 9355–9359.
Iwabe, N., Kuma, K.-i., Kishino, H., Hasegawa, M., and Miyata, T. (1991). *J. Mol. Evol.* **32**, 70–78.
Jacobs, H. T. (1991). *J. Mol. Evol.* **32**, 333–339.
Jacobs, H. T., Elliott, D., Math, V. B., and Farquharson, A. (1988). *J. Mol. Biol.* **202**, 185–217.
James, B. D., Olsen, G. J., Liu, J., and Pace, N. R. (1988). *Cell (Cambridge, Mass.)* **52**, 19–26.
Jang, S. H., and Jaehning, J. A. (1991). *J. Biol. Chem.* **266**, 22,671–22,677.
Janssen, I., Mucke, H., Löffelhardt, W., and Bonhert, H. J. (1987). *Plant Mol. Biol.* **9**, 479–484.
Jeon, K. W. (1987). *Ann. N.Y. Acad. Sci.* **503**, 359–371.
Jess, W., Palm, P., Evers, R., Köck, J., and Cornelissen, A. W. C. A. (1990). *Curr. Genet.* **18**, 547–551.
Jin, L., and Nei, M. (1990). *Mol. Biol. Evol.* **7**, 82–102.
John, P., and Whatley, F. R. (1975a). *Nature (London)* **254**, 495–498.
John, P., and Whatley, F. R. (1975b). *Symp. Soc. Exp. Biol.* **29**, 39–40.
Johnson, A. M., Illana, S., Hakendorf, P., and Baverstock, P. R. (1988). *J. Parasitol.* **74**, 847–860.
Johnson, D. E., and Dutcher, S. K. (1991). *J. Cell Biol.* **113**, 339–346.
Johnson, K. A., and Rosenbaum, J. L. (1990). *Cell (Cambridge, Mass.)* **62**, 615–619.
Joyce, G. F. (1989). *Nature (London)* **338**, 217–224.
Joyce, P. B. M., and Gray, M. W. (1989a) *Nucleic Aids Res.* **17**, 5461–5476.
Joyce, P. B. M., and Gray, M. W. (1989b) *Nucleic Acids Res.* **17**, 7865–7878.
Joyce, P. B. M., Spencer, D. F., Bonen, L., and Gray, M. W. (1988). *Plant Mol. Biol.* **10**, 251–262.
Jukes, T. H. (1990). *Experientia* **46**, 1149–1157.
Jukes, T. H., and Osawa, S. (1990). *Experientia* **46**, 1117–1126.
Karlovsky, P., and Fartmann, B. (1992). *J. Mol. Evol.* **34**, 254–258.
Keha, E. E., Ronft, H., and Kresze, G.-B. (1982). *FEBS Lett.* **145**, 289–292.
Kemmerer, E. C., Lei, M., and Wu, R. (1991). *J. Mol. Evol.* **32**, 227–237.
Kennell, J. C., and Lambowitz, A. M. (1989). *Mol. Cell. Biol.* **9**, 3603–3613.
Kessel, M., and Klink, F. (1980). *Nature (London)* **287**, 250–251.
Kessler, U., Maid, U., and Zetsche, K. (1992). *Plant Mol. Biol.* **18**, 777–780.
Kilejian, A. (1975). *Biochim. Biophys. Acta* **390**, 276–284.
Kimura, M., Arndt, E., Hatakeyama, T., Hatakeyama, T., and Kimura, J. (1989). *Can. J. Microbiol.* **35**, 195–199.
Kishino, H., Miyata, T., and Hasegawa, M. (1990). *J. Mol. Evol.* **31**, 151–160.
Kitakawa, M., Grohmann, L., Graack, H.-R., and Isono, K. (1990). *Nucleic Acids Res.* **18**, 1521–1529.
Kjems, J., and Garrett, R. A. (1990). *J. Mol. Evol.* **31**, 25–32.
Kjems, J., Leffers, H., Garrett, R. A., Wich, G., Leinfelder, W., and Böck, A. (1987). *Nucleic Acids Res.* **15**, 4821–4835.
Köchel, H. G., and Küntzel, H. (1982). *Nucleic Acids Res.* **10**, 4795–4801.

Koerner, T. J., Myers, A. M., Lee, S., and Tzagoloff, A. (1987). *J. Biol. Chem.* **262,** 3690–3696.

Koller, B., and Delius, H. (1984). *Cell (Cambridge, Mass.)* **36,** 613–622.

Koller, B., Fromm, H., Galun, E., and Edelman, M. (1987). *Cell (Cambridge, Mass.)* **48,** 111–119.

Köpke, A. K. E., and Wittmann-Liebold, B. (1989). *Can. J. Microbiol.* **35,** 11–20.

Koslowsky, D. J., Bhat, G. J., Perollaz, A. L., Feagin, J. E., and Stuart, K. (1990). *Cell (Cambridge, Mass.)* **62,** 901–911.

Kostrzewa, M., Valentin, K., Maid, U., Radetzky, R., and Zetsche, K. (1990). *Curr. Genet.* **18,** 465–469.

Kováč, L., Lazowska, J., and Slonimski, P. P. (1984). *Mol. Gen. Genet.* **197,** 420–424.

Kowallik, K. V. (1989). *In* "The Chromophyte Algae: Problems and Perspectives" (J. C. Green, B. S. C. Leadbeater, and W. L. Diver, eds.), pp. 101–124. Clarendon Press, Oxford.

Kraus, M., Götz, M., and Löffelhardt, W. (1990). *Plant Mol. Biol.* **15,** 561–573.

Kreader, C. A., Langer, C. S., and Heckman, J. E. (1989). *J. Biol. Chem.* **264,** 317–327.

Kruse, B., Narasimhan, N., and Attardi, G. (1989). *Cell (Cambridge, Mass.)* **58,** 391–397.

Kubelik, A. R., Kennell, J. C., Akins, R. A., and Lambowitz, A. M. (1990). *J. Biol. Chem.* **265,** 4515–4526.

Kuchino, Y., Ihara, M., Yabusaki, Y., and Nishimura, S. (1982). *Nature (London)* **298,** 684–685.

Kück, U., Choquet, Y., Schneider, M., Dron, M., and Bennoun, P. (1987). *EMBO J.* **6,** 2185–2195.

Kück, U., Godehardt, I., and Schmidt, U. (1990). *Nucleic Acids Res.* **18,** 2691–2697.

Kück, U., and Neuhaus, H. (1986). *Appl. Microbiol. Biotechnol.* **23,** 462–469.

Kudla, J., Igloi, G. L., Metzlaff, M., Hagemann, R., and Kössel, H. (1992). *EMBO J.* **11,** 1099–1103.

Kuhsel, M. G., Strickland, R., and Palmer, J. D. (1990). *Science* **250,** 1570–1573.

Kuiper, M. T. R., Akins, R. A., Holtrop, M. de Vries, H., and Lambowitz, A. M. (1988). *J. Biol. Chem.* **263,** 2840–2847.

Kumano, M., Tomioka, N., and Sugiura, M. (1983). *Gene* **24,** 219–225.

Küntzel, H., and Köchel, H. G. (1981). *Nature (London)* **293,** 751–755.

Küntzel, H., and Noll, H. (1967). *Nature (London)* **215,** 1340–1345.

Kuzmin, E. V., Levchenko, I. V., and Zaitseva, G. N. (1988). *Nucleic Acids Res.* **16,** 4177.

Lake, J. A. (1986a). *Nature (London)* **319,** 626.

Lake, J. A. (1986b). *Nature (London)* **321,** 657–658.

Lake, J. A. (1987a). *Mol. Biol. Evol.* **4,** 167–191.

Lake, J. A. (1987b). *Cold Spring Harbor Symp. Quant. Biol.* **52,** 839–846.

Lake, J. A. (1988). *Nature (London)* **331,** 184–186.

Lake, J. A. (1989). *Can. J. Microbiol.* **35,** 109–118.

Lake, J. A. (1990a). *Nature (London)* **343,** 418–419.

Lake, J. A. (1990b). *In* "The Ribosome: Structure, Function, and Evolution" (W. E. Hill, A. Dahlberg, R. A. Garrett, P. B. Moore, D. Schlessinger, and J. R. Warner, eds.), pp. 579–588. American Society for Microbiology, Washington.

Lake, J. A. (1991). *Trends Biochem. Sci.* **16,** 46–50.

Lake, J. A., Clark, M. W., Henderson, E., Fay, S. P., Oakes, M., Scheinman, A., Thornber, J. P., and Mah, R. A. (1985). *Proc. Natl. Acad. Sci. U.S.A.* **82,** 3716–3720.

Lake, J. A., Henderson, E., Oakes, M., and Clark, M. W. (1984). *Proc. Natl. Acad. Sci. U.S.A.* **81,** 3786–3790.

Lamb, A. J., Clark-Walker, G. D., and Linnane, A. W. (1968). *Biochim. Biophys. Acta* **161,** 415–427.

Lambert, D. H., Bryant, D. A., Stirewalt, V. L., Dubbs, J. M., Stevens, S. E., Jr., and Porter, R. D. (1985). *J. Bacteriol.* **164,** 659–664.

Lambowitz, A. M. (1989). *Cell (Cambridge, Mass.)* **56,** 323–326.

Lambowitz, A. M., and Perlman, P. S. (1990) *Trends Biochem. Sci.* **15,** 440–444.

Lane, D. J., Pace, B., Olsen, G. J., Stahl, D. A., Sogin, M. L., and Pace, N. R. (1985). *Proc. Natl. Acad. Sci. U.S.A.* **82,** 6955–6959.

Lang, B. F., Cedergren, R., and Gray, M. W. (1987). *Eur. J. Biochem.* **169,** 527–537.

Larsen, N., Leffers, H., Kjems, J., and Garrett, R. A. (1986). *System. Appl. Microbiol.* **7,** 49–57.

Lazowska, J., Jacq, C., and Slonimski, P. P. (1980). *Cell (Cambridge, Mass.)* **22,** 333–348.

Leaver, C. J., and Gray, M. W. (1982). *Annu. Rev. Plant Physiol.* **33,** 373–402.

Leaver, C. J., and Harmey, M. A. (1976). *Biochem. J.* **157,** 275–277.

Lechner, K., and Böck, A. (1987). *Mol. Gen. Genet.* **208,** 523–528.

Lechner, K., Heller, G., and Böck, A. (1988). *Nucleic Acids Res.* **16,** 7817–7826.

Lederer, H. (1986). *Nature (London)* **320,** 220.

Lee, J. J., and Corliss, J. O. (1985). *J. Protozool.* **32,** 371–372.

Lee, J. J., Soldo, A. T., Reisser, W., Lee, M. J., Jeon, K. W., and Görtz, H.-D. (1985). *J. Protozool.* **32,** 391–403.

Lee, R. W., Dumas, C., Lemieux, C., and Turmel, M. (1991). *Mol. Gen. Genet.* **231,** 53–58.

Leffers, H., Kjems, J., Østergaard, L., Larsen, N., and Garrett, R. A. (1987). *J. Mol. Biol.* **195,** 43–61.

Lemieux, B., and Lemieux, C. (1985). *Curr. Genet.* **10,** 213–219.

Lemieux, B., Turmel, M., and Lemieux, C. (1985). *BioSystems* **18,** 293–298.

Lenaers, G., Scholin, C., Bhaud, Y., Saint-Hilaire, D., and Herzog, M. (1991). *J. Mol. Evol.* **32,** 53–63.

Levine, N. D. (1985). *In* "Illustrated Guide to the Protozoa" (J. J. Lee, S. H. Hutner, and E. C. Bovee, eds.), pp. 322–374. Society of Protozoologists, Lawrence, KS.

Levings, C. S., III, and Brown, G. G. (1989). *Cell (Cambridge, Mass.)* **56,** 171–179.

Lewin, R. (1986). *Science* **231,** 545–546.

Lewin, R. A. (1976). *Nature (London)* **261,** 697–698.

Lewin, R. A. (1977). *Phycologia* **16,** 217.

Lewin, R. A. (1981). *Ann. N.Y. Acad. Sci.* **361,** 325–328.

Lewin, R. A., and Withers, N. W. (1975). *Nature (London)* **256,** 735–737.

Li, W.-H., and Graur, D. (1991). "Fundamentals of Molecular Evolution." Sinauer, Sunderland, MA.

Liaud, M.-F., Zhang, D. X., and Cerff, R. (1990). *Proc. Natl. Acad. Sci. U.S.A.* **87,** 8918–8922.

Lidholm, J., Szmidt, A. E., Hällgren, J.-E., and Gustafsson, P. (1988). *Mol. Gen. Genet.* **212,** 6–10.

Lindmark, D. G., and Müller, M. (1973). *J. Biol. Chem.* **248,** 7724–7728.

Link, G. (1984). *EMBO J.* **3,** 1697–1704.

Lisowsky, T., and Michaelis, G. (1988). *Mol. Gen. Genet.* **214,** 218–223.

Lockhart, P. J., Howe, C. J., Bryant, D. A., Beanland, T. J., and Larkum, A. W. D. (1992a). *J. Mol. Evol.* **34,** 153–162.

Lockhart, P. J., Penny, D., Hendy, M. D., Howe, C. J., Beanland, T. J., and Larkum, A. W. D. (1992b) *FEBS Lett.* **301,** 127–131.

Lonsdale, D. M. (1984). *Plant Mol. Biol.* **3,** 201–206.

Lonsdale, D. M. (1989). *Biochem. Plants* **15,** 229–295.

Lonsdale, D. M., Hodge, T. P., and Fauron, C. M.-R. (1984). *Nucleic Acids Res.* **12,** 9249–9261.

Lübben, M., Kolmerer, B., and Saraste, M. (1992). *EMBO J.* **11,** 805–812.

Ludwig, B., and Schatz, G. (1980). *Proc. Natl. Acad. Sci. U.S.A.* **77**, 196–200.
Ludwig, M., and Gibbs, S. P. (1985). *Protoplasma* **127**, 9–20.
Ludwig, M., and Gibbs, S. P. (1987). *Ann. N.Y. Acad. Sci.* **503**, 198–211.
MacKay, R. M., Salgado, D., Bonen, L., Stackebrandt, E., and Doolittle, W. F. (1982). *Nucleic Acids Res.* **10**, 2963–2970.
Maerz, M., Wolters, J., Hofmann, C. J. B., Sitte, P., and Maier, U.-G. (1992). *Curr. Genet.* **21**, 73–81.
Mahalingam, R., Seilhamer, J. J., Pritchard, A. E., and Cummings, D. J. (1986). *Gene* **49**, 129–138.
Mahendran, R., Spottswood, M. R., and Miller, D. L. (1991). *Nature (London)* **349**, 434–438.
Maid, U., and Zetsche, K. (1991). *Plant Mol. Biol.* **16**, 537–546.
Maizels, N., and Weiner, A. M. (1987). *Nature (London)* **330**, 616.
Makaroff, C. A., and Palmer, J. D. (1987). *Nucleic Acids Res.* **15**, 5141–5156.
Maloney, A. P., and Walbot, V. (1990). *J. Mol. Biol.* **213**, 633–649.
Manam, S., and Van Tuyle, G. C. (1987) *J. Biol. Chem.* **262**, 10,272–10,279.
Manhart, J. R., and Palmer, J. D. (1990). *Nature (London)* **345**, 268–270.
Manna, E., and Brennicke, A. (1985). *Curr. Genet.* **9**, 505–515.
Mannella, C., Frank, J., and Delihas, N. (1987). *J. Mol. Evol.* **24**, 228–235.
Marchfelder, A., Schuster, W., and Brennicke, A. (1990). *Nucleic Acids Res.* **18**, 1401–1406.
Maréchal, L., Guillemaut, P., Grienenberger, J.-M., Jeannin, G., and Weil, J.-H. (1985). *Nucleic Acids Res.* **13**, 4411–4416.
Maréchal, L., Runeberg-Roos, P., Grienenberger, J. M., Colin, J., Weil, J. H., Lejeune, B., Quetier, F., and Lonsdale, D. M. (1987). *Curr. Genet.* **12**, 91–98.
Maréchal-Drouard, L., Guillemaut, P., Cosset, A., Arbogast, M., Weber, F., Weil, J.-H., and Dietrich, A. (1990). *Nucleic Acids Res.* **18**, 3689–3696.
Maréchal-Drouard, L., Kuntz, M., and Weil, J. H. (1991). In "The Molecular Biology of Plastids" (Vol. 7A in "Cell Culture and Somatic Cell Genetics of Plants") (L. Bogorad and I. K. Vasil, eds.), pp. 169–189. Academic Press, San Diego.
Maréchal-Drouard, L., Weil, J.-H., and Guillemaut, P. (1988). *Nucleic Acids Res.* **16**, 4777–4788.
Margulis, L. (1970). "Origin of Eukaryotic Cells." Yale Univ. Press, New Haven, CT.
Margulis, L. (1975). *Symp. Soc. Exp. Biol.* **29**, 21–38.
Margulis, L. (1981). "Symbiosis in Cell Evolution" Freeman, San Francisco.
Margulis, L., Chase, D., and To, L. P. (1979). *Proc. R. Soc. Lond. B.* **204**, 189–198.
Margulis, L., and Guerrero, R. (1991). *New Scientist* **129**, 46–50.
Margulis, L., and Obar, R. (1985). *BioSystems* **17**, 317–325.
Margulis, L., and Schwartz, K. V. (1988). "Five Kingdoms: An Illustrated Guide to the Phyla of Life on Earth." Freeman, New York.
Margulis, L., To, L. P., and Chase, D. (1981). *Ann. N.Y. Acad. Sci.* **361**, 356–367.
Markowicz, Y., and Loiseaux-de Goër, S. (1991). *Curr. Genet.* **20**, 427–430.
Markowicz, Y., Mache, R., and Loiseaux-de Goër, S. (1988). *Plant Mol. Biol.* **10**, 465–469.
Martin, W., and Cerff, R. (1986). *Eur. J. Biochem.* **159**, 323–331.
Martin, W., Lagrange, T., Li, Y. F., Bisanz-Seyer, C., and Mache, R. (1990). *Curr. Genet.* **18**, 553–556.
Masters, B. S., Stohl, L. L., and Clayton, D. A. (1987). *Cell (Cambridge, Mass.)* **51**, 89–99.
Matheson, A. T., Auer, J., Ramírez, C., and Böck, A. (1990). In "The Ribosome: Structure, Function, and Evolution" (W. E. Hill, A. Dahlberg, R. A. Garrett, P. B. Moore, D. Schlessinger, and J. R. Warner, eds.), pp. 617–635. American Society for Microbiology, Washington.
Matsushita, Y., Kitakawa, M., and Isono, K. (1989). *Mol. Gen. Genet.* **219**, 119–124.
Mattox, K. R., and Stewart, K. D. (1984). "Systematics of the Green Algae." Academic Press, London.

Mayes, S. R., Cook, K. M., and Barber, J. (1990). *FEBS Lett.* **262,** 49–54.
Maynard Smith, J., Dowson, C. G., and Spratt, B. G. (1991). *Nature (London)* **349,** 29–31.
Mayr, E. (1990). *Nature (London)* **348,** 491.
McCarroll, R., Olsen, G. J., Stahl, Y. D., Woese, C. R., and Sogin, M. L. (1983). *Biochemistry* **22,** 5858–5868.
McFadden, G. I. (1990). *J. Cell Sci.* **95,** 303–308.
Mereschkowsky, C. (1910). *Biol. Centr.* **30,** 278–303, 321–347, 353–367.
Michaelis, G., Vahrenholz, C., and Pratje, E. (1990). *Mol. Gen. Genet.* **223,** 211–216.
Michalowski, C., Breunig, K. D., and Bohnert, H. J. (1987). *Curr. Genet.* **11,** 265–274.
Michalowski, C. B., Flachmann, R., Löffelhardt, W., and Bohnert, H. J. (1991a). *Plant Physiol.* **95,** 329–330.
Michalowski, C. B., Löffelhardt, W., and Bohnert, H. J. (1991b). *J. Biol. Chem.* **266,** 11,866–18,870.
Michalowski, C. B., Pfanzagl, B., Löffelhardt, W., and Bohnert, H. J. (1990). *Mol. Gen. Genet.* **224,** 222–231.
Mikelsaar, R. (1987). *J. Mol. Evol.* **25,** 168–183.
Miller, D. L., and Martin, N. C. (1983). *Cell (Cambridge, Mass.)* **34,** 911–917.
Mirabdullaev, L. M. (1985). *Zhurn. Obshch. Biol.* **46,** 483–490.
Moore, L. J., and Coleman, A. W. (1989). *Plant Mol. Biol.* **13,** 459–465.
Morales, M. J., Wise, C. A., Hollingsworth, M. J., and Martin, N. C. (1989). *Nucleic Acids Res.* **17,** 6865–6881.
Morden, C. W., and Golden, S. S. (1989a). *Nature (London)* **337,** 382–385.
Morden, C. W., and Golden, S. S. (1989b). *Nature (London)* **339,** 400.
Morden, C. W., and Golden, S. S. (1991). *J. Mol. Evol.* **32,** 379–395.
Morden, C. W., Wolfe, K. H., de Pamphilis, C. W., and Palmer, J. D. (1991). *EMBO J.* **10,** 3281–3288.
Morin, G. B., and Cech, T. R. (1988). *Nucleic Acids Res.* **16,** 327–346.
Müller, M. (1980). *Symp. Soc. Gen. Microbiol.* **30,** 127–142.
Müller, M. (1985). *J. Protozool.* **32,** 559–563.
Müller, M. (1988). *Annu. Rev. Microbiol.* **42,** 465–488.
Mulligan, R. M. (1991). *Plant Cell* **3,** 327–330.
Mulligan, R. M., Lau, G. T., and Walbot, V. (1988a). *Proc. Natl. Acad. Sci. U.S.A.* **85,** 7998–8002.
Mulligan, R. M., Maloney, A. P., and Walbot, V. (1988b). *Mol. Gen. Genet.* **211,** 373–380.
Mulligan, R. M., Leon, P., and Walbot, V. (1991). *Mol. Cell. Biol.* **11,** 533–543.
Mulligan, R. M., and Walbot, V. (1986). *Trends Genet.* **2,** 263–266.
Muramatsu, T., Nishikawa, K., Nemoto, F., Kuchino, Y., Nishimura, S., Miyazawa, T., and Yokoyama, S. (1988b) *Nature (London)* **336,** 179–181.
Muramatsu, T., Yokoyama, S., Horie, N., Matsuda, A., Ueda, T., Yamaizumi, Z., Kuchino, Y., Nishimura, S., and Miyazawa, T. (1988a). *J. Biol. Chem.* **263,** 9261–9267.
Muto, A., Andachi, Y., Yuzawa, H., Yamao, F., and Osawa, S. (1990). *Nucleic Acids Res.* **18,** 5037–5043.
Myers, A. M., Crivellone, M. D., and Tzagoloff, A. (1987). *J. Biol. Chem.* **262,** 3388–3397.
Myers, A. M., and Tzagoloff, A. (1985). *J. Biol. Chem.* **260,** 15,371–15,377.
Nagata, S., Tsunetsugu-Yokota, Y., Naito, A., and Kaziro, Y. (1983). *Proc. Natl. Acad. Sci. U.S.A.* **80,** 6192–6196.
Nagley, P. (1988). *Trends Genet.* **4,** 46–52.
Nagley, P. (1989). *Trends Genet.* **5,** 67–69.
Nargang, F. E., Pande, S., Kennell, J. C., Akins, R. A., and Lambowitz, A. M. (1992). *Nucleic Acids Res.* **20,** 1101–1108.
Natsoulis, G., Hilger, F., and Fink, G. R. (1986). *Cell (Cambridge, Mass.)* **46,** 235–243.

Nes, W. D., Norton, R. A., Crumley, F. G., Madigan, S. J., and Katz, E. R. (1990). *Proc. Natl. Acad. Sci. U.S.A.* **87,** 7565–7569.

Neumann-Spallart, C., Brandtner, M., Kraus, M., Jakowitsch, M., Bayer, M. G., Maier, T. L., Schenk, H. E. A., and Löffelhardt, W. (1990). *FEBS Lett.* **268,** 55–58.

Neumann-Spallart, C., Jakowitsch, J., Kraus, M., Brandtner, M., Bohnert, H. J., and Löffelhardt, W. (1991). *Curr. Genet.* **19,** 313–315.

Newman, S. M., Derocher, J., and Cattolico, R. A. (1989). *Plant Physiol.* **91,** 939–946.

Newton, K. J. (1988). *Annu. Rev. Plant Physiol. Plant Mol. Biol.* **39,** 503–532.

Nixon, P. J., Gounaris, K., Coomber, S. A., Hunter, C. N., Dyer, T. A., and Barber, J. (1989). *J. Biol. Chem.* **264,** 14,129–14,135.

North, G. (1987). *Nature (London)* **328,** 18–19.

Nugent, J. M., and Palmer, J. D. (1991). *Cell (Cambridge, Mass.)* **66,** 473–481.

Obar, R., and Green, J. (1985). *J. Mol. Evol.* **22,** 243–251.

Ochman, H., and Wilson, A. C. (1987). *J. Mol. Evol.* **26,** 74–86.

Oda, K., Yamato, K., Ohta, E., Nakamura, Y., Takemura, M., Nozato, N., Akashi, K., Kanegae, T., Ogura, Y., Kohchi, T., and Ohyama, K. (1992). *J. Mol. Biol.* **223,** 1–7.

Ohyama, K., Fukuzawa, H., Kohchi, T., Shirai, H., Sano, T., Sano, S., Umesono, K., Shiki, Y., Takeuchi, M., Chang, Z., Aota, S.-i., Inokuchi, H., and Ozeki, H. (1986). *Nature (London)* **322,** 572–574.

Ohyama, K., Kohchi, T., Sano, T., and Yamada, Y. (1988). *Trends Biochem. Sci.* **13,** 19–22.

Okimoto, R., Chamberlin, H. M., Macfarlane, J. L., and Wolstenholme, D. R. (1991). *Nucleic Acids Res.* **19,** 1619–1626.

Okimoto, R., Macfarlane, J. L., Clary, D. O., and Wolstenholme, D. R. (1992). *Genetics* **130,** 471–498.

Okimoto, R., and Wolstenholme, D. R. (1990). *EMBO J.* **9,** 3405–3411.

Olsen, G. J. (1987). *Cold Spring Harbor Symp. Quant. Biol.* **52,** 825–837.

Olsen, G. J., and Woese, C. R. (1989). *Can. J. Microbiol.* **35,** 119–123.

Opperdoes, F. R., and Michels, P. A. M. (1989). *In* "Organelles of Eukaryotic Cells: Molecular Structure and Interaction" (J. M. Tager, A. Azzi, S. Papa, and F. Guerrieri, eds.), pp. 187–195. Plenum, New York.

Osinga, K. A., De Haan, M., Christianson, T., and Tabak, H. F. (1982). *Nucleic Acids Res.* **10,** 7993–8006.

Osinga, K. A., De Vries, E., Van der Horst, G. T. J., and Tabak, H. F. (1984). *Nucleic Acids Res.* **12,** 1889–1900.

Ozeki, H., Ohyama, K., Inokuchi, H., Fukuzawa, H., Kohchi, T., Sano, T., Nakahigashi, K., and Umesono, K. (1987). *Cold Spring Harbor Symp. Quant. Biol.* **52,** 791–804.

Ozeki, H., Umesono, K., Inokuchi, H., Kohchi, T., and Ohyama, K. (1989). *Genome* **31,** 169–174.

Pace, N. R. (1991) *Cell (Cambridge, Mass.)* **65,** 531–533.

Pace, N. R., and Marsh, T. L. (1985). *Origins Life* **16,** 97–116.

Pace, N. R., Olsen, G. J., and Woese, C. R. (1986). *Cell (Cambridge, Mass.)* **45,** 325–326.

Palenik, B., and Haselkorn, R. (1992). *Nature (London)* **355,** 265–267.

Palmer, J. D. (1985a). *Annu. Rev. Genet.* **19,** 325–354.

Palmer, J. D. (1985b). *In* "Molecular Evolutionary Genetics" (R. J. MacIntyre, ed.), pp. 131–240. Plenum, New York/London.

Palmer, J. D. (1987). *Am. Nat.* **130,** (Suppl.), S6–S29.

Palmer, J. D. (1990). *Trends Genet.* **6,** 115–120.

Palmer, J. D. (1991). *In* "The Molecular Biology of Plastids" (Vol. 7A in "Cell Culture and Somatic Cell Genetics of Plants") (L. Bogorad and I. K. Vasil, eds.), pp. 5–53. Academic Press, San Diego.

Palmer, J. D. (1992). *In* "Organelles" (Vol. 6 of "Plant Gene Research") (R. G. Herrmann, ed.), pp. 99–133. Springer-Verlag, Berlin.

Palmer, J. D., and Herbon, L. A. (1988). *J. Mol. Evol.* **28**, 87–97.
Palmer, J. D., Nugent, J. M., and Herbon, L. A. (1987). *Proc. Natl. Acad. Sci. U.S.A.* **84**, 769–773.
Palmer, J. D., and Shields, C. R. (1984). *Nature (London)* **307**, 437–440.
Palmer, J. D., Soltis, D., and Soltis, P. (1992). *Curr. Genet.* **21**, 125–129.
Palmer, J. D., and Thompson, W. F. (1982). *Cell (Cambridge, Mass.)* **29**, 537–550.
Pancic, P. G., Strotmann, H., and Kowallik, K. V. (1991). *FEBS Lett.* **280**, 387–392.
Pang, H., Ihara, M., Kuchino, Y., Nishimura, S., Gupta, R., Woese, C. R., and McCloskey, J. A. (1982). *J. Biol. Chem.* **257**, 3589–3592.
Pape, L. K., Koerner, T. J., and Tzagoloff, A. (1985). *J. Biol. Chem.* **260**, 15,362–15,370.
Pappenheimer, A. M., Dunlop, P. C., Adolph, K. W., and Bodley, J. W. (1983) *J. Bacteriol.* **153**, 1342–1347.
Parisi, M. A., and Clayton, D. A. (1991). *Science* **252**, 965–969.
Parks, T. D., Dougherty, W. G., Levings, C. S., III, and Timothy, D. H. (1984). *Plant Physiol.* **76**, 1079–1082.
Partaledis, J. A., and Mason, T. L. (1988). *Mol. Cell. Biol.* **8**, 3647–3660.
Patterson, D. J., and Sogin, M. L. (1992) *In* "The Origin and Evolution of Prokaryotic and Eukaryotic Cells" (H. Hartman and K. Matsuno, eds.) World Scientific, New Jersey, in press.
Penny, D. (1988). *Nature (London)* **331**, 111–112.
Penny, D., and O'Kelly, C. J. (1991). *Nature (London)* **350**, 106–107.
Perasso, R., Baroin, A., and Adoutte, A. (1990). *In* "Cell Walls and Surfaces, Reproduction, Photosynthesis" (Vol. 1 of "Experimental Phycology") (W. Wiessner, D. G. Robinson, and R. C. Starr, eds.), pp. 1–19. Springer-Verlag, Berlin.
Perasso, R., Baroin, A., Qu, L. H., Bachellerie, J. P., and Adoutte, A. (1989). *Nature (London)* **339**, 142–144.
Perlman, P. S., and Butow, R. A. (1989). *Science* **246**, 1106–1109.
Pfitzinger, H., Weil, J. H., Pillay, D. T. N., and Guillemaut, P. (1990). *Plant Mol. Biol.* **14**, 805–814.
Pietromonaco, S. F., Hessler, R. A., and O'Brien, T. W. (1986). *J. Mol. Evol.* **24**, 110–117.
Pilkington, S. J., Skehel, J. M., Gennis, R. B., and Walker, J. E. (1991a). *Biochemistry* **30**, 2166–2175.
Pilkington, S. J., Skehel, J. M., and Walter, J. E. (1991b). *Biochemistry* **30**, 1901–1908.
Pollard, V. W., Rohrer, S. P., Michelotti, E. F., Hancock, K., and Hajduk, S. L. (1990). *Cell (Cambridge, Mass.)* **63**, 783–790.
Prais, D., Weidner, U., Conzen, C., Azevedo, J. E., Nehls, U., Roehlen, D.-A., Sackmann, U., Sackmann, U., Schneider, R., Werner, S., and Weiss, H. (1991). *Biochim. Biophys. Acta* **1090**, 133–138.
Pring, D. R., and Lonsdale, D. M. (1985) *Int. Rev. Cytol.* **97**, 1–46.
Pritchard, A. E., Seilhamer, J. J., and Cummings, D. J. (1986). *Gene* **44**, 243–253.
Pritchard, A. E., Seilhamer, J. J., Mahalingam, R., Sable, C. L., Venuti, S. E., and Cummings, D. J. (1990a). *Nucleic Acids Res.* **18**, 173–180.
Pritchard, A. E., Sable, C. L., Venuti, S. E., and Cummings, D. J. (1990b). *Nucleic Acids Res.* **18**, 163–171.
Pruitt, K. D., and Hanson, M. R. (1991). *Curr. Genet.* **19**, 191–197.
Pühler, G., Leffers, H., Gropp, F., Palm, P., Klenk, H.-P., Lottspeich, F., Garrett, R. A., and Zillig, W. (1989). *Proc. Natl. Acad. Sci. U.S.A.* **86**, 4569–4573.
Qu, L.-H., Nicoloso, M., and Bachellerie, J.-P. (1988a). *J. Mol. Evol.* **28**, 113–124.
Qu, L.-H., Perasso, R., Baroin, A., Brugerolle, G., Bachellerie, J.-P., and Adoutte, A. (1988b). *BioSystems* **21**, 203–208.
Quagliariello, C., Saiardi, A., and Gallerani, R. (1990). *Curr. Genet.* **18**, 355–363.

Quirk, S. M., Bell-Pedersen, D., and Belfort, M. (1989). *Cell (Cambridge, Mass.)* **56,** 455–465.

Raff, R. A., and Mahler, H. R. (1972). *Science* **177,** 575–582.

Raff, R. A., and Mahler, H. R. (1975). *Symp. Soc. Exp. Biol.* **29,** 41–92.

Ragan, M. A., and Chapman, D. J. (1978). "A Biochemical Phylogeny of the Protists." Academic Press, New York.

Raitio, M., Jalli, T., and Saraste, M. (1987). *EMBO J.* **6,** 2825–2833.

Rapp, W. D., and Stern, D. B. (1992). *EMBO J.* **11,** 1065–1073.

Rasmussen, J., and Hanson, M. R. (1989). *Mol. Gen. Genet.* **215,** 332–336.

Raven, P. H. (1970). *Science* **169,** 641–646.

Reisser, W., Meier, R., Görtz, H.-D., and Jeon, K. W. (1985). *J. Protozool.* **32,** 383–390.

Recipon, H., Perasso, R., Adoutte, A., and Quetier, F. (1992). *J. Mol. Evol.* **34,** 292–303.

Reiter, W.-D., Hüdepohl, U., and Zillig, W. (1990). *Proc. Natl. Acad. Sci. U.S.A.* **87,** 9509–9513.

Reiter, W.-D., Palm, P., Voos, W., Kaniecki, J., Grampp, B., Schulz, W., and Zillig, W. (1987). *Nucleic Acids Res.* **15,** 5581–5595.

Reiter, W.-D., Palm, P., and Zillig, W. (1988). *Nucleic Acids Res.* **16,** 1–19.

Reith, M. (1992). *Plant Mol. Biol.* **18,** 773–775.

Reith, M., and Cattolico, R. A. (1986). *Proc. Natl. Acad. Sci. U.S.A.* **83,** 8599–8603.

Reith, M., and Douglas, S. (1990). *Plant Mol. Biol.* **15,** 585–592.

Reith, M., and Munholland, J. (1991). *FEBS Lett.* **294,** 116–120.

Relman, D. A., Loutit, J. S., Schmidt, T. M., Falkow, S., and Tompkins, L. S. (1990). *N. Engl. J. Med.* **323,** 1573–1580.

Roe, B. A., Ma, D.-P., Wilson, R. K., and Wong, J. F.-H. (1985). *J. Biol. Chem.* **260,** 9759–9774.

Rothschild, L J., Ragan, M. A., Coleman, A. W., Heywood, P., and Gerbi, S. A. (1986). *Cell (Cambridge, Mass.)* **47,** 640.

Rottmann, W. H., Brears, T., Hodge, T. P., and Lonsdale, D. M. (1987). *EMBO J.* **6,** 1541–1546.

Runswick, M. J., Gennis, R. B., Fearnley, I. M., and Walker, J. E. (1989). *Biochemistry* **28,** 9452–9459.

Saitou, N., and Nei, M. (1987). *Mol. Biol. Evol.* **4,** 406–425.

Samac, D. A., and Leong, S. A. (1989). *Mol. Plant–Microbe Interact.* **2,** 155–159.

Sanangelantoni, A. M., Barbarini, D., Di Pasquale, G., Cammarano, P., and Tiboni, O. (1990). *Mol. Gen. Genet.* **221,** 187–194.

Sandman, K., Krzycki, J. A., Dobrinski, B., Lurz, R., and Reeve, J. N. (1990). *Proc. Natl. Acad. Sci. U.S.A.* **87,** 5788–5791.

Sangaré, A., Lonsdale, D., Weil, J.-H., and Grienenberger, J.-M. (1989). *Curr. Genet.* **16,** 195–201.

Sangaré, A., Weil, J.-H., Grienenberger, J.-M., Fauron, C., and Lonsdale, D. (1990). *Mol. Gen. Genet.* **223,** 224–232.

Sapp, J. (1990). *Endocytobiosis Cell Res.* **7,** 5–36.

Saraste, M., Raitio, M., Jalli, T., and Perämaa, A. (1986). *FEBS Lett.* **206,** 154–156.

Scaramuzzi, C. D., Stokes, H. W., and Hiller, R. G. (1992). *Plant Mol. Biol.* **18,** 467–476.

Scherer, S., Herrmann, G., Hirschberg, J., and Böger, P. (1991). *Curr. Genet.* **19,** 503–507.

Schinkel, A. H., Groot Koerkamp, M. J. A., and Tabak, H. F. (1988). *EMBO J.* **7,** 3255–3262.

Schinkel, A. H., Groot Koerkamp, M. J. A., Touw, E. P. W., and Tabak, H. F. (1987). *J. Biol. Chem.* **262,** 12,785–12,791.

Schinkel, A. H., and Tabak, H. F. (1989). *Trends Genet.* **5,** 149–154.

Schlegel, M. (1991). *Eur. J. Protistol.* **27,** 207–219.

Schmidt, R. J., Myers, A. M., Gillham, N. W., and Boynton, J. E. (1984). *Mol. Biol. Evol.* **1,** 317–334.

352

MICHAEL W. GRAY

Schnabel, R., Sonnenbichler, J., and Zillig, W. (1982). *FEBS Lett.* **150**, 400–402.
Schnabel, R., Thomm, M., Gerardy-Schahn, R., Zillig, W., Stetter, K. O., and Huet, J. (1983). *EMBO J.* **2**, 751–755.
Schnare, M. N., and Gray, M. W. (1982). *Nucleic Acids Res.* **10**, 3921–3932.
Schnare, M. N., Heinonen, T. Y. K., Young, P. G., and Gray, M. W. (1985). *Curr. Genet.* **9**, 389–393.
Schnare, M. N., Heinonen, T. Y. K., Young, P. G., and Gray, M. W. (1986). *J. Biol. Chem.* **261**, 5187–5193.
Scholzen, T., and Arndt, E. (1991). *Mol. Gen. Genet.* **228**, 70–80.
Schröder, J., and Klink, F. (1991). *Eur. J. Biochem.* **195**, 321–327.
Schümann, H., and Klink, F. (1989). *System. Appl. Microbiol.* **11**, 103–107.
Schuster, W., and Brennicke, A. (1985). *Curr. Genet.* **9**, 157–163.
Schuster, W., and Brennicke, A. (1987). *EMBO J.* **6**, 2857–2863.
Schuster, W., Hiesel, R., Isaac, P. G., Leaver, C. J., and Brennicke, A. (1986). *Nucleic Acids Res.* **14**, 5943–5954.
Schwartz, R. M., and Dayhoff, M. O. (1978). *Science* **199**, 395–403.
Schwartz, R. M., and Dayhoff, M. O. (1981). *Ann. N.Y. Acad. Sci.* **361**, 260–269.
Schwartzbach, C. J., and Spremulli, L. L. (1989). *J. Biol. Chem.* **264**, 19,125–19,131.
Searcy, D. G. (1987). *Ann. N.Y. Acad. Sci.* **503**, 168–179.
Sederoff, R. R. (1984). *Adv. Genet.* **22**, 1–108.
Seewaldt, E., and Stackebrandt, E. (1982). *Nature (London)* **295**, 618–620.
Seilhamer, J. J., Gutell, R. R., and Cummings, D. J. (1984a). *J. Biol. Chem.* **259**, 5173–5181.
Seilhamer, J. J., Olsen, G. J., and Cummings, D. J. (1984b). *J. Biol. Chem.* **259**, 5167–5172.
Shapleigh, J. P., and Gennis, R. B. (1992). *Mol. Microbiol.* **6**, 635–642.
Shiba, T., Mizote, H., Kaneko, T., Nakajima, T., Kakimoto, Y., and Sano, I. (1971). *Biochim. Biophys. Acta* **244**, 523–531.
Shih, M.-C., Lazar, G., and Goodman, H. M. (1986). *Cell (Cambridge, Mass.)* **47**, 73–80.
Shimada, H., and Sugiura, M. (1991). *Nucleic Acids Res.* **19**, 983–995.
Shinozaki, K., Ohme, M., Tanaka, M., Wakasugi, T., Hayashida, N., Matsubayashi, T., Zaita, N., Chunwongse, J., Obokata, J., Yamaguchi-Shinozaki, K., Ohto, C., Torazawa, K., Meng, B. Y., Sugita, M., Deno, H., Kamogashira, T., Yamada, K., Kusuda, J., Takaiwa, F., Kato, A., Tohdoh, N., Shimada, H., and Sugiura M. (1986). *EMBO J.* **5**, 2043–2049.
Shivji, M. S. (1991). *Curr. Genet.* **19**, 49–54.
Shub, D. A., Gott, J. M., Xu, M.-Q., Lang, B. F., Michel, F., Tomaschewski, J., Pedersen-Lane, J., and Belfort, M. (1988). *Proc. Natl. Acad. Sci. U.S.A.* **85**, 1151–1155.
Sidow, A., and Bowman, B. H. (1991). *Curr. Opin. Genet. Dev.* **1**, 451–456.
Sidow, A., and Wilson, A. C. (1990). *J. Mol. Evol.* **31**, 51–68.
Siemeister, G., Buchholz, C., and Hachtel, W. (1990). *Curr. Genet.* **18**, 457–464.
Siemeister, G., and Hachtel, W. (1989). *Curr. Genet.* **15**, 435–441.
Simpson, A. M., Suyama, Y., Dewes, H., Campbell, D. A., and Simpson, L. (1989). *Nucleic Acids Res.* **17**, 5427–5445.
Simpson, L. (1986). *Int. Rev. Cytol.* **99**, 119–179.
Simpson, L. (1987). *Annu. Rev. Microbiol.* **41**, 363–382.
Simpson, L. (1990). *Science* **250**, 512–513.
Simpson, L., and Shaw, J. (1989). *Cell (Cambridge, Mass.)* **57**, 355–366,
Simpson, L., Simpson, A., and Livingston, L. (1982). *Mol. Biochem. Parasitol.* **6**, 237–350.
Singhal, R. P., and Fallis, P. A. M. (1979) *Progr. Nucleic Acid Res. Mol. Biol.* **23**, 227–263.
Sjöberg, B.-M., Hahne, S., Mathews, C. Z., Mathews, C. K., Rand, K. N., and Gait, M. J. (1986). *EMBO J.* **5**, 2031–2036.
Sloof, P., Van den Burg, J., Voogd, A., Benne, R., Agostinelli, M., Borst, P., Gutell, R., and Noller, H. F. (1985). *Nucleic Acids Res.* **13**, 4171–4190.

Small, I., Suffolk, R., and Leaver, C. J. (1989). *Cell (Cambridge, Mass.)* **58**, 69–76.

Smith, A. E., and Marcker, K. A. (1968). *J. Mol. Biol.* **38**, 241–243.

Smooker, P. M., Kruft, V., and Subramanian, A. R. (1990). *J. Biol. Chem.* **265**, 16,699–16,703.

Snyder, M., Fraser, A. R., LaRoche, J., Gartner-Kepkay, K. E., and Zouros, E. (1987). *Proc. Natl. Acad. Sci. U.S.A.* **84**, 7595–7599.

Sogin, M. L. (1990). *In* "PCR Protocols: A Guide to Methods and Applications" (M. A. Innis, D. H. Gelfand, J. J. Sninsky, and T. J. White, eds.), pp. 307–314. Academic Press, San Diego.

Sogin, M. L. (1991). *Curr. Opin. Genet. Dev.* **1**, 457–463.

Sogin, M. L., Elwood, H. J., and Gunderson, J. H. (1986). *Proc. Natl. Acad. Sci. U.S.A.* **83**, 1383–1387.

Sogin, M. L., Gunderson, J. H., Elwood, H. J., Alonso, R. A., and Peattie, D. A. (1989). *Science* **243**, 75–77.

Somerville, C. C., Jouannic, S., and Loiseaux-de Goër, S. (1992). *J. Mol. Evol.* **34**, 246–253.

Spencer, D. F., Bonen, L., and Gray, M. W. (1981). *Biochemistry* **20**, 4022–4029.

Spencer, D. F., Schnare, M. N., and Gray, M. W. (1984). *Proc. Natl. Acad. Sci. U.S.A.* **81**, 493–497.

Sprague, G. F., Jr. (1991). *Curr. Opin. Genet. Dev.* **1**, 530–533.

Stamper, S. E., Dewey, R. E., Bland, M. M., and Levings, C. S., III (1987). *Curr. Genet.* **12**, 457–463.

Stanier, R. Y. (1970). *Symp. Soc. Gen. Microbiol.* **20**, 1–38.

Stanier, R. Y., and van Niel, C. B. (1962). *Arch. Mikrobiol.* **42**, 17–35.

Starnes, S. M., Lambert, D. H., Maxwell, E. S., Stevens, S. E., Jr., Porter, R. D., and Shively, J. M. (1985). *FEMS Microbiol. Lett.* **28**, 165–169.

Steffens, G. C. M., Buse, G., Oppliger, W., and Ludwig, B. (1983). *Biochem. Biophys. Res. Commun.* **116**, 335–340.

Steinmetz, A. A., Krebbers, E. T., Schwarz, Z., Gubbins, E. J., and Bogorad, L. (1983). *J. Biol. Chem.* **258**, 5503–5511.

Steinmüller, K., Ley, A. C., Steinmetz, A. A., Sayre, R. T., and Bogorad, L. (1989). *Mol. Gen. Genet.* **216**, 60–69.

Stern, D. B., and Gruissem, W. (1987). *Cell (Cambridge, Mass.)* **51**, 1145–1157.

Stern, D. B., and Lonsdale, D. M. (1982). *Nature (London)* **299**, 698–702.

Stern, D. B., and Palmer, J. D. (1984). *Proc. Natl. Acad. Sci. U.S.A.* **81**, 1946–1950.

Stern, D. B., and Palmer, J. D. (1986). *Nucleic Acids Res.* **14**, 5651–5666.

Stewart, K. D., and Mattox, K. R. (1984). *J. Mol. Evol.* **21**, 54–57.

Strauss, S. H., Palmer, J. D., Howe, G. T., and Doerksen, A. H. (1988). *Proc. Natl. Acad. Sci. U.S.A.* **85**, 3898–3902.

Stuart, K. (1991). *Trends Biochem. Sci.* **16**, 68–72.

Stuart, K., Feagin, J. E., and Abraham, J. M. (1989). *Gene* **82**, 155–160.

Stuart, K., and Gelvin, S. (1982). *Mol. Cell. Biol.* **2**, 845–852.

Sturm, N. R., and Simpson, L. (1990). *Cell (Cambridge, Mass.)* **61**, 879–884.

Subramanian, A. R., Smooker, P. M., and Giese, K. (1990). *In* "The Ribosome: Structure, Function, and Evolution" (W. E. Hill, A. Dahlberg, R. A. Garrett, P. B. Moore, D. Schlessinger, and J. R. Warner, eds.), pp. 655–663. American Society for Microbiology, Washington.

Südhof, T. C., Fried, V. A., Stone, D. K., Johnston, P. A., and Xie, X.-S. (1989). *Proc. Natl. Acad. Sci. U.S.A.* **86**, 6067–6071.

Sugiura, M. (1989a) *In* "Molecular Biology" (Vol. 15 of "The Biochemistry of Plants") (A. Marcus, ed.), pp. 133–150. Academic Press, San Diego.

Sugiura, M. (1989b). *Annu. Rev. Cell. Biol.* **5**, 51–70.

Surguchov, A. P. (1987). *Trends Biochem. Sci.* **12**, 335–338.

Suyama, Y. (1985). *Nucleic Acids Res.* **13**, 3273–3284.

Suyama, Y. (1986). *Curr. Genet.* **10**, 411–420.

Suyama, Y., Fukuhara, H., and Sor, F. (1985). *Curr. Genet.* **9**, 479–493.

Suyama, Y., and Jenney, F. (1989). *Nucleic Acids Res.* **17**, 803.

Suyama, Y., Jenney, F., and Okawa, N. (1987). *Curr. Genet.* **11**, 327–330.

Tabak, H. F., Grivell, L. A., and Borst, P. (1983). *CRC Crit. Rev. Biochem.* **14**, 297–317.

Tanaka, Y., Kuroe, K., Angata, K., and Yanagisawa, K. (1990). *Plant Mol. Biol.* **15**, 659–660.

Taylor, F. J. R. (1974). *Taxon* **23**, 229–258.

Taylor, F. J. R. (1979). *Proc. R. Soc. Lond. B* **204**, 267–286.

Taylor, F. J. R. (1987). *Ann. N.Y. Acad. Sci.* **503**, 1–16.

Thorsness, P. E., and Fox, T. D. (1990). *Nature (London)* **346**, 376–379.

Tian, G.-L., Macadre, C., Kruszewska, A., Szczesniak, B., Ragnini, A., Grisanti, P., Rinaldi, T., Palleschi, C., Frontali, L., Slonimski, P. P., and Lazowska, J. (1991a). *J. Mol. Biol.* **218**, 735–746.

Tian, G.-L., Michel, F., Macadre, C., Slonimski, P. P., and Lazowska, J. (1991b). *J. Mol. Biol.* **218**, 747–760.

Timkovich, R., and Dickerson, R. E. (1976). *J. Biol. Chem.* **251**, 4033–4046.

Tomioka, N., and Sugiura, M. (1983). *Mol. Gen. Genet.* **191**, 46–50.

Topper, J. N., and Clayton, D. A. (1989). *Mol. Cell. Biol.* **9**, 1200–1211.

Topper, J. N., and Clayton, D. A. (1990). *Nucleic Acids Res.* **18**, 793–799.

Torazawa, K., Hayashida, N., Obokata, J., Shinozaki, K., and Sugiura, M. (1986) *Nucleic Acids Res.* **14**, 3143.

Trent, J. D., Nimmesgern, E., Wall, J. S., Hartl, F.-U., and Horwich, A. L. (1991). *Nature (London)* **354**, 490–493.

Tu, J., and Zillig, W. (1982). *Nucleic Acids Res.* **10**, 7231–7245.

Turmel, M., Boulanger, J., Schnare, M. N., Gray, M. W., and Lemieux, C. (1991). *J. Mol. Biol.* **218**, 293–311.

Turner, G., and Müller, M. (1983). *J. Parasitol.* **69**, 234–236.

Turner, S., Burger-Wiersma, T., Giovannoni, S. J., Mur, L. R., and Pace, N. R. (1989). *Nature (London)* **337**, 380–382.

Tzagoloff, A., and Myers, A. M. (1986). *Annu. Rev. Biochem.* **55**, 249–285.

Tzagoloff, A., Vambutas, A., and Akai, A. (1989). *Eur. J. Biochem.* **179**, 365–371.

Urbach, E., Robertson, D. L., and Chisholm, S. W. (1992). *Nature (London)* **355**, 267–270.

Uzzell, T., and Spolsky, C. (1974). *Am. Sci.* **62**, 334–343.

Uzzell, T., and Spolsky, C. (1981). *Ann. N.Y. Acad. Sci.* **361**, 481–499.

Vaidya, A. B., Akella, R., and Suplick, K. (1989). *Mol. Biochem. Parasitol.* **35**, 97–108.

Vaidya, A. B., Akella, R., and Suplick, K. (1990). *Mol. Biochem. Parasitol.* **39**, 295–296.

Valentin, K., and Zetsche, K. (1989). *Curr. Genet.* **16**, 203–209.

Valentin, K., and Zetsche, K. (1990a). *Plant Mol. Biol.* **15**, 575–584.

Valentin, K., and Zetsche, K. (1990b). *Mol. Gen. Genet.* **222**, 425–430.

Vambutas, A., Ackerman, S. H., and Tzagoloff, A. (1991). *Eur. J. Biochem.* **201**, 643–652.

van den Boogaart, P., Samallo, J., and Agsteribbe, E. (1982) *Nature (London)* **298**, 187–189.

Van de Peer, Y., Neefs, J.-M., and De Wachter, R. (1990). *J. Mol. Evol.* **30**, 463–476.

van der Spek, H., Arts, G.-J., Zwaal, R. R., van den Burg, J., Sloof, P., and Benne, R. (1991). *EMBO J.* **10**, 1217–1224.

Van Ness, B. G., Howard, J. B., and Bodley, J. W. (1980). *J. Biol. Chem.* **255**, 10,710–10,716.

Van Valen, L. M. (1982). *Nature (London)* **298**, 493,494.

Villanueva, E., Luehrsen, K. R., Gibson, J., Delihas, N., and Fox, G. E. (1985). *J. Mol. Evol.* **22**, 46–52.

von Berg, K.-H. L., and Kowallik, K. V. (1992). *Plant Mol. Biol.* **18**, 83–95.

von Heijne, G. (1986). *FEBS Lett.* **198,** 1–4.

Vossbrinck, C. R., Maddox, J. V., Friedman, S., Debrunner-Vossbrinck, B. A., and Woese, C. R. (1987) *Nature (London)* **326,** 411–414.

Vossbrinck, C. R., and Woese, C. R. (1986). *Nature (London)* **320,** 287–288.

Wächtershäuser, G. (1988). *Mircobiol. Rev.* **52,** 452–484.

Wächtershäuser, G. (1990). *Proc. Natl. Acad. Sci. U.S.A.* **87,** 200–204.

Waddle, J. A., Schuster, A. M., Lee, K. W., and Meints, R. H. (1990). *Plant Mol. Biol.* **14,** 187–195.

Wahleithner, J. A., Macfarlane, J. L., and Wolstenholme, D. R. (1990). *Proc. Natl. Acad. Sci. U.S.A.* **87,** 548–552.

Wahleithner, J. A., and Wolstenholme, D. R. (1988). *Nucleic Acids Res.* **16,** 6897–6913.

Walbot, V. (1991) *Trends Genet.* **7,** 37–39.

Walker, J. E., Runswick, M. J., and Saraste, M. (1982a). *FEBS Lett.* **146,** 393–396.

Walker, J. E., Saraste, M., Runswick, M. J., and Gay, N. J. (1982b). *EMBO J.* **8,** 945–951.

Wallace, D. C. (1982). *Microbiol. Rev.* **46,** 208–240.

Wang, S., and Liu, X-Q. (1991). *Proc. Natl. Acad. Sci. U.S.A.* **88,** 10,783–10,787.

Ward, B. L., Anderson, R. S., and Bendich, A. J. (1981). *Cell (Cambridge, Mass.)* **25,** 793–803.

Warrior, R., and Gall, J. (1985). *Arch. Sci.* **38,** 439–445.

Weber, F., Dietrich, A., Weil, J.-H., and Maréchal-Drouard, L. (1990). *Nucleic Acids Res.* **18,** 5027–5030.

Weeden, N. F. (1981). *J. Mol. Evol.* **17,** 133–139.

Weil, J. H. (1987). *Plant Sci.* **49,** 149–157.

Weiner, A. M. (1988). *Cell (Cambridge, Mass.)* **52,** 155–157.

Weiner, A. M., and Maizels, N. (1987). *Proc. Natl. Acad. Sci. U.S.A.* **84,** 7383–7387.

Weiner, A. M., and Maizels, N. (1990). *Cell (Cambridge, Mass.)* **61,** 917–920.

Wesolowski, M., and Fukuhara, H. (1981). *Mol. Cell. Biol.* **1,** 387–393.

Whatley, F. R. (1981). *Ann. N.Y. Acad. Sci.* **361,** 330–340.

Whatley, J. M. (1981). *Ann. N.Y. Acad. Sci.* **361,** 154–164.

Whatley, J. M. (1983). *Int. Rev. Cytol.,* Suppl. **14,** 329–373.

Whatley, J. M., John, P., and Whatley, F. R. (1979). *Proc. R. Soc. Lond. B* **204,** 165–187.

Whatley, J. M., and Whatley, F. R. (1981). *New Phytol.* **87,** 233–247.

Whitfeld, P. R., and Bottomley, W. (1983). *Annu. Rev. Plant Physiol.* **34,** 279–310.

Whittaker, R. H. (1969). *Science* **163,** 150–160.

Whittaker, R. H., and Margulis, L. (1978). *BioSystems* **10,** 3–18.

Wilcox, L. W., and Wedemayer, G. J. (1985). *Science* **227,** 192–194.

Williamson, D. H., Wilson, R. J. M., Bates, P. A., McCready, S., Perler, F., and Qiang, B.-u. (1985). *Mol. Biochem. Parasitol.* **14,** 199–209.

Williamson, L. R., Plano, G. V., Winkler, H. H., Krause, D. C., and Wood, D. O. (1989). *Gene* **80,** 269–278.

Wilson, A. C., Cann, R. L., Carr, S. M., George, M., Gyllensten, U. B., Helm-Bychowski, K. M., Higuchi, R. G., Palumbi, S. R., Prager, E. M., Sage, R. D., and Stoneking, M. (1985). *Biol. J. Linn. Soc.* **26,** 375–400.

Wilson, R. J. M., Fry, M., Gardner, M. J., Feagin, J. E., and Williamson, D. H. (1992). *Curr. Genet.* **21,** 405–408.

Wilson, R. J. M., Gardner, M. J., Feagin, J. E., and Williamson, D. H. (1991). *Parasitol. Today* **7,** 134–136.

Winker, S., and Woese, C. R. (1991). *System. Appl. Microbiol.* **14,** 305–310.

Wintz, H., Grienenberger, J.-M., Weil, J.-H., and Lonsdale, D. M. (1988). *Curr. Genet.* **13,** 247–254.

Wintz, H., and Hanson, M. R. (1991). *Curr. Genet.* **19,** 61–64.

Wissinger, B., Hiesel, R., Schuster, W., and Brennicke, A. (1988). *Mol. Gen. Genet.* **212**, 56–65.

Wissinger, B., Schuster, W., and Brennicke, A. (1990). *Mol. Gen. Genet.* **224**, 389–395.

Wissinger, B., Schuster, W., and Brennicke, A. (1991). *Cell (Cambridge, Mass.)* **65**, 473–482.

Witt, D., and Stackebrandt, E. (1988). *Arch. Microbiol.* **150**, 244–248.

Wittmann-Liebold, B., Köpke, A. K. E., Arndt, E., Krömer, W., Hatakeyama, T., and Wittmann, H.-G. (1990). *In* "The Ribosome: Structure, Function, and Evolution" (W. E. Hill, A. Dahlberg, R. A. Garrett, P. B. Moore, D. Schlessinger, and J. R. Warner, eds.), pp. 598–616. American Society for Microbiology, Washington.

Woese, C. R. (1981). *Sci. Am.* **244**, 98–122.

Woese, C. R. (1987). *Microbiol. Rev.* **51**, 221–271.

Woese, C. R. (1990). *Science* **247**, 789.

Woese, C. R. (1991). *In* "Evolution at the Molecular Level" (R. K. Selander, A. G. Clark, and T. S. Whittam, eds.), pp. 1–24. Sinauer, Sunderland, MA.

Woese, C. R., and Fox, G. E. (1977). *Proc. Natl. Acad. Sci. U.S.A.* **74**, 5088–5090.

Woese, C. R., Gibson, J., and Fox, G. E. (1980). *Nature (London)* **283**, 212–214.

Woese, C. R., Kandler, O., and Wheelis, M. L. (1990). *Proc. Natl. Acad. Sci. U.S.A.* **87**, 4576–4579.

Woese, C. R., Kandler, O., and Wheelis, M. L. (1991). *Nature (London)* **351**, 528–529.

Woese, C. R., Magrum, L. J., and Fox, G. E. (1978). *J. Mol. Evol.* **11**, 245–252.

Woese, C. R., and Olsen, G. J. (1986). *System. Appl. Microbiol.* **7**, 161–177.

Woese, C. R., Pace, N. R., and Olsen, G. J. (1986). *Nature (London)* **320**, 402.

Woese, C. R., Stackebrandt, E., Weisburg, W. G., Paster, B. J., Madigan, M. T., Fowler, V. J., Hahn, C. M., Blanz, P., Gupta, R., Nealson, K. H., and Fox, G. E. (1984). *System. Appl. Microbiol.* **5**, 315–326.

Woese, C. R., and Wolfe, R. S. (1985). *In* "Archaebacteria" (Vol. 8 of "The Bacteria: A Treatise on Structure and Function") (C. R. Woese and R. S. Wolfe, eds.), pp. 561–564. Academic Press, San Diego.

Wolf, K. (1987a). *In* "Gene Structure in Eukaryotic Microbes" (Vol. 22, Special Publications of the Society for General Microbiology) (J. R. Kinghorn, ed.), pp. 41–67. IRL Press, Oxford/Washington.

Wolf, K. (1987b). *In* "Gene Structure in Eukaryotic Microbes" (Vol. 22, Special Publications of the Society for General Microbiology) (J. R. Kinghorn, ed.), pp. 69–91. IRL Press, Oxford/Washington.

Wolf, K., and Del Giudice, L. (1988). *Adv. Genet.* **25**, 185–308.

Wolfe, K. H., Li, W.-H., and Sharp, P. M. (1987). *Proc. Natl. Acad. Sci. U.S.A.* **84**, 9054–9058.

Wolfe, K. H., Morden, C. W., and Palmer, J. D. (1991). *Curr. Opin. Genet. Dev.* **1**, 523–529.

Wolff, G., and Kück, U. (1990). *Curr. Genet.* **17**, 347–351.

Wolstenholme, D. R., Macfarlane, J. L., Okimoto, R., Clary, D. O., Wahleithner, J. A. (1987). *Proc. Natl. Acad. Sci. U.S.A.* **84**, 1324–1328.

Wolters, J., and Erdmann, V. A. (1986). *J. Mol. Evol.* **24**, 152–166.

Wolters, J., and Erdmann, V. A. (1989). *Can. J. Microbiol.* **35**, 43–51.

Xu, B., and Clayton, D. A. (1992). *Nucleic Acids Res.* **20**, 1053–1059.

Xu, M.-Q., Kathe, S. D., Goodrich-Blair, H., Nierzwicki-Bauer, S. A., and Shub, D. A. (1990). *Science* **250**, 1566–1570.

Xue, Y., Davies, D. R., and Thomas, C. M. (1990). *Mol. Gen. Genet.* **221**, 195–198.

Yang, D., Oyaizu, Y., Oyaizu, H., Olsen, G. J., and Woese, C. R. (1985). *Proc. Natl. Acad. Sci. U.S.A.* **82**, 4443–4447.

Yokoi, F., Vassileva, A., Hayashida, N., Torazawa, K., Wakasugi, T., and Sugiura, M. (1990). *FEBS Lett.* **276**, 88–90.

Young, E. G., and Hanson, M. R. (1987). *Cell (Cambridge, Mass.)* **50**, 41–49.

Young, I. G., Rogers, B. L., Campbell, H. D., Jaworowski, A., and Shaw, D. C. (1981). *Eur. J. Biochem.* **116**, 165–170.

Zaita, N., Torazawa, K., Shinozaki, K., and Sugiura, M. (1987). *FEBS Lett.* **210**, 153–156.

Zillig, W. (1986). *Nature (London)* **320**, 220.

Zillig, W. (1991). *Curr. Opin. Genet. Dev.* **1**, 544–551.

Zillig, W., Klenk, H.-P., Palm, P., Leffers, H., Pühler, G., Gropp, F., and Garrett, R. A. (1989). *Endocytobiosis Cell Res.* **6**, 1–25.

Zillig, W., Palm, P., Klenk, H.-P., Pühler, G., Gropp, F., and Schleper, C. (1991). *In* "General and Applied Aspects of Halophilic Microorganisms" (F. Rodriguez-Valera, ed.), pp. 321–332. Plenum, New York.

Zillig, W., Palm, P., Reiter, W.-D., Gropp, F., Pühler, G., and Klenk, H.-P. (1988). *Eur. J. Biochem.* **173**, 473–482.

Zillig, W., Stetter, K. O., Schnabel, R., and Thomm, M. (1985). *In* "Archaebacteria" (Vol. 8 of "The Bacteria: A Treatise on Structure and Function") (C. R. Woese and R. S. Wolfe, eds.), pp. 499–524. Academic Press, San Diego.

Zimmer, M., Lückemann, G., Lang, B. F., and Wolf, K. (1984). *Mol. Gen. Genet.* **196**, 473–481.

NOTE ADDED IN PROOF: Comparison of elongation factor EF-1α sequences has provided new evidence that extremely thermophilic, sulfur-metabolizing bacteria (eocytes) are the closest surviving relatives of the eukaryotes [Rivera, M. C., and Lake, J. A. (1992). *Science* **257**, 74–76] (see Section III,A). The results of recent phylogenetic analyses of plastid protein sequences have been interpreted as supporting a monophyletic origin of all plastids, with subsequent lateral transfer of the *rbcLS* operon from a purple bacterium to a rhodophyte [Morden, C. W., Delwiche, C. F., Kuhsel, M., and Palmer, J. D. (1992). *BioSystems,* in press] (see Section IV,F). Complete genetic maps have now been determined for mtDNAs from *Allomyces macrogynus* (a saprophytic fungus) [Paquin, B., and Lang, B. F. (1992). *In* "Genetic Maps—1992" (S. J. O'Brien, ed.), 6th ed. Cold Spring Harbor Laboratory, Cold Spring Harbor, NY, in press.], *Prototheca wickerhamii* (green algal class Chlorococcales) [Burger, G., Lang, B. F., Wolff, G., and Kück, U., (1992). *In* "Genetic Maps—1992" (S. J. O'Brien, ed.), 6th ed. Cold Spring Harbor Laboratory, Cold Spring Harbor, NY, in press.], and *Phytophthora infestans* (an oomycete) [Lang, B. F., and Forget, L., (1992). *In* "Genetic Maps—1992" (S. J. O'Brien, ed.), 6th ed. Cold Spring Harbor Laboratory, Cold Spring Harbor, NY, in press.]. Three aspects of these new data on gene content (Tables II and III) are particularly relevant to the issue of mitochondrial origins (see Section V,C): (i) *P. infestans* mtDNA contains almost exactly the same set of ribosomal protein genes as is found in liverwort mtDNA, while *P. wickerhamii* mtDNA contains a smaller subset of these; (ii) *P. wickerhamii* mtDNA contains homologs of *orf25* and *orfB*, two genes that had previously been found only in land plant mtDNA; and (iii) *P. wickerhamii* mtDNA encodes a 5S rRNA gene that in primary sequence and potential secondary structure is a specific relative of the 5S rRNA encoded by land plant mtDNA. The latter observation constitutes the first solid evolutionary connection between the mitochondrial genomes of multicellular land plants and at least some unicellular green algae.

Index

A

Achyla, mitochondrial genomes in, 92–93
Achyla ambisexualis, mitochondrial
  genomes in, 59, 91–92
Acipenser transmontanus, animal mtDNA
  and, 202–204
Agaricus, mitochondrial genomes in,
  103–105
Algae
  ciliates and, 58
  endosymbiont hypothesis and, 278,
    296–297
  eukaryotes, 276–277
  mitochondria, 327
  molecular biology, 265
  nuclear genome, 252
  phylogenies, 273–275
  plastids, 254–255, 261–262, 279
  sequences, 271–272
Alleles
  higher plant mitochondrial genomes and,
    131–134
  mitochondrial genomes in fungi and, 107,
    109
Amino acids
  animal mtDNA and, 180–186, 192, 206
    transcription, 220–221
  ciliates and, 36–38, 41
  endosymbiont hypothesis and, 249
    mitochondria, 304, 307, 309
    plastids, 268–269, 278
  higher plant mitochondrial genomes and,
    149, 159, 163–164
  kinetoplastid mtDNA and, 71, 82
  mitochondrial genomes in fungi and,
    118–119
Anacystis nidulans
  endosymbiont hypothesis and, 269, 297
  mitochondrial genomes of, 43, 50, 58

Anacystis tumefaciens, mitochondrial
  genomes of, 43, 58
Animal mitochondrial DNA, *see*
  Mitochondrial DNA, animal
Antibiotic resistance, ciliates and, 2–3
Antibiotic sensitivity, endosymbiont
  hypothesis and, 267, 306
Antibodies, higher plant mitochondrial
  genomes and, 131
Apicomplexa, endosymbiont hypothesis
  and, 326
Apocytochrome *b*
  ciliates and, 12, 41
  kinetoplastid mtDNA and, 71
  mitochondrial genomes in fungi and
    DNA diversity, 92, 96–98, 100, 102–103,
    105
  generation of mtDNA diversity, 107,
    109
Apodachyla, mitochondrial genomes in fungi
  and, 91–92
Arabidopsis thaliana, endosymbiont
  hypothesis and, 250, 278
Archaebacteria, endosymbiont hypothesis
  and, 235
  future prospects, 331
  lineages of life, 238, 240–246
  mitochondria, 280, 307–309, 313, 327
  nuclear genome, 247–251
  organelles, 329
  plastids, 253, 265, 267–269
  sequences, 270
Ascaris suum
  animal mtDNA and, 188, 191, 202
    nucleotide bias, 204–205
    tRNA, 197–198
  endosymbiont hypothesis and, 290
Ascomycetes
  endosymbiont hypothesis and, 282, 290,
    300

372

INDEX

Prochlorophytes, endosymbiont hypothesis
and, 255, 269
Progenote, endosymbiont hypothesis and,
235, 238, 246
Prokaryotes
animal mtDNA and, 193, 195–196, 198
replication, 225
transcription, 221
ciliates and, 43, 45, 59
endosymbiont hypothesis and, 233,
235–237
future prospects, 332
lineages of life, 238, 240, 243–244, 246
mitochondria, 298, 306, 309, 312–313
nuclear genome, 247
organelles, 330
plastids, 255, 258, 267–268, 278
mitochondrial genomes in fungi and, 120
Promoters
animal mtDNA and, 202, 229–230
regulation, 227
replication, 223
transcription, 218–221
endosymbiont hypothesis and, 267
gene expression, 298–299, 301, 333, 336
genome organization, 289, 297
higher plant mitochondrial genomes and,
161
kinetoplastid mtDNA and, 78–79
mitochondrial genomes in fungi and, 108,
115
Protein
animal mtDNA and, 174, 180, 208–209,
229–231
genes, 181–189
nucleotide bias, 204–206
regulation, 226, 229
replication, 225
rRNA, 189–191
sequences, 204
transcription, 218–222
tRNA, 193, 199
ciliates and, 4, 11, 58–59
genetic code, 53, 55
Paramecium aurelia, 12, 35–38
endosymbiont hypothesis and, 234
gene expression, 298–299, 301–302,
304, 306–307
gene structure, 308–313
gene transfer, 278–279, 324–325
genome organization, 292–295, 298

lineages of life, 242, 246
mitochondria, 281–282, 329
nuclear genome, 247–249
organelles, 329
Plasmodium, 326
plastids, 253, 263, 265, 267–268
sequences, 268–273
higher plant mitochondrial genomes and,
162–164
abnormal phenotypes, 131–134
gene location, 157–159
structure, 137, 146
kinetoplastid mtDNA and, 66, 69, 76, 82
mitochondrial genomes in fungi and, 96,
98, 107
Protein-coding genes
endosymbiont hypothesis and
gene expression, 298, 304
gene structure, 308–309
gene transfer, 325
genome organization, 289–290, 293,
295–297
lineages of life, 239
mitochondria, 288
nuclear genome, 247, 250
plastids, 267
higher plant mitochondrial genomes and,
157–159, 162
kinetoplastid mtDNA and, 70–73
Proteolysis, ciliates and, 6
Protistan mitochondria, endosymbiont
hypothesis and, 305–306
Protistan mtDNA, endosymbiont hypothesis
and, 294–298, 316
Protists, endosymbiont hypothesis and, 251,
278, 283, 286, 324, 332
Protoplastids, endosymbiont hypothesis
and, 277–278
Protoplasts, higher plant mitochondrial
genomes and, 151, 153
Prototheca wickerhamii, endosymbiont
hypothesis and, 296–297, 324
Protozoa
ciliates and, 4, 12, 36
endosymbiont hypothesis and, 252, 273,
275, 294, 330
psb G, ciliates and, 36–38
PstI sites, mitochondrial genomes in fungi
and, 110, 116
Pyrenomonas salina, endosymbiont
hypothesis and, 276–277